QM² Quantitative Methods in Mathematics

Fifth Edition
Revised

Charlotte T. Sukta
Saint Laurence High School

Joseph J. Sukta
Moraine Valley Community College

KENDALL/HUNT PUBLISHING COMPANY
4050 Westmark Drive   Dubuque, Iowa 52002

Cover: "David and Goliath",
A 16th century mural on the Goliathhaus,
Regensburg, Germany
Photo by André J. Sukta

Copyright © 1997, 1998, 1999, 2001, 2006 by Kendall/Hunt Publishing Company
Revised Printing 2008

ISBN 978-0-7575-5002-7

All rights reserved. No part of this publication may be reproduced,
stored in a retrieval system, or transmitted, in any form or by any means,
electronic, mechanical, photocopying, recording, or otherwise,
without the prior written permission of the copyright owner.

Printed in the United States of America
10 9 8 7 6 5 4

*Dedicated to André
and our Parents*

TABLE OF CONTENTS

Preface	IX
How To Solve It	XI

I. Review of Basic Math Skills
1.1 Operations with Signed Numbers	1
1.2 Order of Operations	20
1.3 Ratio and Proportion	35
1.4 Proportion Applications	46
1.5 Solving Linear Equations	59
Project: Proportion Applications	76
Project: Linear Equations and Proportions in Radiology	77
Chapter 1 Review	79

II. Personal Finance
2.1 Simple Interest: Savings Accounts and Loans	82
2.2 Compound Interest: Savings Account and Loans	94
2.3 Amortization	114
2.4 Remaining Balance on Simple Interest Loans	132
Project: Car Purchase	155
Chapter 2 Review	157

III. More Personal Finance
3.1 The Actual Monthly Payment	161
3.2 Strategies to Save on Interest Charges	181
3.3 Closing Costs	202
3.4 Annuities that Grow	218
3.5 Annuities that Decay	232
3.6 Annuities Revisited	241
Project: Purchasing a Residence	255
Project: A Personal Retirement Plan	257
Chapter 3 Review	258

IV. Geometry and Home Renovation
4.1 Perimeter and Circumference	263
4.2 Area	283
4.3 Perimeter and Area of Irregular Shapes	299
4.4 Volume	312
4.5 Do – It – Yourself	323
Project: National Debt	338

Project: Sweat Equity	339
Chapter 4 Review	343

V. Statistics
5.1 Descriptive Statistics	348
5.2 Descriptive Statistics with Quantitative Data	366
5.3 Central Tendency	379
5.4 IQR Boxplot and Standard Deviation	388
5.5 Standard Normal Curve	399
5.6 Standard Score Based on an Area	410
5.7 Converting to Standard Scores	415
5.8 Data (x) Based on a Probability	423
Project: Graphs for Qualitative Data	430
Project: Descriptive Statistics for Quantitative Data	431
Project: Normal Probability Distribution	432
Chapter 5 Review	433

VI. Probability
6.1 Probability	436
6.2 Odds	455
6.3 Probability Distributions	465
6.4 Decision Trees, Decision-Making under Uncertainty	481
6.5 Venn Diagrams	493
6.6 "OR", The Addition Rule	507
6.7 "AND", The Multiplication Rule	520
Project: Raffle	528
Project: Decision Tree	530
Project: Venn Diagrams and Probability	532
Chapter 6 Review	534

Appendices
Solutions to Chapter Review
Chapter 1	A1
Chapter 2	A6
Chapter 3	A13
Chapter 4	A20
Chapter 5	A31
Chapter 6	A42

Solutions to Odd Numbered Problems
    Chapter 1   B1
    Chapter 2   B13
    Chapter 3   B21
    Chapter 4   B34
    Chapter 5   B50
    Chapter 6   B67

Tax Returns   C1

Index   D1

Standard Normal Distribution Table   **Inside Back Cover**

Preface

In the legend of David and Goliath, Goliath is the champion of Gath and portrayed as an intimidating figure. His was over six and a half feet tall and was clad in bronze armor. He wore a bronze helmet, a bronze corselet, bronze greaves (knee and leg guards) and carried a bronze saber called a scimitar. He arrogantly challenged his opponents to one-on-one combat for their freedom. Upon seeing this giant even the most seasoned soldiers retreated in fear.

David, a mere youth, decided to accept Goliath's challenge. He tried on his brother's armor, but his inexperience made it difficult for him to move, so he discarded the armor. Armed with only a staff, five smooth stones taken from a brook, a sling, and determination, he set out to defeat Goliath. **David was victorious!**

Many people view Mathematics as an unbeatable opponent, a "Goliath" of a course. You, like David, can be victorious, but in your mathematical endeavors. Set out with confidence and a willingness to try. Take the risk. You, too, can defeat your Goliath and succeed. Our book aids you in learning math for your future. Daily, in our lives, we encounter a variety of problems. The problem solving approach used throughout our text will give you the tools necessary to analyze and solve these problems. Use these tools and problem solving techniques.

Mathematics is the vehicle we use to develop problem solving skills. The goal of our book is to develop these problem solving skills and to characterize mathematics as a valuable ally. The problem solving skills used in Mathematics are necessary for survival in today's ever changing technical society.

The student who studies our text will find that each chapter reinforces George Polya's successful approach to problem solving. The first step in problem solving encourages you to approach any situation patiently and rationally. Take the time to collect the information and your thoughts, rather than react to a situation. **UNDERSTAND** the problem. An old adage states "A picture is worth a thousand words." Once you have the picture in your mind and understand both what the data represents and what is asked for, then you are ready to develop a strategy. The second step of problem solving is to decide on the direction to pursue. This means develop a **PLAN**. Look for patterns, formulas and recall similar problems. The third step is to take action. Carry out the plan. **SOLVE** the problem. The final step is to **CHECK** the result. Is your solution reasonable and is question answered? There are times that the problem solving process may have to be repeated. Through the continued application of these four steps, you will be able to calmly approach futuristic mathematical and life problems with confidence.

The use of a scientific calculator will make smooth work of your calculations. Get familiar with the built in functions of your calculator, but use this tool cautiously. Always check your answers for reasonableness. Compare the solution with your estimate of the result. Only round your final answer. It is best to store the numbers from the intermediate steps in the memory of the calculator if they are needed for continued calculations. We caution you not to blindly rely on this tool of technology. Reflect on your results to determine whether they make sense.

We are confident that you will view Mathematics in a new light. Much success in your encounters and endeavors with Mathematics in your life.

Charlotte T. Sukta

Joseph J. Sukta

HOW TO SOLVE IT

First, understand and picture the problem	**UNDERSTAND AND PICTURE THE PROBLEM** A. What is the question? What is the data? What is the unknown? B. Identify sufficient information needed to solve the problem. C. Draw a figure. Write suitable notation. D. Separate the various parts of the problem.
Second, find the connection between the data and the unknown. You may consider an auxiliary problem. Devise a plan for the solution.	**DEVISE A PLAN** A. Have you seen this type of problem before? B. Do you know a related problem? Do you have a theorem that is useful? C. Is there a problem similar to yours which was solved before? Can you use its result? Can you use its method? D. Restate the problem. Go back to definitions. E. If you cannot solve the proposed problem, try to solve a related problem. Imagine a more general problem. Solve part of the problem. Derive something useful from the data. F. Did you use all pertinent data?
Third, carry out your plan.	**EXECUTE THE PLAN** Carry out your plan of the solution. Check each step.
Fourth, examine the solution obtained.	**CHECK YOUR WORK** A. Check the result? B. Is the result reasonable? C. Was the question answered?

CHAPTER 1: REVIEW OF BASIC MATH SKILLS

1.1 SIGNED NUMBERS
Objective: Perform arithmetic operations with signed numbers.

The Real Number Line
The number line consists of real numbers. Draw a straight line and arbitrarily place a zero on this line. Zero is not the start of the number line and does not have to be in the center of the number line. To the left of zero are the negative values and to the right of zero are the positive values.

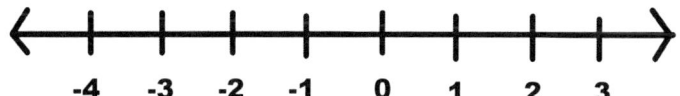

Distance
The real numbers without their negative and positive signs are specific distances from zero. Distance is a unit of measure. The units of measure may be feet, yards, meters, kilometers, etc. Distance cannot be a negative number. A football field is 100 yards in length, never –100 yards. Distance is always a non-negative value. On the number line, the negative sign communicates direction. The negative sign indicates that the number is to the left of zero.

The number line below shows that the value –400 is 400 units to the left of zero, while the value +300 is 300 units to the right of zero. Since 400 units are greater than 300 units, the point –400 is further from zero than the point 300.

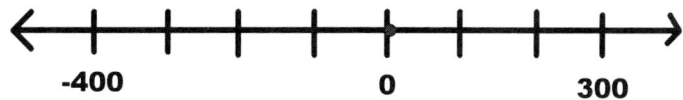

Signed Numbers
Zero, positive and negative numbers are used to model many real life situations. For example, positive and negative numbers are used to keep track of deposits and withdrawals in checking and savings accounts. Positive values represent deposits, money added to the account, and negative values represent withdrawals, money subtracted from the account.

Another interpretation of signed numbers is to record changes (gains, no change, or losses) over time. Positive values represent an increase (gain), zero indicates no change, and negative values represent a decrease (loss). Changes occur in altitude, weight, stock prices, revenue, etc. The following graph exhibits change over a period of time. The mortality rate for cancer patients increased from 1978 to 1990. From 1990 to 2002 the mortality rate for cancer patients decreased.

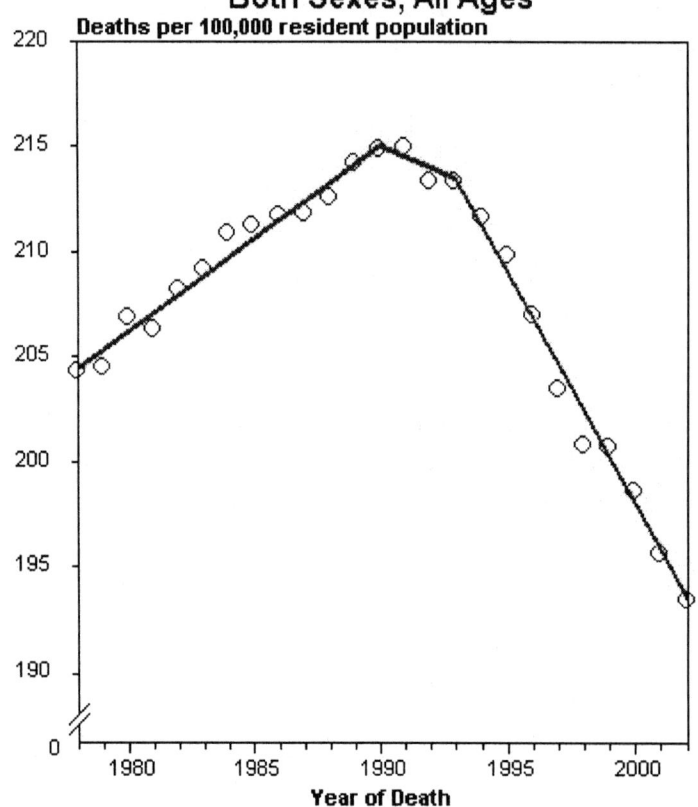

Signed numbers may also be used to represent time. A negative value indicates the past, zero is the present, and a positive value represents the future. The following table summarizes various interpretations of negative, zero, and positive values.

Negative (−) Decrease or decay	Zero (0) Values is an initial value	Positive (+) Increase or growth
Account withdrawal (Account balance decreases.)	Account inactivity (Account balance remains the same.)	Account deposit (Account balance increases.)
Time in the past	Time is the present	Time in the future
Weight loss	No weight change	Weight gain
Population decreases from an initial value	Population's initial value	Population increases from an initial value

Addition Rules
Case 1: Signed numbers with the same sign:
Consider deposits (increases) of $400 and $300 into a checking account. The effect of the two deposits is:

$$\begin{array}{r} \$400 \\ + \ \underline{\$300} \\ \$700 \end{array}$$

The account has increased by + $700, or $700. The positive sign (+) is not needed to represent an increase.

Consider two checks (withdrawals or decreases) written on the account. The first check is $200 and the second check is $140.
The effect of the two checks on the account is:

$$\begin{array}{r} -\$200 \\ + \ \underline{-\$140} \\ -\$340 \end{array}$$

The account has decreased by $340, represented by – $340. $340 has been withdrawn from the account. The negative sign (–) is needed to represent a decrease when words indicating a loss are omitted.

The Addition of values with the same sign is summarized by Addition Rule #1.

> **Addition Rule #1**
> If the signs of the numbers are the same, add the numbers and the sign remains the same.

Example 1: Evaluate 250 + 140.
Solution:
Step 1: Understand and picture the problem.
   Problem involves the addition of signed numbers with the same sign.

Step 2: Develop a plan.
   Apply Addition Rule #1 (same sign).

Step 3: Execute the plan.
   (+250) + (+140) = +390

Step 4: Check your work.
   1. Check the numbers and your calculations.
   2. Based on gains and losses, do the results seem reasonable?
   3. Was the question answered?

Example 2: Evaluate (−250) + (−442).
Solution:
Step 1: Understand and picture the problem.
Problem involves the addition of signed numbers with the same sign.

Step 2: Develop a plan.
Apply Addition Rule #1 (same sign).

Step 3: Execute the plan.
(−250) + (−442) = −692

Step 4: Check your work.
1. Check the numbers and your calculations.
2. Based on gains and losses, do the results seem reasonable?
3. Was the question answered?

Case 2: Signed numbers with different signs:
What is the unique sum of opposite numbers? 5 and (−5) are one pair of opposite numbers. The sum of 5 and (−5) or any pair of opposites is always zero.

Now, consider a deposit (increase) of $500 into an account and a check for $200 (decrease) written on the account. Find the change in the account balance after making the deposit and writing the check.
The $500 deposit is written as +$500 and the $200 check (withdrawal) is written as −$200. Combining the deposit and check:

$500 + (−$200) =	This is the problem.
($300 + $200) + (−$200) =	Rewrite 500 as 300 + 200.
$300 + [$200 + (−$200)] =	Pair the opposite values.
$300 + $0 =	Simplify, sum of opposites is zero.
$300	The result is $300.

While this method is mathematically correct, it is rather tedious. Another way to express this problem is:

$$\begin{array}{r} \mathbf{\$500} \\ \mathbf{+(\text{-}\$200)} \\ \hline \mathbf{\$300} \end{array}$$

The account has increased by +$300, or $300. The numbers 500 and 200 are subtracted resulting in an increase of 300. The positive sign (+) may be omitted.

Next Consider a check (decrease) written on an account for $400 and a deposit (increase) of $300 into the account. Find the effect of the deposit and the check on the account.

The $400 check is written as –$400 and the $300 deposit is written as +$300, or $300. Combining the deposit and check:

(–$400) + (+$300) =	This is the problem.
[(–$100) + (–$300)] + (+$300) =	Rewrite –$400 as [(–$100) + (–$300)].
(–$100) + [(–$300) + (+$300)] =	Pair the opposite values.
–$100 + $0 =	Simplify, sum of opposites is zero.
–$100	The result is –$100.

While this method is mathematically correct, it is again tedious. Another way to express this problem is:

$$-\$400$$
$$+\underline{\$300}$$
$$-\$100$$

The account has decreased by $100, represented by (– $100). The numbers 400 and 300 are subtracted resulting in a decrease of 100. The negative sign (–) must be used to represent a decrease.

The addition of values with different signs is summarized by Addition Rule #2.

Addition Rule #2
If the signs of the numbers are different, disregard the signs. Subtract the larger value minus the smaller value. The final result inherits the sign of the value furthest from zero on the number line.

Example 3: Evaluate (–250) + (+140).
Solution:
Step 1: Understand and picture the problem.
   The problem involves the addition of signed numbers with different signs.

Step 2: Develop a plan.
   Apply Addition Rule #2 (different signs).

Step 3: Execute the plan.
   (–250) + (+140) = –110 Addition Rule #2. Without regard for the signs, subtract the larger value minus the smaller value, 250 – 140 = 110. The result inherits the sign of the number furthest from zero. On the number line, –250 is further from zero than 140. The answer is –110.
   (–250) + (+140) = –110.

Step 4: Check your work.
1. Check the numbers and your calculations.
2. Based on gains and losses, do the results seem reasonable?
3. Was the question answered?

Example 4: Evaluate (+250) + (–140).
Solution:
Step 1: Understand and picture the problem.
The problem involves the addition of signed numbers with different signs.

Step 2: Develop a plan.
Apply Addition Rule #2 (different signs).

Step 3: Execute the plan.
(250) + (–140) = 110   Addition Rule #2. Without regard for the signs, subtract the larger value minus the smaller value, 250 – 140 = 110. The result inherits the sign of the number furthest from zero. On the number line, 250 is further from zero than –140. The answer is 110.
(250) + (–140) = 110.

Step 4: Check your work.
1. Check the numbers and your calculations.
2. Based on gains and losses, do the results seem reasonable?
3. Was the question answered?

Summary: Addition Rules for Signed Numbers:
1. If the signs of the numbers are the same, add the numbers and the sign remains the same.
2. If the signs of the numbers are different, disregard the signs. Subtract the larger value minus the smaller value. The final result inherits the sign of the value furthest from zero on the number line.

Subtraction Rules
Subtraction is the inverse operation of addition, it "undoes" the addition. For example, start with the number 18:
$$18 + 6 = 24, \quad \text{add 6 to 18 and get 24.}$$
$$24 - 6 = 18, \quad \text{subtract 6 from 24 and get 18.}$$
Notice that the addition of 18 and 6 was "undone" by the subtraction of 6 from the sum, 24. The original value of 18 was returned.

Addition and Subtraction are inverse operations. Subtraction is defined as adding the "opposite".

> **Subtraction Rule: Subtraction is adding the opposite.**
> 1. Write the first number with no changes.
> 2. Change the subtraction symbol to an addition symbol **AND** replace the second number with its opposite.
> 3. Follow the rules for the addition of signed numbers.

Every subtraction problem can be rewritten as an addition problem by adding the opposite. Hence, subtracting a value and adding the opposite value are equivalent expressions. For example, *subtracting* a *positive* six (+6) is the same as *adding* a *negative* six (–6).

Consider the problem 24 – 6 = 18.

24 – 6 =	Original problem, Positive 24 minus Positive 6
(+24)	1. Write the first number with no changes, Positive 24.
(+24) + (–6) =	2. Change the subtraction symbol to an addition symbol **AND** replace the second number with its opposite. **POSITIVE 24 PLUS NEGATIVE 6.**
+18 or 18	3. Follow the rules for the addition of signed numbers with different signs. **POSITIVE 24 PLUS NEGATIVE 6 equals 18.**

Thus the subtraction problem, 24 – 6 is the same as (+24) + (–6).

Example 5: Evaluate 100 – 44.
Solution:
Step 1: Understand and picture the problem.
  Subtraction of signed numbers requires rewriting the problem as an addition problem using the opposite of the second number and then applying the addition rules.

Step 2: Develop a plan.
  Apply the subtraction rule for signed numbers.

Step 3: Execute the plan.

100 – 44	Original problem, **POSITIVE 100 MINUS POSITIVE 44.**
+100	1. Write the first number with no changes, Positive 100.
100 + (–44)	2. Change the subtraction symbol to an addition symbol **AND** replace the second number with its opposite. **POSITIVE 100 PLUS NEGATIVE 44.**
+56 or 56	3. Follow the rules for the addition of signed numbers. with different signs. Without regard for the signs, subtract the larger value minus the smaller value, 100 – 44 = 56. The result inherits the sign of the number furthest from zero, 56.

Therefore, 100 – 44 = 100 + (–44) = 56.

Step 4: Check your work.
1. Check the numbers and your calculations.
2. Based on gains and losses, do the results seem reasonable?
3. Was the question answered?

Example 6: Evaluate −100 − 44.
Solution:
Step 1: Understand and picture the problem.
Subtraction of signed numbers requires rewriting the problem as an addition problem using the opposite of the second number and then applying the addition rules.

Step 2: Develop a plan.
Apply the subtraction rule for signed numbers.

Step 3: Execute the plan.
−100 − 44	Original problem, **NEGATIVE 100 MINUS POSITIVE 44**.
−100	1. Write the first number with no changes, Negative 100.
−100 + (−44)	2. Change the subtraction symbol to an addition symbol **AND** replace the second number with its opposite. **NEGATIVE 100 PLUS NEGATIVE 44.**
−144	3. Follow the rules for the addition of signed numbers, with the same signs. Add the numbers and keep the same sign.

Therefore, −100 − 44 = −144.

Step 4: Check your work.
1. Check the numbers and your calculations.
2. Based on gains and losses, do the results seem reasonable?
3. Was the question answered?

Example 7 Evaluate −100 − (−44).
Solution:
Step 1: Understand and picture the problem.
Subtraction of signed numbers requires rewriting the problem as an addition problem using the opposite of the second number and then applying the addition rules.

Step 2: Develop a plan.
Apply the subtraction rule for signed numbers.

Step 3: Execute the plan.
    −100 − (−44)   Original problem, **NEGATIVE 100 MINUS NEGATIVE 44.**
    −100   1. Write the first number with no changes, Negative 100.
    −100 + (44)   2. Change the subtraction symbol to an addition symbol **AND** replace the second number with its opposite. **NEGATIVE 100 PLUS POSITIVE 44.**
    −56   3. Follow the rules for the addition of signed numbers, with different signs. Without regard for the signs, subtract the larger value minus the smaller value, 100 − 44 = 56. The result inherits the sign of the number furthest from zero, −56.
    Therefore, −100 − (−44) = −100 + (44) = −56.

Step 4: Check your work.
    1. Check the numbers and your calculations.
    2. Based on gains and losses, do the results seem reasonable?
    3. Was the question answered?

Example 8: Evaluate 100 − (−44).
Solution:
Step 1: Understand and picture the problem.
    Subtraction of signed numbers requires rewriting the problem as an addition problem using the opposite of the second number and then applying the addition rules.

Step 2: Develop a plan.
    Apply the subtraction rule for signed numbers.

Step 3: Execute the plan.
    100 − (−44)   Original problem, **POSITIVE 100 MINUS NEGATIVE 44.**
    100   1. Write the first number with no changes, Positive 100.
    100 + (44)   2. Change the subtraction symbol to an addition symbol **AND** replace the second number with its opposite. **POSITIVE 100 PLUS POSITIVE 44.**
    144   3. Follow the rules for the addition of signed numbers, with the same signs. Add the numbers and keep the same sign.
    Therefore, 100 − (−44) = 144.

Step 4: Check your work.
    1. Check the numbers and your calculations.
    2. Based on gains and losses, do the results seem reasonable?
    3. Was the question answered?

A problem can have both addition and subtraction as demonstrated in the following two examples.

Example 9: Evaluate $120+(-24)-48+55-(-79)+(66)$.

Solution:

Step 1: Understand and picture the problem.
Addition uses the numbers as given. Subtraction of signed numbers requires rewriting the problem as an addition problem using the opposite of the subtracted number.

Step 2: Develop a plan.
1. Rewrite the problem using only addition.
2. Find the sum of all positive terms.
3. Find the sum of all negative terms.
4. Apply the addition rule for signed numbers to the sums.

Step 3: Execute the plan.
1. $120+(-24)-48+55-(-79)+(66)=120+(-24)+(-48)+55+79+66$
2. $120+55+79+66=320$
3. $(-24)+(-48)=-72$
4. $320+(-72)=248$

Thus $120+(-24)-48+55-(-79)+(66)=248$

Step 4: Check your work.
1. Check the numbers and your calculations.
2. Based on gains and losses, do the results seem reasonable?
3. Was the question answered?

Example 10: Evaluate $-144+(-66)-(-72)+24-55+(-48)$.

Solution:

Step 1: Understand and picture the problem.
Addition uses the numbers as given. Subtraction of signed numbers requires rewriting the problem as an addition problem using the opposite of the subtracted number.

Step 2: Develop a plan.
1. Rewrite the problem using only addition.
2. Find the sum of all positive terms.
3. Find the sum of all negative terms.
4. Apply the addition rule for signed numbers to the sums.

Step 3: Execute the plan.
1. $-144+(-66)-(-72)+24-55+(-48) = -144+(-66)+72+24+(-55)+(-48)$
2. $72 + 24 = 96$
3. $-144+(-66)+(-55)+(-48) = -313$
4. $-313 + 96 = -217$

Thus $-144+(-66)-(-72)+24-55+(-48) = -217$

Step 4: Check your work.
 1. Check the numbers and your calculations.
 2. Based on gains and losses, do the results seem reasonable?
 3. Was the question answered?

Multiplication | Division Rules

Multiplication

Multiplication can be interpreted as repeated addition of the same number. This interpretation works in most cases.

Case 1: Consider 3 times 5, the product of two positive factors. One value may be written as often as the other number specifies. Thus, *3 times 5* may be written as either
1) 3 written 5 times: $3 + 3 + 3 + 3 + 3 = 15$.
2) 5 written 3 times: $5 + 5 + 5 = 15$.

Case 2: Consider the product of one positive factor and one negative factor, such as $(-3) \times 5$. This problem can only be written one way. The negative value is written as often as the positive number specifies:
$(-3) \times 5$ becomes (-3) written 5 times: $(-3) + (-3) + (-3) + (-3) + (-3) = -15$.

Similarly $(-5) \times 3$ becomes (-5) written 3 times: $(-5) + (-5) + (-5) = -15$.

Case 3: Two negative factors, such as $(-5) \times (-3)$, cannot be written as an addition problem. -5 cannot be written -3 times nor can -3 be written -5 times.

To interpret $(-5) \times (-3)$, a different model is needed. Consider the following scenario involving a storage tank for liquids connected to a pipeline. See the following figure:

Suppose that 5 gallons of liquid is put into the tank every second. This is written as $5\frac{\text{gallons}}{\text{second}}$. At the end of 3 seconds, how much liquid is in the tank? At the end of 3 seconds, there is more liquid in the tank and the result is positive.

$$\left(+5 \; \frac{\text{gallons}}{\cancel{\text{second}}}\right) \; times \; (+3 \; \cancel{\text{seconds}}) = (+15 \text{ gallons}).$$

The positive product means that 15 gallons of liquid has been added to the tank. The level of liquid in the tank has risen 15 gallons above its initial position.

Let's perform the "inverse" operation and take liquid out of the tank. Use the same figure except close the top valve and open the bottom valve.

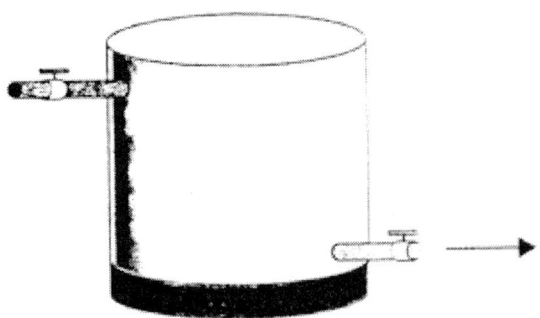

Suppose the liquid is removed from the tank at a rate of 5 gallons every second. This removal is represented by the value, $-5 \; \frac{\text{gallons}}{\text{second}}$. How much liquid has been removed from the tank at the end of three seconds?

$$\left(-5 \; \frac{\text{gallons}}{\cancel{\text{second}}}\right) \; times \; (+3 \; \cancel{\text{seconds}}) = (-15 \text{ gallons}).$$

The negative product means 15 gallons has been removed from the tank. The level of the liquid has fallen (decreased) 15 gallons below its initial position.

Finally consider the product, $\left(-5 \frac{\text{gallons}}{\text{second}}\right) \times (-3 \text{ seconds})$. The negative sign on the $\left(-5 \frac{\text{gallons}}{\text{second}}\right)$ still means to remove 5 gallons of liquid every second. The negative sign on the –3 seconds means to go "backwards" (into the past) in time.

Since time travel is only a Science Fiction topic, this is an unusual request. However, this time travel can be simulated with a video camera. Position a video camera inside the storage tank. Filming the tank as the liquid drains simulates the following problem:

$$\left(-5 \frac{\text{gallons}}{\cancel{\text{second}}}\right) \text{ times } \left(+3 \cancel{\text{seconds}}\right) = (-15 \text{ gallons}).$$

When the tape is played, the level of the liquid is seen falling within the storage tank. Now simulate the –3 hours by viewing the past. Play the tape in *reverse*. The tape will show the tank filling up. The level of the liquid is rising within the tank. Thus $\left(-5 \frac{\text{gallons}}{\cancel{\text{second}}}\right) \text{ times } \left(-3 \cancel{\text{seconds}}\right) = (+15 \text{ gallons}).$

The level of the liquid in the tank has risen 15 gallons above the initial position at the start of the tape. This scenario reflects the multiplication rules for signed numbers. The multiplication of two negative numbers results in a **positive** product.

> **Multiplication Rule #1:**
> The product of two factors with the same sign is positive.
> **Multiplication Rule #2:**
> The product of two factors with different signs is negative.

Example 11: Evaluate $6 \times (-3)$.
Solution:
Step 1: Understand and picture the problem.
  Multiplication problem has two factors of different signs.

Step 2: Develop a plan.
  Using Multiplication Rule #2, the product is negative.

Step 3: Execute the plan.
6 × (–3) = –18.

Step 4: Check your work.
1. Check the numbers and your calculations.
2. Does the result seem reasonable?
3. Was the question answered?

Example 12: Evaluate (–4) × (–5).
Solution:
Step 1: Understand and picture the problem.
Multiplication problem has two factors with the same signs.

Step 2: Develop a plan.
Using Multiplication Rule #1, the product is positive.

Step 3: Execute the plan.
(–4) × (–5) = 20

Step 4: Check your work.
1. Check the numbers and your calculations.
2. Does the result seem reasonable?
3. Was the question answered?

Division
Division is the inverse operation of Multiplication. It "undoes" the multiplication. For example, start with the number 5:

$$5 \times 6 = 30, \quad \textit{Multiply 5 by 6 to get 30.}$$
$$30 \div 6 = 5, \quad \textit{Divide 30 by 6 to get 5.}$$

Notice that the multiplication of 5 by 6 was "undone" by the division of 6 into the product, 30. The original value of 5 was returned. Multiplication and Division are inverse operations. Division can also be defined as multiplying by the "reciprocal". Reciprocals are usually fractions. The reciprocal of a number is the number inverted (upside down). $5 = \frac{5}{1}$ and $\frac{1}{5}$ are reciprocals. $\frac{2}{3}$ and $\frac{3}{2}$ are reciprocals.

The following algorithm describes the process of rewriting a division problem as a multiplication problem.
1. Write the first number with no changes.
2. Change the division symbol to a multiplication symbol **AND** replace the second number with its reciprocal.
3. Follow the rules for the multiplication of signed numbers.

Every division problem can be rewritten as a multiplication problem. Hence, dividing by a value and multiplying by the reciprocal value are equivalent.

For example, *dividing* by 2 is the same as *multiplying* by $\frac{1}{2}$. Consider the problem $(-24) \div 2 = -12$.

$(-24) \div 2 =$ Original problem, **NEGATIVE 24 DIVIDED BY POSITIVE 2.**

(-24) 1. Write the first number with no changes, Negative 24.

$(-24) \times \left(\frac{1}{2}\right) =$ 2. Change the division symbol to a multiplication symbol

                    AND replace the second number with its reciprocal.

                    NEGATIVE 24 TIMES POSITIVE $\frac{1}{2}$.

$\frac{-24}{2} = -12$ 3. Follow the rules for the multiplication of signed

                   numbers. **NEGATIVE 24 TIMES POSITIVE** $\frac{1}{2}$ **equals –12.**

                   (Signed numbers multiplication rule #2.)

Thus the division problem, $(-24) \div 2$ is the same as the multiplication problem $(-24) \times \left(\frac{1}{2}\right)$.

Consider multiplying more than two factors. Multiply $(-3) \times (-4) \times (-2)$.

$(-3) \times (-4) \times (-2) =$ the product of $(-3) \times (-4)$ which is 12 times (-2). Then $12 \times (-2) = -24$. Thus, $(-3) \times (-4) \times (-2) = -24$.

A multiplication problem with an odd number of negative signs produces a negative answer. The above example consists of three negative signs resulting in a negative value.

Consider multiplying more factors. Multiply $(-3) \times (-4) \times (-2) \times (-3)$.

$(-3) \times (-4) \times (-2) \times (-3) =$ the product of $(-3) \times (-4)$ which is 12 times $(-2) \times (-3)$. Then $12 \times (-2)$ is -24 times (-3). Finally $(-24) \times (-3) = 72$.

Thus $(-3) \times (-4) \times (-2) \times (-3) = 72$.

A multiplication problem with an even number of negative signs produces a positive answer. The above example consists of four negative signs resulting in a positive value.

The product | quotient of several factors results in the general rule for the multiplication | division of signed numbers.

> **General Multiplication | Division Rule #1:**
> The product | quotient of an even count of negative factors is positive.
>
> **General Multiplication | Division Rule #2:**
> The product | quotient of an odd count of negative factors is negative.

The count of positive factors does not affect the sign of the answer. It is the count of negative factors that determines the sign of the result. An odd count of negative factors produces a negative result, while an even count of negative factors produces a positive result. Factors are the values being multiplied. In (2)(3)(4), there are 3 factors, numbers being multiplied. While in (2)(3)(4)(5), there are 4 factors in the multiplication problem.

Example 13: Evaluate $\left(\dfrac{-24}{25}\right) \times \left(\dfrac{100}{-48}\right) \times \left(\dfrac{-8}{-64}\right)$.

Solution:
Step 1: Understand and picture the problem.
    The multiplication problem involves four negative signs. The solution will be positive.

Step 2: Develop a plan.
    Apply the General Multiplication | Division rule. The count of negative values is even. The answer is positive.

Step 3: Execute the Plan.

$$\left(\frac{-24}{25}\right)\left(\frac{100}{-48}\right)\left(\frac{-8}{-64}\right) = \left(\frac{24}{25}\right)\left(\frac{100}{48}\right)\left(\frac{8}{64}\right) = \left(\frac{1}{1}\right)\left(\frac{4}{2}\right)\left(\frac{1}{8}\right) = \frac{1}{4} = 0.25$$

Step 4: Check your Work.
    1. Check the numbers and your calculations.
    2. Based on the General Multiplication | Division Rule, do the results seem reasonable?
    3. Was the question answered?

Example 14: Evaluate $(-24) \times (-48) \times (-27) \times \left(\dfrac{-45}{-9}\right)$.

Solution:
Step 1: Understand and picture the problem.
  The multiplication problem involves five negative sings. The solution will be negative.

Step 2: Develop a plan.
  Apply the General Multiplication | Division rule. The count of negative values is odd. The solution is negative.

Step 3: Execute the Plan.

$$(-24) \times (-48) \times (-27) \times \left(\dfrac{-45}{-9}\right) = (-24) \times (-48) \times (\overset{3}{\cancel{-27}}) \times \left(\dfrac{-45}{\underset{1}{\cancel{-9}}}\right) = -155{,}520.$$

Step 4: Check your Work.
  1. Check the numbers and your calculations.
  2. Based on the General Multiplication | Division Rule, do the results seem reasonable?
  3. Was the question answered?

 (c) 2006 JupiterImages Corporation	**Historical Perspective** **Brahmagupta** (about 598 – 670) India

Brahmagupta's understanding of numbers went far beyond that of others during the period. In his work, the *Brahmasphutasiddhanta,* he defined zero as the result of subtracting a number from itself. He also gives arithmetical rules in terms of fortunes (positive numbers) and debts (negative numbers):
- A debt subtracted from zero is a fortune.
- A fortune subtracted from zero is a debt.
- The product of zero multiplied by a debt or fortune is zero.
- The product or quotient of two fortunes is one fortune.
- The product or quotient of two debts is one fortune.
- The product or quotient of a debt and a fortune is a debt.
- The product or quotient of a fortune and a debt is a debt.

These became the Rules for working with signed numbers presented in this section.

? Cognitive Problems ?

1. What is the opposite of –3?
2. What is the reciprocal of –3?
3. The terms, sum difference, product and quotient, apply to what operation? Give an example of each.
4. Explain the difference between a negative sign and a minus sign.
5. Give at least one example of an application using a:
   a. negative number   b. zero   c. positive number.
6. Name the inverse operations of:
   a. subtraction   b. multiplication   c. addition   d. division.
7. Identify the four basic arithmetic operations.
8. When performing arithmetic with two numbers, which two operations allow switching the order of the two values without changing the result? Note: This is called the Commutative Property.
9. Write and answer your own real life word problem that uses positive and negative numbers.
10. Subtracting 14 from a number is the same as adding what number?
11. Subtracting (–9) from a number is the same as adding what number?
12. Dividing by $\frac{1}{3}$ is the same as multiplying by what number?
13. Select a pair of opposite numbers and find their sum. What is the result? Make a general statement concerning the sum of opposite numbers.
14. Find the product of a number and its reciprocal. What is the result? Make a general statement concerning the product of a number and its reciprocal.
15. What is the difference between the opposite of a number and its reciprocal? Give an example of each.

(c) 2006 JupiterImages Corp.

Exercise 1.1
Perform the following operations on signed numbers.

1. 20 + 56
2. (–20) + 56
3. 20 + (–56)
4. (–20) + (–56)
5. 20 – 56
6. (–20) – 56
7. 20 – (–56)

8. (–20) – (–56)
9. (–934) + 87
10. 87 – (–94)
11. (–23) – 45

12. (–36) – (–92)
13. (–861) + (–48)
14. 95 + (–35)

15. (–3) + (+5) – (–8)   16. 18 – (–7) – 9

17. Andre's checking account has a balance of $258.31. He wrote checks for $32.70, $124.65, $420.88 and $1,257.00 and made a deposit of $2,500.00. What is the new balance of his checking account?

18. Charlotte keeps track of the daily gains and losses of her stock investment. During the last two weeks, her stock has fluctuated in the following manner, +1.2, –0.8, –0.4, +0.2, –0.1, +0.4, +0.2, –0.2, +0.1, and –0.2. What is the overall change in her stock?

19. Charisma keeps track of her weight on a daily basis. She has an electronic scale that measures weight to the tenth of a pound. Last week she experienced weight changes of –1.8, +0.4, –0.3, –0.6, +1.2, –0.2 and –0.4 pounds. What was the overall change in her weight?

20. Nick went to the racetrack and bet on 10 races. His betting yielded the following results, –$4, +$3.40, –$12, –$2, –$6, +$0.40, –$12, +$6.40, –$4, and +$18.60. What was the overall result of his betting?

21. (–7)(+8)   22. (+7)(+8)   23. (–7)(–8)   24. 8(–7)

25. (–13)(9)   26. (–5)(7)   27. $\dfrac{36}{4}$ 28. $\dfrac{-56}{-14}$

29. $\dfrac{24}{-3}$ 30. $\dfrac{-5.7}{3}$ 31. $\dfrac{54}{-1.6}$ 32. $\dfrac{-2.8}{8}$

33. (–3)(–2)(4)   34. (–7)(3)(0)   35. (–1.2)(5)(–8)

36. $\dfrac{3}{8}(-96)$ 37. $\left(\dfrac{15}{16}\right)\left(\dfrac{-48}{125}\right)$ 38. $\left(\dfrac{-12}{13}\right)\left(\dfrac{52}{-96}\right)$

39. $(-48)\left(\dfrac{15}{28}\right)\left(\dfrac{49}{-35}\right)(120)$ 40. $(-144)\left(\dfrac{-32}{35}\right)(-250)\left(\dfrac{-7}{20}\right)(-81)$

1.2 ORDER OF OPERATIONS
Objective: Perform the Order of Operations on mathematical expressions.

Order is necessary in society.
Just as there is an "order" to our daily activities, arithmetic computations follow a specific "order". Imagine driving to school and suddenly the "Rules of the Road" change. Suppose at one intersection the green light "stops" vehicles and the red light mandates "proceed with caution". Yet, at another intersection, the red light "stops" vehicles, while the green light means "proceed with caution". Without specific standards, there would be chaos and many life threatening accidents.

(c) 2006 JupiterImages Corp.

In light of setting standards, evaluate the following mathematical expression: $3 + 4 \times 5$. Does the expression $3 + 4 \times 5$ mean:

[3 + 4] × 5 add 4 and 3 = 7 7 × 5 multiply 7 and 5 = 35 35? with a result of 35?	**OR**	3 + [4 × 5] multiply 4 and 5 = 20 3 + 20 add 3 and 20 = 23 23? with a result of 23?

Neither solution has an arithmetic error, so it appears that either answer should be correct. However, only ***ONE*** solution is correct. Just like the traffic light mandates only one interpretation, arithmetic problems mandate only ***ONE*** interpretation. Otherwise, computations performed by accountants, pharmacists, nuclear physicists, or anyone studying the same problem, could result in different solutions.

For instance, one accountant may compute that $10,000 is owed to the government in taxes, while another accountant using the same figures, but following a different "order of operations", determines that a refund of $500 is due. Second, consider the consequences of a pharmacist giving different dosages of the same prescription medicine. An over-dose or under-dose of a medication could be life threatening or lethal. As a third example, a nuclear physicist could generate explosive and catastrophic results at a nuclear power plant as a consequence of setting the nuclear reactor incorrectly. Without an "Order of Operations, there would be total chaos. Mathematics would not exist. The "order" in which a problem is solved must be standardized to generate a single and unique solution to any computation.

Returning to the original problem, 3 + 4 × 5, the only difference between the two solutions is the "**order**" in which the problem was solved. A hierarchy, a ranking of the operations, mandates the standard that is followed. This ranking is called the *"ORDER OF OPERATIONS"*.

Order of Operations: Addition | Subtraction
The order of operations has four levels. The lowest rank of these four levels is Addition | Subtraction. Addition | Subtraction is usually identified as Addition. All subtraction problems can be rewritten as addition problems. **Subtracting** a number is equivalent to **adding the opposite** of that number. For example, 10 − 6 = 4 is equivalent to 10 + (−6) = 4. **All addition and subtraction operations are equally ranked and performed from left to right.**

Let's study a problem involving only addition and subtraction: **10 − 6 + 5**. Find the solution to 10 minus 6 plus 5. The two operations, as they appear from left to right, are Subtraction (minus) followed by Addition (plus). These are the same rank. The standard procedure mandates performing equally ranked operations from left to right. This means whichever operation appears first, from left to right, is performed first. Use this guideline and work from left to right to simplify:

10 − 6 + 5	All operations are the same rank.
[10 − 6] + 5	10 minus 6 equals 4. Replace 10 − 6 with 4.
4 + 5	4 plus 5 equals 9. Replace 4 + 5 with 9.
9	The result is 9.

Thus, the value of 10 − 6 + 5 = 9.

ORDER OF OPERATIONS
Rank 1:
Rank 2:
Rank 3:
Rank 4: Addition | Subtraction (left to right).

Example 1: Simplify 48 + 2 − 12 + 3 (Operations of the same rank).
Solution:
Step 1: Understand and picture the problem.
    1. Question: Simplify the expression by performing the appropriate calculations.
    2. There is a correct order to evaluate any expression.
    3. The operations are the same rank.

Step 2: Devise a plan.
Evaluate by following the order of operations, Rank 4: addition | subtraction (left to right).

Step 3: Execute the plan.
The operations are the same rank, so each operation is performed from the left to the right.

[48 + 2] − 12 + 3	48 plus 2 equals 50. Replace 48 + 2 with 50.
50 − 12 + 3	Perform the next operation, 50 minus 12 equals 38. Replace 50 − 12 with 38.
38 + 3	38 plus 3 equals 41. Replace 38 + 3 with 41.
41	The result is 41.

Thus, the value of 48 + 2 − 12 + 3 = 41.

Step 4: Check your work.
1. Check the numbers and your calculations.
2. Was the order of operations followed?
3. Is the answer reasonable?
4. Was the question answered?

Order of Operations: Multiplication | Division
Outranking Addition | Subtraction is Multiplication | Division. Multiplication | Division is third in the hierarchy of the "Order of Operations". They are also equally ranked and must be dealt with in the "order" they appear from left to right. Thus, the "Order of Operations" becomes:

> **ORDER OF OPERATIONS**
> Rank 1:
> Rank 2:
> Rank 3: Multiplication | Division (left to right).
> Rank 4: Addition | Subtraction (left to right).

Recall that Multiplication | Division is usually identified as Multiplication. All division problems can be rewritten as multiplication problems. **Dividing** by a number is equivalent to **multiplying by the reciprocal** of that number. For example, $10 \div 2 = 5$ is equivalent to $10 \times \frac{1}{2} = 5$. Peruse the following example of Multiplication | Division performed from left to right.

Example 2: Simplify 16 ÷ 2 × 4 (Operations of the same rank).
Solution:
Step 1: Understand and picture the problem.
   1. Question: Simplify the expression by performing the appropriate calculations.
   2. There is a correct order to evaluate any expression.
   3. The Multiplication | Division operations are the same rank.

Step 2: Devise a plan.
   Follow the order of operations.
   Rank 3: Multiplication | Division (left to right).
   Rank 4: Addition | Subtraction (left to right).

Step 3: Execute the plan.
   The operations are the same rank, each operation is worked from left to right.
   [16 ÷ 2] × 4   16 divided by 2 equals 8. Replace 16 ÷ 2 with 8.
   8 × 4   8 times 4 equals 32. Replace 8 × 4 with 32.
   32   The result is 32.
   Thus, the value of 16 ÷ 2 × 4 = 32.

Step 4: Check your work.
   1. Check the numbers and your calculations.
   2. Was the order of operations followed?
   3. Is the answer reasonable?
   4. Was the question answered?

Returning to the problem presented at the start of this section, 3 + 4 × 5, the correct solution is sequence B. This method produced the solution of 23. Based on the "Order of Operations", Multiplication must be performed before Addition. For verification, key this problem into a scientific calculator.

Example 3: Simplify 120 − 30 + 10 ÷ 5 × 2.
Solution:
Step 1: Understand and picture the problem.
   1. Question: Simplify the expression by performing the appropriate calculations.
   2. There is a correct order to evaluate any expression.
   3. The operations have different ranks. Rank 3 and rank 4 operations comprise this problem.

Step 2: Devise a plan.
Follow the order of operations.
Rank 3: Multiplication | Division (left to right).
Rank 4: Addition | Subtraction (left to right).
By the hierarchy of the Order of Operations the Multiplication | Division must be performed before the Addition | Subtraction, from left to right.

Step 3: Execute the plan on $120 - 30 + 10 \div 5 \times 2$.

$120 - 30 + [10 \div 5] \times 2$ From left to right, do the first rank 3 operation, division. 10 divided by 5 equals 2. Replace $10 \div 5$ with 2.

$120 - 30 + [2 \times 2]$ Do the second rank 3 operation, multiply. 2 times 2 equals 4. Replace 2×2 with 4. (Note: No more rank 3 operations.)

$[120 - 30] + 4$ From left to right, do the rank 4 operation, subtract. 120 minus 30 equals 90. Replace $120 - 30$ with 90.

$90 + 4$ Do the final rank 4 operation, add. 90 plus 4 equals 94. Replace $90 + 4$ with 94.

94   The result is 94.

Thus, the value of $120 - 30 + 10 \div 5 \times 2 = 94$.

Step 4: Check your work.
1. Check the numbers and your calculations.
2. Was the order of operations followed?
3. Is the answer reasonable?
4. Was the question answered?

Order of Operations: Exponents (Powers)
The next ranked operation is exponents. The exponent is an abbreviation for a multiplication problem involving the same factor. For example:
$2^4 = 2 \times 2 \times 2 \times 2 = 16$
The factor, 2, is called the "base". The number, 4, indicating how many times the "base" is written is called the "exponent" or "power". In the example, 2^4, the base is the factor 2 and the exponent (power) is 4. Hence, four factors of 2 are written and multiplied to produce 16.

Exponents outrank both Multiplication | Division and Addition | Subtraction. Thus, the "Order of Operations" becomes:

> **ORDER OF OPERATIONS**
> Rank 1:
> Rank 2: Exponents, powers (left to right).
> Rank 3: Multiplication | Division (left to right).
> Rank 4: Addition | Subtraction (left to right).

Example 4: Simplify $10 - 4^3 \div 32 \times 5 + 3$ (several ranks).
Solution:
Step 1: Understand and picture the problem.
  1. Question: Simplify the expression by performing the appropriate calculations.
  2. There is a correct order to evaluate any expression.
  3. The operations have different ranks. Ranks 2, 3 and 4 comprise this problem.

Step 2: Devise a plan.
  Follow the order of operations.
  Rank 2: Exponents (left to right).
  Rank 3: Multiplication | Division (left to right).
  Rank 4: Addition | Subtraction (left to right).
  First compute the rank 2 operations. That means all exponents are evaluated from left to right. Then perform all rank 3 operations. That means all Multiplication | Division are evaluated from left to right. Finally perform all rank 4 operations. That means all Addition | Subtraction are computed from left to right.

Step 3: Execute the plan on $10 - 4^3 \div 32 \times 5 + 3$.
  Work from left to right.

  $10 - [4^3] \div 32 \times 5 + 3$ Perform rank 2 operation, power. 4^3 equals 64. Replace 4^3 with 64.

  $10 - [64 \div 32] \times 5 + 3$ From left to right, perform the rank 3 operation, divide. 64 divided by 32 equals 2. Replace $64 \div 32$ with 2.

  $10 - [2 \times 5] + 3$ Perform rank 3 operation, multiply. 2 times 5 equals 10. Replace 2×5 with 10.

  $[10 - 10] + 3$ From left to right, perform rank 4 operation, subtract. 10 minus 10 equals 0. Replace $10 - 10$ with 0.

$0 + 3$	Do rank 4 operation, add, 0 plus 3 equals 3.
	Replace $0 + 3$ with 3.
3	Thus, the value of $10 - 4^3 \div 32 \times 5 + 3 = 3$.

Step 4: Check your work.
1. Check the numbers and your calculations.
2. Was the order of operations followed?
3. Is the answer reasonable?
4. Was the question answered?

INVERSE RELATIONSHIPS

Roots are the same rank as exponents (powers). Roots and exponents have a unique relationship. Their relationship is called an *inverse* relationship. An inverse relationship means one operation is used to "undo" the other operation. Addition and Subtraction are inverse operations. Focus on how the addition and subtraction of 3 to the initial value of 5 returns the value of 5 in the following situation:

$$5 + 3 = 8 \quad \textbf{5 } add \textbf{ 3} \longrightarrow \textbf{8}$$
$$8 - 3 = 5 \quad \textbf{8 } subtract \textbf{ 3} \longrightarrow \textbf{5.}$$

The *addition of* 3 to 5 is "undone" by the *subtraction of* 3 from their sum, 8. The number is once again 5.

Multiplication and Division are also inverse operations. Focus on the effect of the multiplication and division of 4 to the initial value of 6 in the following situation:

$$\textbf{6} \times 4 = 24 \quad \textbf{6 } multiplied\ by\ \textbf{4} \longrightarrow 24$$
$$24 \div 4 = 6 \quad 24\ divided\ by\ 4 \longrightarrow \textbf{6.}$$

The *multiplication of* 4 by 6 is "undone" by the *division of* 4 to the product 24. The number is once again 6.

Exponents and Roots

Exponents and Roots are also inverse operations. For instance, $2^3 = 8$. The 3 is the exponent or power, the 2 is the base and the result is 8. This problem is read as "two to the third power is 8". It can also be read as "2 cubed equals 8".

To investigate roots, rewrite $2^3 = 8$ as $\sqrt[3]{8} = 2$. In $\sqrt[3]{8}$, the 3 is called the index, the $\sqrt{}$ is called the radical sign, and the 8 is called the radicand. The root problem is read "3rd root of 8" or the "cube root of 8". It asks, what single factor may be written three times to obtain the product 8?

Focus on raising 2 to the third power followed by taking the third root of the result, this process returns the value of 2 in the following situation.

$$2^3 = 8 \quad \textbf{2 to the } \textit{third power} \longrightarrow 8$$
$$\sqrt[3]{8} = 2 \quad \textbf{8 to the } \textit{third root} \longrightarrow \textbf{2.}$$

Exponents and roots "undo" each other. Exponents and roots are inverse operations.

Study these examples of exponents.

Base	Exponent	Problem	Read	Evaluated
3	2	3^2	3 squared	9
2	3	2^3	2 cubed	8
5	4	5^4	5 to the fourth power	625
2	5	2^5	2 to the fifth power	32

Study these examples of roots.

Index	Radicand	Problem	Read	Evaluated
2	9	$\sqrt{9}$	Square root of 9	3
3	8	$\sqrt[3]{8}$	Cube root of 8	2
4	625	$\sqrt[4]{625}$	Fourth root of 625	5
5	32	$\sqrt[5]{32}$	Fifth root of 32	2

For example, find the $\sqrt[4]{81}$ (the 4th root of 81). The problem $\sqrt[4]{81}$ translates into the inverse problem, $(\text{base})^4 = 81$. In other words, what number can be used as a factor, base value, written four times yielding the result of 81? Thus $\sqrt[4]{81} = 3$, because $3^4 = 81$ or $3 \times 3 \times 3 \times 3 = 81$.

Another root problem is, find $\sqrt{81}$ (the square root of 81). The $\sqrt{81}$ translates into the inverse problem, $(\text{base})^2 = 81$. In a square root problem, the index is understood to be 2. In other words, what number will be used as a factor twice resulting in the product 81? Thus $\sqrt{81} = 9$, because $9 \times 9 = 81$.

Technology Note:
Use a scientific calculator to evaluate these problems. Roots may also be entered as fractional exponents that are reciprocals of the index. Thus,

$$\sqrt[3]{8} = 8^{(1/3)}, \quad \sqrt[4]{81} = 81^{(1/4)}, \text{ and } \sqrt{81} = 81^{(1/2)}.$$

Caution: Remember to enclose the fractional exponent within parenthesis.

28

The Order of Operations becomes

> **ORDER OF OPERATIONS**
> Rank 1:
> Rank 2: Exponents, Powers | Roots (left to right).
> Rank 3: Multiplication | Division (left to right).
> Rank 4: Addition | Subtraction (left to right).

Example 5: Simplify $5 + \sqrt[4]{625} \times 4 - 3$.
Solution:
Step 1: Understand and picture the problem.
  1. Question: Simplify the expression by performing the appropriate calculations.
  2. There is a correct order to evaluate any expression.
  3. The operations have different ranks. Ranks 2, 3 and 4 comprise this problem.

Step 2: Devise a plan.
  Follow the order of operations.
  Rank 2: Exponents (Powers) | Roots (left to right).
  Rank 3: Multiplication | Division (left to right).
  Rank 4: Addition | Subtraction (left to right).
  First, compute the rank 2 operations. All Exponents | Roots are evaluated from left to right. Next perform all rank 3 operations. All Multiplication | Division are evaluated from left to right. Finally, perform all rank 4 operations. All Addition | Subtraction are computed from left to right.

Step 3: Execute the plan on $5 + \sqrt[4]{625} \times 4 - 3$.
  According to the "Order of Operations" perform the rank 2 operation, which is the root, first. By using the inverse strategy, $(base)^4 = 625$, the base is 5. The fourth root of 625 is 5.

  $5 + [\sqrt[4]{625}] \times 4 - 3$ From left to right, do the rank 2 operation, root. The fourth root of 625 equals 5. Replace $\sqrt[4]{625}$ with 5.

  $5 + [5 \times 4] - 3$ Do the rank 3 operation, multiply. 5 times 4 equals 20. Replace 5×4 with 20.

  $[5 + 20] - 3$ From left to right, do the rank 4 operation, add. 5 plus 20 equals 25. Replace $5 + 20$ with 25.

25 − 3	Do the rank 4 operation, subtract. 25 minus 3 equals 22. Replace 25 − 3 with 22.
22	Thus, $5 + \sqrt[4]{625} \times 4 - 3 = 22$.

Step 4: Check your work.
1. Check the numbers and your calculations. Remember when using the exponent key, the cube root is entered as the power $\left(\dfrac{1}{3}\right)$, the fourth root as the power $\left(\dfrac{1}{4}\right)$, the fifth root as the power $\left(\dfrac{1}{5}\right)$, etc.
2. Was the order of operations followed?
3. Is the answer reasonable?
4. Was the question answered?

Order of Operations: Grouping Symbols

The highest rank in the "Order of Operations" is grouping symbols, parenthesis, braces, and brackets. All operations within a set of grouping symbols must be performed before any computations outside of the grouping symbols. Grouping symbols make it possible to change the order of a problem. It controls which rank operation is performed first. However, the order of operations must still be followed within any grouping symbols. The "Order of Operations" becomes:

> **ORDER OF OPERATIONS**
> Rank 1: Grouping Symbols.
> Rank 2: Exponents, Powers | Roots (left to right).
> Rank 3: Multiplication | Division (left to right).
> Rank 4: Addition | Subtraction (left to right).

Review the introductory problem to this section, $3 + 4 \times 5 = 23$. If parentheses are introduced, then the "order" and the solution may change. For example, inspect the new problem $(3 + 4) \times 5$. The parenthesis has the highest priority. Follow the ORDER OF OPERATIONS:

$(3 + 4) \times 5$	Perform the rank 1 operation, parenthesis. 3 plus 4 equals 7. Replace 3 + 4 with 7.
7×5	Perform the rank 3 operation, multiplication. 7 times 5 equals 35. Replace 7 × 5 with 35.
35	The result is 35. Thus, $(3 + 4) \times 5 = 35$.

Example 6: Simplify $10 + (48 - 12 \times 2) \div 4 \times (-5)$.
Solution:
Step 1: Understand and picture the problem.
  1. Question: Simplify the expression by performing the appropriate calculations.
  2. There is a correct order to evaluate any expression.
  3. The evaluation process involves all four ranks.

Step 2: Devise a plan.
  Follow the order of operations.
  Rank 1: Grouping Symbols.
  Rank 2: Exponents, Powers | Roots (left to right).
  Rank 3: Multiplication | Division (left to right).
  Rank 4: Addition | Subtraction (left to right).
  First, compute all operations within any set of grouping symbols. These operations within parenthesis must follow the order of operations.
  When there are no more grouping symbols, compute the rank 2 operations. Evaluate all Exponents | Roots from left to right. Next perform all rank 3 operations. Multiply | Divide from left to right. Finally, perform all rank 4 operations. Add | Subtract from left to right.

Step 3: Execute the plan on $10 + (48 - 12 \times 2) \div 4 \times (-5)$.
  First perform the operations following the ranks within the parenthesis. Then proceed with the ranks in order.

$10 + (48 - 12 \times 2) \div 4 \times (-5)$	Perform the rank 1 operation, grouping symbols. Follow the hierarchy within the parenthesis, perform the rank 3 operation of multiplication. 12 times 2 equals 24. Replace 12×2 with 24.
$10 + (48 - 24) \div 4 \times (-5)$	Do the rank 4 operation of subtraction within the parenthesis. 48 minus 24 equals 24. Replace $48 - 24$ with 24.
$10 + (24) \div 4 \times (-5)$	From left to right, perform the rank 3 operation, division. 24 divided by 4 equals 6. Replace $24 \div 4$ with 6.
$10 + 6 \times (-5)$	Do the rank 3 operation, multiplication. 6 times (-5) equals -30. Replace $6 \times (-5)$ with -30.

   10 + (−30)   Do the rank 4 operation, addition.
               10 plus (−30) equals −20. Replace
               10 + (−30) with −20.
 −20   The result is −20.
   Thus, 10 + (48 − 12 × 2) ÷ 4 × (−5) = −20.

Step 4: Check your work.
   1. Check the numbers and your calculations.
   2. Was the order of operations followed?
   3. Is the answer reasonable?
   4. Was the question answered?

What happens when there is more than one set of grouping symbols in the same problem? For example, evaluate:

120 + 5[24 − (8 + 2 × 3)] − 12.

The operations are performed on the innermost set of grouping symbols and then expanded to the next set of grouping symbols. In other words, you start inside, following the hierarchy, and work your way out. **Thus, the complete "Order of Operations" is:**

> ### ORDER OF OPERATIONS (*FINAL*)
>
> Rank 1: Grouping Symbols (innermost first).
> Rank 2: Exponents, Powers | Roots (left to right).
> Rank 3: Multiplication | Division (left to right).
> Rank 4: Addition | Subtraction (left to right).

Example 7: Simplify 120 + 5[24 − (8 + 2 × 3)] − 12.
Solution:
Step 1: Understand and picture the problem.
   1. Question: Simplify the expression by performing the appropriate calculations.
   2. There is a correct order to evaluate any expression.
   3. The evaluation process involves all four ranks.

Step 2: Devise a plan.
   Follow the order of operations.
   Rank 1: Grouping Symbols (innermost first).
   Rank 2: Exponents, Powers | Roots (left to right).

Rank 3: Multiplication | Division (left to right).
Rank 4: Addition | Subtraction (left to right).
First, compute all operations within any set of grouping symbols, starting with the innermost set of grouping symbols and follow the order of operations. Next compute the rank 2 operations. Evaluate Exponents (Powers) | Roots from left to right. Then perform all rank 3 operations. Multiply | Divide from left to right. Finally, perform all rank 4 operations. Add | Subtract from left to right.

Step 3: Execute the plan on $120 + 5[24 - (8 + 2 \times 3)] - 12$.
According to the "Order of Operations", the evaluation starts with the innermost grouping symbols $(8 + 2 \times 3)$.

$120 + 5[24 - (8 + 2 \times 3)] - 12$	Perform the rank 1 operation of the innermost parenthesis. Do the rank 3 operation, multiplication. 2 times 3 equals 6. Replace 2×3 with 6.
$120 + 5[24 - (8 + 6)] - 12$	Continue within the parenthesis and perform the rank 4 operation, 8 plus 6 equals 14. Replace $8 + 6$ with 14.
$120 + 5[24 - 14] - 12$	Still within the brackets, perform the rank 4 operation, 24 minus 14 equals 10. Replace $24 - 14$ with 10.
$120 + 5 [10] - 12$	Perform the rank 3 operation, multiplication. 5 times 10 equals 50. Replace $5[10]$ with 50.
$120 + 50 - 12$	Do the rank 4 operation, addition. 120 plus 50 equals 170. Replace $120 + 50$ with 170.
$170 - 12$	Do the Rank 4 operation, subtraction. 170 minus 12 equals 158. Replace 170 minus 12 with 158.
158	The final value is 158.

Thus, $120 + 5[24 - (8 + 2 \times 3)] - 12 = 158$.

Step 4: Check your work.
1. Check the numbers and your calculations.
2. Was the order of operations followed?
3. Is the answer reasonable?
4. Was the question answered?

? Cognitive Problems 1.2 ?

1. Explain, in your own words, the necessity for the "Order of Operations". Give an example of a problem that could arise if the "Order of Operations" did not exist.

2. Find the error in the following problems. Explain the error and correct it.
   A. $7 - 3 + 2 =$
      $7 - 5 =$
      2
   B. $15 \div 3 \times 5 =$
      $15 \div 15 =$
      1
   C. $4(3 + 5 \times 2) =$
      $4(8 \times 2) =$
      $4(16) =$
      64

3. Give an example of the inverse operation for:
   A. Addition.
   B. Division.
   C. Square Root.
   D. Taking a number to the 5th power.

4. Explain the difference between the power and the root of a number.

5. Create your own problem which uses all four ranked operations. Evaluate your problem and then use a scientific calculator to check your work.

6. What is a mathematical expression?
   (Hint: All the examples in this section are mathematical expressions.)

(c) 2006 JupiterImages Corp.

Exercise 1.2
Perform the indicated operations. Round to hundredths when necessary.

1. $3 + 4 \times 5$
2. $(3 + 4) \times 5$
3. $3 + (-4) \times 5$

4. $(3 + (-4)) \times 5$
5. $24 - 12 \div 4 \times 3$
6. $(24 - 12) \div 4 \times 3$

7. $(24 - 12) \div (4 \times 3)$
8. $24 - 12 \div (4 \times 3)$
9. $24 - 12 \div 4 \times (-3)$

10. $(24 - 12) \div 4 \times (-3)$
11. $(24 - 12) \div (4 \times (-3))$
12. $24 - 12 \div (4 \times (-3))$

13. $4^3 - 36 \div 2^2 \times 3$
14. $4^3 - 36 \div (2^2 \times 3)$
15. $(4^3 - 36) \div 2^2 \times 3$

16. $4^3 - 36 \div (2 \times 3)^2$
17. $(-4)^3 - 36 \div 2^2 \times 3$
18. $(-4)^3 - 36 \div (2^2 \times 3)$

19. $((-4)^3 - 36) \div 2^2 \times 3$
20. $(-4)^3 - 36 \div (2 \times 3)^2$

33

21. $\sqrt{64} \times \sqrt[3]{64} + \sqrt{81} \div \sqrt[4]{81}$

22. $\sqrt{64} \times (\sqrt[3]{64} + \sqrt{81} \div \sqrt[4]{81})$

23. $7 + 2[50 - 3(4 + 6)]$

24. $7 + 2[50 - 3 \times 4 + 6]$

25. $7 + 2[(50 - 3) \times 4 + 6]$

26. $7 + 2[(50 - 3) \times (4 + 6)]$

27. $7 + 2[50 - (3 \times 4 + 6)]$

28. $(7 + 2) \times [50 - 3(4 + 6)]$

29. $7 + 2[50 - (3 \times (-4) + 6)]$

30. $(7 + 2) \times [50 - 3((-4) + 6)]$

31. $(7 + 2) \times [(50 - 3) \times (-4) + 6]$

32. $(7 + 2) \times [(50 - 3) \times ((-4) + 6)]$

33. $(7 + 2) \times [50 - (3 \times (-4) + 6)]$

34. $\dfrac{18 - 6 \times 2}{24}$

35. $\dfrac{(18 - 6) \times 2}{24}$

36. $\dfrac{100 + 4 \times 6 - 4^3}{2^3 + 2^2}$

37. $\dfrac{100 + (4 \times 6 - 4)^3}{2^3 + 2^2}$

38. $\dfrac{(100 + 4) \times 6 - 4^3}{(2^3 + 2)^2}$

39. $\dfrac{(100 + 4) \times 6 - 4^3}{2^3 + 2^2}$

40. $\dfrac{(100 + 4) \times 6 - (-4)^3}{(2^3 + 2)^2}$

1.3 RATIO AND PROPORTION
Objective: Solve proportion problems.

A ratio is a fraction. $\frac{3}{4}$ is an example of a ratio. A proportion is the equality of two ratios. $\frac{6}{8} = \frac{3}{4}$ is an example of a proportion. The fundamental property of proportions is that their cross products are equal.

Fundamental Property of Proportions

Given the proportion $\frac{a}{b} = \frac{c}{d}$, then their cross products

$\left(\frac{a}{b} \times \frac{c}{d}\right)$ produces $ad = bc$.   Where $b \neq 0, d \neq 0$.

Consider the proportion, $\frac{6}{8} = \frac{3}{4}$. The ratios are equal. The cross product $\frac{6}{8} \times \frac{3}{4}$ produces $6(4) = 8(3)$, which is $24 = 24$. Similarly, $\frac{6}{8} \neq \frac{3}{5}$, because $\frac{6}{8} \times \frac{3}{5}$ produces $6(5) \neq 8(3)$, which is $30 \neq 24$. A proportion must consist of two equal ratios.

Given any three values of a proportion, the "Fundamental Property of Proportions" permits finding the fourth value. Consider the proportion $\frac{x}{18} = \frac{2}{9}$. The cross product, $\frac{x}{18} \times \frac{2}{9}$ produces $9x = 36$. Solving for "x" requires using the "Multiplication | Division Property of Equality".

Multiplication | Division Principle of Equality:

If $a = b$, then $ac = bc$ and $\frac{a}{c} = \frac{b}{c}$

for all real numbers a, b, c with $c \neq 0$.

To solve $9x = 36$ for "x", the factor "9" must be removed. Undo the multiplication of "9" by multiplying by the multiplicative inverse or reciprocal, "$\frac{1}{9}$" (This is equivalent to dividing by "9"). The Multiplication | Division Principle of Equality

mandates that both sides of the equation must be multiplied by "$\frac{1}{9}$" (or divided by "9"). Thus,
$$9x = 36$$
$$\left(\frac{1}{9}\right)9x = \left(\frac{1}{9}\right)36$$
$$\left(\frac{1}{\cancel{9}}\right)(\cancel{9})x = \left(\frac{1}{\cancel{9}}\right)(\cancel{36})^4$$
$$x = 4.$$

Example 1: Solve for x, $\frac{24}{x} = \frac{36}{15}$.

Solution:

Step 1: Understand and picture the problem.
Question: Solve for x in the proportion.

Step 2: Determine a plan.
1. Write the proportion.
2. Use the Fundamental Principal of Proportions to write the cross product of the proportion.
3. Use the Multiplication | Division Principle of Equality to solve for x.

Step 3: Execute the plan.

1. *Write the proportion.* $\quad \frac{24}{x} = \frac{36}{15}$

2. *Write the cross products.* $\quad \frac{24}{x} \times \frac{36}{15}, \quad 24(15) = 36x, \quad 360 = 36x.$

3. *Multiply by the reciprocal of 36.* $\quad \left(\frac{1}{36}\right)360 = \left(\frac{1}{36}\right)36x$

$$\left(\frac{1}{\cancel{36}}\right)(\cancel{360})^{10} = \left(\frac{1}{\cancel{36}}\right)(\cancel{36})x, \quad 10 = x.$$

Step 4: Check your work.
1. Check the numbers and your calculations.
2. Does the result seem reasonable?
3. Was the question answered?

Example 2: Solve for x, $\dfrac{5}{24} = \dfrac{x}{60}$.

Solution:
Step 1: Understand and picture the problem.
  Question: Solve for x in the proportion.

Step 2: Determine a plan.
  1. Write the proportion.
  2. Use the Fundamental Principal of Proportions to write the cross product of the proportion.
  3. Use the Multiplication | Division Principle of Equality to solve for x.

Step 3: Execute the plan.
  1. *Write the proportion.* $\dfrac{5}{24} = \dfrac{x}{60}$.

  2. *Write the cross products.* $\dfrac{5}{24} \diagdown\!\!\!\!\diagup \dfrac{x}{60}$, $\quad 5(60) = 24x, \quad 300 = 24x$.

  3. *Multiply by the reciprocal of 24* $\quad \left(\dfrac{1}{24}\right)300 = \left(\dfrac{1}{24}\right)24x$

  $\left(\dfrac{1}{24}\right)(300) = \left(\dfrac{1}{\cancel{24}}\right)(\cancel{24})x, \quad \dfrac{300}{24} = x, \quad 12.5 = x.$

Step 4: Check your work.
  1. Check the numbers and your calculations.
  2. Does the result seem reasonable?
  3. Was the question answered?

	Percents
50%	Percent means divide by 100. 50% means 50 divided by 100 = $\dfrac{50}{100} = \dfrac{1}{2}$. Percentages are used in many professions. There are three components in percent problems: the part, the percent and the whole amount.

Here are some sample questions involving percent.
1. Find the part when the percent and the whole amount are given.
   For example: 3% of a sample of 2,200 people experienced headaches after being given a new prescription drug for allergies. How may people in this group experienced headaches?

2. Find the percent by comparing the part to the whole amount.
   For example: In a survey, 360 people were interviewed. 160 had their own cell phone. What percent of people have their own cell phone?

3. Find the whole amount when the percent and the part are known.
   For example, in a medical study of a new prescription allergy medicine, 72 or 6% of the participants experienced drowsiness. How many people participated in the study?

Proportions may be used to solve these percent problems. Each percent problem is solved using the following proportion model.

Proportion Model for Percent Problems
$$\frac{\text{Part}}{\text{Whole}} = \frac{a}{100} = \frac{b}{c}$$
Where: a is the percentage without the % symbol
b is part of a quantity
c is the whole amount and $c \neq 0$.
100 replaces the % symbol.

Consider the percent problem: The sales tax is 6%. How much sales tax is charged on a $299 digital camera purchase?

Solution Method 1: The percentage is 6 (a = 6) and the total cost is $299 (c = 299). Substituting these values into the proportion model for percents produces
$$\frac{\text{Part}}{\text{Whole}} = \frac{6}{100} = \frac{b}{299}.$$

(Note: % means divide by 100. 6% means 6 divided by $100 = \frac{6}{100}$).

Use the cross product and solve for b;

Write the proportion. $\quad \frac{\text{Part}}{\text{Whole}} = \frac{6}{100} = \frac{b}{299}.$

Write the cross products. $\quad \frac{6}{100} \times \frac{b}{299}, \quad 6(299) = 100b, \quad 1{,}794 = 100\,b.$

Multiply by the reciprocal of 100. $\quad \frac{1}{100}(1{,}794) = \frac{1}{100}(100)b.$

$\left(\frac{1}{100}\right)(1{,}794) = \left(\frac{1}{\cancel{100}}\right)(\cancel{100})b, \quad \frac{1{,}794}{100} = 17.94 = b.$ The sale tax is $17.94.

If the question was to find the total cost of the digital camera, then the selling price and the sales tax would be added. The total cost of the digital camera is $299.00 + $17.94 = $316.94.

Solution Method 2: Another way to calculate total cost of the camera is to add the percent cost of the camera (100%) and the percent of the sales tax (6%). The total cost of the camera (tax included) is 100% + 6% = 106%. Using proportions

Write the proportion. $\quad \dfrac{Part}{Whole} = \dfrac{106}{100} = \dfrac{b}{299}.$

Write the cross products. $\quad \dfrac{106}{100} \times \dfrac{b}{299}, \quad 106(299) = 100b, \quad 31{,}694 = 100b.$

Multiply by the reciprocal of 100. $\quad \dfrac{1}{100}(31{,}694) = \dfrac{1}{100}(100)b.$

$\left(\dfrac{1}{100}\right)(31{,}946) = \left(\dfrac{1}{100}\right)(100)b, \quad \dfrac{31{,}694}{100} = 316.94 = b.$

The total cost of the camera is $316.94. (Note: Same answer found previously.)

Example 3: 3% of a sample of 2,200 people experienced headaches after being given a new prescription drug for allergies. How may people in this study experienced headaches?
Solution
Step 1: Understand and picture the problem.
    Question: Solve for the part.
    a = 3 (percent) because percent means divide by 100.
    c = 2,200 (whole) because it is the total number of people tested.

Step 2: Determine a plan.
    1. Use the Proportion Model for Percent Problems.
    2. Use the Fundamental Principal of Proportions to write the cross product of the proportion.
    3. Use the Multiplication | Division Principle of Equality.

Step 3: Execute the plan.
    1. *Write the proportion.* $\quad \dfrac{Part}{Whole} = \dfrac{a}{100} = \dfrac{b}{c}, \quad \dfrac{Part}{Whole} = \dfrac{3}{100} = \dfrac{b}{2{,}200}.$

2. *Write the cross products.* $\dfrac{3}{100} \diagup\!\!\!\!\!\diagdown \dfrac{b}{2{,}200}$, $\quad 3(2{,}200) = 100(b), \quad 6{,}600 = 100b.$

3. *Multiply by the reciprocal of 100.* $\quad \dfrac{1}{100}(6{,}600) = \dfrac{1}{100}(100b),$

$$\left(\dfrac{1}{\cancel{100}}\right)\overset{66}{\cancel{(6{,}600)}} = \left(\dfrac{1}{\cancel{100}}\right)\overset{1}{\cancel{(100)}}b, \quad 66 = b.$$

66 people in the medical study experienced headaches.

Step 4: Check your work.
   1. Check the numbers and your calculations.
   2. Does the result seem reasonable?
   3. Was the question answered?

Example 4: 3% of a sample of 2,200 people experienced headaches after being given a new prescription drug for allergies. How may people in this study did not experience headaches?
Solution Method 1:
Step 1: Understand and picture the problem.
   Question: Solve for the part.
   a = 3 (percent) because percent means divide by 100.
   c = 2,200 (whole) because it is the total number of people tested.

Step 2: Determine a plan.
   1. Use the Proportion Model for Percent Problems.
   2. Use the Fundamental Principal of Proportions to write the cross product of the proportion.
   3. Use the Multiplication | Division Principle of Equality.
   4. After finding the number of people who did experience a headache, subtract that value from the total number in the study to find the number of people who did not experience a headache.

Step 3: Execute the plan.
   1. *Write the proportion.* $\quad \dfrac{\text{Part}}{\text{Whole}} = \dfrac{a}{100} = \dfrac{b}{c}, \quad \dfrac{\text{Part}}{\text{Whole}} = \dfrac{3}{100} = \dfrac{b}{2{,}200}$

   2. *Write the cross products.* $\dfrac{3}{100} \diagup\!\!\!\!\!\diagdown \dfrac{b}{2{,}200}$, $\quad 3(2{,}200) = 100(b), \quad 6{,}600 = 100b.$

3. *Multiply by the reciprocal of 100.* $\quad \dfrac{1}{100}(6,600) = \dfrac{1}{100}(100b)$,

$$\left(\dfrac{1}{\cancel{100}}\right)\cancel{(6,600)}^{66} = \left(\dfrac{1}{\cancel{100}}\right)(\cancel{100})^{1} b, \quad 66 = b.$$

66 people in the medical study experienced headaches.

4. 2,200 − 66 = 2,134
   Thus, 2,134 people in the medical study did not experience headaches.

Step 4: Check your work.
   1. Check the numbers and your calculations.
   2. Does the result seem reasonable?
   3. Was the question answered?

Solution Method 2: An alternate way of solving Example 4 is to subtract the percent of patients who experienced headaches (3%) from 100%. The percent of patients who did not experience headaches is 100% − 3% = 97%.

Write the proportion. $\quad \dfrac{\text{Part}}{\text{Whole}} = \dfrac{97}{100} = \dfrac{b}{2,200}$

Write the cross products. $\quad \dfrac{97}{100} \times \dfrac{b}{2,200}, \quad 97(2,200) = 100b, \quad 213400 = 100\,b.$

Multiply by the reciprocal of 100. $\quad \dfrac{1}{100}(213,400) = \dfrac{1}{100}(100)b$,

$$\left(\dfrac{1}{100}\right)(213,400) = \left(\dfrac{1}{\cancel{100}}\right)(\cancel{100})^{1} b, \quad \dfrac{213,400}{100} = 2,134 = b.$$

The number of patients who did not experience headaches is 2,134.
Note: This is the same answer found previously.

Example 5: In a survey, 360 people interviewed. 160 people in this group had their own cell phone. What percent of people have their own cell phone?
Solution:
Step 1: Understand and picture the problem.
   Question: Solve for the percent of the part compared to the whole.

b = 160 (part) because it represents some of the 360 people surveyed.
c = 360 (whole) because it is the total number of people interviewed.

Step 2: Determine a plan.
1. Use the Proportion Model for Percent Problems.
2. Use the Fundamental Principal of Proportions to write the cross product of the proportion.
3. Use the Multiplication | Division Principle of Equality.

Step 3: Execute the plan.
1. *Write the proportion.* $\dfrac{\text{Part}}{\text{Whole}} = \dfrac{a}{100} = \dfrac{b}{c}$, $\dfrac{\text{Part}}{\text{Whole}} = \dfrac{a}{100} = \dfrac{160}{360}$.

2. *Write the cross products.* $\dfrac{a}{100} \times \dfrac{160}{360}$, $360(a) = 100(160)$, $360a = 16{,}000$.

3. *Multiply by the reciprocal of 360.* $\dfrac{1}{360}(360a) = \dfrac{1}{360}(16{,}000)$,

$\left(\dfrac{1}{360}\right)(360)a = \left(\dfrac{1}{360}\right)(16{,}000)$, $a = \dfrac{16{,}000}{360} \approx 44.444444$.

About 44.4% of the people surveyed possess a cell phone.

Step 4: Check your work.
1. Check the numbers and your calculations.
2. Does the result seem reasonable?
3. Was the question answered?

Example 6: In a medical study of a new prescription allergy medicine, 72 or 6% of the participants experienced drowsiness. How many people participated in the study?

Solution:

Step 1: Understand and picture the problem.
Question: Given a percent and a part, determine the whole amount.
a = 6 (percent) and b = 72 (part of the people in the medical study).

Step 2: Determine a plan.
1. Use the Proportion Model for Percent Problems.
2. Use the Fundamental Principal of Proportions to determine the cross product of the proportion.
3. Use the Multiplication | Division Principle of Equality.

Step 3: Execute the plan.
1. *Write the proportion.* $\dfrac{\text{Part}}{\text{Whole}} = \dfrac{a}{100} = \dfrac{b}{c}$, $\dfrac{\text{Part}}{\text{Whole}} = \dfrac{6}{100} = \dfrac{72}{c}$.

2. *Write the cross products.* $\dfrac{6}{100} \times \dfrac{72}{c}$, $6(c) = 100(72)$, $6c = 7{,}200$.

3. *Multiply by the reciprocal of 6.* $\dfrac{1}{6}(6c) = \dfrac{1}{6}(7{,}200)$,

$\left(\dfrac{1}{6}\right)(6)c = \left(\dfrac{1}{6}\right)(7{,}200)$, $c = 1{,}200$.

1,200 people participated in the medical study.

Step 4: Check your work.
1. Check the numbers and your calculations.
2. Does the result seem reasonable?
3. Was the question answered?

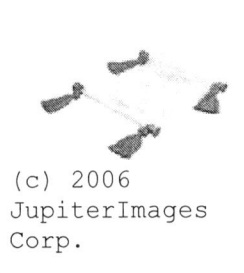 (c) 2006 JupiterImages Corp.	**Historical Perspective** **Eratosthenes of Syrene (276 – 194 BC) Greece**	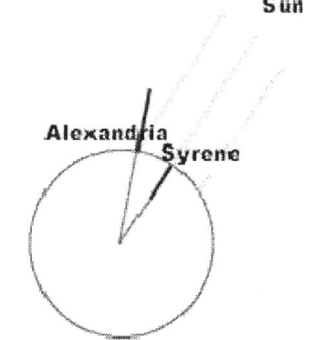

Eratosthenes estimated the circumference of the Earth using shadows and proportions. In Syrene, Egypt, at noon on the day of the summer solstice, the longest daylight day of the year, Eratosthenes placed a stick into the ground. Since the sun was directly overhead of Syrene, there was no shadow. Another stick was placed into the ground at Alexandria, a city to the North of Syrene. In Alexandria, a shadow was cast at noon of the summer solstice. The angle of the shadow in Alexandria was about 7^0. Using ratios, Eratosthenes set up the following proportion:

$$\dfrac{7^0}{360^0} = \dfrac{\text{Distance between Syrene and Alexandria}}{\text{Circumference of the Earth}}$$

Eratosthenes measured the distance between Syrene and Alexandria as 5,000 stadia. Solving the proportion produced.

$$\frac{7^0}{360^0} = \frac{5{,}000 \text{ stadia}}{x}$$

The cross products $\frac{7}{360} \diagup \frac{5{,}000}{x}$ produces $7x = 360(5{,}000) = 1{,}800{,}000$ stadia.

Solving for x produces $\left(\frac{1}{\cancel{7}}\right)(\cancel{7})x = \left(\frac{1}{7}\right)(1{,}800{,}000) = \frac{1{,}800{,}000}{7} \approx 257{,}143$ stadia.

A stadium, (singular for stadia) is estimated to be 172 yards (516 feet).

Using dimensional analysis; $\frac{257{,}143 \;\cancel{\text{stadia}}}{1}\left(\frac{516 \text{ feet}}{1 \;\cancel{\text{stadia}}}\right) \approx 132{,}685{,}788$ feet

A mile is 5,280 feet, thus $\frac{132{,}685{,}788 \;\cancel{\text{feet}}}{1}\left(\frac{1 \text{ mile}}{5{,}280 \;\cancel{\text{feet}}}\right) \approx 25{,}130$ miles.

Using today's scientific methods, the circumference of the Earth measures 24,855 miles. Eratosthenes was off by **ONLY 1%** !

Cognitive Problems 1.3

1. Compose a percent problem solving for the whole amount.
2. Compose a percent problem solving for the part.
3. Compose a percent problem using the part and whole amount, solving for the percent.
4. A discounted item has a price of $124.55 including a 12% discount. Find the original price of this item.

(c) 2006 JupiterImages Corp.

Exercise 1.3
Solve for x.

1. $\frac{x}{12} = \frac{2}{8}$ 2. $\frac{x}{15} = \frac{1}{6}$ 3. $\frac{12}{x} = \frac{8}{12}$ 4. $\frac{6}{x} = \frac{3}{8}$

5. $\frac{5}{9} = \frac{x}{12}$ 6. $\frac{3}{7} = \frac{x}{35}$ 7. $\frac{5}{8} = \frac{12}{x}$ 8. $\frac{9}{7} = \frac{27}{x}$

9. A digital camera costs $399.98 and the sales tax is 7%
   a. What is the sales tax?
   b. What is the total price of the digital camera?

10. A computer costs $1,399.00 and the sales tax is 6.5%. What is the total cost of the computer?

11. A car costs $22,167.48 and is discounted 4%. What is the sales price of the car?

12. A stuffed pizza costs $19.50. A 20% discount coupon reduces the price. What is the cost of the pizza?

13. In a survey 198 out of 250 people interviewed had received a speeding ticket at least one time. What percent of people had received speeding tickets?

14. In a survey 428 out of 720 people interviewed expressed opposition to a library tax increase. What percent of people are opposed to the tax increase?

15. In a medical study of a new skin cleansing cream, 18 of 400 people participating experienced a rash. What percent of the participants experienced a rash?

16. In a medical study of a new cold medicine, 10 out of 250 people experienced fatigue. What percent of the participants did not experience fatigue?

17. In an opinion poll, 252 or 63% of people interviewed, were opposed to the upcoming library referendum. How many people were interviewed?

18. In an opinion poll, 98 or 28% of people interviewed, were opposed to the upcoming high school referendum. How many people were interviewed?

19. In an opinion poll, 163 or 65.2% of people interviewed, were in favor of the fire protection district's upcoming referendum. How many people were interviewed?

20. In a medical study 154 or 35% of people tested experienced fatigue after being treated with the new allergy formula. How many people participated in the study?

1.4 PROPORTION APPLICATIONS

Objective: Solve application proportion problems.

International Banking, global businesses, vacationers in a foreign country, etc, routinely uses ratios and proportions. Proportions are used to convert dollars into the national money and national money into dollars.

(c)2006 JupiterImages Corp.

Nursing, pharmacy, chemical engineering, dieticians, photography, etc. are professions that routinely use ratios and proportions. Ratios and proportions are used to calculate mixture amounts. If the mixture, such as a prescription drug, is too strong or too weak, the results could be catastrophic or lethal.

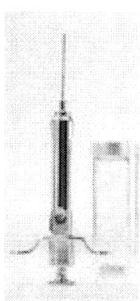
(c) 2006 JupiterImages Corp.

Navigators, architects, surveyors, civil engineers, etc. are also professionals that routinely use ratios and proportions. Ratios and proportions help determine the actual length of an object based on a map or blueprint measurement.

(c) 2006 JupiterImages Corp.

Monetary conversion.
International businesses require converting dollars into another country's currency. Proportions are used for this conversion.

Proportion Model for Monetary Conversion
$$\frac{\$1 \text{ US}}{\text{foreign equivalent}} = \frac{\text{dollar amount (US)}}{\text{foreign equivalent amount}}$$
Where: Foreign equivalent can be pounds, francs, yens, rubles, euros, etc. Dollar amount is the price in dollars. Foreign equivalent amount is the price in the foreign currency.

Example 1: Given that $1 US is equivalent to 0.88 euros. What is $500 US worth in euros?

Solution:

Step 1: Understand and picture the problem.
    1. Question: How many euros equal 500 dollars?

2. The ratio is $\dfrac{1 \text{ US dollar}}{0.88 \text{ Euro}}$.
3. The dollar amount is $500.
4. Let x represent the number of Euros.

Step 2: Determine a plan.
1. Use the Proportion Model for Monetary Conversion.
2. Use the Fundamental Principal of Proportions to determine the cross product of the proportion.
3. Use the Multiplication | Division Principle of Equality to solve for x.

Step 3: Execute the plan.
1. $\dfrac{\$1 \text{ US}}{\text{foreign equivalent}} = \dfrac{\text{dollar amount (US)}}{\text{foreign equivalent amount}}$.

   Write the proportion. $\quad \dfrac{1}{0.88} = \dfrac{500}{x}$.

2. *Write the cross products.* $\quad \dfrac{1}{0.88} \diagup \dfrac{500}{x}, \quad 1(x) = 0.88(500)$.

3. x = 440, $500 (US) equals 440 euros.

Step 4: Check your work.
1. Check the numbers and your calculations.
2. Does the result seem reasonable?
3. Was the question answered?

Example 2: Given that $1 US is equivalent to 1.35 Swiss francs. What is 1,200 Swiss francs worth in US dollars?
Solution:
Step 1: Understand and picture the problem.
1. Question: How many US dollars equal 1,200 Swiss francs?
2. The ratio is $\dfrac{1 \text{ US dollar}}{1.35 \text{ Swiss Francs}}$.
3. The foreign currency amount is 1,200 Swiss francs.
4. Let x represent the amount of US dollars.

Step 2: Determine a plan.
1. Use the Proportion Model for Monetary Conversion.
2. Use the Fundamental Principal of Proportions to determine the cross product of the proportions.
3. Use the Multiplication | Division Principle of Equality to solve for x.

47

Step 3: Execute the plan.

1. $$\frac{\$1 \text{ US}}{\text{foreign equivalent}} = \frac{\text{dollar amount (US)}}{\text{foreign equivalent amount}}.$$

   Write the proportion. $\quad \dfrac{1}{1.35} = \dfrac{x}{1,200}.$

2. *Write the cross products.*

   $\dfrac{1}{1.35} \diagup\!\!\!\!\diagdown \dfrac{x}{1,200}, \quad 1(1,200) = 1.35(x), \quad 1,200 = 1.35x.$

3. *Multiply by the reciprocal of 1.35.*

   $\dfrac{1}{1.35}(1,200) = \dfrac{1}{1.35}(1.35)x, \quad \dfrac{1,200}{1.35} = x, \quad 888.8888889 \approx x.$

   Thus 1,200 Swiss francs is approximately $888.89 US.

Step 4: Check your work.
   1. Check the numbers and your calculations.
   2. Does the result seem reasonable?
   3. Was the question answered?

Example 3: $20 US will purchase a 606 ruble item in Russia. How many rubles are equivalent to the dollar?
Solution:
Step 1: Understand and picture the problem.
   1. Question: How many rubles equal $1 US?
   2. The US dollar amount is 20 dollars.
   3. The ruble amount is 606 rubles.
   4. Let x represent the amount of rubles.

Step 2: Determine a plan.
   1. Use the Proportion Model for Monetary Conversion.
   2. Use the Fundamental Principal of Proportions to determine the cross product of the proportions.
   3. Use the Multiplication | Division Principle of Equality to solve for x.

Step 3: Execute the plan.

1. $$\frac{\$1 \text{ US}}{\text{foreign equivalent}} = \frac{\text{dollar amount (US)}}{\text{foreign equivalent amount}}.$$

   Write the proportion. $\quad \dfrac{1}{x} = \dfrac{20}{606}.$

2. *Write the cross products.* $\frac{1}{x} \asymp \frac{20}{606}$,   1(606) = 20(x),   606 = 20x.

3. *Multiply by the reciprocal of 20.*

$$\frac{1}{20}(606) = \frac{1}{20}(20)x, \quad \frac{606}{20} = x, \quad 30.3 = x.$$

There are 30.3 rubles per US dollar.

Step 4: Check your work.
    1. Check the numbers and your calculations.
    2. Does the result seem reasonable?
    3. Was the question answered?

Geometric Proportions: Map and Blueprint Reading
Another application of proportions is comparing distances. These distances are usually dimensions read from a map or blueprint and have to be converted to "actual" dimensions. The scale of the map or blueprint is given in the legend.

Proportion Model for Geometric Problems

$$\frac{1}{\text{scale}} = \frac{\text{measured distance}}{\text{actual distance}}$$

Where: **Scale** is the conversion factor taken from the legend.
    Measured distance is the distance measured between the two points on the map or blueprint.
    Actual distance is the "real life" distance.

Example 4: A pilot plans to fly from Chicago, IL to St. Louis, MO. The scale on the map indicates one inch equals 12 miles. The measured distance between Chicago, IL and St. Louis, MO is 24.75 inches. How far is it from Chicago, IL to St. Louis, MO?

Solution:
Step 1: Understand and picture the problem.
    1. Question: How many miles is it from Chicago, IL to St. Louis, MO?
    2. The scale is 1 inch equals 12 miles or $\frac{1 \text{ inch}}{12 \text{ miles}}$.
    3. The measured distance on the map is 24.75 inches.
    4. Let "x" represent the "real life" distance.

Step 2: Determine a plan.
   1. Use the Proportion Model for Geometric Problems.
   2. Use the Fundamental Principal of Proportions to determine the cross product of the proportions.
   3. Use the Multiplication | Division Principle of Equality to solve for x.

Step 3: Execute the plan.
   1. *Write the proportion.* $\dfrac{1}{\text{scale}} = \dfrac{\text{measurement}}{\text{actual}}$ $\dfrac{1}{12} = \dfrac{24.75}{x}$.

   2. *Write the cross products.* $\dfrac{1}{12} \diagup\!\!\!\!\!= \dfrac{24.75}{x}$, $1(x) = 12(24.75)$.

   3. $x = 297$, The distance from Chicago, IL to St. Louis, MO is 297 miles.

Step 4: Check your work.
   1. Check the numbers and your calculations.
   2. Does the result seem reasonable?
   3. Was the question answered?

Example 5: An architect needs to represent a distance of 61.5 feet (length of a house) on a blueprint. The scale is one inch equals 4 feet. How long of a line does the architect draw on the blueprint to represent this distance.
Solution:
Step 1: Understand and picture the problem.
   1. Question: What line length should be drawn on the blueprint to represent a distance of 61.5 feet.
   2. The scale is 1 inch equals 4 feet or $\dfrac{1 \text{ inch}}{4 \text{ feet}}$.
   3. The real life distance 61.5 feet.
   4. Let "x" represent the length of the line on the blueprint.

Step 2: Determine a plan.
   1. Use the Proportion Model for Geometric Problems.
   2. Use the Fundamental Principal of Proportions to determine the cross product of the proportion.
   3. Use the Multiplication | Division Principle of Equality.

Step 3: Execute the plan.
   1. *Write the proportion.* $\dfrac{1}{\text{scale}} = \dfrac{\text{measurement}}{\text{actual}}$, $\dfrac{1}{4} = \dfrac{x}{61.5}$.

   2. *Write the cross products.* $\dfrac{1}{4} \diagup\!\!\!\!\!= \dfrac{x}{61.5}$, $4(x) = 1(61.5)$, $4x = 61.5$.

3. *Multiply by the reciprocal of 4.*

$$\left(\frac{1}{4}\right)(4x) = \left(\frac{1}{4}\right)(61.5), \quad \left(\frac{1}{\cancel{4}}\right)(\cancel{4})x = \left(\frac{1}{4}\right)(61.5), \quad x = \frac{61.5}{4} = 15.375.$$

The line drawn on the blueprint is 15.375 inches long.

Step 4: Check your work.
    1. Check the numbers and your calculations.
    2. Does the result seem reasonable?
    3. Was the question answered?

For Your Health
Walking and jogging are ways to burn calories. Ratio and proportions are used to determine the distance needed to burn the calories from the food we eat.

Proportion Model for Burning Calories

$$\frac{\text{Activity}}{\text{Unit Distance or Unit Time}} = \frac{\text{Calories in Food}}{\text{Distance or Time}}$$

Where:
Activity represents the number of calories your body burns performing the activity.
Unit Distance or Unit Time represents 1 mile or 1 hour
Calories in Food is the number of calories in a particular food you have eaten.
Distance or Time is the distance (miles) or time (hours) required to burn the calories of the food you have eaten.

Example 6: Walking burns 70 calories per mile. A Burger King Whopper contains 670 calories. How far would you have to walk to burn these calories?
Solution:
Step 1: Understand and picture the problem.
    1. Question: What walking distance is required to burn the calories from a Burger King Whopper?
    2. Walking burns $\frac{70 \text{ calories}}{1 \text{ mile}}$.
    3. A Burger King Whopper has 670 calories.
    4. Let "x" represent the distance in miles needed to burn the calories in a Whopper.

52

Step 2: Determine a plan.
1. Use the Proportion Model for Burning Calories.
2. Use the Fundamental Principal of Proportions to determine the cross product of the proportion.
3. Use the Multiplication | Division Principle of Equality.

Step 3: Execute the plan.
1. *Write the proportion.*

$$\frac{\text{Activity}}{\text{Unit Distance or Unit Time}} = \frac{\text{Calories in Food}}{\text{Distance or Time}}, \quad \frac{70}{1} = \frac{670}{x}.$$

2. *Write the cross products.* $\quad \frac{70}{1} \diagup \frac{670}{x}, \quad 70(x) = 1(670).$

3. *Multiply by the reciprocal of 70.*

$$\left(\frac{1}{70}\right)(70x) = \left(\frac{1}{70}\right)(670), \quad \left(\frac{1}{\cancel{70}}\right)(\cancel{70})x = \left(\frac{1}{70}\right)(670), \quad x = \frac{670}{70} \approx 9.57.$$

It takes walking 9.57 miles to burn the calories from eating a Burger King Whopper.

Step 4: Check your work.
1. Check the numbers and your calculations.
2. Does the result seem reasonable?
3. Was the question answered?

Example 7: Leisurely freestyle swimming burns 563 calories in 1 hour. The Taco Bell taco supreme has 260 calories. How long would you have to swim to burn these calories?
Solution:
Step 1: Understand and picture the problem.
1. Question: How long do you need to swim to burn the calories from a Taco Bell taco supreme?
2. Swimming Freestyle burns $\frac{563 \text{ calories}}{1 \text{ hour}}$.
3. A Taco Bell taco supreme has 260 calories.
4. Let "x" represent the time in hours needed to burn the calories in a taco supreme.

Step 2: Determine a plan.
    1. Use the Proportion Model for Burning Calories.
    2. Use the Fundamental Principal of Proportions to determine the cross product of the proportion.
    3. Use the Multiplication | Division Principle of Equality.

Step 3: Execute the plan.
    1. *Write the proportion.*
$$\frac{\text{Activity}}{\text{Unit Distance or Unit Time}} = \frac{\text{Calories in Food}}{\text{Distance or Time}}, \quad \frac{563}{1} = \frac{260}{x}.$$

    2. *Write the cross products.* $\quad 563(x) = 1(260).$

    3. *Multiply by the reciprocal of 563.*

$$\left(\frac{1}{563}\right)(563x) = \left(\frac{1}{563}\right)(260), \quad \left(\frac{1}{563}\right)(563)x = \left(\frac{1}{563}\right)(260),$$

$$x = \frac{260}{563} \approx 0.46.$$

It requires approximately 0.46 hours to burn the calories from eating a Taco Bell taco supreme.

Note: $0.46 \text{ hours} \left(\frac{60 \text{ minutes}}{1 \text{ hour}}\right) \approx 27.6 \text{ minutes}$

Step 4: Check your work.
    1. Check the numbers and your calculations.
    2. Does the result seem reasonable?
    3. Was the question answered?

Mixtures

Nursing, pharmacy, dietician, photography, etc. are professions that use ratios and proportions. If the mixture is too strong or too weak, the results could be catastrophic or lethal. The proportion model for diluting mixture problems is

---

Proportion Model for Diluting Mixture Problems

$$\frac{FS}{OS} = \frac{OA}{FA} \quad \text{and } OS \neq 0, FA \neq 0$$

Where: **FS** is the **F**inal **S**trength (percent) needed of the final solution.
           OS is the **O**riginal **S**trength (percent) of the original solution.
           OA is the **O**riginal **A**mount of the original solution.
           FA is the **F**inal **A**mount needed of the final solution.

However, it is easier to work with the cross product when solving mixtures dilution problems.

Cross Product Model for Diluting Mixture Problems

$$\frac{FS}{OS} \times \frac{OA}{FA} \quad \text{yields} \quad (FS)(FA) = (OS)(OA)$$

Where: FS is the Final Strength (percent) needed of the final solution.
OS is the Original Strength (percent) of the original solution.
OA is the Original Amount of the original solution.
FA is the Final Amount needed of the final solution.

Example 8: A nurse needs to prepare 12 ml of a solution that is 8% strong from a solution that is 20% strong. How much 20% solution and distilled water are needed to make the required solution?

Solution:

Step 1: Understand and picture the problem.
1. 12 ml is the final amount needed (FA = 12).
2. 8% is the final strength (percent) of final solution (FS = 8).
3. 20% is the original strength (percent) (OS = 20).
4. Question: Solve for the remaining unknown, OA, the amount of 12% solution used to make the diluted solution.

Step 2: Determine a plan.
1. Use the Cross Product Model for Diluting Mixture Problems.
2. Use the Multiplication | Division Principle of Equality.

Step 3: Execute the plan.
1. *Write the cross products.*
   (FS)(FA) = (OS)(OA),   8(12) = 20(x)   96 = 20x.
2. *Multiply by the reciprocal of 20.*

$$\frac{1}{20}(96) = \frac{1}{20}(20x), \quad \frac{1}{20}(96) = \left(\frac{1}{20}\right)(20x), \quad \frac{96}{20} = x, \quad 4.8 = x.$$

Thus, 4.8 ml of 20% solution is used to make 12 ml of the 8% solution. This must be added to 7.2 ml of distilled water. (Note: 12 − 4.8 = 7.2)

Step 4: Check your work.
1. Check the numbers and your calculations.
2. Does the result seem reasonable?
3. Was the question answered?

Example 9: A pharmacist needs to prepare 250 ml of a solution that is 42% strong from a pure solution. How much pure solution and distilled water is needed to make the required solution?

Solution:

Step 1: Understand and picture the problem.
1. 250 ml is the final amount needed (FA = 250).
2. 42% is the final strength (percent) needed (FS = 42).
3. Pure is 100%, the original strength (percent) of the original solution being diluted (OS = 100).
4. Problem: Solve for the remaining unknown, OA, the amount of the original solution.

Step 2: Determine a plan.
1. Use the Cross Product Model for Diluting mixture Problems.
2. Use the Multiplication | Division Principle of Equality.

Step 3: Execute the plan.
1. *Write the cross products.*
   (FS)(FA) = (OS)(OA),   42(250) = 100(x)   10,500 = 100x.
2. *Multiply by the reciprocal of 100.*

$$\frac{1}{100}(10,500) = \frac{1}{100}(100)x, \quad \left(\frac{1}{\cancel{100}}\right)(\cancel{10,500}^{105}) = \left(\frac{1}{\cancel{100}}\right)(\cancel{100}^{1})x, \quad 105 = x.$$

Thus, 105 ml 100% (pure) solution must be added to 145 ml of distilled water to make 250 ml of 42% solution. (Note: 250 – 105 = 145)

Step 4: Check your work.
1. Check the numbers and your calculations.
2. Does the result seem reasonable?
3. Was the question answered?

(c) 2006 JupiterImages Corp.

Optional Student Project
"Proportion Application" can be found on page 76.

? Cognitive Problems 1.4 ?

1. Compose a proportion problem for a dietician or baker.
2. Explain the significance of 100% used for a pure solution.
3. Describe the problems that can result if a medication is too weak or too strong.
4. Give at least one consequence of what can happen if a building contractor misreads a blueprint.
5. When would 0% be used in a mixture problem?

Exercise 1.4

1. Given that $1 US is equivalent to 1.4 Canadian dollars. What is $300 US worth in Canadian dollars?

2. Given that $1 US is equivalent to 10.72 Mexican pesos. What is $240 US worth in Mexican peso?

3. Given that $1 US is equivalent to 3.86 Polish zlotys. What is 2,000 Polish zlotys worth in US dollars?

4. Given that $1 US is equivalent to 28.38 Czech korunas. What is 480 Czech korunas worth in US dollars?

5. Given that $1 US is equivalent to 4.46 Israeli shekels. What is 5,500 Israeli shekels worth in US dollars?

6. $70 US will purchase a 49 dinar item in Jordan. How many dinars are equivalent to $1 US?

7. $600 US will purchase a 34,600 rupee item in Pakistan. How many rupees are equivalent to $1 US?

8. $120 will purchase a 7.25 krone item in Norway. How many krones are equivalent to $1 US?

9. On a blueprint, 1 inch equals 4 feet. If the length of the living room measures 4.25 inches by 3.75 inches, what are the actual dimensions of the living room?

10. On a blueprint, 1 inch equals 4 feet. If the width of the house measures 7.75 inches, what is the actual width of the house?

11. On a map, 1 in equals 75 miles. The measured distance from New York City, NY to Boston, MA is 3 in. What is the actual distance from New York City, NY to Boston, MA?

12. On a map, 1 inch equals 300 miles. The measured distance from Los Angeles, CA to New York City, NY is 9.25 inches. What is the actual distance from Los Angeles, CA to New York City, NY?

13. On a blueprint, 1 inch equals 8 feet. The building being drawn has a length of 244 feet. What is the length of the building on the blueprint?

14. On a blueprint, 1 inch equals 8 feet. The building being drawn has a length of 366 feet. What is the length of the building on the blueprint?

15. On a blueprint, 1 inch equals 4 feet. The house being drawn has a length of 52 feet and a width of 31 feet. How many inches is the length and width of the house on the blueprint?

16. On a blueprint, 1 inch equals 12 feet. The diameter of a circular oil storage tank is 54 feet. What is the diameter of the oil storage tank on the blueprint?

17. Walking burns 70 calories per mile. A McDonald's Big Mac contains 590 calories. How far would you have to walk to burn the calories from eating a Big Mac?

18. Jogging burns 105 calories per mile. A McDonald's Egg McMuffin contains 275 calories. How far would you have to jog to burn the calories from eating an Egg McMuffin?

19. Power walking burns 225 calories per mile. An order of five Burger King's french toast sticks contains 390 calories. How far would you have to power walk to burn the calories from eating five Burger King's french toast sticks?

20. Swimming freestyle burns 400 calories per mile. Wendy's big bacon classic contains 580 calories. How far would you have to swim freestyle to burn the calories in Wendy's big bacon classic?

21. Playing tennis burns 493 calories per hour. A 6 inch Subway BMT contains 415 calories. How long would you have to play tennis to burn the calories in a 6 inch Subway BMT?

22. Skateboarding burns 352 calories per hour. Six pieces of KFC's honey bbq contain 607 calories. How long would you have to skateboard to burn the calories in six pieces of KFC bbq?

23. Noncompetitive soccer burns 493 calories per hour. Burger King's enormous omelet sandwich contains 730 calories. How long would you have to play noncompetitive soccer to burn the calories in this breakfast sandwich?

24. Competitive soccer burns 704 calories per hour. Burger King's enormous omelet sandwich contains 730 calories. How long would you have to play competitive soccer to burn the calories in this breakfast sandwich?

25. A pharmacist needs to prepare 350 ml of a medicine that is 20% strong. How much 50% solution and distilled water is needed to fill this prescription?

26. A pharmacist needs to prepare 240 ml of a medicine that is 12% strong. How much pure solution and distilled water is needed to fill this prescription?

27. A nurse needs to prepare 10 cc of a medicine that is 75% strong. How much pure solution and distilled water is needed to fill this order?

28. A nurse needs to prepare 8 cc a medicine that is 48% strong. How much 80% solution and distilled water is needed to fill this order?

1.5 SOLVING LINEAR EQUATIONS
Objective: Solve linear equations.

Thoughts and relationships are expressed in Mathematics, just as they are in English. An equation is a form of communication that expresses a problem to be solved. Each equation is a sentence that has its own set of grammar rules. One rule is that every equation must contain an equal sign. There are many types of equations, but only linear equations will be studied here. Study the following examples of linear equations. List their attributes.

$$3x + 2 = 6$$
$$x = 3$$
$$y = -4$$
$$12 = 7x - 4$$
$$y = 2x + 5$$
$$4x + 2y = 36$$

As a comparison, peruse the following samples of nonlinear equations. How are they different from the above linear equations?

$$4x^3 = 12$$
$$0 = 5x^2 + 3x - 5$$
$$x^{-3} = 1000$$
$$xy = 24$$

Notice that the variables of a linear equation are raised only to the first power. The variables in a linear equation may be added or subtracted, but they may not be multiplied or divided by other variables.

It is necessary to solve linear equations to accomplish many mathematical tasks. The process of finding the solution of a linear equation is an algorithm. Solving linear equations requires finding the solution by isolating the chosen variable. This means that the chosen variable is only on one side of the equal sign and a number is on the other side of the equal sign, like $x = 5$ or $5 = x$.

The process to accomplish this is similar to keeping a scale in balance. When a weight is placed on one side, an equal weight must be placed on the other side in order to keep the scale in balance. The same principle applies when a weight is removed from one side of the scale. The same weight must be removed from the other side of the scale to keep the scale balanced.

(c) 2006 JupiterImages Corp.

The equal sign in an equation serves the same function as the pivot on a balanced beam scale. When a number is added to one side of the equation, the same value must be added to the other side in order to keep the equation in balance. Since subtraction is equivalent to adding the opposite, this principle also applies to subtraction. The rule is called the Addition | Subtraction Principle of Equality.

> **Addition | Subtraction Principle of Equality**:
>
> If $a = b$, Then $a + c = b + c$
> for all real numbers a, b, c

For example, solve $x + 8 = 14$. The objective is accomplished by isolating the x. Undo the addition of 8 to the x on the left side of the equation. Add the opposite, –8, (called the additive inverse). By the Addition | Subtraction Principle of Equality, –8 must be added to both the left and right sides of the equation. Thus,

$x + 8 = 14$ *Original problem.*
$x + 8 + (-8) = 14 + (-8)$ *Addition | Subtraction Prin. for Equality, add a negative eight (–8) to both sides of the equation.*
$x = 6$ *Simplify the expressions on both sides of the equal sign. (Note: $x + 0 = x$ on the left side of the equation.)*

The solution to the equation $x + 8 = 16$ seems to be 6. The solution may be checked by substituting the number, 6, into the original equation. Substitute 6 for x into the original equation and perform the arithmetic to verify its validity.

$x + 8 = 14$ *Original problem.*
$6 + 8 = 14$ *Substitute 6 for x.*
$14 = 14$ *Calculate 6 + 8.*

It does work, the solution to the equation $x + 8 = 14$ is 6.

Example 1: Solve $x - 21 = -12$
Solution:
Step 1: Understand and Picture the Problem.
    1. Question: Solve the equation: x minus 21 equals negative 12.
    2. Identify the problem as a linear equation.
    3. Isolate the variable, x, on the left side of the equation by removing the subtraction of 21 from x.

Step 2: Develop a plan.
Use the Addition | Subtraction Principle of Equality to isolate the variable. Add 21 to both sides of the equation to undo the subtraction of 21 to the variable.

Step 3: Execute the plan.

$x - 21 = -12$	*Original problem.*
$x - 21 + 21 = -12 + 21$	*Add 21 to both sides of the equation.*
$x = 9$	*Simplify both sides of the equation. (Note: $x + 0 = x$ on the left side of the equation.)*

Step 4: Check your work.

$x - 21 = -12$	*Original problem.*
$9 - 21 = -12$	*Substitute 9 for x and simplify.*
$-12 = -12$	*It works, $x = 9$.*

The solution to the equation $x - 21 = -12$ is 9. The solution is reasonable and the question is answered.

The same concept applies to equations that contain multiplication or division. The Multiplication | Division Principle of Equality is:

Multiplication | Division Principle of Equality:

If $a = b$, Then $ac = bc$
For all real numbers a, b, c with $c \neq 0$.

For example, solve $6x = 27$. The objective is accomplished by isolating the x. Undo the multiplication of 6 to the variable. Multiply 6 by its reciprocal, $\frac{1}{6}$ (or divide by 6). Use the Multiplication | Division Principle of Equality. Both sides of the equation must be multiplied by $\frac{1}{6}$ (or divided by 6). Thus,

$6x = 27$	*Original problem.*
$\left(\frac{1}{6}\right) 6x = \left(\frac{1}{6}\right) 27$	*Multiplication	Division Principle of Equality, multiply both sides of the equation by $\left(\frac{1}{6}\right)$.*

$$\left(\frac{1}{\cancel{6}}\right)(\cancel{6})x = \left(\frac{1}{\cancel{6}}\right)(\cancel{27})$$

Simplify both sides of the equation.

$$x = \frac{9}{2} = 4.5$$

(Note: 1x = x on the left side of the equation.)

The solution is checked by substituting the solution into the original equation to verify it works.

6x = 27	*Original problem.*
6(4.5) = 27	*Substitute 4.5 for x and simplify.*
27 = 27	

It does work, the solution to the equation 6x = 27 is 4.5.

Example 2: Solve $-8 = \frac{4}{5}x$

Solution:

Step 1: Understand and Picture the Problem.
1. Question: Find the solution to the equation: Negative 8 equals four fifths times x.
2. Identify the problem as a linear equation.
3. Isolate the variable, x, on the right side of the equation by removing the multiplication of x by $\frac{4}{5}$.

Step 2: Develop a plan.
1. Use the multiplicative inverse, reciprocal, to undo the multiplication by $\frac{4}{5}$.
2. Use the Multiplication | Division Principle of Equality to isolate the variable. Multiply both sides of the equation by $\frac{5}{4}$, which is the reciprocal of $\frac{4}{5}$.

Step 3: Execute the plan.

$$-8 = \frac{4}{5}x \qquad \textit{Original problem.}$$

$$\left(\frac{5}{4}\right)(-8) = \left(\frac{5}{4}\right)\left(\frac{4}{5}\right)(x)$$ *Multiply both sides of the equation by $\frac{5}{4}$.*

$$\left(\frac{5}{\cancel{4}}\right)(\cancel{-8}^{-2}) = \left(\frac{\cancel{5}^1}{\cancel{4}^1}\right)\left(\frac{\cancel{4}^1}{\cancel{5}^1}\right)x$$ *Simplify both sides of the equation.*

$$-10 = x$$ *(Note: $1x = x$ on the right side of the equation.)*

Step 4: Check your work.

$$-8 = \frac{4}{5}x$$ *Original problem.*

$$-8 = \frac{4}{5}(-10)$$ *Substitute -10 for x and simplify.*

$$-8 = -8$$ *It does work, $x = -10$.*

The solution to the equation $-8 = \frac{4}{5}x$ is -10. The solution is reasonable and the question is answered.

Both the Addition and Multiplication Principles for Equality can apply to the same equation. For example, solve the linear equation $3x + 5 = 17$. To accomplish the objective, focus on the variable and read the problem. "Three times x plus five equals seventeen". The variable x is multiplied by 3 and then 5 is added to this product. **To isolate the x, the inverse operations must be used in the reverse order.** This is accomplished by first subtracting 5 and then multiplying by $\frac{1}{3}$ to both sides of the equation. The 5 must be removed first. This reversal of steps is similar to wrapping and unwrapping a present. To wrap a present, first wrapping paper, then ribbon is applied to the present. To open the present, the ribbon is removed first, then the wrapping paper. It is this taking apart process that mandates that the addition be undone first to solve an equation. Thus,

$3x + 5 = 17$ *Original problem.*
$3x + 5 + (-5) = 17 + (-5)$ *Add -5, to both sides of the equation, the opposite of 5.*
$3x = 12$ *Simplify.*

$$\left(\frac{1}{3}\right)(3)(x) = 12\left(\frac{1}{3}\right)$$ *Multiply both sides of the equation by $\frac{1}{3}$, the reciprocal of 3.*

$$\left(\frac{1}{\cancel{3}}\right)(\cancel{3})x = \left(\frac{1}{\cancel{3}}\right)(\cancel{12}^{4})$$ *Simplify.*

$$x = 4$$ *(Note $1x = x$ on the left side of the equation.)*

This solution can be checked by substituting the result into the original equation to verify that it works.

$3x + 5 = 17$	*Original problem.*
$3(4) + 5 = 17$	*Substitute 4 for x.*
$12 + 5 = 17$	*Perform the math.*
$17 = 17$	*It does work, $x = 4$.*

The solution to the equation $3x + 5 = 17$ is 4. The solution is reasonable and the question is answered.

Example 3: Solve $14 = -2x - 8$
Solution:
Step 1: Understand and Picture the Problem.
   1. Question: Find the solution to the equation: 14 equals negative 2 times x minus 8.
   2. Identify the problem as a linear equation.
   3. The last operation performed to the variable is the first operation that has to be removed (the unwrapping process).
      a. The removal of the subtraction has to be performed first.
      b. The removal of the multiplication is performed next.

Step 2: Develop a plan.
   1. The subtraction of 8 is undone by adding 8.
   2. Use the Addition | Subtraction Principle of Equality, add 8 to both sides of the equation to undo the subtraction to the variable.
   3. The multiplication by -2 is undone by using the multiplicative inverse, $\frac{-1}{2}$, or divide by (-2).
   4. Use the Multiplication | Division Principle of Equality, multiply both sides of the equation by $\frac{-1}{2}$ to undo the multiplication to the variable.

Step 3: Execute the plan.

$14 = -2x - 8$ — *Original problem.*

$14 + 8 = -2x - 8 + 8$ — *Use the Addition | Subtraction Principle of Equality, add 8 to both sides of the equation.*

$22 = -2x$ — *Simplify.*

$\left(\dfrac{-1}{2}\right)(22) = \left(\dfrac{-1}{2}\right)(-2x)$ — *Use the Multiplication | Division Principle of Equality, multiply both sides of the equation by $\dfrac{-1}{2}$.*

$\left(\dfrac{-1}{\cancel{2}}\right)(\cancel{22}^{11}) = \left(\dfrac{-1}{\cancel{2}}\right)(\cancel{-2}^{(-1)})x$ — *Simplify.*

$-11 = x$ — *(Note: $(-1)(-1)x = 1x = x$ on the right side of the equation.)*

Step 4: Check your work.

$14 = -2x - 8$ — *Original problem.*
$14 = -2(-11) - 8$ — *Substitute –11 for x.*
$14 = 22 - 8$ — *Perform the math.*
$14 = 14$ — *It does work, x = –11.*

The solution to the equation $14 = -2x - 8$ is –11. The solution is reasonable and the question is answered.

Another variation of a linear equation involves grouping symbols. In this type of problem, the grouping symbols must be removed before solving for the variable. Consider the equation $5(3x - 12) = -30$. Read the problem: "5 times the difference of 3x and 12 equals –30". The equation must be simplified by multiplying the terms within the parenthesis by the factor 5. (Remember the Distributive Property?)

$5(3x - 12) = -30$
$15x - 60 = -30$ — *Distributive Property.*

Now the Addition Principle and the Multiplication Principle can be applied to find a solution to the equation. Completing the solution:

$15x - 60 = -30$ — *Original problem.*
$15x - 60 + 60 = -30 + 60$ — *Addition | Subtraction Principle of Equality, add 60 to both sides of the equation.*
$15x = 30$ — *Simplify.*

$$\left(\frac{1}{15}\right)(15x) = \left(\frac{1}{15}\right)(30)$$ *Multiplication | Division Principle of Equality, multiply both sides of the equation by $\frac{1}{15}$.*

$$\left(\frac{1}{\cancel{15}}\right)(\cancel{15})x = \left(\frac{1}{\cancel{15}}\right)(\cancel{30})^2$$ *Simplify.*

$$x = 2$$ *(Note: 1x = x on the left side of the equation.)*

Checking the solution in the original equation,

15x − 60 = −30	*Original problem.*
15(2) − 60 = −30	*Substitute 2 for x.*
30 − 60 = −30	*Simplify.*
−30 = −30	*It does work, x = 2.*

The solution to the equation 15x − 60 = −30 is 2. The solution is reasonable and the question is answered.

Example 4: Solve 4(3x + 6) + 2 = 2.
Solution:
Step 1: Understand and Picture the Problem.
    1. Question: Find the solution to the equation: 4 times the sum of 3 times x and 6 added to 2 equals 2.
    2. Identify the problem as a linear equation.
    3. First the parenthesis must be removed by using the Distributive Property. Then simplify the expression on the left side of the equation by combining like terms.
    4. The addition to the variable must be undone first.

Step 2: Develop a plan.
    1. Use the Distributive Property, to undo the parenthesis in the linear equation. Multiply all the terms within the parenthesis by 4.
    2. Combine like terms.
    3. Use the Addition | Subtraction Principle for Equality, to undo the addition to the variable.
    4. Use the Multiplication | Division Principle for Equality to undo the Multiplication to the variable.

Step 3: Execute the plan.

$$4(3x + 6) + 2 = 2 \quad \textit{Original problem.}$$
$$12x + 24 + 2 = 2 \quad \textit{Use the Distributive Property, multiply each term by 4.}$$
$$12x + 26 = 2 \quad \textit{Simplify by combining like terms.}$$
$$12x + 26 - 26 = 2 - 26 \quad \textit{Use the Addition | Subtraction Principle of Equality, subtract 26 from both sides of the equation.}$$
$$12x = -24 \quad \textit{Simplify by combining like terms.}$$
$$\left(\frac{1}{12}\right)(12x) = \left(\frac{1}{12}\right)(-24) \quad \textit{Use the Multiplication | Division Principle of Equality, multiply both sides of the equation by } \frac{1}{12}.$$

$$\left(\frac{1}{\cancel{12}}\right)(\cancel{12})x = \left(\frac{1}{\cancel{12}}\right)(\cancel{-24}) \quad \textit{Simplify.}$$

$$x = -2$$

Step 4: Check your work.

$$4(3x + 6) + 2 = 2 \quad \textit{Original problem.}$$
$$4(3(-2) + 6) + 2 = 2 \quad \textit{Substitute –2 for x.}$$
$$4(-6 + 6) + 2 = 2 \quad \textit{Perform the math using the order}$$
$$4(0) + 2 = 2 \quad \textit{of operations.}$$
$$0 + 2 = 2$$
$$2 = 2 \quad \textit{It does work, x = –2.}$$

The solution to the equation $4(3x + 6) + 2 = 2$ is –2. The solution is reasonable and the question is answered.

Yet another variation to the linear equation involves variables appearing on both sides of the equal sign. In this problem, use the Addition Principle of Addition to move the variable to only one side of the equal sign in the equation. Usually the variable with the smaller coefficient is removed to simplify the problem. (Remember that a coefficient is a factor of the variable, as 3 is the coefficient of 3z.) For example, consider the equation: $5z + 12 = 3z - 20$, the 3z is removed from the right-hand side because between 5z and 3z, 3 is the smaller of the two coefficients, 3 and 5.

Example 5: Solve $5z + 12 = 3z - 20$.
Solution:
Step 1: Understand and Picture the Problem.
   1. Question: Find the solution to the equation: 5 times z plus 12 equals 3 times z minus 20.
   2. Identify the problem as a linear equation.
   3. Variables are on both sides of the equal sign.
   4. The variable must be on only one side of the equal sign.

Step 2: Develop a plan.
   1. Use the Addition | Subtraction Principle of Equality to remove the smaller of the variable terms from one side of the equation. Rewrite the equation.
   2. Subtract $3z$ from both sides of the equation.
   3. Use the Addition | Subtraction Principle of Equality to undo the addition to the variable.
   4. Use the Multiplication | Division Principle of Equality to undo the multiplication, to the variable.

Step 3: Execute the plan.

$5z + 12 = 3z - 20$	*Original problem.*
$5z + 12 - 3z = 3z - 20 - 3z$	*Use the Addition	Subtraction Principle of Equality, subtract $3z$ from both sides of the equation.*
$2z + 12 = -20$	*Simplify by combining like terms.*
$2z + 12 - 12 = -20 - 12$	*Use the Addition	Subtraction Principle of Equality, subtract 12 from both sides of the equation.*
$2z = -32$	*Simplify by combining like terms.*
$\left(\dfrac{1}{2}\right)(2z) = \left(\dfrac{1}{2}\right)(-32)$	*Use the Multiplication	Division Principle of Equality, multiply both sides of the equation by $\dfrac{1}{2}$.*
$\left(\dfrac{1}{\cancel{2}}\right)(\cancel{2})z = \left(\dfrac{1}{\cancel{2}}\right)(\cancel{-32}^{-16})$	*Simplify.*
$z = -16$	

Step 4: Check your work.

$$5z + 12 = 3z - 20 \quad \textit{Original problem.}$$
$$5(-16) + 12 = 3(-16) - 20 \quad \textit{Substitute -16 for each z.}$$
$$-80 + 12 = -48 - 20 \quad \textit{Perform the math using the order of operations.}$$
$$-68 = -68 \quad \textit{It does work, z = -16.}$$

The solution to the equation $5z + 12 = 3z - 20$ is -16.

Summary of Steps for Solving Linear Equations (Final)
1. Use the Distributive Property, to remove the parenthesis in the linear Equation and simplify, if possible.
2. Use the Addition Property of Equations to get the variable on one side of the equal (=) sign.
3. Use the Addition Property of Equations to isolate the variable.
4. Use the Multiplication Property of Equations to get 1x.
5. Check the result.

Applications of Linear Equations

One major factor in the selling price of a house is its area. The larger the living area, the more expensive the house.

Example 6: Suppose the formula defining the selling price of a townhouse in a suburb is: Selling Price equals $130.80 times (Living Area) plus $24,955.70.

$$P = \$130.80(A) + \$24,955.70.$$

where: P = Selling Price, A = Living Area

Find the living area (square feet) of a townhouse costing $150,000.

Solution:

Step 1: Understand and Picture the Problem.
1. Question: Find the square footage of a $150,000 townhouse.
2. The equation used is P = $130.80(A) + $24,955.70.
3. Identify the problem as a linear equation to be solved.
4. Selling price is $150,000.
5. Read the problem. The last operation performed, is the first operation that has to be removed (the unwrapping process).
6. The removal of the addition has to be performed first.
7. Isolate the "Living Area" on the right side of the equation by undoing the addition of $24,955.70, then the multiplication of $130.80.

Step 2: Develop a plan.
1. The addition of 24,955.70 is undone by subtracting 24,955.70.
2. Use the Addition | Subtraction Principle of Equality, subtract 24,955.70 from both sides of the equation.
3. The multiplication by 130.80 is undone by using the multiplicative inverse, $\frac{1}{130.80}$, or divide by (130.80).
4. Use the Multiplication | Division Principle of Equality, multiply both sides of the equation by $\frac{1}{130.80}$ to undo the multiplication to the variable.

Step 3: Execute the plan.
150,000 = 130.80(A) + 24,955.70   *Original problem.*

150,000 − 24,955.70 = 130.80(A) + 24,955.70 − 24,955.70   *Use the Addition | Subtraction Principle of Equality, subtract 130.80 from both sides of the equation.*

125,044.30 = 130.80(A)   *Simplify.*

$\frac{1}{130.80}(125,044.30) = 130.80\left(\frac{1}{130.80}\right)(A)$ *Use the Multiplication | Division Principle of Equality, multiply both sides of the equation by $\frac{1}{130.80}$.*

$\frac{1}{130.80}(125,044.30) = \frac{1}{\cancel{130.80}}(\cancel{130.80})(A)$ *Simplify.*

$\frac{125,044.30}{130.80} \approx 955.996 = A$

The area of the townhouse is approximately 956 square feet.

Step 4: Check your work.
$150,000 = $130.80(A) + $24,955.70   *Original problem.*
$150,000 = $130.80(956) + $24,955.70   *Substitute 996 for A.*
$150,000 = $125,044.80 + $24,955.70   *Perform the math.*
$150,000 ≈ $150,000.50   *It does work, the difference is due to rounding the living area.*

The solution to the equation $150,000 = $130.80(A) + $24,955.70 is approximately 956 square feet.

Many scientific formulas use degrees Celsius for their temperature measurements. Since many people are familiar with degrees Fahrenheit, a formula is needed to convert degrees Fahrenheit to degrees Celsius. This formula is

$$C = \frac{5}{9}(F - 32)$$

Example 7: Water boils at 100 °C. What is the boiling point of water in degrees Fahrenheit?
Solution:
Step 1: Understand and Picture the Problem.
1. Question: Find the solution to the equation: 100 degrees Celsius equals $\frac{5}{9}$ times the difference between degrees Fahrenheit and 32 degrees.
2. Identify the problem as solving a linear equation.
3. Celsius is 100 degrees.
4. First the parenthesis must be removed by using the Distributive Property. Then simplify the expression on the right side of the equation.
5. Isolate the "F" on the right side of the equation by removing the subtraction, then remove the multiplication.

Step 2: Develop a plan.
1. Use the Distributive Property, to undo the parenthesis in the linear equation. Multiply all the terms within the parenthesis by $\frac{5}{9}$.
2. Use the Addition | Subtraction Principle for Equality, to undo the subtraction to the variable.
3. Use the Multiplication | Division Principle for Equality to undo the Multiplication to the variable.

Step 3: Execute the plan.

$100 = \frac{5}{9}(F - 32)$ *Original problem.*

$100 = \frac{5}{9}F - \frac{160}{9}$ *Use the Distributive Property, multiply each term by $\frac{5}{9}$.*

$100 + \frac{160}{9} = \frac{5}{9}F - \frac{160}{9} + \frac{160}{9}$ *Use the Addition | Subtraction Principle of Equality, add $\frac{160}{9}$ to both sides of the equation.*

$$\frac{900}{9} + \frac{160}{9} = \frac{5}{9}F \qquad\qquad Simplify\ to \qquad \frac{1060}{9} = \frac{5}{9}F,$$

$$\left(\frac{9}{5}\right)\left(\frac{1060}{9}\right) = \left(\frac{9}{5}\right)\left(\frac{5}{9}\right)F \qquad ,,,,,,,,,,,Use\ the\ Multiplication\ |\ Division$$

Principle of Equality, multiply both sides of the equation by $\frac{9}{5}$.

$$\left(\frac{\cancel{9}}{\cancel{5}}\right)\left(\frac{\cancel{1060}^{212}}{\cancel{9}}\right) = \left(\frac{\cancel{9}}{\cancel{5}}\right)\left(\frac{\cancel{5}}{\cancel{9}}\right)F \qquad ,,,,,,,,,,,Simplify. \quad 212 = F$$

The boiling point of water is $212^\circ F$.

Step 4: Check your work.

$$100 = \frac{5}{9}(F - 32) \qquad\qquad Original\ problem.$$

$$100 = \frac{5}{9}(212 - 32) \qquad\qquad Substitute\ 212\ for\ F.$$

$$100 = \frac{5}{9}(180) \qquad\qquad Perform\ the\ math.$$

$$100 = \frac{5}{\cancel{9}}(\cancel{180}^{20}) \qquad 100 = 100 \quad It\ does\ work,\ F = 212.$$

The solution to the equation $100 = \frac{5}{9}(F - 32)$ is 212 degrees Fahrenheit.

The solution is reasonable and the question is answered.

 (c) 2006 JupiterImages Corp.	**Historical Perspective** **Diophantus (about 200 – 284) Greece**

Diophantus, is credited as the "Father of Algebra". He is best known for his "*Arithmetica*", a work on the solution of algebraic equations. The "*Arithmetica*", is the most outstanding work on algebra in Greek mathematics. It is a collection of 130 problems giving numerical solutions. Diophantus represented the unknown number he was solving for by a letter.

(c) 2006 JupiterImages Corp.

Student Project
"Linear Equations and Proportions Used in Radiology" can be found on page 77.

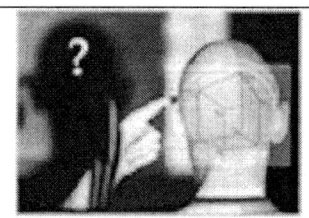
(c) 2006 JupiterImages Corp.

Chapter Review
Chapter review problems can be found on page 79.

❓ Cognitive Problems 1.5 ❓

1. State equivalent operations and give an example of each.
   a. Adding the opposite.   b. Multiplying by the reciprocal.
2. What are the attributes of a linear equation?
3. What is the sum of a pair of additive inverses? Give at least two examples.
4. What is the product of a pair of multiplicative inverses? Give at least two examples.
5. What is meant by finding the solution to an equation?
6. When undoing addition to a variable, why is its opposite value, the additive inverse, added to both sides of the equation?
7. When undoing a multiplication to a variable, why is it multiplied by its multiplicative inverse?
8. Why is it incorrect to add 3 to both sides of the equation, $-3x = 12$, when solving for x?
9. Why is $x = 6 + 2x$ not a solution. What is the solution to this linear equation?
10. Why does the Multiplication | Division Principle of Equality state $0 \neq 0$?

(c) 2006 JupiterImages Corp.

Exercise 1.5
Solve the following equations.

1. $x + 6 = 21$
2. $-36 = x + 10$
3. $17 + x = 11$
4. $49 + x = 21$
5. $x + \dfrac{3}{8} = 10$
6. $48 + x = \dfrac{201}{4}$

7. $12.4 = x - 7.9$
8. $x + 48.3 = 97.4$
9. $5x = 120$

10. $-3w = 42$
11. $144 = 36y$
12. $-4.8x = -240$

13. $-x = \dfrac{43}{8}$
14. $\dfrac{3}{4}x = 72$
15. $3.6z = -25.56$

16. $8.4x = 98.28$
17. $3x + 12 = 42$
18. $-3 = -2x + 7$

19. $48 - y = 79$
20. $18a - 24 = 30$
21. $5x + \dfrac{15}{8} = 10$

22. $-3x - \dfrac{35}{8} = \dfrac{17}{4}$
23. $4.2b + 9.4 = 25.36$
24. $29.3 = 2.7 - 7.6a$

25. $4(x + 7) = 44$
26. $6(x - 3) = -96$
27. $106 = -2(4x - 5)$

28. $5(-6x + 3) = 150$
29. $\dfrac{5}{8}(16x - 8) = 18$
30. $\dfrac{11}{16}(48x + 32) = -44$

31. $7.2(3.6z + 4) = 41.76$
32. $2.4(8.5x - 3.4) = 8.16$

33. $17w + 24 = 15w - 48$
34. $-7t - 19 = -12t + 21$

35. $4(2x + 6) = 6x + 6$
36. $-3(8x - 4) = 18x - 30$

37. $-2(8x - 5) = 3(-5x + 11)$
38. $9(3z - 4) = -3(5z - 37)$

39. $-3(6x - 12) + 20 = 5(4x - 2) + 47$
40. $8(5x - 3) + 14 = 3(4x + 6) + 7$

41. Suppose the selling price of a townhouse is linear and the equation is Selling Price = $130.80(Living Area) + $24,955.70. Find the living area of a $250,000 townhouse.

42. Suppose the selling price of a townhouse is linear and the equation is Selling Price = $130.80(Living Area) + $24,955.70. Find the living area of a $175,000 townhouse.

43. Suppose the selling price of a home is linear and the equation is Selling Price = $288.50(Living Area) + $49,790. Find the living area of a $500,000 home.

44. Suppose the selling price of a home is linear and the equation is
Selling Price = $288.50(Living Area) + $49,790. Find the living area of a $1,000,000 home.

45. The melting point of Gold is 1,068 $°C$. Find the melting point of Gold in degrees Fahrenheit.

46. The melting point of Lead is 327 $°C$. Find the melting point of Lead in degrees Fahrenheit.

47. The boiling point of Liquid Oxygen is –183 $°C$. Find the boiling point of Liquid Oxygen in degrees Fahrenheit.

48. The boiling point of Liquid Hydrogen is –253 $°C$. Find the boiling point of Liquid Hydrogen in degrees Fahrenheit.

49. The melting point of Iron is 1,538 $°C$. Find the melting point of Iron in degrees Fahrenheit.

50. The boiling point of Iron is 2,861 $°C$. Find the boiling point of Iron in degrees Fahrenheit.

51. The melting point of Silver is 1,761 $°F$. Find the melting point of Silver in degrees Centigrade.

52. The melting point of Tin is 450 $°F$. Find the melting point of Tin in degrees Centigrade.

53. The melting point of Zinc 787 $°F$. Find the melting point of Zinc in degrees Centigrade.

54. The melting point of Copper is 1,981 $°F$. Find the melting point of Copper in degrees Centigrade.

55. The melting point of Aluminum is 1,220 $°F$. Find the melting point of Aluminum in degrees Centigrade.

56. The melting point of Nickel is 2,651 $°F$. Find the melting point of Nickel in degrees Centigrade.

PROJECT: PROPORTION APPLICATIONS

Objective: Determine the time or distance required to burn calories.

(c) 2006 JupiterImages Corp.

(c) 2006 JupiterImages Corp.

THIRTY MINUTES OF BRISK WALKING BENEFITS HEALTH

The Meals
Breakfast
A Starbucks cappuccino (180 calories), a Burger King enormous omelet sandwich (730 calories) and large hash browns (390 calories).
Lunch
A Starbucks cafe latte (260 calories), a KFC original recipe breast (400 calories), leg (140 calories), thigh (250 calories), 2 biscuits (180 calories each), and a double chocolate chip cake (320 calories) for dessert.
Dinner
A McDonald's Big Mac (590 calories), a large order of fries (540 calories), a large coke (310 calories), 2 baked apple pies (260 calories each), and an M&M McFlurry (630 calories) for dessert.
Nightcap
5 ounces of Baileys Irish Cream (468 calories)

Exercise Options
1. Walking burns $\dfrac{70 \text{ calories}}{\text{mile}}$.
2. Jogging burns $\dfrac{105 \text{ calories}}{\text{mile}}$.
3. Basketball burns $\dfrac{563 \text{ calories}}{\text{hour}}$.
4. Ice Hockey burns $\dfrac{690 \text{ calories}}{\text{hour}}$.
5. Running (treadmill) burns $\dfrac{493 \text{ calories}}{\text{hour}}$.
6. Volleyball burns $\dfrac{281 \text{ calories}}{\text{hour}}$.

Project
1. Determine the total number of calories ingested by the person who ate the above meals.
2. Determine the amount of time or distance required to burn all the calories for each given exercise option.

PROJECT: LINEAR EQUATIONS AND PROPORTIONS USED IN RADIOLOGY

Objective: Solve problems used by radiologic technologists.

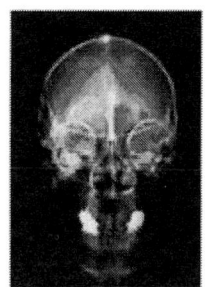
(c) 2006 JupiterImages Corp.

The X–Ray Machine

Two settings on an x-ray machine are the milliamperage (mA) dial and the timer. The milliamperage setting controls the amount of radiation used to take the x-ray. The timer controls the amount of time in seconds "s" the patient is exposed to the radiation. Together the two settings determine the density, the blackness of the radiograph (x-ray photo).

The relationship between milliamps and seconds is defined by the linear equation:

Milliamps × seconds = milliampere-seconds, which abbreviates to:
$$mA \times s = mAs$$

Thus the quantity of radiation is measured in milliampere-seconds (mAs). For example, consider a radiograph made with settings of 200mA for 0.4 seconds. The exposure of the radiograph is
Milliamps × seconds = milliampere-seconds
$$mA \times s = mAs$$
$$(200 \text{ mA})(0.4s) = 80 \text{ mAs}$$
The radiograph was produced by an exposure of 80 milliampere-seconds.

Another relationship used in Radiology compares the exposure used to produce (mAs) a radiograph to the distance between the x-ray tube and the film in inches (in). This relationship is defined by the proportion.

$$\frac{\text{New mAs}}{\text{Old mAs}} = \frac{(\text{New Distance})^2}{(\text{Old Distance})^2}$$

For example, if 100 mAs produces a satisfactory radiograph at a distance of 40 inches, what mAs would be required to produce a satisfactory radiograph at a distance of 50 inches?

$$\frac{\text{New mAs}}{\text{Old mAs}} = \frac{(\text{New Distance})^2}{(\text{Old Distance})^2}$$

Let: x = New mAs, 100 = Old mAs, 50 = New Distance, 40 = Old Distance
then the proportion is

$$\frac{x}{100} = \frac{(50)^2}{(40)^2} \rightarrow \frac{x}{100} = \frac{2{,}500}{1{,}600} \rightarrow \frac{x}{100} \diagup \frac{2{,}500}{1{,}600} \rightarrow 1{,}600x = 250{,}000 \rightarrow$$

$$\frac{1}{1{,}600}(1{,}600)x = \frac{1}{1{,}600}(250{,}000) \rightarrow x = \frac{250{,}000}{1{,}600} = 156.25 \quad x = 156.25 \text{ mAs}$$

Use the linear and proportional relationships defined above to solve the following problems.

I. A satisfactory radiograph is made using 240 mAs at a distance of 30 inches, what mAs would be required to produce a satisfactory radiograph at a distance of 36 in?

II. A satisfactory radiograph is made using 240 mAs, what is the time of exposure for the x-ray using 800 mA?

III. A satisfactory radiograph is made using 90 mAs at a distance of 48 inches, what mAs would be required to produce a satisfactory radiograph at a distance of 36 in?

IV. A satisfactory radiograph is made using 90 mAs and 0.16875 seconds, what is the mA setting for the x-ray?

CHAPTER 1 REVIEW

Section 1.1: Signed Numbers

1. Sergei went to the dog track and bet on 8 races. His betting yielded the following results: –$6.00, $4.42, –$24.00, $7.24, $0.80, $16.60, –$12.00, $2.55. What was the overall result of his betting.

2. $48\left(\dfrac{-1}{6}\right)(72)\left(\dfrac{1}{-2}\right) =$

3. $(-8)+12-6-(-7)+(-2) =$

4. $\dfrac{-7}{18}\left(\dfrac{35}{42}\right)\left(\dfrac{-6}{-21}\right)\left(\dfrac{2}{5}\right) =$

5. Elena's checking account started with a balance of $1,248.55. Today she wrote checks in the amount of $27.16, $46.18, $256.79, $1,480.18. She also deposited checks in the amount of $748.16 and $24.55, but took $100.00 out in cash. What is the balance of her checking account.

Section 1.2: Order of Operations

6. $\sqrt{4}\left[7-3(4)\right]+8 =$

7. $\sqrt{4}\left[7-3^2(4)\right]+8 =$

8. $\sqrt{4}\left[7-3(4)\right]^2 +8 =$

9. $\sqrt{4}\left[(7-3)(4)\right]+8 =$

10. $\sqrt{4}\left[7-3^3(4)\right]+8 =$

Section 1.3: Ratio and Proportion

11. Solve for x: $\dfrac{3.5}{6} = \dfrac{21}{x}$

12. Solve for x: $\dfrac{x}{3} = \dfrac{20}{24}$

80

13. The sales tax is 8.25%. What is the sales tax on a $24,286.44 car?

14. 60 out of 1,250 students surveyed did not have an e-mail account. What percent of students do have e-mail accounts?

15. In an opinion poll, 434 or 35% of those people surveyed, expressed confidence in the president's economic policies. How many people were interviewed?

16. In a medical study 400 patients were given a new formula for chemotherapy. 4.5% of those patients suffered extreme nausea. How many patients suffered extreme nausea?

Section 1.4: Proportion Applications

17. Given that $1 US is equivalent to 7.42 South African rands. What is South African's equivalent of $324 US?

18. Given that $1 US is equivalent to 0.88 euros. What is the US dollar equivalent of 1,248 euros?

19. The map legend sets 1 inch equal to 80 miles. The actual distance between two cities is 670 miles. How far apart are the two cities on the map?

20. A blueprint sets 1 inch equal to 6 feet. A house measures 11.5 inches by 9.75 inches. What are the dimensions of the house?

21. For lunch a person had a Pizza Hut Personal Pan Italian Sausage pizza containing 740 calories and a large coke containing 310 calories. An intense weightlifting session burns 422 calories per hour. How long would the weightlifting session have to last to burn the calories consumed at lunch?

22. For lunch a person had a Pizza Hut Personal Pan Italian Sausage pizza containing 740 calories and a large coke containing 310 calories. Jumping Rope burns 704 calories per hour. How long would the person have to jump rope to burn the calories consumed at lunch?

23. A pharmacist need to prepare 108 ml of a solution that is 16% strong. How much 24% solution and distilled water need to be mixed to prepare this solution?

24. A chemist needs to prepare 40 ml of a solution that is 8% strong from a solution that is pure. How much pure solution and distilled water are needed to make the required solution?

Section 1.5: Solving Linear Equations

25. 24x − 36 = 79.2

26. 72 + 6x = 48 + 9x

27. 4(7x − 3) = 6(3x + 8)

28. 3(3x − 5) = 4(2x + 9) + 5

29. 8(4x − 3) − 12 = 5(4x + 6) + 18

30. Suppose the formula for the selling price of a home in Beverly Hills, CA is
    Selling Price = $283(Living Area) + $143,650
    Find the living area of a $5,250,000 home.

31. The melting point of an unknown is 348 °C. Find the melting point of this unknown in degrees Fahrenheit.

32. The melting point of an unknown is 1,979 °F. Find the melting point of this unknown in degrees Centigrade.

CHAPTER 2: PERSONAL FINANCE

2.1 SIMPLE INTEREST: SAVINGS ACCOUNTS AND LOANS
Objective: Solve problems involving simple interest.

Money may be borrowed to purchase a car or home. Money may be deposited into a savings account for future needs. Whether borrowing money or depositing money, a service charge is associated with these transactions. This service charge is called interest. Interest is the amount of money paid by the person who borrows the money or by the holder who invests the money.

(c) 2006 JupiterImages Corp.

The amount of interest paid is related to three factors; the **principal**, **interest rate**, and **time**. The principal is the amount of money borrowed or deposited. The interest rate is the annual rate written as a percentage. Time is expressed in years and refers to the time it takes to repay the loan or to the time that the principal remains in the savings account or other investment.

SIMPLE INTEREST
The relationship of principal, interest rate, and time for simple interest is defined by

> **SIMPLE INTEREST**
> **I = PRT**
> where: I is the money paid in interest.
> P is the money borrowed or saved.
> R is the annual interest rate written as a decimal.
> T is the time in years.

Example 1: Find the amount of interest paid on a $1,000 deposit at 6% simple interest for 2 years.
Solution:
 Step 1: Understand and picture the problem.
  1. Question: Find the amount of interest paid.
  2. List the given data: P = $1,000, R = 6%, T = 2 years.
  3. Identify "interest" as the unknown.

 Step 2: Develop a plan.
  1. Choose the appropriate formula.
  2. Convert the given data into the appropriate units: rate must be written as a decimal.
  3. Substitute these values into the formula and simplify.

Step 3: Execute the plan.
   1. The formula is: I = PRT
   2. Convert the data: R = 6% = 0.06 (Note: Percent means per hundred.)
   3. Substitute and simplify: I = ($1,000)(0.06)(2) = $120.
      The interest for 2 years is $120.

Step 4: Check your work.
   1. Check the numbers and your calculations
   2. Is $120 in interest reasonable?
   3. Was the question answered?

Example 2: Find the amount of interest paid on a $1,000 deposit at 6% simple interest for 7 months.
Solution:
Step 1: Understand and picture the problem.
   1. Question: Find the amount of interest paid.
   2. List the given data: P = $1,000, R = 6%, T = 7 months.
   3. Identify "interest" as the unknown.

Step 2: Develop a plan.
   1. Choose the appropriate formula.
   2. Convert the given data into the appropriate units:
      a. Rate must be written as a decimal.
      b. Time must be converted to years.
   3. Substitute these values into the formula and simplify.

Step 3: Execute the plan.
   1. The formula is: I = PRT.
   2. Convert the data:
      a. R = 6% = 0.06
      b. T = 7 months = $(7 \text{ months})\left(\dfrac{1 \text{ year}}{12 \text{ months}}\right) = \dfrac{7}{12}$ year
   3. Substitute and simplify: I = ($1,000)(0.06)$\left(\dfrac{7}{12}\right)$ = $35

   The interest for 7 months is $35.

Step 4: Check your work.
   1. Check the numbers and your calculations.
   2. Is $35 in interest reasonable?
   3. Was the question answered?

Example 3: Find the amount of interest paid on a $5,000 loan at $7\frac{1}{4}$% simple interest for 18 months.
Solution:
  Step 1: Understand and picture the problem.
    1. Question: Find the amount of interest paid.
    2. List the given data: P = $5,000., R = $7\frac{1}{4}$%, T = 18 months.
    3. Identify "interest" as the unknown.

  Step 2: Develop a plan.
    1. Choose the appropriate formula.
    2. Convert the given data into the appropriate units:
       a. Rate must be written as a decimal.
       b. Time must be converted into years.
    3. Substitute these values into the formula and simplify.

  Step 3: Execute the plan.
    1. The formula is: I = PRT.
    2. Convert the data:
       a. R = $7\frac{1}{4}$% = 7.25% = 0.0725.

       b. T = 18 months = $(18 \text{ months})\left(\dfrac{1 \text{ year}}{12 \text{ months}}\right) = \dfrac{3}{2}$ year = 1.5 years.

    3. Substitute and simplify: I = ($5,000)(0.0725)(1.5) = $543.75.

  Step 4: Check your work.
    1. Check the numbers and your calculations.
    2. Is $543.75 in interest reasonable?
    3. Was the question answered?

It is also possible to use the simple interest formula, I = PRT, to solve for the other variables. This type of problem is demonstrated in Example 4.

Example 4: Find the rate of interest, if $220 is paid in simple interest on a $2000 loan for 24 months.
Solution:
  Step 1: Understand the problem.
    1. Question: Find the rate of interest paid.
    2. List the given data: I = $220, P = $2,000, T = 24 months.

3. Identify the "rate of interest" as the unknown.

Step 2: Devise a plan.
1. Choose the appropriate formula: I = PRT.
2. Convert the given data to the units used in the formula.
3. Substitute these values into the formula and solve.

**Remember how to solve a linear equation?
If review is necessary, see Section 1.5.**

(c) 2006 JupiterImages Corp.

Step 3: Execute the plan.
1. The formula is: I = PRT.
2. Convert the data:

$$T = 24 \text{ months} = \left(\overset{2}{\cancel{24 \text{ months}}}\right)\left(\frac{1 \text{ year}}{\underset{1}{\cancel{12 \text{ months}}}}\right) = 2 \text{ years}.$$

3. Substitute, simplify and solve for R.
   I = PRT.
   220 = (2,000)(R)(2)
   220 = 4,000R

Multiply both sides of the equation by the reciprocal of 4,000.

$$\left(\frac{1}{\underset{200}{\cancel{4000}}}\right)\left(\overset{11}{\cancel{220}}\right) = \left(\frac{1}{\underset{1}{\cancel{4000}}}\right)\left(\overset{1}{\cancel{4000}}\right)R$$

$$\frac{11}{200} = 0.055 = R \quad \text{The rate of interest is 5.5\%.}$$

Step 4: Check your work.
1. Check the numbers and your calculations.
2. Is 5.5% a reasonable rate of interest?
3. Was the question answered?

Similarly, the simple interest formula I = PRT can be solved for the variables P or T. (See problems 7 through 10 of the exercises at the end of this section.)

Future Value of a Simple Interest Account

The future value of a simple interest savings account (A) is the total amount of money in the account after a predetermined time. Similarly, the future value of a simple interest loan (A) is the total amount of money required to repay the loan. In both cases, this amount is equal to the amount of money borrowed or invested (P) plus the service charge (I). The formula for the future value at simple interest is
$$A = P + I.$$

Look back at Example 2, the future amount is $1,035. This includes the original deposit of $1,000 and the service charge of $35. Similarly, the total amount of money to repay the loan in Example 3 is $5,543.75. This includes the original loan of $5,000 and the service charge of $543.75. Again, the formula that defines the future value of a loan is:
$$A = P + I$$
$$A = P + PRT \quad \text{(Substitute I = PRT)}$$
$$A = P(1 + RT) \quad \text{(Factor out the P)}$$

Future Value of a Simple Interest Account:
$$A = P(1 + RT)$$
where: A is the future value of the account.
P is the principal, the amount of money borrowed or deposited.
R is the annual interest rate written as a decimal.
T is the time in years.

Example 5: Find the future value of a $1,000 deposit at 6% simple interest for 7 months.
Solution:
 Step 1: Understand and picture the problem.
  1. Question: Find the future value of the loan.
  2. List the given data: P = $1,000, R = 6%, T = 7 months.
  3. Identify the "future value of the loan" as the unknown.

 Step 2: Devise a plan.
  1. Choose the appropriate formula.
  2. Convert the given data to the units used in the formula.
  3. Substitute these values into the formula and simplify.

 Step 3: Execute the plan.
  1. The formula is: $A = P(1 + RT)$.
  2. Convert the data:
   a. R = 6% = 0.06
   b. T = 7 months = $(7 \text{ months})\left(\dfrac{1 \text{ year}}{12 \text{ months}}\right) = \dfrac{7}{12}$ year.

3. Substitute and simplify: $A = P(1 + RT)$

$$A = \$1{,}000 \left(1 + (0.06)\left(\frac{7}{12}\right)\right)$$

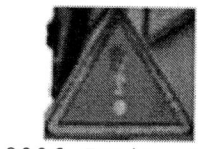
(c) 2006 JupiterImages Corp.

Remember the Order of Operations?
If review is necessary, see Section 1.2 of your text.

$$A = \$1{,}000 \, (1 + 0.035)$$
$$A = \$1{,}000(1.035) = \$1{,}035$$

The future value of the savings account is $1,035 in 7 months.

Step 4: Check your work.
1. Check the numbers and your calculations.
2. Is $1,035 a reasonable future value for this investment?
3. Was the question answered?

Example 6: Find the future value of a $5,000 loan that is due in 18 months at $7\frac{1}{4}\%$ simple interest.
Solution:
Step 1: Understand the problem.
1. Question: Find the future value of the account.
2. List the given data: $P = \$5{,}000$, $R = 7\frac{1}{4}\%$, $T = 18$ months.
3. Identify the "future value of an account" as the unknown.

Step 2: Devise a plan.
1. Choose the appropriate formula.
2. Convert the given data to the units used in the formula.
3. Substitute these values into the formula and simplify.

Step 3: Execute the plan.
1. The formula is: $A = P(1 + RT)$.
2. Convert the data:
   a. $R = 7\frac{1}{4}\% = 7.25\% = 0.0725$.

b. $T = 18 \text{ months} = \left(\cancel{18 \text{ months}}^{3}\right)\left(\dfrac{1 \text{ year}}{\cancel{12 \text{ months}}_{2}}\right) = \dfrac{3}{2} \text{ year} = 1.5 \text{ years}$

   3. Substitute and simplify: $A = P(1 + RT)$
$$A = \$5,000(1 + (0.0725)(1.5))$$
$$A = \$5,000(1 + 0.10875)$$
$$A = \$5,000(1.10875) = \$5,543.75$$
The future value (amount to repay the loan) in 1.5 years is $5,543.75.

Step 4: Check your work.
   1. Check the numbers and your calculations.
   2. Is $5,543.75 a reasonable amount to repay the loan?
   3. Was the question answered?

The future value simple interest formula, $A = P(1 + RT)$, may be used to solve for the other variables.

Example 7: Find the interest rate for a simple interest account that has a present value of $10,000 and will have a future value of $13,400 in 4 years.

Solution:
Step 1: Understand and picture the problem.
   1. Question: Find the rate of interest paid.
   2. List the given data: $A = \$13,400$, $P = \$10,000$, $T = 4$ years.
   3. Identify the "interest rate" as the unknown.

Step 2: Devise a plan.
   1. Choose the appropriate formula.
   2. Substitute these values into the formula and solve for R.

Step 3: Execute the plan.
   1. The formula is: $A = P(1 + RT)$.
   2. Substitute the appropriate values into the formula and solve for R
$$\$13,400 = \$10,000(1 + R(4))$$
Simplify (Distribute) $\$13,400 = \$10,000 + 40,000R$
Subtract $10,000 from both sides of the equation and simplify.
$$\$13,400 - \$10,000 = \$10,000 + 40,000R - \$10,000$$
$$\$3,400 = 40,000R$$

Multiply both sides of the equation by the reciprocal of 40,000.

$$\left(\frac{1}{40{,}000}\right)(3{,}400) = \left(\frac{1}{40{,}000}\right)(40{,}000)R$$

$$\frac{17}{200} = R = 0.085$$

The interest rate is 8.5%.

Step 4: Check your work.
1. Check the numbers and your calculations.
2. Is 8.5% a reasonable interest rate?
3. Was the question answered?

Similarly, A = P(1 + RT) may be solved for the variable T. (Problems 17 and 18 of the exercises at the end of this section.)

Present Value of a Simple Interest Account

The present value (P) of a simple interest problem is equivalent to the initial amount of money invested or borrowed. In Example 5, the future value was $1,035 on a 7 month savings account at 6% simple interest. The initial deposit or present value was $1,000. The formula that defines the present value of simple interest is a manipulation of the future value formula; A = P(1 + RT). In solving for the "Present Value", P, both sides of the equation are multiplied by $\frac{1}{1+RT}$ producing

---

Present Value of a Simple Interest Account:

$$P = \frac{A}{1+RT}$$

where: A is the future value of the account.
P is the principal (present value), the money borrowed or invested.
R is the annual interest rate written as a decimal.
T is the time in years.

Example 8: Find the present value of a 6% simple interest account if 7 months later the account contains $1,035.

Solution:
Step 1: Understand and picture the problem.
1. Question: Find the "present value", initial investment.
2. List the given data: A = $1,035, Rate = 6%, Time = 7 months.
3. Identify "present value" as the unknown.

Step 2: Devise a plan.
  1. Choose the appropriate formula.
  2. Convert the given data to the units used in the formula.
  3. Substitute these values into the formula and simplify.

Step 3: Execute the plan.
  1. The formula is: $P = \dfrac{A}{1+RT}$.
  2. Convert the data:
     a. 6% = 0.06 = R.
     b. 7 months = $(7 \text{ months})\left(\dfrac{1 \text{ year}}{12 \text{ months}}\right) = \dfrac{7}{12}$ year = T.
  3. Substitute and simplify: $P = \dfrac{A}{1+RT}$
     $$P = \dfrac{\$1{,}035}{1+(0.06)\left(\dfrac{7}{12}\right)} = \dfrac{\$1{,}035}{1.035} = \$1{,}000.$$

  The present value is $1,000.

Step 4: Check your work.
  1. Check the numbers and your calculations.
  2. Is $1,000 a reasonable present value?
  3. Was the question answered?

Example 9: Find the amount of money initially borrowed (present value) if after 18 months at $7\dfrac{1}{4}$% simple interest it costs $5,543.75 to repay the loan.

Solution:
Step 1: Understand the problem.
  1. Question: Find the present value.
  2. List is the given data: A = $5,543.75, Rate = $7\dfrac{1}{4}$%, Time = 18 months.
  3. Identify the "present value" as the unknown.

Step 2: Devise a plan.
  1. Choose the appropriate formula.
  2. Convert the given data to the units used in the formula.
  3. Substitute these values into the formula and simplify.

Step 3: Execute the plan.
1. The formula is: $P = \dfrac{A}{1+RT}$.
2. Convert the data:
   a) $7\dfrac{1}{4}\% = 0.0725 = R$.

   b) $18 \text{ months} = \left(\cancel{18 \text{ months}}^{3}\right)\left(\dfrac{1 \text{ year}}{\cancel{12 \text{ months}}_{2}}\right) = \dfrac{3}{2} \text{ year} = 1.5 \text{ years} = T$.

3. Substitute and simplify: $P = \dfrac{A}{1+RT}$

$$P = \dfrac{\$5{,}543.75}{1+(0.0725)(1.5)} = \dfrac{\$5{,}543.75}{1.10875} = \$5{,}000.00.$$

The present value of the loan is $5,000.

Step 4: Check your work.
1. Check the numbers and calculations.
2. Is $5,000 a reasonable present value?
3. Was the question answered?

(c) 2006 JupiterImages Corp.

Historical Perspective
Uruk (about 3000 BC)
What is now Saudi Arabia.

We may never know when finance began, because financial contracts are as old as written language. It appears writing was invented for the purpose of recording financial deals. The first archaeological traces of financial activity appear in the urban civilizations in the Near East. Buried in the ruins of Uruk were clay tablets covered with figures (jars, loaves and animals.) The people of Uruk, appear to be the origin of the practice of lending money at **interest**. They had a system for recording contractual obligations and a numerical system that could specify particular quantities of goods.

?   Cognitive Problems   **?**

1. Identify the variables in the simple interest formula, $I = PRT$. What are the units of each variable?

2. Identify the variables in the future value of the simple interest formula, $A = P(1 + RT)$. What are the units of each variable?
3. Identify the variables in the present value of the simple interest formula, $P = \dfrac{A}{1+RT}$. What are the units of each variable?
4. What does the term "percent" mean?
5. Explain the difference between 4% and 40%.
6. Explain the difference between 0.5% and 0.5.
7. The annual percentage rate must be written as a decimal value before any calculations are performed. Explain how to convert any percent into a decimal.
8. Explain how to change $6\dfrac{3}{8}\%$ into a decimal.

Exercise 2.1

(c) 2006 JupiterImages Corp.

1. Find the interest paid on a $2,400 deposit at 5.5% simple interest for 1 year.

2. Find the interest paid on a $4,800 deposit at $6\dfrac{3}{8}\%$ simple interest for 2 years.

3. Find the interest paid on a $2,100 loan at $6\dfrac{7}{8}\%$ simple interest for 17 months.

4. Find the interest paid on a $5,000 deposit at 4.8% simple interest for 10 months.

5. Find the rate of simple interest, if $25.50 in interest is paid on a $1,800 loan for 3 months.

6. Find the rate of simple interest, if $357.20 in interest is paid on a 2 year loan for $3,800.

7. How many years must $2,000 stay in a 6% simple interest account before it earns $660 in interest?

8. How many months must $4,800 stay in a $4\dfrac{1}{8}\%$ simple interest account before it earns $892.50 in interest?

9. A 6.5% simple interest loan for 8 months had an interest charge of $234. How much money was borrowed?

10. A 4.8% simple interest loan for 5 months had an interest charge of $25. How much money was borrowed?

11. Find the amount of money needed to repay a $12,000 loan at 5.25% simple interest for 2 years.

12. Find the amount of money needed to repay a $3,600 loan at $7\frac{1}{4}$% simple interest for 30 months.

13. Find the future value of a $6\frac{5}{8}$% simple interest savings account after 42 months. The initial deposit was $4,800.

14. Find the future value of a 7.1% simple interest savings account after 18 years. The initial deposit was $6,600.

15. Find the rate of interest, if the future value of a savings account is $2,760 when $2,400 was deposited 3 years ago?

16. Find the rate of interest, if the future value of a savings account is $925 when $800 was deposited 30 months ago?

17. How many months must $2,400 stay in a 5% simple interest account before its future value is $2,760?

18. How many years must $800 stay in a 6.25% simple interest account before its future value is $925?

19. After 4 years, an account that earns 3.5% simple interest has grown to $9,120. How much money was initially deposited (present value) into the account?

20. It cost $15,220.80 to repay a loan. The loan was for 3 years at 1.9% simple interest. How much money (present value) was borrowed?

2.2 COMPOUND INTEREST: SAVINGS ACCOUNTS AND LOANS
Objective: Solve problems involving compound interest.

Simple interest is not the only type of interest. Many loans and savings accounts are calculated using compound interest. Once asked, "What is the most powerful force on earth?" Albert Einstein replied without hesitation, "Compound interest." ("You Can Make a Million", Reader's Digest, July 1996.)

$e = mc^2$

The following chart is an excerpt taken from a bank's weekly interest rate sheet.

Certificate of Deposit (CD)

	Minimum Balance To Open	Interest Rate	Annual Percentage Yield
Passbook (Compounded Quarterly)	$100	4.00%	4.06%
Statement Savings (Compounded Daily)	$100	4.00%	4.08%
Christmas Club (Simple Interest)	$ 3	4.00%	4.00%

Compound interest on a loan or savings account involves calculating the interest after a periodic amount of time. The account is then credited with the interest for that period. The next period's interest is calculated on the new total in the account. The time intervals (periods) for compound interest may be once a year (annually), twice a year (semi-annually), four times a year (quarterly), 12 times a year (monthly), 52 times a year (weekly), or even daily. In the case of daily compound interest, some banks define a year as 360 days and other banks use 365 days. For problems in this textbook, a year will be defined as 365 days. Study the following problem to comprehend the difference between compound interest and simple interest.

Example 1: Calculate and compare the balance in the following saving accounts. The account contains $1,000 for 1 year and earns:
A. 4% simple interest.
B. 4% interest compounded quarterly.

Solution:
Step 1: Understand and picture the problem.

Simple Interest	Interest Compounded Quarterly
1. Question: Find the future value of the simple interest account. 2. List the given data. P = $1,000, Rate = 4%. T = 1 year. 3. Identify the "future value of the account" as the unknown balance.	1. Question: Find the future value of the compound interest account. 2. List the given data. P = $1,000, T = 1 year, Rate = 4% compounded quarterly. 3. Identify the "future value of the account" as the unknown balance.

Step 2: Develop a plan.

Simple Interest	Interest Compounded Quarterly
1. Choose the appropriate formula. 2. Convert the given data to the appropriate units. 3. Substitute these values into the formula and simplify.	1. Choose the appropriate formula. 2. Convert the given data to the appropriate units. 3. Substitute these values into the formula and simplify. 4. Repeat these calculations with the new principal for each quarter until the quarters total one year.

Step 3: Execute the plan.

Simple Interest	Interest Compounded Quarterly
1. The formula is: A = P(1 + RT). 2. Convert the data: 4% = 0.04 = R. 3. Substitute and simplify: P = $1,000 R = 0.04 T = 1. A = P(1 + RT) A = $1,000(1 + (.04)(1)) A = $1,000(1.04) = $1,040. The future value of the account is $1,040 for 1 year.	1. The formula is: A = P(1 + RT) for each quarter with a new principal. 2. Convert the data: 4% = 0.04 = R. Convert time (T) to quarterly. $T = \dfrac{1}{4}$ year = 0.25 year. 3. Substitute and solve: **Months 1 to 3 (first quarter):** P = $1,000 R = 0.04 T = 0.25. A = P(1 + RT) A = $1,000(1 + (.04)(.25)) A = $1,000(1.01) A = $1,010. This $1,010 is the future value at the end of the first quarter and the new principal (P) for the next quarter.

Simple Interest	Interest Compounded Quarterly
	4. **Repeat for the 2nd, 3rd, and 4th quarters: Months 4 to 6 (second quarter)** $P = \$1{,}010$ $R = 0.04$ $T = 0.25$. Substitute and solve: $A = P(1 + RT)$ $A = \$1{,}010(1 + (.04)(.25))$ $A = \$1{,}010(1.01)$ $A = \$1{,}020.10$. This $\$1{,}020.10$ is the future value at the end of the second quarter and the new principal (P) for the next quarter.
The future value of the simple interest account is $\$1{,}040$ for 1 year.	
Comment on Example 1: Simple interest involves a single calculation as displayed in this column. Compound interest repeats the calculations as displayed in the right column. Interest is calculated each period using the previous period's principal. Each year four periodic calculations are performed for quarterly compound interest.	**Months 7 to 9 (third quarter)** $P = \$1{,}020.10$ $R = 0.04$ $T = 0.25$. $A = P(1 + RT)$ $A = \$1{,}020.10(1+(.04)(.25))$ $A = \$1{,}020.10(1.01)$ $A \approx \$1{,}030.30$. This $\$1{,}030.30$ is the future value at the end of the third quarter and the new principal (P) for the next quarter.
	Months 10 to 12 (fourth quarter) $P = \$1{,}030.30$ $R = 0.04$ $T = 0.25$. $A = P(1 + RT)$ $A = \$1{,}030.30(1+(.04)(.25))$ $A = \$1{,}030.30(1.01)$ $A \approx \$1{,}040.60$. The future value of the compound quarterly interest account is $\$1{,}040.60$ for one year.

Step 4: Check your work.

Simple Interest	Interest Compounded Quarterly
1. Check the numbers and calculations. 2. Is $\$1{,}040$ a reasonable future value for this account? 3. Was the question answered?	1. Check the numbers and calculations. 2. Is $\$1{,}040.60$ a reasonable future value for this account? 3. Was the question answered?

Note: The quarterly compound interest account pays $1,040.60 and the simple interest account pays $1,040.00. Even though the time period is short, 1 year, sixty cents is the net difference between these two accounts. Imagine the effect compounding may have on an account over many years.

Future Value of Compound Interest
Since the previous example was interest compounded quarterly, it required four calculations per year. Imagine if the previous problem involved calculating daily compound interest, it would require 365 calculations per year. This would be a very tedious task. The need for a "future value" formula for compound interest is evident. The following formula computes the same future value with one single calculation.

FUTURE VALUE OF COMPOUND INTEREST

$$A = P\left(1 + \frac{R}{n}\right)^{nT}$$

where: A is the future value of the account.
P is the principal, the initial value of the account.
R is the annual interest rate.
T is the time in years.
n is the number of compound periods per year.

n	Type of Compounding
1	Annual
2	Semi-annual
4	Quarterly
12	Monthly
52	Weekly
360 or 365 * dependent on the bank	Daily

---

* For this course, use 365 for n when calculating daily compound interest.

Example 2: Find the future value of a savings account that started with $1,000 at 4% compounded quarterly after one year.
Solution:
  Step 1: Understand and picture the problem.
    1. Question: Find the future value of the compound interest account.
    2. List the given data: P = $1,000,   R = 4% compounded quarterly,
        T = 1 year,   n = 4 pay periods per year.
    3. Identify the "future value" as the unknown.

Step 2: Devise a plan.
    1. Choose the appropriate formula.
    2. Convert the given data to the appropriate units.
    3. Substitute these values into the formula and simplify.

Step 3: Execute the plan.
    1. The formula is: $A = P\left(1 + \dfrac{R}{n}\right)^{nT}$.
    2. Convert the data: 4% = 0.04 = R.
    3. Substitute and simplify: P = $1,000, R = 0.04, T = 1, n = 4.

$$A = P\left(1 + \dfrac{R}{n}\right)^{nT}$$

$$A = \$1{,}000\left(1 + \dfrac{0.04}{4}\right)^{(4\times 1)} = \$1{,}000(1 + 0.01)^4 = \$1{,}000(1.01)^4$$

A = $1,000(1.04060401) ≈ $1040.60.
The future value of the compound interest account is $1,040.60.

Step 4: Check your work.
    The result matches Example 1. The work is reasonable and correct.

Example 3: Find the future value of a savings account that opened with $5,000 at $5\tfrac{1}{4}$% compounded monthly for 18 years.

Solution:
Step 1: Understand and picture the problem.
    1. Question: Find the future value of the compound interest account.
    2. List the given data: P = $5,000, Rate = $5\tfrac{1}{4}$% compounded monthly, T = 18 years, n = 12 pay periods per year.
    3. "Future value of a compound interest account" is the unknown.

Step 2: Devise a plan.
    1. Choose the appropriate formula.
    2. Convert the given data to the appropriate units.
    3. Substitute these values into the formula and simplify.

Step 3: Execute the plan.
    1. The formula is: $A = P\left(1 + \dfrac{R}{n}\right)^{nT}$.

2. Convert the rate: $5\frac{1}{4}\% = 0.0525 = R$.
3. Substitute and simplify. **Do not round any numbers used in the calculations. Store the values in your calculator's memory. See your calculator's manual for instructions on how to store values in memory. Only the final answer is rounded to the nearest penny.** (Round up, when the digit in the thousandths place is 5 or larger.)
P = $5,000, R = 0.0525, T = 18 n = 12.

$$A = P\left(1+\frac{R}{n}\right)^{nT}.$$

$$A = \$5,000\left(1+\frac{0.0525}{12}\right)^{(12\times 18)} = \$5,000(1+0.004375)^{216}.$$

$A = \$5,000(1.004375)^{216} \approx \$5,000(2.567515814) \approx \$12,837.58.$
The future value of the account is $12,837.58.

Step 4: Check your work.
1. Check the numbers and calculations.
2. Is $12,837.58 a reasonable future value for this account?
3. Was the question answered?

Example 4: Find the future value of a savings account that opened with $1,000 at 4% compounded daily for three years.
Solution:
  Step 1: Understand and picture the problem.
    1. Question: Find the future value of a compound interest account
    2. List the given data: P = $1,000, Rate = 4% compounded daily,
                            T = 3 years,   n = 365 pay periods per year.
    3. Identify the "future value of compound interest" as the unknown.

  Step 2: Devise a plan.
    1. Choose the appropriate formula.
    2. Convert the given data to the appropriate units.
    3. Substitute these values into the formula and simplify.

  Step 3: Execute the plan.
    1. The formula is: $A = P\left(1+\frac{R}{n}\right)^{nT}.$
    2. Convert the data: 4% = 0.04 = R.
    3. Substitute and simplify: P = $1,000, R = 0.04, T = 3   n = 365.

$$A = P\left(1 + \frac{R}{n}\right)^{nT}.$$

$$A = \$1,000\left(1 + \frac{0.04}{365}\right)^{(365 \times 3)} \approx \$1,000(1 + 0.000109589)^{1095}$$

$$A = \$1,000(1.000109589)^{1095} \approx \$1,000(1.127489438) \approx \$1,127.49.$$

The future value of the account is $1,127.49. Only round the final result to the nearest penny.

Step 4: Check your work.
1. Check the numbers and calculations.
2. Is $1,127.49 a reasonable future value for this account?
3. Was the question answered?

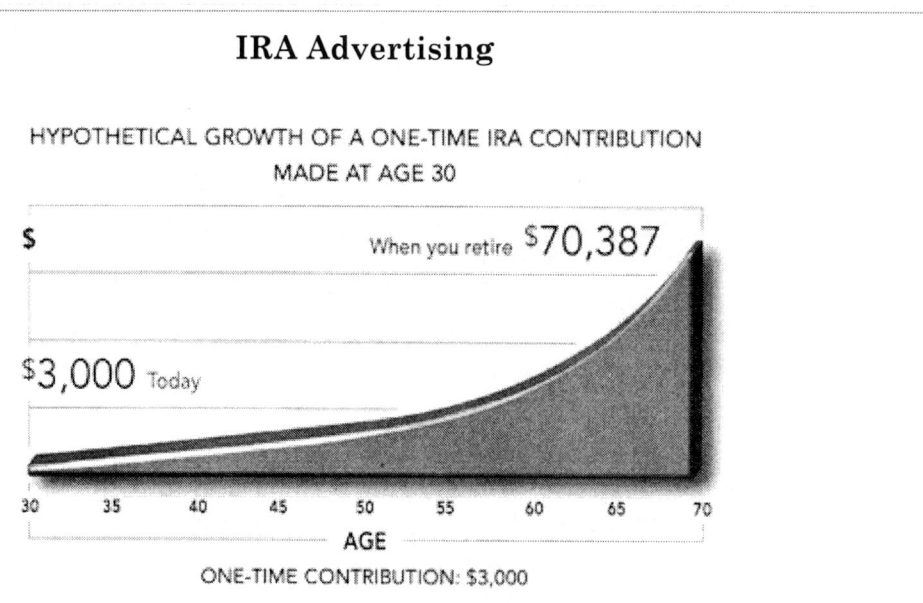

This hypothetical growth example is for illustrative purposes only and does not represent the performance of any security. Please keep in mind that investing involves risk, including the risk of loss. The chart assumes a one-time contribution beginning at age 30 and held through age 70 (41 years). It also assumes a constant annual rate of return of 8% (compounded annually). This example is further explained in the following example.

Example 5: Find the future value of a savings account that opened with $3,000 at 8% compounded annually for forty-one years.
Solution:
  Step 1: Understand and picture the problem.
    1. Question: Find the future value of a compound interest account
    2. List the given data: P = $3,000,   Rate = 8% compounded daily,
                             T = 41 years,   n = 1 pay period per year.
    3. Identify the "future value of compound interest" as the unknown.

  Step 2: Devise a plan.
    1. Choose the appropriate formula.
    2. Convert the given data to the appropriate units.
    3. Substitute these values into the formula and simplify.

  Step 3: Execute the plan.
    1. The formula is: $A = P\left(1 + \dfrac{R}{n}\right)^{nT}$.
    2. Convert the data: 8% = 0.08 = R.
    3. Substitute and simplify: P = $3,000, R = 0.08, T = 41 n = 1.

    $$A = P\left(1 + \dfrac{R}{n}\right)^{nT}$$

    $$A = \$3{,}000\left(1 + \dfrac{0.08}{1}\right)^{(1 \times 41)} \approx \$3{,}000(23.46248322)$$

    A ≈ $70,387.45.
    The future value of the account is $70,387.45. Only round the final result to the nearest penny.

  Step 4: Check your work.
    1. Check the numbers and calculations.
    2. Was the question answered?
    3. The result agrees with the graph preceding this example.

Continuous Compound Interest
Some banks offer accounts that are compound continuous interest. This savings account grows continuously as if it were a well tended plant in a greenhouse. The formula for continuous compound interest is:

> **FUTURE VALUE OF**
> **CONTINUOUS COMPOUND INTEREST**
> $$A = Pe^{RT}$$
> where: A is the future value of the account.
> P is the principal, the initial value of the account.
> R is the annual interest rate as a decimal.
> T is the time in years.
> **Note:** Like π, e is a constant (e $\approx$ 2.718281828...).
> Do not enter the value for e into your
> calculator. Use the e^x key on your calculator,
> where x represents R $\times$ T.

Example 6: Find the future value of a savings account that opened with $1,000 at 4% compounded continuously for 3 years.

Solution:

Step 1: Understand and picture the problem.
   1. Question: Find the future value of a continuous interest account.
   2. List the given data: P = $1,000,   e, use the e^x key on your calculator,
                                  T = 3 years,   Rate = 4% compounded continuously.
   3. Identify the "future value of compound interest" as the unknown.

Step 2: Devise a plan.
   1. Choose the appropriate formula.
   2. Convert the given data to the appropriate units.
   3. Substitute these values into the formula and simplify.

Step 3: Execute the plan.
   1. The formula is: $A = Pe^{RT}$.
   2. Convert the rate: 4% = 0.04 = R.
   3. Substitute and simplify: P = $1,000, use the e^x key on your calculator, R = 0.04, T = 3.
$$A = Pe^{RT} = \$1,000(e)^{(0.04 \times 3)} = \$1,000(e)^{0.12}.$$
A = $1,000(1.127496852) $\approx$ $1,127.50. Only the final result is rounded.
The future value is $1,127.50.

Step 4: Check your work.
   1. Check the numbers and calculations.
   2. Is $1,127.50 a reasonable future value for this account?
   3. Was the question answered?

Present Value of Compound Interest Accounts

In a related problem, how much money must be deposited (today) into an account, earning 4.75% interest compounded semi-annually, to earn $2,000 in 3 years? This formula is called the "Present Value of Compound Interest". (Compare this formula with the "Future Value of Compound Interest" formula. Notice the similarities in the variables used and the computations required.)

PRESENT VALUE OF COMPOUND INTEREST

$$P = \frac{A}{\left(1 + \frac{R}{n}\right)^{nT}}$$

where: A is the future value of the account.
P is the principal, the present (initial) value of the account.
R is the annual interest rate as a decimal.
T is the time in years.
n is the number of compound periods per year.

Example 7: Find the present value of a savings account that grows to $2,000 in 3 years compounded semi-annually at 4.75%.

Solution:

Step 1: Understand and picture the problem.
 1. Question: Find the present value of a compound interest account.
 2. List the given data: A = $2,000,
  Rate = 4.75% compounded semi-annually,
  n = 2 pay periods per year, T = 3 years.
 3. Identify the "present value of a compound interest account" as the unknown.

Step 2: Devise a plan.
 1. Choose the appropriate formula.
 2. Convert the given data to the appropriate units.
 3. Substitute these values into the formula and simplify.

Step 3: Execute the plan.

 1. The formula is: $P = \dfrac{A}{\left(1 + \dfrac{R}{n}\right)^{nT}}$.

 2. Convert the rate: 4.75% = 0.0475 = R.
 3. Substitute and simplify: A = $2,000,   R = 0.0475,   n = 2,   T = 3.

$$P = \frac{A}{\left(1+\frac{R}{n}\right)^{nT}}$$

$$P = \frac{\$2{,}000}{\left(1+\frac{0.0475}{2}\right)^{(2\times3)}} = \frac{\$2{,}000}{(1+0.02375)^6} = \frac{\$2{,}000}{(1.02375)^6} = \frac{\$2{,}000}{1.151233685}$$

Use your calculator's memory function. Do not rekey 1.151233685, recall it from memory.

$P \approx \$1{,}737.27.$ Only round the final result.

The present value of the account is $1,737.27.

Step 4: Check your work.
1. Check the numbers and calculations.
2. Is $1,737.27 a reasonable present value for this account?
3. Was the question answered?

Example 8: Find the present value of a savings account that grows to $12,000 in 5 years compounded monthly at 6%.
Solution:
Step 1: Understand and picture the problem.
1. Question: Find the present value of a compound interest account.
2. List the given data: A = $12,000,   Rate = 6% compounded monthly,
                        n = 12 pay periods per year,   T = 5 years.
3. Identify the "present value of a compound interest account" as the unknown.

Step 2: Devise a plan.
1. Choose the appropriate formula.
2. Convert the given data to the appropriate units.
3. Substitute these values into the formula and simplify.

Step 3: Execute the plan.
1. The formula is: $P = \dfrac{A}{\left(1+\dfrac{R}{n}\right)^{nT}}$.

2. Convert the rate: 6% = 0.06 = R.
3. Substitute and simplify: A = $12,000,   R = 0.06,   n = 12,   T = 5.

$$P = \frac{A}{\left(1+\frac{R}{n}\right)^{nT}}$$

$$P = \frac{\$12{,}000}{\left(1+\dfrac{0.06}{12}\right)^{(12\times 5)}} = \frac{\$12{,}000}{(1+0.005)^{60}} = \frac{\$12{,}000}{(1.005)^{60}} = \frac{\$12{,}000}{1.348850153}.$$

P ≈ $8,896.47.   Only round the final result.
The present value of the account is $8,896.47.

Step 4: Check your work
1. Check the numbers and calculations.
2. Is $8,896.47 a reasonable present value for this account?
3. Was the question answered?

Continuous Compound Interest Revisited
Present value can also be found for continuous compound interest. (Compare this formula with the "Future Value of Continuous Compound Interest" formula. Notice the similarities in the variables used and the computations required.)

**PRESENT VALUE FOR
CONTINUOUS COMPOUND INTEREST**

$$P = \frac{A}{e^{RT}}$$

where: P is the principal, the present (initial) value of the account.
A is the future value of the account.
R is the annual interest rate.
T is the time in years.
Note: Like π, e is a constant (e ≈ 2.718281828...).
Do not enter the value for e into your calculator. Use the e^x key on your calculator, where x represents R × T.

Example 9: Find the present value of a savings account that grows to $12,000 in 5 years compounded continuously at 6%.
Solution:
  Step 1: Understand and picture the problem.
    1. Question: Find the present value of a continuous compound interest account.
    2. List the given data: A = $12,000,   use the e^x key on your calculator,
                Rate = 6%,   T = 5 years.
    3. Identify the "present value of a continuous compound interest account" as the unknown.

Step 2: Devise a plan.
1. Choose the appropriate formula.
2. Convert the given data to the appropriate units.
3. Substitute these values into the formula and simplify.

Step 3: Execute the plan.
1. The formula is: $P = \dfrac{A}{e^{RT}}$.
2. Convert the rate: 6% = 0.06 = R.
3. Substitute and simplify: A = $12,000
   e, use the e^x key on your calculator   R = 0.06   T = 5.
   $P = \dfrac{A}{e^{RT}} = \dfrac{\$12{,}000}{e^{(0.06 \times 5)}} = \dfrac{\$12{,}000}{e^{0.3}} = \dfrac{\$12{,}000}{1.349858808} \approx \$8{,}889.82$.
   Only the final result is rounded.
   The present value of the account is $8,889.82.

Step 4: Check your work.
1. Check the numbers and calculations.
2. Is $8,889.82 a reasonable present value for this account?
3. Was the question answered?

ANNUAL PERCENTAGE YIELD (APY) FOR SAVINGS AND ANNUAL PERCENTAGE RATE (APR) FOR LOANS

The Annual Percentage Yield (APY) and the Annual Percentage Rate (APR) are effective rates. Effective rates convert compound interest rates to simple interest rates. Use this conversion to display the simple interest rate that earns the same amount of money (interest) as the compound interest rate. Thus saving accounts and loans with varied interest rates can be easily compared using the APY and APR. APY compares savings accounts and APR compares loans. The federal government mandates that all saving accounts display both the compound interest rate and its equivalent simple interest rate (APY). Similarly the federal government mandates that all loans display both the compound interest rate and its equivalent simple interest rate (APR). Consider the following chart which is an excerpt taken from a bank's weekly interest rate sheet for certificates of deposit (CD).

(c) 2006 JupiterImages Corp.

Term	Compound Interest (Daily)	Annual Percentage Yield
1 Year	4.00%	4.08%
2 Years	4.16%	4.25%
3 Years	5.35%	5.50%
4 Years	6.30%	6.50%
5 Years	7.23%	7.50%

In this excerpt, the 4.00% compound daily interest on the 1 year CD is equivalent to 4.08% simple interest (APY). The 4.16% compound daily interest on the 2 year CD is equivalent to 4.25% simple interest (APY), etc.

The Annual Percentage Yield (APY) and Annual Percentage Rate (APR) formulas are used for converting compound rates to simple interest rates.

APY and APR Formulas	
For compound interest accounts (excluding continuous) APY (Savings) or APR (Loans) $= \left(1 + \dfrac{R}{n}\right)^n - 1$ where: Annual Percentage Yield (APY) Annual Percentage Rate (APR) R is the annual interest rate. n is the number of compound periods per year	**For continuous compound interest** APY (Savings) or APR (Loans) $= e^R - 1$ where: Annual Percentage Yield (APY) Annual Percentage Rate (APR) R is the annual interest rate. **Note:** Like π, e is a constant ($e \approx 2.718281828...$). Do not enter the value for e into your calculator, use the e^x key on your calculator, where x represents $R \times T$.

The same formula is used to calculate APY and APR. The difference between APY and APR is their application. APY is used to describe savings accounts. The bankd is adding money to your savings account. The consumer wants the **HIGHEST APY** in this setting. APR is used to describe loans. The consumer is paying the interest on the money they borrowed. The consumer wants the **LOWEST APR** in this setting.

Example 10: Find the Annual Percentage Yield (APY) for the savings account that earns 3% compounded quarterly.
Solution:
 Step 1: Understand and picture the problem.
    1. Question: Find the APY.
    2. List the given data: Rate = 3% compounded quarterly,
       n = 4 pay periods per year.
    3. Identify the "APY" as the unknown.

 Step 2: Devise a plan.
    1. Choose the appropriate formula.
    2. Convert the given data to the units used in the formula.
    3. Substitute these values into the formula and simplify.

Step 3: Execute the plan.
1. The formula is: $APY = \left(1 + \frac{R}{n}\right)^n - 1$.
2. Convert the rate: 3% = 0.03 = R.
3. Substitute and simplify: R = 0.03, n = 4.
$$APY = \left(1 + \frac{R}{n}\right)^n - 1.$$
$$APY = \left(1 + \frac{0.03}{4}\right)^4 - 1 = (1.0075)^4 - 1 = (1.030339191) - 1.$$
$APY \approx 0.030339191 \approx 3.03\%$.   Only the final result is rounded.
The APY is 3.03%.

Step 4: Check your work.
1. Check the numbers and calculations.
2. The answer is reasonable, it matches the APY in the chart.
3. Was the question answered?

Example 11: Choose the savings account that is most profitable for the depositor.
A. 5.9% simple interest.
B. $5\frac{7}{8}$% interest compounded semi-annually.
C. 5.8% interest compounded daily.
D. 5.76% interest compounded continuously.

Solution:
Step 1: Understand and picture the problem.
1. Question: Find the APY for each account.
2. List the given data:
    A. Rate = 5.9% simple interest.
    B. Rate = $5\frac{7}{8}$% compounded semi-annually,
        n = 2 pay periods per year.
    C. Rate = 5.8% compounded daily,
        n = 365 pay periods per year.
    D. Rate = 5.76% compounded continuously.
3. Identify "APY" as the unknown.

Step 2: Devise a plan.
1. Choose the appropriate formula.
2. Convert the given data to the appropriate units.

3. Substitute these values into the formula and simplify.
4. Compare the APY and choose the **largest** APY value.
   (Note: Since this is a savings account, the goal is to get the most money.)

Step 3: Execute the plan.

A. Rate is 5.9% simple interest. This is already a simple interest rate and does not require a conversion. APY = 0.059.
B. $5\frac{7}{8}\% = 0.05875 = R$ Substitute and simplify: R = 0.05875   n = 2. $APY = \left(1 + \frac{R}{n}\right)^n - 1.$ $APY = \left(1 + \frac{0.05875}{2}\right)^2 - 1 = (1.029375)^2 - 1 = 1.059612891 - 1.$ $APY \approx 0.059612891 \approx 5.96\%.$
C. 5.8% = 0.058 = R   Substitute and simplify: R = 0.058   n = 365. $APY = \left(1 + \frac{R}{n}\right)^n - 1.$ $APY = \left(1 + \frac{0.058}{365}\right)^{365} - 1 = (1.000158904)^{365} - 1 = 1.059710113 - 1.$ $APY \approx 0.059710113 \approx 5.97\%.$
D. 5.76% = 0.0576 = R   Substitute and simplify: R = 0.0576.   e, use the e^x key on your calculator. $APY = e^R - 1$ $APY = e^{0.0576} - 1 \approx 1.059291194 - 1 \approx 0.059291194 \approx 5.93\%.$

4. In summary,

CD Rate	APY
5.9% simple	5.9%
$5\frac{7}{8}\%$ compounded semi-annually	5.96%
5.8% compounded daily	5.97% Largest APY
5.76% compounded continuously	5.93%

The 5.8% compounded daily account is the most profitable for the saver, because an APY of 5.97% is the greatest percentage.

110

Step 4: Check your work.
    1. Check the numbers and calculations.
    2. Are the answers reasonable?
    3. Was the question answered?

Example 12: Choose the loan that is most advantageous for the borrower, person applying for a loan.
  A. 5.9% simple interest.
  B. $5\frac{7}{8}$% interest compounded semi-annually.
  C. 5.8% interest compounded daily.
  D. 5.76% interest compounded continuously.

Solution:
Step 1: Understand and picture the problem.
    1. Question: Find the APR for each account.
    2. List the given data:
      A. Rate = 5.9% simple interest.
      B. Rate = $5\frac{7}{8}$% compounded semi-annually, n = 2.
      C. Rate = 5.8% compounded daily, n = 365.
      D. Rate = 5.76% compounded continuously.
    3. Identify "APR" as the unknown.

Step 2: Devise a plan.
    1. Choose the appropriate formula.
    2. Convert the given data to the appropriate units.
    3. Substitute these values into the formula and simplify.
    4. Compare the APR and choose the **smallest** APR value.
      (Note: This is a loan. The goal is to get the best deal by paying the least amount of money in interest.)

Step 3: Execute the plan.
See Example 11. The calculations are the same. The summary,

Loan Rate	APR
5.9% simple	5.9% Lowest APR
$5\frac{7}{8}$% compounded semi-annually	5.96%
5.8% compounded daily	5.97%
5.76% compounded continuously	5.93%

The 5.9% simple interest account is the most advantageous for the borrower, because 5.9% is the smallest APR.

Step 4: Check your work.
1. Check the numbers and calculations.
2. Are the answers reasonable?
3. Was the question answered?

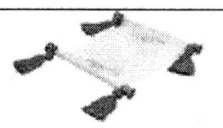 (c) 2006 JupiterImages Corp.	**Historical Perspective** **John Napier (1550 – 1617) Scotland** **Jacob Bernoulli (1654 – 1705) Switzerland**
The number *e* first comes into mathematics in a curious way. In 1618, in an appendix to Napier's work on logarithms, a table appeared giving the natural logarithms of various numbers. That these numbers were logarithms to base *e* was not recognized. *e* was first "discovered" through the study of continuous compound interest. In 1683 Jacob Bernoulli studied the problem. He tried to find the limit of $(1 + 1/n)^n$ as n approaches infinity. He demonstrated that *e* had to lie between 2 and 3. This is considered as the first approximation for *e*.	

? Cognitive Problems ?

1. How is simple interest different from compound interest?
2. Explain the meaning of Annual Percentage Yield, (APY).
3. Explain the meaning of Annual Percentage Rate, (APR).
4. In a savings situation, who would want
   a. the smallest APY. b. the largest APY.
5. In a loan situation, who would want
   a. the smallest APR. b. the largest APR.
6. Explain the difference between present value and future value of an account.

(c) 2006 JupiterImages Corp.

Exercise 2.2

1. Find the future value of a savings account after 3 years, that is opened with $2,400 at 5.1% interest compounded annually.

2. Find the future value of a savings account after 3 years, that is opened with $2,400 at 5.1% interest compounded semi-annually.

3. Find the future value of a savings account after 18 years, that is opened with $5,000 at 6.6% interest compounded monthly.

4. Find the future value of a savings account after 18 years, that is opened with $5,000 at 6.6% interest compounded daily.

5. Find the future value of a certificate of deposit (CD) after 2 years, that is opened with $5,000 at $6\frac{3}{8}$% interest compounded continuously.

6. Find the future value of a certificate of deposit (CD) after 2 years, that is opened with $5,000 at $6\frac{3}{8}$% interest compounded quarterly.

7. Find the future value of a certificate of deposit (CD) after 2 years, that is opened with $5,000 at $6\frac{3}{8}$% interest compounded weekly.

8. Find the present value of a money market account that grows to $13,400 in 4 years, if the account is compounded quarterly at $7\frac{1}{8}$%.

9. Find the present value of a savings account that grows to $5,400 in 18 months, if the account is compounded semi-annually at 4.2%.

10. Find the present value of a savings account that grows to $6,500 in 2 years, if the account is compounded daily at 4.4%.

11. Find the present value of a savings account that grows to $4,400 in 4 years, if the account is compounded continuously at 4.8%.

12. Find the present value of a savings account that grows to $5,000 in 3 years, if the account is compounded monthly at 4.4%.

13. Find the Annual Percentage Yield (APY) for a 6.25% interest account compounded quarterly.

14. Find the Annual Percentage Yield (APY) for a 6.25% interest account compounded monthly.

15. Find the Annual Percentage Rate (APR) for a $6\frac{5}{8}$% loan compounded semi-annually.

16. Find the Annual Percentage Rate (APR) for a $6\frac{5}{8}$% loan compounded continuously.

17. Find the Annual Percentage Yield (APY) for the following savings accounts:
    a. $4\frac{5}{8}$% interest compounded annually.
    b. 4.62% interest compounded weekly.
    c. 4.61% interest compounded daily.
    d. 4.6% interest compounded continuously.
    Which account is most profitable for the savings account holder?

18. Find the Annual Percentage Rate (APR) for the following loans:
    a. $4\frac{5}{8}$% interest compounded annually.
    b. 4.62% interest compounded weekly.
    c. 4.61% interest compounded daily.
    d. 4.6% interest compounded continuously.
    Which account is most advantageous for the borrower?

19. Find the Annual Percentage Rate (APR) for the following loans:
    a. $4\frac{3}{8}$% interest compounded semi-annually.
    b. 4.37% interest compounded quarterly.
    c. 4.35% interest compounded monthly.
    d. 4.32% interest compounded continuously.
    Which account is most advantageous for the borrower?

20. Find the Annual Percentage Yield (APY) for the following accounts:
    a. $4\frac{3}{8}$% interest compounded semi-annually.
    b. 4.37% interest compounded quarterly.
    c. 4.35% interest compounded monthly.
    d. 4.32% interest compounded continuously.
    Which account is most profitable for the account holder?

2.3 AMORTIZATION

Objective: Calculate the periodic payment for a conventional (simple interest) loan using the amortization formula.

Car Loans

Most people will apply for a loan when they purchase a car. In the majority of these loans it is impossible to repay the loan and interest in a single payment. Thus, the repayment is made periodically, such as monthly payments.

(c) 2006 JupiterImages Corp.

The following excerpt is taken from a car loan contract. According to the "Truth in Lending Act" of 1968, the federal government requires full disclosure of the terms of a loan. Study the following portion of an auto loan contract and find the

1. Annual Percentage Rate, interest rate on the loan.
2. Finance Charge, amount paid in interest.
3. Amount Financed, the loan amount.
4. Total of Payments = (number of payments) × (monthly payment).
5. Total Sales = Total of payments + Down Payment.
6. Number of payments.
7. Monthly Payment.

ANNUAL PERCENTAGE RATE	FINANCE CHARGE	AMOUNT FINANCED	TOTAL OF PAYMENTS	TOTAL SALES PRICE Total purchase on credit including your down payment of $1,389.50 is $16,256.06
8.75%	$2,361.06	$12,505.50	$14,866.56	
NUMBER OF PAYMENTS 48		**MONTHLY PAYMENT** $309.72		

The terms of the car loan are:
1. R = ANNUAL PERCENTAGE RATE (APR), 8.75% which is 0.0875.
2. The finance charge (Interest) is $2,361.06.
3. P = Amount of the loan is $12,505.50 (AMOUNT FINANCED).
4. Total payments (48 times $309.72) is $14,866.56.
5. Total sales (payments and down payment) is $14,866.56 + $1,389.50 = $16,256.06.
6. T = Time of the loan, 4 years. NUMBER OF PAYMENTS is 48 months.

$$\frac{48 \text{ months}}{1} \left(\frac{1 \text{ year}}{12 \text{ months}} \right) = \frac{48}{12} \text{ years} = 4 \text{ years}$$

7. The monthly payment is $309.72.

> **Question 1**: What formula was used to calculate the monthly payment? Can the future value for simple interest formula be used to calculate the monthly payment?

Answer: Future Value for Simple Interest is A = P(1 + RT).
Using the data from contract P = $12,505.50,   R = 8.75%,   T = 4 years.
A = $12,505.50 (1 + 0.0875(4)) = $12,505.50(1.35) = $16,882.425.
Dividing this amount into 48 equal monthly payments yields $\frac{\$16,882.425}{48} \approx \$351.72$.
Rounded to the nearest penny.
This amount does not match the contract. The future value for simple interest formula is **NOT** used to calculate the monthly payment.

(c)2006 Jupiter Images Corp.

BEWARE! Some financiers will try to finance your car loan using this formula.

> **Question 2**: What formula was used to calculate the monthly payment? Can the future value for compound interest formula be used to calculate the monthly payment?

Answer: Future Value for Compound Interest is $A = P\left(1 + \frac{R}{n}\right)^{nT}$.

Using the data from contract P = $12,505.50,   R = 8.75%,   n = 12,   T = 4 years.

$A = \$12,505.50\left(1 + \frac{0.0875}{12}\right)^{12(4)} \approx \$12,505.50(1.007291667)^{48}$.

$A \approx \$12,505.50(1.417266658) \approx \$17,723.62819$.

Dividing this amount into 48 equal monthly payments yields
$\frac{\$17,723.62819}{48} \approx \$369.24$. Rounded to the nearest penny.

This amount does **NOT** match the contract. The future value for compound interest formula is not used to calculate the monthly payment.

Amortization

What formula was used to calculate the monthly payment? This formula is called the amortization formula. What does amortization mean? It means the zeroing out or the gradual reduction of debt using periodic payments. The amortization formula yields the amount of the periodic payment necessary to repay a loan over a specified time period. The periodic payment is defined by the formula:

> **AMORTIZATION OF A LOAN**
>
> $$P = \frac{\left(\dfrac{AR}{n}\right)\left(1 + \dfrac{R}{n}\right)^{nT}}{\left(1 + \dfrac{R}{n}\right)^{nT} - 1}$$
>
> where: A is the amount of the loan.
> P is the fixed periodic payment.
> R is the annual interest rate.
> n is the number of compound periods in a year.
> T is the number of years.

Example 1: Calculate the monthly payment on a $500 loan for 4 months that charges 8% interest compounded monthly.

Solution:

Step 1: Understand and picture the problem.
1. Question: Find the monthly payment.
2. List the given data: A = $500, Rate = 8% compounded monthly, n = 12 pay periods per year, Time = 4 months = $\dfrac{4}{12}$.
3. Identify the "amortization of a loan" as the unknown.

Step 2: Devise a plan.
1. Use the appropriate formula.
2. Convert the given data into the appropriate units.
3. Perform the intermediate calculations that will make the formula easier to manipulate. Determine $\dfrac{AR}{n}$, $\left(1 + \dfrac{R}{n}\right)^{nT}$, $\left(1 + \dfrac{R}{n}\right)^{nT} - 1$, and **store these values in your calculator's memory.** (Note: See your calculator's manual for directions on how to store numbers in multiple memory locations.)
4. Substitute these values into the formula and simplify.

Note: Do not round the intermediate calculations. Rounding these intermediate values will cause a rounding error in your final answer. Rounding is only performed on the final result. Instead store these values in your calculator's memory. Consult your owner's manual if you are unfamiliar with the memory function on your calculator.

Step 3: Execute the plan.

1. The formula is: $P = \dfrac{\left(\dfrac{AR}{n}\right)\left(1+\dfrac{R}{n}\right)^{nT}}{\left(1+\dfrac{R}{n}\right)^{nT} - 1}$.

2. Convert the data:
   a. $8\% = 0.08 = R$.
   b. Time = 4 months = $(4 \text{ months})\left(\dfrac{1 \text{ year}}{12 \text{ months}}\right) = \dfrac{4}{12} = \dfrac{1}{3}$ year = T.

3. Intermediate calculations: $A = \$500 \quad R = 0.08 \quad T = \dfrac{1}{3} \quad n = 12$.

   a. $\dfrac{AR}{n} = \dfrac{\$500(0.08)}{12} \approx \$3.3333333$.
   (Store in calculator's memory 1.)

   b. $\left(1+\dfrac{R}{n}\right)^{nT} = \left(1+\dfrac{0.08}{12}\right)^{\left(12 \times \frac{1}{3}\right)} \approx (1.0066666667)^4 \approx 1.02693452$.
   (Store in calculator's memory 2.)

   c. $\left(1+\dfrac{R}{n}\right)^{nT} - 1 = \left(1+\dfrac{0.08}{12}\right)^{\left(12 \times \frac{1}{3}\right)} - 1 \approx (1.0066666667)^4 - 1 \approx$ 0.02693452. **(Store in calculator's memory 3.)**

4. Substitute, recall the appropriate stored values from memory following the order of operations to simplify. (Note: See your calculator's manual for directions on how to recall numbers from multiple memory locations.)

   $P = \dfrac{(\$3.33333333)(1.02693452)}{0.02693452} \approx \$127.0902546 \approx \$127.09$.

   The monthly payment is $127.09.
   (Note: Only the final result is rounded.)

Step 4: Check your work.
  1. Check the numbers and your calculations.
  2. Is the answer reasonable?
  3. Was the question answered?

The following table shows the amortization of the $500 loan at 8% using four monthly payments of $127.09.

Payment #	Payment	Interest	Principal	Balance
---	---	---	---	$500.00
1	$127.09	$3.33	$123.76	$376.24
2	$127.09	$2.51	$124.58	$251.66
3	$127.09	$1.68	$125.41	$126.25
4	$127.09	$0.84	$126.25	$ 0.00

Study the breakdown of each payment into interest and principal. Like many words in the English language, the term "principal" has two interpretations. First, principal (P) is the current balance of the loan used to calculate interest. (Let I = PRT, P = $500, R = 0.08 and T = $\frac{1}{12}$ representing 1 month.) Second, principal is used to describe the amount of the payment that is applied to the loan, monthly payment minus monthly interest.

The principal, the current balance of the loan, is $500.
The principal for the first payment is $123.76.
The principal for the second payment is $124.58.
The principal for the third payment is $125.41
The principal for the final payment is $126.25.

The following process uses these formulas.
1. Interest = Principal × Rate × Time   (I = PRT).
2. Principal = Payment – Interest.
3. Current Balance = Previous Balance – Principal.

Money problems are rounded to two decimal places. Rounding is performed at the end of each formula. For the first month:

1. **$500**(0.08)$\left(\frac{1}{12}\right)$ = $3.33   PRT = I, where P = $500 is the current balance.

2. $127.09 – $3.33 = $123.76   Payment minus interest is principal. $123.76 is the payment applied to paying off the loan.

3. $500 – $123.76 = $376.24   Previous loan balance minus the principal ($123.76 applied to the loan) equals the remaining loan balance, $376.24.

For the second month, this process is repeated, using the new balance, $376.24.

1. **$376.24**(0.08)$\left(\frac{1}{12}\right)$ = $2.51   PRT = I, where P = $376.24 is the current balance.

2. $127.09 – $2.51 = $124.58   Payment minus interest is principal. $124.58 is the payment applied to paying off the loan.

3. $376.24 – $124.58 = $251.66   Previous loan balance minus the principal ($124.58 applied to the loan) equals the remaining loan balance, $251.66.

For the third month, the same process is repeated, using another new balance, $251.66.

1. **$251.66**$(0.08)\left(\dfrac{1}{12}\right) = \$1.68$ PRT = I, where P = $251.66 is the current balance.
2. $127.09 − $1.68 = $125.41   Payment minus interest is principal. $125.41 is the payment applied to paying off the loan.
3. $251.66 − $125.41 = $126.25   Previous loan balance minus the principal ($125.41 applied to the loan) equals the remaining loan balance, $126.25.

For the fourth month, the process is repeated, using the new balance, $126.25.

1. **$126.25**$(0.08)\left(\dfrac{1}{12}\right) = \$0.84$ PRT = I, where P = $126.25 is the current balance.
2. $127.09 − $0.84 = $126.25   Payment minus interest is principal. $126.26 is the payment applied to paying off the loan.
3. $126.25 − $126.25 = $0.00   Previous loan balance minus the principal ($125.26 applied to the loan) equals the remaining loan balance, $0.00.

The balance of the loan is zeroed out. The loan is repaid in four equal payments of $127.09.

Example 2: Let's revisit the loan contract presented at the beginning of this section. A used car costs $13,895 cash ("Out-the-door" price, all taxes, options and dealer fees included) or 10% down (cash/trade) with the remainder financed at 8.75% for 48 months. Find the monthly payment.

(c) 2006 JupiterImages Corp.

Solution:
  Step 1: Understand and picture the problem.
    1. Question: Find the monthly payment.
    2. List the given data:
      "Out-the-door" price of the car = $13,895   n = 12 pay periods per year
      Time = 48 months   Rate = 8.75% compounded monthly, with 10% down.
    3. Identify the "amortization of a loan" as the unknown.

  Step 2: Devise a plan.
    1. Calculate the down payment: Down Payment = 0.10 *times* price.
    2. Calculate the amount of money financed:
      Amount financed = price *minus* down payment.

3. Use the amortization formula to calculate the monthly payment.
4. Convert the given data into the appropriate units.
5. Perform the intermediate calculations that will make the formula easier to manipulate. Determine $\dfrac{AR}{n}$, $\left(1+\dfrac{R}{n}\right)^{nT}$, $\left(1+\dfrac{R}{n}\right)^{nT}-1$, and store these values in your calculator's memory.
6. Substitute these values into the formula and simplify.

Step 3: Execute the plan.
1. Calculate the down payment: Down = 0.10($13,895.00) = $1,389.50.
2. Calculate the amount financed:
   Amount financed = $13,895.00 − $1,389.50 = $12,505.50.
3. The amortization formula is: $P = \dfrac{\left(\dfrac{AR}{n}\right)\left(1+\dfrac{R}{n}\right)^{nT}}{\left(1+\dfrac{R}{n}\right)^{nT}-1}$.

4. Convert the data:
   a. 8.75% = 0.0875 = R.
   b. 48 months = $(\cancel{48 \text{ months}}^{4})\left(\dfrac{1 \text{ year}}{\cancel{12 \text{ months}}_{1}}\right)$ = 4 years = T.

5. Perform the intermediate calculations:
   a. $\dfrac{AR}{n} = \dfrac{\$12{,}505.50(0.0875)}{12} \approx \$91.1859375$.
   (Store in calculator's memory 1.)

   b. $\left(1+\dfrac{R}{n}\right)^{nT} = \left(1+\dfrac{0.0875}{12}\right)^{(12\times 4)} \approx (1.007291667)^{48} \approx 1.417266658$.
   (Store in calculator's memory 2.)

   c. $\left(1+\dfrac{R}{n}\right)^{nT}-1 = \left(1+\dfrac{0.0875}{12}\right)^{(12\times 4)}-1 \approx (1.007291667)^{48}-1 \approx$ 0.417266658 **(Store in calculator's memory 3.)**

6. Substitute, using the order of operations, recall the appropriate stored values from memory and simplify. (Note: The denominator is one less than the second factor of the numerator.)

$$P = \frac{(\$91.1859375)(1.417266658)}{0.417266658} \approx \$309.7175068.$$

Note: The monthly payment is $309.72, which agrees with the loan contract presented at the beginning of this section. This confirms the use of the amortization formula in the calculation of periodic payments.

Step 4: Check your work.
1. Check the numbers and your calculations.
2. Is the answer reasonable?
3. Was the question answered?

Example 3: A new motorcycle costs $10,500 cash ("Out-the-door" price, all taxes, options and dealer fees included) or 15% down (cash/trade) with the remainder financed at 7.25% for 60 months. Calculate the monthly payment.

Solution:
Step 1: Understand and picture the problem.
1. Question: Find the monthly payment.
2. List the given data:
"Out-the-door" price of the motorcycle = $10,500,
n = 12 pay periods per year,   Time = 60 months,
Rate = 7.25% compounded monthly, with 15% down.
3. Identify the "amortization of a loan" as the unknown.

Step 2: Devise a plan.
1. Calculate the down payment: Down Payment = 0.15 *times* price.
2. Calculate the amount of money financed:
Amount financed = price *minus* down payment.
3. Use the amortization formula.
4. Convert the given data into the appropriate units.
5. Perform the intermediate calculations that will make the formula easier to manipulate. Determine $\frac{AR}{n}$, $\left(1+\frac{R}{n}\right)^{nT}$, $\left(1+\frac{R}{n}\right)^{nT}-1$, and store these values in your calculator's memory.
6. Substitute these values into the formula and simplify.

Step 3: Execute the plan.
1. Calculate the down payment:
Down payment = 0.15($10,500.00) = $1,575.00.

2. Calculate the amount financed:
   Amount financed = $10,500.00 − $1,575.00 = $8,925.00.

3. The amortization formula is: $P = \dfrac{\left(\dfrac{AR}{n}\right)\left(1+\dfrac{R}{n}\right)^{nT}}{\left(1+\dfrac{R}{n}\right)^{nT} - 1}$.

4. Convert the data:
   a. 7.25% = 0.0725 = R.
   b. 60 months = $(\cancel{60 \text{ months}})\left(\dfrac{1 \text{ year}}{\cancel{12 \text{ months}}}\right)$ = 5 years = T.

5. Perform the intermediate calculations:
   a. $\dfrac{AR}{n} = \dfrac{\$8{,}925.00(0.0725)}{12} \approx \$53.921875$.
   (Store in calculator's memory 1.)

   b. $\left(1+\dfrac{R}{n}\right)^{nT} = \left(1+\dfrac{0.0725}{12}\right)^{(12\times 5)} \approx (1.006041667)^{60} \approx 1.435350885$.
   (Store in calculator's memory 2.)

   c. $\left(1+\dfrac{R}{n}\right)^{nT} - 1 = \left(1+\dfrac{0.0725}{12}\right)^{(12\times 5)} - 1 \approx (1.00604041667)^{60} - 1 \approx$ 0.435350885. **(Store in calculator's memory 3.)**

6. Substitute, using the order of operations, recall the appropriate stored values from memory and simplify. (Note: The denominator is one less than the second factor of the numerator.)
   $P = \dfrac{(\$53.921875)(1.43530885)}{0.435350885} \approx \$177.7803001$.
   The monthly payment is $177.78.

Step 4: Check your work.
1. Check the numbers and your calculations.
2. Is the answer reasonable?
3. Was the question answered?

Example 4: A new pickup truck costs $18,028 cash ("Out-the-door" price, all taxes, options and dealer fees included) or 20% down (cash/trade) with the remainder financed at 3.9% for 36 months. Find the monthly payment.

Solution:
Step 1: Understand and picture the problem.
  1. Question: Find the monthly payment.
  2. List the given data:
     "Out-the-door" price of the truck = $18,028,
     n = 12 pay periods per year,   Time = 36 months,
     Rate = 3.9% compounded monthly, with 20% down.
  3. Identify the "amortization of a loan" as the unknown.

Step 2: Devise a plan.
  1. Calculate the down payment: Down Payment = 0.20 *times* price.
  2. Calculate the amount of money financed:
     Amount financed = price *minus* down payment.
  3. Use the amortization formula.
  4. Convert the given data into the appropriate units.
  5. Perform the intermediate calculations that will make the formula easier to manipulate. Determine $\dfrac{AR}{n}$, $\left(1+\dfrac{R}{n}\right)^{nT}$, $\left(1+\dfrac{R}{n}\right)^{nT}-1$, and store these values in your calculator's memory.
  6. Substitute these values into the formula and simplify.

Step 3: Execute the plan.
  1. Calculate the down payment:
     Down payment = 0.20($18,028) = $3,605.60.
  2. Calculate the amount financed:
     Amount financed = $18,028.00 − $3,605.60 = $14,422.40.
  3. The amortization formula is: $P = \dfrac{\left(\dfrac{AR}{n}\right)\left(1+\dfrac{R}{n}\right)^{nT}}{\left(1+\dfrac{R}{n}\right)^{nT}-1}$.
  4. Convert the data:
     a. 3.9% = 0.039 = R.

123

124

b. 36 months = $\left(\cancel{36 \text{ months}}^{3}\right)\left(\dfrac{1 \text{ year}}{\cancel{12 \text{ months}}_{1}}\right)$ = 3 years = T.

5. Perform the intermediate calculations:

   a. $\dfrac{AR}{n} = \dfrac{\$14{,}422.40(0.039)}{12} \approx \$46.8728.$

   (Store in calculator's memory 1.)

   b. $\left(1+\dfrac{R}{n}\right)^{nT} = \left(1+\dfrac{0.039}{12}\right)^{(12\times 3)} \approx (1.00325)^{36} \approx 1.123906189.$

   (Store in calculator's memory 2.)

   c. $\left(1+\dfrac{R}{n}\right)^{nT} - 1 = \left(1+\dfrac{0.039}{12}\right)^{(12\times 3)} - 1 \approx (1.00325)^{36} - 1 \approx 0.123906189.$

   (Store in calculator's memory 3.)

6. Substitute, using the order of operations, recall the appropriate stored values from memory and simplify. (Note: The denominator is one less than the second factor of the numerator.)

   $P = \dfrac{(\$46.8728)(1.123906189)}{0.123906189} \approx \$425.1654463.$ Only round the final result. The monthly payment is $425.17.

Step 4: Check your work.
1. Check the numbers and your calculations.
2. Is the answer reasonable?
3. Was the question answered?

(c) 2006 JupiterImages Corp.

BUYER BEWARE!

A loan based on the amortization formula is also referred to as a simple interest loan. There are two formulas used to calculate the periodic payment on a simple interest loan!
1. The future value of simple interest divided by the number of payments.
2. The amortization formula

Simple Interest Formulas for Loans	
Future Value $$\text{Payment} = \frac{P(1+RT)}{nT}$$ Where: P is the Amount of the loan R is the annual interest rate. n is the number of compound periods in a year. T is the number of years.	**Amortization** $$\text{Payment} = \frac{\left(\frac{AR}{n}\right)\left(1+\frac{R}{n}\right)^{nT}}{\left(1+\frac{R}{n}\right)^{nT} - 1}$$ where: A is the amount of the loan. R is the annual interest rate. n is the number of compound periods in a year. T is the number of years.

When purchasing a car, the financial manager will inform you that the monthly payment is calculated using the simple interest formula. The problem arises in which simple interest loan formula is used to calculate the monthly payment.

All loan payments consist of two parts Principal and Interest (P&I). Interest is the portion of the payment taken by the lender as a service charge for the loan. Principal is the remainder of the payment applied to the balance of the loan. It is in the calculation of the interest payment where the two formulas differ.

Future Value of Simple Interest	**Amortization**
The interest payment is based on the loan amount and remains the same for every payment.	The interest payment is based on the remaining balance of the loan and decreases with each payment.

Simple interest loans based on the amortization formula will ALWAYS be cheaper than simple interest loans based on the future value of simple interest formula.

Revisiting the loan contract presented at the beginning of this section. The $12505.50 loan at 8.75% for 4 years has monthly payments of

1. Using the future value of simple interest formula $\left[\text{Payment} = \frac{P(1+RT)}{nT}\right]$, the montly payment is $351.72.

2. Using the amortization formula $\left[\text{Payment} = \frac{\left(\frac{AR}{n}\right)\left(1+\frac{R}{n}\right)^{nT}}{\left(1+\frac{R}{n}\right)^{nT} - 1}\right]$, the monthly payment is $309.72.

The P&I of a loan will be covered in detail in the next section.

Example 5: A new pickup truck costs $18,028 cash ("Out-the-door" price, all taxes, options and dealer fees included) or 20% down (cash/trade) with the remainder financed at 3.9% for 36 months. The simple interest (amortization) monthly payment is $425.17 (See Example 4). Find the simple interest (future value) monthly payment.

Solution:
Step 1: Understand and picture the problem.
　　1. Question: Find the monthly payment.
　　2. List the given data:
　　　　"Out-the-door" price of the truck = $18,028,
　　　　n = 12 pay periods per year,　Time = 36 months,
　　　　Rate = 3.9% compounded monthly, with 20% down.
　　3. Identify the monthly payment as the unknown.

Step 2: Devise a plan.
　　1. Calculate the down payment: Down Payment = 0.20 *times* price.
　　2. Calculate the amount of money financed:
　　　　Amount financed = price *minus* down payment.
　　3. Use the future value of simple interest divided by the number of payments.
　　4. Convert the given data into the appropriate units.
　　5. Substitute these values into the formula and simplify.

Step 3: Execute the plan.
　　1. Calculate the down payment:
　　　Down payment = 0.20($18,028) = $3,605.60.
　　2. Calculate the amount financed:
　　　Amount financed = $18,028.00 − $3,605.60 = $14,422.40 .
　　3. The future value of simple interest divided by the number of payments is $\dfrac{P(1+RT)}{nT}$
　　4. Convert the data:
　　　a. 3.9% = 0.039 = R.

b. 36 months = $\left(\cancel{36 \text{ months}}^{3}\right)\left(\dfrac{1 \text{ year}}{\cancel{12 \text{ months}}_{1}}\right) = 3$ years = T.

5. Substitute, recall the appropriate stored values from memory and simplify

Payment = $\dfrac{\$14,422.40(1+(0.039)(3))}{12(3)} \approx \$447.50$.   Only round the final result. The monthly payment is $447.50.

Note: The monthly payment for simple interest (future value, $447.40) is more expensive than the monthly payment for simple interest (amortization, $425.17).

Step 4: Check your work.
1. Check the numbers and your calculations.
2. Is the answer reasonable?
3. Was the question answered?

Zero Percent Financing
A new enticement to purchasing a vehicle is to offer zero percent, 0%, financing. The lending institution does not charge interest for the loan. The monthly payment for a zero percent interest loan is the loan amount divided by the number of payments.

MONTHLY PAYMENT FOR A ZERO PERCENT INTEREST LOAN

$$P = \dfrac{A}{nT}$$

where: A is the amount of the loan.
P is the fixed periodic payment.
n is the number of compound periods in a year.
T is the number of years.

The following example demonstrates a zero (no) interest loan.
Example 6: A new pickup truck costs $18,028 cash ("Out-the-door" price, all taxes, options and dealer fees included) or 20% down (cash/trade) with the remainder financed at 0% for 36 months. Find the monthly payment.

Solution:
Step 1: Understand and picture the problem.
1. Question: Find the monthly payment.
2. List the given data:
"Out-the-door" price of the truck = $18,028,

n = 12 pay periods per year,   Time = 36 months,
Rate = 0% compounded monthly, with 20% down.

Step 2: Devise a plan.
1. Calculate the down payment: Down Payment = 0.20 *times* price.
2. Calculate the amount of money financed:
   Amount financed = price *minus* down payment.
3. Use the appropriate formula to calculate the monthly payment.
4. Convert the given data into the appropriate units.
5. Substitute these values into the appropriate formula and simplify.

Step 3: Execute the plan.
1. Calculate the down payment:
   Down payment = 0.20($18,028) = $3,605.60.
2. Calculate the amount financed:
   Amount financed = $18,028.00 − $3,605.60 = $14,422.40.
3. $P = \dfrac{A}{nT}$

4. 36 months = $\left(\cancel{36\text{ months}}^{3}\right)\left(\dfrac{1\text{ year}}{\cancel{12\text{ months}}_{1}}\right) = 3$ years = T.

5. Substitute and simplify.
   $P = \dfrac{\$14{,}422.40}{12(3)} = \dfrac{\$14{,}422.40}{36} \approx \$400.62$ Only round the final result. The monthly payment is $400.62.

Step 4: Check your work.
1. Check the numbers and your calculations.
2. Is the answer reasonable?
3. Was the question answered?

❓ Cognitive Problems ❓

1. Give an example besides buying a vehicle that requires a loan?
2. Does rounding for the intermediate calculations of the amortization formula produce a different monthly payment? Give an example.
3. When purchasing a car, why should the buyer be concerned when told that the simple interest formula is being used to calculate the periodic payment?

Exercise 2.3

1. Calculate the monthly payment on a car loan. The "out-the-door" price is $26,999 cash or a down payment of $3,000. The remainder of the balance is financed at 9% for 5 years. Use
   A. Future value of simple interest.
   B. Amortization.

2. Calculate the monthly payment on a car loan. Te "out-the-door" price is $26,999 cash or a down payment of $3,000. The remainder of the balance is financed at 0% for 5 years

3. Calculate the monthly payment on a car loan. The "out-the-door" price is $19,790 cash or a down payment of $2,000. The remainder of the balance is financed at 0% for 3 years.

4. Calculate the monthly payment on a car loan. The "out-the-door" price is $19,790 cash or a down payment of $2,000. The remainder of the balance is financed at 8.75% for 3 years. Use
   A. Future value of simple interest.   B. Amortization.

5. Calculate the monthly payment on a car loan. The "out-the-door" price is $46,700 cash or a down payment of 10% (cash/trade) and the remainder of the balance is financed at 9.2% for 6 years. Use
   A. Future value of simple interest.
   B. Amortization.

6. Calculate the monthly payment on a car loan. The "out-the-door" price is $46,700 cash or a down payment of 10% (cash/trade) and the remainder of the balance is financed at 0% for 6 years

130

7. Calculate the monthly payment on an "off-road" vehicle loan. The "out-the-door" price is $18,400 cash or no down payment and the $18,400 is financed at 0% for 4 years.

8. Calculate the monthly payment on an "off-road" vehicle loan. The "out-the-door" price is $18,400 cash or a down payment of 5% (cash/trade) and the remainder of the balance is financed at 7.9% for 4 years. Use
   A. Future value of simple interest.
   B. Amortization.

9. Calculate the monthly payment on a car loan. The "out-the-door" price is $12,480 cash or a down payment of $500. The remainder of the balance is financed at 6.8% for 3 years. Use
   A. Future value of simple interest.
   B. Amortization.

(c) 2006 JupiterImages Corp.

10. Calculate the monthly payment on a motorcycle loan. The "out-the-door" price is $12,480 cash or a down payment of $500. The remainder of the balance is financed at 0% for 3 years.

11. Calculate the monthly payment on a motorcycle loan. The "out-the-door" price is $10,500 cash or a down payment of $500. The remainder of the balance is financed at 0% for 2 years.

12. Calculate the monthly payment on a water craft loan. The "out-the-door" price is $10,500 cash or a down payment of $500. The remainder of the balance is financed at 6.4% for 2 years. Use
    A. Future value of simple interest.
    B. Amortization.

13. Calculate the monthly payment on a truck loan. The "out-the-door" price is $22,680 cash or a down payment of 10% and the remainder of the balance is financed at 5.75% for 4 years. Use
    A. Future value of simple interest.
    B. Amortization.

14. Calculate the monthly payment on a truck loan. The "out-the-door" price is $22,680 cash or a down payment of 10% and the remainder of the balance is financed at 0% for 4 years.

15. Calculate the monthly payment on a truck loan. The "out-the-door" price is $16,200 cash or 0% down and the $16,200 is financed at 0% for 5 years.

16. Calculate the monthly payment on a truck loan. The "out-the-door" price is $16,200 cash or a down payment of 5% and the remainder of the balance is financed at 5.95% for 5 years. Use
    A. Future value of simple interest.
    B. Amortization.

17. Calculate the monthly payment on a boat loan. The "out-the-door" price is $170,000 cash or a down payment of 15% and the remainder of the balance is financed at 7% for 15 years. Use
    A. Future value of simple interest.
    B. Amortization.

(c) 2006 JupiterImages Corp.

18. Calculate the monthly payment on a boat loan. The "out-the-door" price is $170,000 cash or a down payment of 15% and the remainder of the balance is financed at 0% for 15 years.

2.4 REMAINING BALANCE ON SIMPLE INTEREST LOANS

Objective: Calculate the remaining balance on a loan, the amount of money needed to pay off the loan early.

REMAINING BALANCE

Suppose after 2 years, a person with a 4 year loan, wants to trade-in their car and purchase another car. In order to get the title to the car from the loaning institution, the remaining balance on the loan must be paid. The remaining balance is **NOT** always the monthly payment times the number of payments remaining. It may be less than that amount. One way to find out is to call the bank and ask customer service for the remaining balance. However, how do you know they are giving you the correct amount? In this section, the methods used to calculate the remaining balance on a simple interest loan are presented. Both types of simple interest loans: future value of simple interest and amortization are presented.

REMAINING BALANCE ON A SIMPLE INTEREST LOAN (FUTURE VALUE)

In the previous section, a 48 month simple interest loan (future value) for $12,505.50 at 8.75% has a monthly payment of $351.72. The following table shows the breakdown of this loan. Each payment is displayed with a breakdown of the principal and the interest (P & I). **Note, the principal and the interest remain constant.** It also shows the remaining balance after each payment and the eventual zeroing out of the debt.

(c) 2006 JupiterImages Corp.

PAYMENT #	PAYMENT	INTEREST	PRINCIPAL	BALANCE
				$12,505.50
1	$351.72	$91.19	$260.53	$12,244.97
2	$351.72	$91.19	$260.53	$11,984.44
3	$351.72	$91.19	$260.53	$11,723.91
4	$351.72	$91.19	$260.53	$11,463.38
5	$351.72	$91.19	$260.53	$11,202.85
6	$351.72	$91.19	$260.53	$10,942.32
7	$351.72	$91.19	$260.53	$10,681.79
8	$351.72	$91.19	$260.53	$10,421.26
9	$351.72	$91.19	$260.53	$10,160.73
10	$351.72	$91.19	$260.53	$ 9,900.20
11	$351.72	$91.19	$260.53	$ 9,639.67
12	$351.72	$91.19	$260.53	$ 9,379.14
13	$351.72	$91.19	$260.53	$ 9,118.61
14	$351.72	$91.19	$260.53	$ 8,858.08

PAYMENT #	PAYMENT	INTEREST	PRINCIPAL	BALANCE
15	$351.72	$91.19	$260.53	$ 8,597.55
16	$351.72	$91.19	$260.53	$ 8,337.02
17	$351.72	$91.19	$260.53	$ 8,076.49
18	$351.72	$91.19	$260.53	$ 7,815.96
19	$351.72	$91.19	$260.53	$ 7,555.43
20	$351.72	$91.19	$260.53	$ 7,294.90
21	$351.72	$91.19	$260.53	$ 7,034.37
22	$351.72	$91.19	$260.53	$ 6,773.84
23	$351.72	$91.19	$260.53	$ 6,513.31
24	$351.72	$91.19	$260.53	$ 6,252.78
25	$351.72	$91.19	$260.53	$ 5,992.25
26	$351.72	$91.19	$260.53	$ 5,731.72
27	$351.72	$91.19	$260.53	$ 5,471.19
28	$351.72	$91.19	$260.53	$ 5,210.66
29	$351.72	$91.19	$260.53	$ 4,950.13
30	$351.72	$91.19	$260.53	$ 4,689.60
31	$351.72	$91.19	$260.53	$ 4,429.07
32	**$351.72**	**$91.19**	**$260.53**	**$ 4,168.54**
33	$351.72	$91.19	$260.53	$ 3,908.01
34	$351.72	$91.19	$260.53	$ 3,647.48
35	$351.72	$91.19	$260.53	$ 3,386.95
36	$351.72	$91.19	$260.53	$ 3,126.42
37	$351.72	$91.19	$260.53	$ 2,865.89
38	$351.72	$91.19	$260.53	$ 2,605.36
39	$351.72	$91.19	$260.53	$ 2,344.83
40	$351.72	$91.19	$260.53	$ 2,084.30
41	$351.72	$91.19	$260.53	$ 1,823.77
42	$351.72	$91.19	$260.53	$ 1,563.24
43	$351.72	$91.19	$260.53	$ 1,302.71
44	$351.72	$91.19	$260.53	$ 1,042.18
45	$351.72	$91.19	$260.53	$ 781.65
46	$351.72	$91.19	$260.53	$ 521.12
47	$351.72	$91.19	$260.53	$ 260.59
48	$351.72	$91.19	$260.53	$ 0.06

Note: Because of rounding errors, the final payment may be different from the other payments by a few cents. In this case the 48th payment is be the monthly payment plus the six cents, $351.72 + $0.06 = $351.78. This amount completely pays off the loan.

Starting with a balance of $12,505.50 payment #1 begins the process of paying off the loan.

1. Calculate the interest, I = PRT = $(\$12{,}505.50)(0.0875)\left(\dfrac{1}{12}\right) \approx \$91.19$,

   Interest = $91.19.
2. Find the principal, the amount of money applied to paying off the loan
   Principal = Payment − Interest = $351.72 − $91.19 = $260.53.
3. Finally, find the remaining balance after the first payment,
   New Balance = Old Balance − Principal = $12,505.50 − $260.53 = $12,244.97.

PAYMENT #	PAYMENT	INTEREST	PRINCIPAL	BALANCE
—	—	—	—	$12,505.50
1	$351.72	$91.19	$260.53	$12,244.97

The interest is the same every month. Payment #2 continues the process of paying off the loan.
1. Interest = $91.19.
2. Find the principal, the amount of money applied to paying off the loan. This amount is the same as the previous month.
   Principal = Payment − Interest = $351.72 − $91.19 = $260.53.
3. Finally, find the remaining balance after the second payment,
   New Balance = Old Balance − Principal = $12,244.97 − $260.53 = $11,984.44.

PAYMENT #	PAYMENT	INTEREST	PRINCIPAL	BALANCE
—	—	—	—	$12,505.50
1	$351.72	$91.19	$260.53	$12,244.97
2	$351.72	$91.19	$260.53	$11,984.44

There is an easier way of finding the remaining balance of a simple interest loan without having to calculate a new balance every month. Use the following formula to calculate the remaining balance on a simple interest loan (future value).

REMAINING BALANCE FOR A SIMPLE INTEREST LOAN (FUTURE VALUE)

$$B = A - m\left[P - \dfrac{A(R)}{n}\right]$$

where: B is the remaining balance on the loan.
A is the amount of the loan.
m is the number of payments made.
P is the fixed periodic payment.
R is the annual interest rate as a decimal.
n is the number of compound periods in a year.

To demonstrate the validity of this formula, let's calculate the remaining balance for the above loan after making payments for 2 years and 8 months (32 payments).

Example 1: Find the remaining balance on a 48 month simple interest (future value) loan after 2 years and 8 months. The loan is for $12,505.50 at 8.75% interest. The monthly payments are $351.72.

(c) 2006 JupiterImages Corp.

Solution:
Step 1: Understand and picture the problem.
   1. Question: Find the remaining balance of the simple interest loan.
   2. List the given data:
      A = $12,505.50,   Life of loan = 48 months,
      n = 12 pay periods per year,   Time = 2 years and 8 months,
      Rate = 8.75%,   P = $351.72.
   3. Identify the "Remaining balance of a simple interest loan" as the unknown.

Step 2: Devise a plan.
   1. Use the appropriate formula.
   2. Convert the given data into the appropriate units.
   3. Substitute these values into the formula and simplify.

Step 3: Execute the plan.
   1. The formula is: $B = A - m\left[P - \dfrac{A(R)}{n}\right]$.

   2. Convert the data:
      a. Payments have been made for 2 years and 8 months:

      $2 \text{ yrs} + 8 \text{ months} = (2 \cancel{\text{years}})\left(\dfrac{12 \text{ months}}{1 \cancel{\text{year}}}\right) + 8 \text{ months} =$

      24 months + 8 months = 32 months, m = 32.
      b. 8.75% = 0.0875 = R.

   3. Substitute, recall the appropriate stored values from memory and simplify.

   $B = \$12,505.50 - 32\left[\$351.72 - \dfrac{\$12,505.50(0.0875)}{12}\right] = \$4,168.41$

   The remaining balance on the loan is $4,168.41.

Step 4: Check your work.
Compare the calculated remaining balance to the balance in the table (highlighted in bold print). The result matches within pennies. Rounding errors of a penny may occur every month. Since the remaining balance is calculated after 32 months, the figure should be within ± $0.32 of the table value. The formula's remaining balance is $4,168.41. The table's remaining balance is $4,168.54 The difference of −$0.13 is reasonable and acceptable.

REMAINING BALANCE ON A SIMPLE INTEREST LOAN (AMORTIZATION)

In the previous section, a 48 month simple interest loan for $12,505.50 at 8.75% has a monthly payment of $309.72. The following table shows the breakdown of this loan. Each payment is displayed with a breakdown of the principal and the interest (P & I). **Note, the principal and the interest do not remain constant** but the monthly payment does remain the same. It also shows the remaining balance after each payment and eventually zeroing out of the debt.

(c) 2006 JupiterImages Corp.

PAYMENT #	PAYMENT	INTEREST	PRINCIPAL	BALANCE
—	—	—	—	$12,505.50
1	**$309.72**	**$91.19**	**$218.53**	**$12,286.97**
2	**$309.72**	**$89.59**	**$220.13**	**$12,066.84**
3	$309.72	$87.99	$221.73	$11,845.11
4	$309.72	$86.37	$223.35	$11,621.76
5	$309.72	$84.74	$224.98	$11,396.78
6	$309.72	$83.10	$226.62	$11,170.16
7	$309.72	$81.45	$228.27	$10,941.89
8	$309.72	$79.78	$229.94	$10,711.95
9	$309.72	$78.11	$232.61	$10,480.34
10	$309.72	$76.42	$233.30	$10,247.04
11	$309.72	$74.72	$235.00	$10,012.04
12	$309.72	$73.00	$236.72	$ 9,775.32
13	$309.72	$71.28	$238.44	$ 9,536.88
14	$309.72	$69.54	$240.18	$ 9,296.70
15	$309.72	$67.79	$241.93	$ 9,054.77
16	$309.72	$66.02	$243.70	$ 8,811.07
17	$309.72	$64.25	$245.47	$ 8,565.60
18	$309.72	$62.46	$247.26	$ 8,318.34
19	$309.72	$60.65	$249.07	$ 8,069.27
20	$309.72	$58.84	$250.88	$ 7,818.39
21	$309.72	$57.01	$252.71	$ 7,565.68
22	$309.72	$55.17	$254.55	$ 7,311.13

PAYMENT #	PAYMENT	INTEREST	PRINCIPAL	BALANCE
23	$309.72	$53.31	$256.41	$ 7,054.72
24	$309.72	$51.44	$258.28	$ 6,796.44
25	$309.72	$49.56	$260.16	$ 6,536.28
26	$309.72	$47.66	$262.06	$ 6,274.22
27	$309.72	$45.75	$263.97	$ 6,010.25
28	$309.72	$43.82	$265.90	$ 5,744.35
29	$309.72	$41.89	$267.83	$ 5,476.52
30	$309.72	$39.93	$269.79	$ 5,206.73
31	$309.72	$37.97	$271.75	$ 4,934.98
32	**$309.72**	**$35.98**	**$273.74**	**$ 4,661.24**
33	$309.72	$33.99	$275.73	$ 4,385.51
34	$309.72	$31.98	$277.74	$ 4,107.77
35	$309.72	$29.95	$279.77	$ 3,828.00
36	$309.72	$27.91	$281.81	$ 3,546.19
37	$309.72	$25.86	$283.86	$ 3,262.33
38	$309.72	$23.79	$285.93	$ 2,976.40
39	$309.72	$21.70	$288.02	$ 2,688.38
40	$309.72	$19.60	$290.12	$ 2,398.26
41	$309.72	$17.49	$292.23	$ 2,106.03
42	$309.72	$15.36	$294.36	$ 1,811.67
43	$309.72	$13.21	$296.51	$ 1,515.16
44	$309.72	$11.05	$298.67	$ 1,216.49
45	$309.72	$ 8.87	$300.85	$ 915.64
46	$309.72	$ 6.68	$303.04	$ 612.60
47	$309.72	$ 4.47	$305.25	$ 307.35
48	$309.72	$ 2.24	$307.48	–$ 0.06

Note: Because of rounding errors, the final payment may be different from the other payments by a few cents. In this case the 48th payment is be the monthly payment minus the six cents, $309.72 –$0.06 = $309.66. This amount completely pays off the loan.

Starting with a balance of $12,505.50 payment #1 begins the process of paying off the loan.

1. Calculate the interest, I = PRT = $(\$12{,}505.50)(0.0875)\left(\dfrac{1}{12}\right) \approx \$91.19$

   Interest = $91.19.
2. Find the principal, the amount of money applied to paying off the loan
   Principal = Payment – Interest = $309.72 – $91.19 = $218.53

3. Finally, find the remaining balance after the first payment,
   New Balance = Old Balance − Principal = $12,505.50 − $218.53 = $12,286.97.

PAYMENT #	PAYMENT	INTEREST	PRINCIPAL	BALANCE
—	—	—	—	$12,505.50
1	$309.72	$91.19	$218.53	$12,286.97

The interest is **NOT** the same every month. Payment #2 continues the process of paying off the loan.

1. Find the interest, $I = PRT = (\$12,286.97)(0.0875)\left(\dfrac{1}{12}\right) \approx \$89.59$,

   Interest = $89.59.
2. Find the principal, the amount of money applied to paying off the loan.
   Principal = Payment − Interest = $309.72 − $89.59 = $220.13.
3. Finally, find the remaining balance after the second payment,
   New Balance = Old Balance − Principal = $12,286.97 − $220.13 = $12,066.84.

PAYMENT #	PAYMENT	INTEREST	PRINCIPAL	BALANCE
—	—	—	—	$12,505.50
1	$309.72	$91.19	$218.53	$12,286.97
2	$309.72	$89.59	$220.13	$12,066.84

Each month, the interest is calculated on the remaining (new) balance. As payments are made on the loan, the amount of money the bank takes in interest decreases and the principal increases. There is an easier way of finding the remaining balance of a simple interest loan without having to calculate a new balance every month. Use the following formula to calculate the remaining balance on a simple interest loan.

REMAINING BALANCE FOR A SIMPLE INTEREST (AMORTIZATION) LOAN

$$B = P \left(\dfrac{\left(1 + \dfrac{R}{n}\right)^x - 1}{\left(\dfrac{R}{n}\right)\left(1 + \dfrac{R}{n}\right)^x} \right)$$

where: B is the remaining balance on the loan.
P is the fixed periodic payment.
R is the annual interest rate as a decimal.
n is the number of compound periods in a year.
x is the number of remaining payments.

To demonstrate the validity of this formula, let's calculate the remaining balance for the above loan after making payments for 2 years and 8 months (32 payments).

Example 2: Find the remaining balance on a 48 month simple interest (amortization) loan after 2 years and 8 months. The loan is for $12,505.50 at 8.75% interest. The monthly payments are $309.72.

(c) 2006 JupiterImages Corp.

Solution:
Step 1: Understand and picture the problem.
    1. Question: Find the remaining balance of the simple interest loan.
    2. List the given data:
        A = $12,505.50,   Life of loan = 48 months,
        n = 12 pay periods per year,   Time = 2 years and 8 months, (32 months),
        Rate = 8.75%,   P = $309.72.
    3. Identify the "Remaining balance of a simple interest loan" as the unknown.

Step 2: Devise a plan.
    1. Use the appropriate formula.
    2. Convert the given data into the appropriate units.
    3. Perform the intermediate calculations that will make the formula easy to manipulate. Determine $\frac{R}{n}$, $\left(1+\frac{R}{n}\right)^x$, $\left(1+\frac{R}{n}\right)^x - 1$ and store these values in your calculator's memory.
    4. Substitute these values into the formula and simplify.

Step 3: Execute the plan.
    1. The formula is: $B = P\left(\dfrac{\left(1+\frac{R}{n}\right)^x - 1}{\left(\frac{R}{n}\right)\left(1+\frac{R}{n}\right)^x}\right)$.

    2. Convert the data:
       a. Payments have been made for 2 years and 8 months:
        2 yrs + 8 months = $(2 \cancel{\text{years}})\left(\dfrac{12 \text{ months}}{1 \cancel{\text{year}}}\right)$ + 8 months =

        24 months + 8 months = 32 months.
        x = 48 − 32 = 16 months of remaining payments.
       b. 8.75% = 0.0875 = R.

3. Perform the intermediate calculations:
   a. $\dfrac{R}{n} = \dfrac{0.0875}{12} \approx 0.0072917.$
      (Store in your calculator's memory 1.)
   b. $\left(1+\dfrac{R}{n}\right)^x = \left(1+\dfrac{0.0875}{12}\right)^{16} \approx 1.1232692.$
      (Store in your calculator's memory 2.)
   c. $\left(1+\dfrac{R}{n}\right)^x - 1 = \left(1+\dfrac{0.0875}{12}\right)^{16} - 1 \approx 0.1232692.$
      (Store in your calculator's memory 3.)
4. Substitute, use the order of operations to recall the appropriate stored values from memory and simplify. (Note: The numerator is one less than the second factor of the denominator.)

$$B = \$309.72\left(\dfrac{0.1232692}{(0.0072917)(1.1232692)}\right) \approx \$4{,}661.366159$$

The remaining balance on the loan is $4,661.37.

Step 4: Check your work.
Compare the calculated remaining balance to the balance in the table (highlighted in bold print). The result matches within pennies. Rounding errors of a penny may occur every month. Since the remaining balance is calculated after 32 months, the figure should be within ± $0.32 of the table value. The formula's remaining balance is $4,661.37. The table's remaining balance is $4,661.24 The difference of $0.13 is reasonable and acceptable.

Rule of 78

There is another method used to calculate the remaining balance on a simple interest (amortization) loan. This method is titled the "Rule of 78". It uses a different calculation to determine the monthly interest. Reconsider the car loan presented at the beginning of this section.

A 48 month simple interest (amortization) loan for $12,505.50 at 8.75% has a monthly payment of $309.72. The following table shows the breakdown of this loan using the "Rule of 78". This breakdown displays how much of each payment applies to the interest and the principal. Note, while the interest and principal do not remain constant throughout the loan, the monthly payment does remain the same. It also shows the remaining balance after each payment.

(c) 2006 JupiterImages Corp.

PAYMENT #	PAYMENT	INTEREST	PRINCIPAL	BALANCE
				$12,505.50
1	$309.72	$96.37	$213.35	$12,292.15
2	$309.72	$94.36	$215.36	$12,076.79
3	$309.72	$92.35	$217.37	$11,859.42
4	$309.72	$90.35	$219.37	$11,640.05
5	$309.72	$88.34	$221.38	$11,418.67
6	$309.72	$86.33	$223.39	$11,195.28
7	$309.72	$84.32	$225.40	$10,969.88
8	$309.72	$82.32	$227.40	$10,742.48
9	$309.72	$80.31	$229.41	$10,513.07
10	$309.72	$78.30	$231.42	$10,281.65
11	$309.72	$76.29	$233.43	$10,048.22
12	$309.72	$74.29	$235.43	$ 9,812.79
13	$309.72	$72.28	$237.44	$ 9,575.35
14	$309.72	$70.27	$239.45	$ 9,335.90
15	$309.72	$68.26	$241.46	$ 9,094.44
16	$309.72	$66.25	$243.47	$ 8,850.97
17	$309.72	$64.25	$245.47	$ 8,605.50
18	$309.72	$62.24	$247.48	$ 8,358.02
19	$309.72	$60.23	$249.49	$ 8,108.53
20	$309.72	$58.22	$251.50	$ 7,857.03
21	$309.72	$56.22	$253.50	$ 7,503.53
22	$309.72	$54.21	$255.51	$ 7,348.02
23	$309.72	$52.20	$257.52	$ 7,090.50
24	$309.72	$50.19	$259.53	$ 6,830.97
25	$309.72	$48.18	$261.54	$ 6,569.43
26	$309.72	$46.16	$263.54	$ 6,305.89
27	$309.72	$44.17	$265.55	$ 6,040.34
28	$309.72	$42.16	$267.56	$ 5,772.78
29	$309.72	$40.15	$269.57	$ 5,503.21
30	$309.72	$38.15	$271.57	$ 5,231.64
31	$309.72	$36.14	$273.58	$ 4,958.06
32	**$309.72**	**$34.13**	**$275.59**	**$ 4,682.47**
33	$309.72	$32.12	$277.60	$ 4,404.87
34	$309.72	$30.12	$279.60	$ 4,125.27
35	$309.72	$28.11	$281.61	$ 3,843.66
36	$309.72	$26.10	$283.62	$ 3,560.04
37	$309.72	$24.09	$285.63	$ 3,274.41
38	$309.72	$22.08	$287.64	$ 2,986.77
39	$309.72	$20.08	$289.64	$ 2,697.13
40	$309.72	$18.07	$291.65	$ 2,405.48

PAYMENT #	PAYMENT	INTEREST	PRINCIPAL	BALANCE
41	$309.72	$16.06	$293.66	$ 2,111.82
42	$309.72	$14.05	$295.67	$ 1,816.15
43	$309.72	$12.05	$297.67	$ 1,518.48
44	$309.72	$10.04	$299.68	$ 1,218.80
45	$309.72	$ 8.03	$301.69	$ 917.11
46	$309.72	$ 6.02	$303.70	$ 613.41
47	$309.72	$ 4.02	$305.70	$ 307.71
48	$309.72	$ 2.01	$307.71	$ 0.00

The "Rule of 78" which is used to calculate the interest charge each month is determined by the different formula. Rather than get involved with each monthly interest charge, lets study the remaining balance formula for the Rule of 78.

The formula for calculating the remaining balance on a simple interest (amortization) loan using the "Rule of 78" is

> **REMAINING BALANCE FOR A SIMPLE INTEREST LOAN USING THE RULE OF 78**
>
> $$B = x(m) - \left(\frac{x(x+1)}{k(k+1)}\right)(k \times m - a)$$
>
> where: B is the remaining balance on the loan.
> x is the number of remaining payments.
> m is the monthly payment.
> k is the total number of payments.
> a is the amount of the loan.

Let's calculate the remaining balance for the above loan after 2 years and 8 months (32 payments).

Example 3: A car loan uses the "Rule of 78" to calculate interest. Find the remaining balance on a 48 month simple interest loan after 2 years and 8 months. The loan is for $12,505.50 at 8.75% interest. The monthly payments are $309.72.

(c) 2006 JupiterImages Corp.

Solution:
Step 1: Understand and picture the problem.
1. Question: Find the remaining balance of the simple interest loan using the "Rule of 78".
2. List the given data: a = $12,505.50, n = 48 months, m = $309.72, Time = 2 years and 8 months.

3. Identify the "Remaining balance of a simple interest loan" as the unknown.

Step 2: Devise a plan.
1. Use the appropriate formula.
2. Calculate the number of remaining payments.
3. Substitute these values into the formula and simplify.

Step 3: Execute the plan.
1. The formula is: $B = x(m) - \left(\dfrac{x(x+1)}{k(k+1)}\right)(k \times m - a)$.

2. Convert the data:
Payments have been made for 2 years and 8 months:

2 yrs + 8 months = $(2 \text{ years})\left(\dfrac{12 \text{ months}}{1 \text{ year}}\right)$ + 8 months =

24 months + 8 months = 32 months.
x = 48 − 32 = 16 months of remaining payments.

3. Substitute and simplify:
where: k = 48 months in the loan.
x = 16 for the sixteen remaining payments.
m = $309.72 for the monthly payment.
a = $12,505.50 for the original loan..

$B = 16(\$309.72) - \dfrac{16(17)}{48(49)}(48 \times \$309.72 - \$12,505.50)$.

$B = \$4955.52 - \dfrac{272}{2,352}(\$2,361.06) = \$4955.52 - \$273.0477551$.

B ≈ $4,682.472245.
The remaining balance on the loan is $4,682.47.

Step 4: Check your work.
Compare the calculated remaining balance to the balance in the table (highlighted in bold print). The result matches. (Note: Rounding errors of a penny may occur every month. Since the remaining balance is calculated after 32 months, the figure should be within ± $0.32 of the table value.) The result is reasonable and correct.

The following excerpt is taken from a loan agreement for the car described in Example 4. According to the "Truth in Lending Act" of 1968, the federal government requires full disclosure of the terms of a loan. Look for these features in the following contract:

a. **Annual Percentage Rate**, interest rate on the loan.
b. **Finance Charge**, amount paid in interest.
c. **Amount Financed**, the loan amount.
d. **Total Payment** is the (number of payments) × (monthly payment).
e. **Total Sales Price** is the total payment + down payment.
f. **Number of payments** is the number of payments made.
g. **Amount of Payment** is the amount of each monthly payment.
h. **No Prepayment Penalty clause.**
i. **Rebate for Prepayment** is the "Rule of 78" clause.

ANNUAL PERCENTAGE RATE	FINANCE CHARGE	AMOUNT FINANCED	TOTAL PAYMENT	TOTAL SALES PRICE
8.75%	$2,361.06	$12,505.50	$14,866.56	The total purchase price is $16,256.06

Number of Payments	Amount of Payments
48	$309.72

Prepayment You have the right to prepay the unpaid balance in full or in part at anytime without penalty.

REBATE FOR PREPAYMENT: In the event of prepayment of the contract before the maturity of the final installment, the buyer receives a rebate of unearned finance charge computed by the Rule of 78.

Remaining Balance for a Zero Percent Interest Loan

The money saved by prepaying a "conventional loan" or "rule of 78 loan" is interest charges. A "zero percent interest loan" has no money paid in interest charges. With no interest charges, there is no money saved by prepaying the loan. The remaining balance for a "zero percent interest loan" is the remaining payments left in the loan.

REMAINING BALANCE FOR A ZERO PERCENT INTEREST LOAN

$$B = x(m)$$

where: B is the remaining balance on the loan.
x is the number of remaining payments.
m is the monthly payment.

Example 4: A truck loan has "Zero Percent Interest". Find the remaining balance on a 36 month loan after 1 year and 5 months. The loan is for $14,422.40 at 0% interest. The monthly payments are is $400.62.

Solution:
Step 1: Understand and picture the problem.
1. Question: Find the remaining balance of a zero percent interest loan.
2. List the given data: a = $14,420.40,   n = 36 months,   m = $400.62, Time = 1 year and 5 months.
3. Identify the "Remaining balance of a simple interest loan" as the unknown.

Step 2: Devise a plan.
1. Use the appropriate formula.
2. Calculate the number of remaining payments.
3. Substitute these values into the formula and simplify.

Step 3: Execute the plan.
1. The formula is: $B = x(m)$.
2. Convert the data:
Payments have been made for 1 year and 5 months:

$$1 \text{ yr} + 5 \text{ months} = (1 \cancel{\text{year}}) \left(\frac{12 \text{ months}}{1 \cancel{\text{year}}} \right) + 5 \text{ months} =$$

12 months + 5 months = 17 months.
x = 36 − 17 = 19 months of remaining payments.

3. Substitute and simplify:
where: x = 19 for the nineteen remaining payments.
m = $400.62 for the monthly payment.
$B = 19(\$400.62) = \$7,611.78$.

The remaining balance on the loan is $7,611.78.

Step 4: Check your work
1. Check the numbers and your calculations.
2. Is the answer reasonable?
3. Was the question answered?

NO PREPAYMENT PENALTY

In Example 2, the remaining balance for the 4 year car loan after 2 years, 8 months (paid 32 of 48 payments) was $4,661.37. This is the amount it costs to repay the loan early and get the title of the car. This figure is less than making the remaining 16 payments of $309.72.

$$16(\$309.72) = \$4,955.52$$

There is a savings of $294.15.

$$\$4,955.52 - \$4,661.37 = \$294.15$$

The money saved is the amount of interest required for the remainder (16 payments) of the loan. Check the payment table for the amortization loan presented earlier in this section. The sum of the interest payments from month 33 ($32.12) to month 48 ($2.01) totals the money saved by paying the loan off early.

These savings occur **ONLY** if the loan contract has a **"NO PREPAYMENT PENALTY"** clause. If the contract **does not** have a **"NO PREPAYMENT PENALTY"** clause, then it will cost the borrower all the remaining payments. In this case, 16($309.72) = $4,955.52 gets the title of the car with no savings for repaying the loan early. **(Consumer Tip: Make sure your loan has a "NO PREPAYMENT PENALTY" clause.)**

The following excerpt is taken from a loan agreement for the car described in Example 2. According to the "Truth in Lending Act" of 1968, the federal government requires full disclosure of the terms of a loan. Study the contract for

a. **Annual Percentage Rate**, interest rate on the loan.
b. **Finance Charge**, amount paid in interest.
c. **Amount Financed**, the loan amount.
d. **Total Payment** is the (number of payments) × (monthly payment).
e. **Total Sales Price** is the total payment + down payment.
f. **Number of payments** is the number of payments made.
g. **Amount of Payment** is the amount of each monthly payment.
h. **No Prepayment Penalty clause.**

ANNUAL PERCENTAGE RATE	FINANCE CHARGE	AMOUNT FINANCED	TOTAL PAYMENT	TOTAL SALES PRICE
8.75%	$2,361.06	$12,505.50	$14,866.56	The total purchase price is $16,256.06
Number of Payments 48			Amount of Payments $309.72	

Prepayment You have the right to prepay the unpaid balance in full or in part at anytime without penalty.

⚠️**Caution!** Because of the Federal Disclosure Act of 1968, all loan contracts must disclose the terms of the loan. The **"No Prepayment Penalty"** clause is **NOT** a standard clause in all loans. If it is not part of your contract, have it added to your loan agreement. The significance of the **"No Prepayment Penalty"** clause is demonstrated in the next three examples.

Example 5: A loan has a "No Prepayment Penalty" clause. Find the amount of money needed to repay the remaining balance of a 60 month simple interest (future value) loan after 1 year and 3 months. The loan is for $8,925 at 7.25% interest. The monthly payments are $202.67.

Solution:
Step 1: Understand and picture the problem.
  1. Question: Find the remaining balance of the simple interest loan.
  2. List the given data: A = $8,925, Life of loan = 60 months,
     n = 12 pay periods per year, Time = 1 year and 3 months,
     Rate = 7.25%, P = $202.67 monthly.
  3. Identify the "Remaining balance of a simple interest (future value) loan" as the unknown.

Step 2: Devise a plan.
  1. Use the appropriate formula.
  2. Convert the given data into the appropriate units.
  3. Substitute these values into the formula and simplify.

Step 3: Execute the plan.
  1. The formula is: $B = A - m\left[P - \dfrac{A(R)}{n}\right]$.

  2. Convert the data:
     a. Payments have been made for 1 years and 3 months:

     1 yr + 3 months = $\left(1 \cancel{\text{year}}\right)\left(\dfrac{12 \text{ months}}{1 \cancel{\text{year}}}\right)$ + 3 months =

     12 months + 3 months = 15 months, m = 15.
     b. 7.25% = 0.0725 = R.

  3. Substitute, recall the appropriate stored values from memory and simplify.

147

$$B = \$8{,}925 - 15\left[\$202.67 - \frac{\$8925(0.0725)}{12}\right] \approx \$6{,}693.78$$

The remaining balance on the loan is $6,693.78.

Step 4: Check your work.
1. Check the numbers and your calculations.
2. Is the answer reasonable?
3. Was the question answered?

Example 6: A loan has a "No Prepayment Penalty" clause. Find the amount of money needed to repay the remaining balance of a 60 month simple interest (amortization) loan after 1 year and 3 months. The loan is for $8,925 at 7.25% interest. The monthly payments are $177.78

Solution:
Step 1: Understand and picture the problem.
1. Question: Find the remaining balance of the simple interest loan.
2. List the given data: A = $8,925, Life of loan = 60 months, n = 12 pay periods per year, Time = 1 year and 3 months, Rate = 7.25%, P = $177.78 monthly.
3. Identify the "Remaining balance of a simple interest loan" as the unknown.

Step 2: Devise a plan.
1. Use the appropriate formula.
2. Convert the given data into the appropriate units.
3. Perform the intermediate calculations that will make the formula easy to manipulate. Determine $\frac{R}{n}$, $\left(1+\frac{R}{n}\right)^x$, $\left(1+\frac{R}{n}\right)^x - 1$ and store these three values in your calculator's memory.
4. Substitute these values into the formula and use the order of operations to simplify.

Step 3: Execute the plan.
1. The formula is: $B = P\left(\dfrac{\left(1+\frac{R}{n}\right)^x - 1}{\left(\frac{R}{n}\right)\left(1+\frac{R}{n}\right)^x}\right)$.

2. Convert the data:
   a. Payments have been made for 1 year and 3 months:

   1 year + 3 months = $(1 \text{ years})\left(\dfrac{12 \text{ months}}{1 \text{ year}}\right)$ + 3 months =

   12 months + 3 months = 15 months.
   x = 60 − 15 = 45 months of remaining payments.
   b. 7.25% = 0.0725 = R.
3. Perform the intermediate calculations:
   a. $\dfrac{R}{n} = \dfrac{0.0725}{12} \approx 0.0060416667.$

   (Store in your calculator's memory 1.)

   b. $\left(1+\dfrac{R}{n}\right)^x = \left(1+\dfrac{0.0725}{12}\right)^{45} \approx 1.31134982.$

   (Store in your calculator's memory 2.)

   c. $\left(1+\dfrac{R}{n}\right)^x - 1 = \left(1+\dfrac{0.0725}{12}\right)^{45} - 1 \approx 0.31134982.$

   (Store in your calculator's memory 3.)

4. Substitute, recall the appropriate stored values from memory and use the order of operations to simplify. (Note: The numerator is one less than the second factor of the denominator.)

   $B = \$177.78\left(\dfrac{0.31134982}{(0.0060416667)(1.31134982)}\right) \approx \$6{,}986.444281.$

   The remaining balance on the loan is $6,986.44.

Step 4: Check your work.
   1. Check the numbers and your calculations.
   2. Is the answer reasonable?
   3. Was the question answered?

Example 7: This loan does not have a "No Prepayment Penalty" clause. Find the amount of money needed to repay the remaining balance of a 60 month simple interest loan after 1 year and 3 months. The loan is for $8,925 at 7.25% interest. The monthly payments are $177.78.

Solution:

Step 1: Understand and picture the problem.
1. Question: Find the remaining balance of the simple interest loan.
2. List the given data: A = $8,925,   Life of loan = 60 months,
   n = 12 pay periods per year,   Time = 1 year and 3 months,
   Rate = 7.25%,   P = $177.78 monthly.
3. Identify the number of remaining payments as the unknown.

Step 2: Devise a plan.
1. Calculate the number of remaining payments.
2. Find the product of the "Number of Remaining Payments" and the "Monthly Payment" as the "Remaining Balance".

Step 3: Execute the plan.
1. Payments have been made for 1 year and 3 months:

   $$1 \text{ year} + 3 \text{ months} = (1 \cancel{\text{years}}) \left(\frac{12 \text{ months}}{1 \cancel{\text{year}}} \right) + 3 \text{ months} =$$

   12 months + 3 months = 15 months.
   60 – 15 = 45 months of remaining payments.
2. 45($177.78) = $8,000.10. The amount to repay the loan is $8,000.10.

Step 4: Check your work.
1. Check the numbers and your calculations.
2. Is the answer reasonable?
3. Was the question answered?

Note: The largest remaining balance to repay a loan occurs when the contract DOES NOT HAVE a "No Prepayment Penalty Clause".

TIPS FOR PURCHASING A MEANS OF TRANSPORTATION (CAR, TRUCK, ETC.)

1. **Negotiate price starting with the "Out – the – Door" price.**
   All taxes, dealer prep, options, etc. have been included in the "Out – the – Door" price. After the down payment is subtracted, this is the amount of money you are borrowing and should appear on your loan contract as the amount financed.
2. **Shop around for the lowest interest rate.**
   You do not have to finance through the dealer. Your bank or credit union may offer a lower rate.
3. **Make sure the monthly payment is calculated using the amortization (simple interest) formula.** The future value (simple interest) is always more expensive.

4. **Make sure the contract contains a "No Prepayment Penalty" Clause.**
   If you repay your loan early, you will not pay interest on the remainder of the loan.
5. **If possible, purchase at the end of the month.**
   There may be an inhouse dealer incentive program in progress.

Optional Student Project
Project "Car Purchase" is found on page 155.

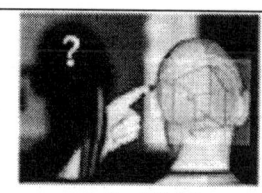

Chapter Review
Chapter review problems can be found on page 157.

Cognitive Problems

1. Two methods were used to calculate the interest and remaining balance on a simple interest loan, the "amortization method" and the "amortization method with the rule of 78". Explain how these methods differ.
2. If loans are repaid early, which method of calculating interest is in the lender's favor, the "amortization method" or the "amortization method with the rule of 78".
3. Explain the significance of the "No Prepayment Penalty" clause.
4. What is meant by rounding errors?
5. Is it possible for a bank to make an error when calculating the remaining balance on your loan? Explain.

Exercise 2.4

1. A simple interest (future value) loan of $23,999 at 9% for 5 years has a monthly payment of $579.98. Find the remaining balance after 2 years and 4 months:
   a. The contract has a "No Prepayment Penalty Clause".
   b. The contract does not have a "No Prepayment Penalty Clause".

2. A simple interest (amortization) loan of $23,999 at 9% for 5 years has a monthly payment of $498.18. Find the remaining balance after 2 years and 4 months:
   a. The contract has a "No Prepayment Penalty Clause".
   b. The contract has a "No Prepayment Penalty Clause" and the "Rule of 78".
   c. The contract does not have a "No Prepayment Penalty Clause".

3. A zero percent interest loan of $23,999 for 5 years has a monthly payment of $399.98 Find the remaining balance after 2 years and 4 months.

4. A zero percent interest loan of $17,790 at for 3 years
   has a monthly payment of $494.17. Find the remaining balance after 1 year and 9 months.

(c) 2006 JupiterImages Corp.

5. A simple interest (amortization) loan of $17,790 at 8.75% for 3 years has a monthly payment of $563.65. Find the remaining balance after 1 year and 9 months:
   a. The contract has a "No Prepayment Penalty Clause".
   b. The contract has a "No Prepayment Penalty Clause" and the "Rule of 78".
   c. The contract does not have a "No Prepayment Penalty Clause".

6. A simple interest (future value) loan of $17,790 at 8.75% for 3 years has a monthly payment of $623.89. Find the remaining balance after 1 year and 9 months:
   a. The contract has a "No Prepayment Penalty Clause".
   b. The contract does not have a "No Prepayment Penalty Clause".

7. A simple interest (amortization) loan of $42,030 at 9.2% for 6 years has a monthly payment of $761.79. Find the remaining balance after 4 years and 8 months:
   a. The contract has a "No Prepayment Penalty Clause".
   b. The contract has a "No Prepayment Penalty Clause" and the "Rule of 78".
   c. The contract does not have a "No Prepayment Penalty Clause".

(c) 2006 JupiterImages Corp.

8. A zero percent interest loan of $42,030 for 6 years has a monthly payment of $583.75. Find the remaining balance after 4 years and 8 months.

153

9. A simple interest (future value) loan of $42,030 at 9.2% for 6 years has a monthly payment of $905.98. Find the remaining balance after 4 years and 8 months:
   a. The contract has a "No Prepayment Penalty Clause".
   b. The contract does not have a "No Prepayment Penalty Clause".

10. A zero percent interest loan of $18,400 for 4 years has a monthly payment of $383.33. Find the remaining balance after 3 years:

11. A simple interest (amortization) loan of $17,480 at 7.9% for 4 years has a monthly payment of $425.92. Find the remaining balance after 3 years:
    a. The contract has a "No Prepayment Penalty Clause".
    b. The contract has a "No Prepayment Penalty Clause" and the "Rule of 78".
    c. The contract does not have a "No Prepayment Penalty Clause".

12. A simple interest (future) loan of $17,480 at 7.9% for 4 years has a monthly payment of $479.24. Find the remaining balance after 3 years:
    a. The contract has a "No Prepayment Penalty Clause".
    b. The contract does not have a "No Prepayment Penalty Clause".

13. A simple interest (amortization) loan of $11,980 at 6.8% for 3 years has a monthly payment of $368.81. Find the remaining balance after 1 year and 6 months:
    a. The contract has a "No Prepayment Penalty Clause".
    b. The contract has a "No Prepayment Penalty Clause" and the "Rule of 78".
    c. The contract does not have a "No Prepayment Penalty Clause".

(c) 2006 JupiterImages Corp.

14. A zero percent interest loan of $11,980 for 3 years has a monthly payment of $332.78. Find the remaining balance after 1 year and 6 months.

15. A simple interest (future value) loan of $11,980 at 6.8% for 3 years has a monthly payment of $400.66. Find the remaining balance after 1 year and 6 months.
    a. The contract has a "No Prepayment Penalty Clause".
    b. The contract does not have a "No Prepayment

16. A zero percent interest loan of $20,412 for 4 years has a monthly payment of $425.25. Find the remaining balance after 2 years and 11 months.

17. A simple interest (future value) loan of $20,412 at 5.75% for 4 years has a monthly payment of $516.14. Find the remaining balance after 2 years and 11 months:
    a. The contract has a "No Prepayment Penalty Clause".
    b. The contract does not have a "No Prepayment Penalty Clause".

18. A simple interest (amortization) loan of $20,412 at 5.75% for 4 years has a monthly payment of $477.04. Find the remaining balance after 2 years and 11 months:
    a. The contract has a "No Prepayment Penalty Clause".
    b. The contract has a "No Prepayment Penalty Clause" and the "Rule of 78".
    c. The contract does not have a "No Prepayment Penalty Clause".

19. A zero percent interest loan of $16,200 for 5 years has a monthly payment of $270.00 Find the remaining balance after 4 years.

20. A simple interest (amortization) loan of $15,390 at 5.75% for 5 years has a monthly payment of $297.17. Find the remaining balance after 4 years:
    a. The contract has a "No Prepayment Penalty Clause".
    b. The contract has a "No Prepayment Penalty Clause" and the "Rule of 78".
    c. The contract does not have a "No Prepayment Penalty Clause".

21. A simple interest (future value) loan of $15,390 at 5.75% for 5 years has a monthly payment of $297.17. Find the remaining balance after 4 years:
    a. The contract has a "No Prepayment Penalty Clause".
    b. The contract does not have a "No Prepayment Penalty Clause".

PROJECT: VEHICLE PURCHASE
Objective: Compare various options for financing a vehicle.

CHOOSE YOUR MEANS OF TRANSPORTATION		
	 (c) 2006 JupiterImages Corp.	
Vehicle A The "Out-the-door" price (all taxes and fees included) for this vehicle is $36,792.40.	**Vehicle B** The "Out-the-door" price (all taxes and fees included) for this vehicle is $50,485.50.	**Vehicle C** The "Out-the-door" price (all taxes and fees included) for this vehicle is $28,545.50.

Part I Monthly Payments
A. Choose one of the vehicles displayed above.
B. For only your vehicle, find the monthly payment for the following finance plans. Show all formulas used and your calculations. Be neat in your work.

Deal #1: The loan requires a 10% down payment. Finance the remainder of the amount at 0.9% for 2 years using simple interest.
   1. Future value   2. Amortization

Deal #2: The loan requires a 10% down payment. Finance the remainder of the amount at 1.9% for 3 years using simple interest.
   1. Future value   2. Amortization

Deal #3: The loan requires a 10% down payment. Finance the remainder of the amount at 2.9% for 4 years using simple interest.
   1. Future value   2. Amortization

Deal #4: The loan requires a 10% down payment. Finance the remainder of the amount at 3.9% for 5 years using simple interest.
   1. Future value   2. Amortization

Deal #5: The manufacturer is offering a $3,000 rebate. With the rebate, the "out-the-door" price of the car/motorcycle is reduced by $3,000.00 and then the down payment is calculated. The loan requires a 10% down payment. Finance the remainder of the amount at 4.9% for 5 years using simple interest.
    1. Future value   2. Amortization

Deal #6: Finance the "out the door" price of the vehicle at 0% down and 0% interest for 5 years. (Hint: Use the price of the vehicle for the loan amount.)

Part II Remaining Balance

Select from deals 1 through 5, the plan that produces the most inexpensive monthly payment. Use only that plan to calculate the remaining balance after 1 year 4 months.

A. The loan contract has a "No Prepayment Penalty" clause.
    The loan is simple interest.
    1. Future value
    2. Amortization
    3. Rule of 78 (Use the amortized monthly payment in your calculation.)

B. The loan contract does not have a "No Prepayment Penalty" clause.
    The loan is simple interest.
    1. Future value
    2. Amortization
    3. Rule of 78 (Use the amortized monthly payment in your calculation.)

C. Find the remaining balance for the zero percent interest loan from Deal #6 after 1 year 4 months.

CHAPTER 2 REVIEW

Section 2.1: Simple Interest

1. Find the interest paid on a $4,855 deposit at 3.2% simple interest for 7 months.

2. Find the interest paid on a $9,700 deposit at $4\frac{5}{8}$% simple interest for 2 years.

3. Find the rate of simple interest, if $470.40 in interest is paid on a $2,400 loan for 4 years.

4. How many years must $1,650 stay in a 6% simple interest account before it earns $222.75 in interest?

5. Find the future value of a $6\frac{1}{8}$% *6.125* simple interest savings account after 42 months. The initial deposit was $5,500.

6. After 18 years, an account that earns 3.5% simple interest has grown to $20,212. How much money was initially deposited (present value) into the account?

Section 2.2: Compound Interest

7. Find the future value of a savings account after 4 years, that is opened with $3,600 at 5.3% interest compounded quarterly.

8. Find the future value of a savings account after 4 years, that is opened with $3,600 at 5.3% interest compounded continuously.

9. Find the present value of a money market account that grows to $14,800 in 8 years, if the account is compounded quarterly at $7\frac{1}{8}$%. *7.125*

10. Find the present value of a savings account that grows to $7,400 in 3 years, if the account is compounded continuously at 4.9%.

11. Find the Annual Percentage Rate (APR) for the following accounts. Which account has the best loan rate?
    a. $7\frac{7}{8}$% *7.875%* interest compounded semi-annually.
    b. 7.85% interest compounded monthly.
    c. 7.8% interest compounded continuously.

12. Find the Annual Percentage Yield (APY) for the following accounts. Which account has the most profitable savings rate?
   a. $7\frac{7}{8}$% interest compounded semi-annually.
   b. 7.85% interest compounded monthly.
   c. 7.8% interest compounded continuously.

Section 2.3: Car Loans

13. Calculate the monthly payment on a simple interest car loan. The "out-the-door" price is $9,999 cash or a down payment of $2,000. The remainder of the balance is financed at 6.4% for 5 years.
    a. Future value   b. Amortization

14. Calculate the monthly payment on a car loan. The "out-the-door" price is $9,999 cash or a down payment of $2,000. The remainder of the balance is financed at 0% for 5 years.

15. Calculate the monthly payment on a car loan. The "out-the-door" price is $22,790 cash or a down payment of $1,500. The remainder of the balance is financed at 0% for 4 years

16. Calculate the monthly payment on a simple interest car loan. The "out-the-door" price is $22,790 cash or a down payment of $1,500. The remainder of the balance is financed at 4.2% for 4 years
    a. Future value   b. Amortization

17. Calculate the monthly payment on a car loan. The "out-the-door" price is $22,790 cash or a down payment of $0. The remainder of the balance is financed at 0% for 4 years

18. Calculate the monthly payment on a simple interest car loan. The "out-the-door" price is $64,100 cash or a down payment of 10% (cash/trade) and the remainder of the balance is financed at 2.4% for 6 years
    a. Future value   b. Amortization

19. Calculate the monthly payment on a simple interest motorcycle loan. The "out-the-door" price is $8,480 cash or a down payment of 5%. The remainder of the balance is financed at 4.5% for 3 years.
    a. Future value   b. Amortization

20. Calculate the monthly payment on a simple interest truck loan. The "out-the-door" price is $18,555 cash or a down payment of 10% and the remainder of the balance is financed at 5.5% for 4 years.
    a. Future value   b. Amortization

21. Calculate the monthly payment on a simple interest watercraft loan. The "out-the-door" price is $5,880 cash or a down payment of 10% and the remainder of the balance is financed at 3.75% for 2 years.
    a. Future value   b. Amortization

Section 2.4: Remaining Balance

22. Using the simple interest vehicle loan from problem 13. Calculate the remaining balance after 2 years 7 months. The contract does have a "No Prepayment Penalty" clause.
    a. Future value   b. Amortization   c. Rule of 78

23. Using the vehicle loan from problem 14. Calculate the remaining balance after 2 years 7 months.

24. Using the car loan from problem 15. Calculate the remaining balance after 1 year 2 months.

25. Using the simple interest car loan from problem 16. Calculate the remaining balance after 1 year 2 months. The contract does have a "No Prepayment Penalty" clause.
    a. Future value   b. Amortization   c. Rule of 78

26. Using the car loan from problem 17. Calculate the remaining balance after 1 year 2 months.

27. Using the simple interest car loan from problem 18. Calculate the remaining balance after 4 years and 8 months. The contract does not have a "No Prepayment Penalty" clause.
    a. Future value   b. Amortization   c. Rule of 78

28. Using the simple interest motorcycle loan from problem 19. Calculate the remaining balance after 1 year 1 month. The contract does not have a "No Prepayment Penalty" clause.
    a. Future value   b. Amortization   c. Rule of 78

29. Using the simple interest truck loan from problem 20. Calculate the remaining balance after 3 years. The contract does have a "No Prepayment Penalty" clause.
    a. Future value
    b. Amortization
    c. Rule of 78

30. Using the simple interest watercraft loan from problem 21. Calculate the remaining balance after 10 months. The contract does not have a "No Prepayment Penalty" clause.
    a. Future value
    b. Amortization
    c. Rule of 78

CHAPTER 3: MORE PERSONAL FINANCE

3.1 HOME PURCHASE
Objective: Calculate the complete monthly payment for the purchase of a residence. This is the sum of the money applied toward the principal and interest, P & I, on the mortgage and the money held in escrow to pay the property taxes and insurance.

Mortgages
Most people apply for a mortgage when they purchase a house, townhouse, or condominium. Similar to a car loan, the loan is repaid periodically, usually monthly, but for a longer period of time, like 15 or 30 years. In addition to repaying the loan, most mortgages have additional fees attached to them.

There are different types of mortgages:
- **Fixed rate mortgage**: The interest rate is the same throughout the life of the loan. Thus, the periodic payment remains the same throughout the life of the loan. These payments can be made monthly or bi-weekly (every 2 weeks).
- **Balloon mortgage**: Loan payments are required for a predetermined number of years, after which the balance is due. For example, a loan may be amortized based on a 30 year term, but the balance of the loan is due at the end of 10 years. This forces the borrower to refinance the loan at the end of 10 years at a new rate. This new rate could be higher than the original rate of ten years ago.
- **Adjustable rate mortgage (ARM)**: The interest rate is subject to change according to market conditions after a predetermined period of time (1 year, 3 years, or 5 years). As the rate changes, so does the monthly payment. ARM's are attractive to first-time home buyers. The interest rates are usually lower than the current fixed rate mortgage. However, when comparing the adjustable rate to the fixed rate, the mortgage caps of the ARM must be considered. These caps limit the rate increase rate of the ARM. The ARM may not exceed the rates stated by the cap. However, the rate may rise substantially after each adjustment period and over the life of the loan.
- **Interest-only mortgage**: The borrower only pays the interest on the loan for a fixed period. The loan balance does not change during this time. At the end of this period, the interest – only loan converts to a fixed rate mortgage

for the remainder of the term of the mortgage. For example, in a 30 year interest–only mortgage with a 10 year interest–only period, the loan balance remains unchanged for the first 10 years and the balance is amortized over the remaining 20 years.

In this section, the fixed rate mortgage and interest–only mortgages will be studied. All mortgages have two common features. The ***principal***, the amount of money borrowed, and the ***interest***, the service charge you pay for the use of the money borrowed.

The principal and interest (P & I) portion of the mortgage payment is calculated with the same amortization formula used to calculate the periodic payment of a car loan.

Amortization Formula

$$P = \frac{\left(\dfrac{AR}{n}\right)\left(1 + \dfrac{R}{n}\right)^{(nT)}}{\left(1 + \dfrac{R}{n}\right)^{(nT)} - 1}$$

where: P is the fixed periodic payment.
A is the amount borrowed.
R is the annual interest rate as a decimal.
n is the number of compound periods per year.
T is the time in years.

Another part of the mortgage payment is the ***escrow*** fee. The escrow account is a fund set aside by the lender to pay
- Property taxes.
- Homeowner's hazard insurance.
- Private mortgage insurance, PMI, if the down payment is less than 20%.

The annual property taxes, the insurance premium and the private mortgage insurance are divided into 12 payments and added to the principal and interest (P & I) payment. The private mortgage insurance is usually 0.24% to 0.84% of the amount financed. When the property taxes and insurance premiums are due, money is taken from the escrow account to pay them. Thus, the actual monthly payment on any type of mortgage involves both P & I and escrow.

Example 1: Find the monthly payment on a $105,000 condominium that requires 5% down with the remainder of the balance financed at a fixed rate of 8.5% for 30 years. The property taxes are $2,000 annually, the homeowners insurance is $250 annually, and the private mortgage insurance (PMI) is 0.48%.

(c) 2006 JupiterImages Corp

Solution:
Step 1: Picture and understand the problem.
1. Question: What is the monthly payment on the condominium?
2. List the given data:
   Price of the residence = $105,000.00,
   N = 12 pay periods annually,   Rate = 8.5%,
   Down payment rate = 5%,   Time = 30 years,
   Taxes = $2,000.00 annually,   Insurance = $250.00 annually,
   PMI = 0.48% of amount financed annually.
3. Identify the unknowns as the amortization formula and escrow payment.

Step 2: Devise a Plan.
1. Calculate the down payment: Down payment = (0.05)(Price of residence).
2. Determine the amount financed:
   Amount financed = price of residence − down payment.
3. Select the amortization formula to calculate the monthly payment on the loan, (Principal & Interest):

$$P = \frac{\left(\dfrac{AR}{n}\right)\left(1+\dfrac{R}{n}\right)^{(nT)}}{\left(1+\dfrac{R}{n}\right)^{(nT)} - 1}.$$

4. Convert the given data into the appropriate units.
5. Find the intermediate values that make the formula easier to manipulate.

Determine $\dfrac{AR}{n}$, $\left(1+\dfrac{R}{n}\right)^{(nT)}$, $\left(1+\dfrac{R}{n}\right)^{(nT)} - 1$ and store these values in your calculator's memory without rekeying the data.

6. Substitute the appropriate values into the formula and use the order of operations to simplify.
7. Find the amount of money placed in escrow.
   a. PMI = 0.0048 (Amount financed).
   b. Escrow = $\dfrac{\text{Taxes} + \text{Insurance} + \text{PMI}}{12}$.

8. Determine the monthly payment for the residence:
   Payment = P & I **plus** Escrow.

Step 3: Execute the Plan.
1. Down Payment = 0.05($105,000.00) = $5,250.00.
2. Amount financed:
   A = $105,000.00 − $5,250.00 = $99,750.00.
3. Monthly Payment: $P = \dfrac{\left(\dfrac{AR}{n}\right)\left(1+\dfrac{R}{n}\right)^{(nT)}}{\left(1+\dfrac{R}{n}\right)^{(nT)} - 1}$.
4. Rate: 8.5% = 0.085 = R,   n = 12 pay periods annually,   T = 30 years.
5. $\dfrac{AR}{n} = \dfrac{(\$99{,}750)(.085)}{12} = \$706.5625$.

   (Store in calculator's memory 1 without rekeying the data.)

   $\left(1+\dfrac{R}{n}\right)^{(nT)} = \left(1+\dfrac{0.085}{12}\right)^{(12\times30)} \approx 12.69249879$.

   (Store in calculator's memory 2 without rekeying the data.)

   $\left(1+\dfrac{R}{n}\right)^{(nT)} - 1 = \left(1+\dfrac{0.085}{12}\right)^{(12\times30)} - 1 \approx 11.69249879$.

   (Store in calculator's memory 3 without rekeying the data.)

6. Substitute, recall the appropriate stored values from memory and simplify. (Note: The denominator is one less than the second factor of the numerator.)

   $P = \dfrac{(\$706.5625)(12.69249879)}{11.69249879} \approx \$766.99$.

7a. PMI = 0.0048($99,750) = $478.80.

 b. Escrow = $\dfrac{\text{Taxes} + \text{Insurance} + \text{PMI}}{12}$ =
   $\dfrac{\$2{,}000 + \$250 + \$478.80}{12} = \$227.40$.

8. Total Monthly Payment: $766.99 + $227.40 = $994.39.
   The actual monthly payment is $994.39.

Step 4: Check Your Work.
1. Check the numbers and your calculations.
2. Is $994.39 a reasonable monthly payment for this condominium?
3. Was the question answered?

Bi-weekly Mortgages

Another way to finance a home is to apply for a bi-weekly mortgage. Rather than make payments monthly, the bi-weekly mortgage payment is due every two weeks. By paying your mortgage payment every two weeks, instead of once a month, you will make 26 payments over the course of a year. (1 year = 52 weeks.) **These payments are one half of the monthly payment.** You make an extra monthly payment per year $\left(\dfrac{26 \text{ payments}}{\text{year}} \times \dfrac{1}{2} \text{monthly amount} = \dfrac{13 \text{ monthly payments}}{\text{year}}\right)$. This extra payment will shorten the time it takes to repay your loan.

Example 2: Find the bi-weekly payment on a $105,000 condominium that requires 5% down with the remainder of the balance financed at a fixed rate of 8.5% for 30 years. The property taxes are $2,000, the homeowners insurance is $250 annually and the private mortgage insurance (PMI) is 0.48%.

(c) 2006 JupiterImages Corp

Solution:
Step 1: Picture and understand the problem.
    1. Question: What is the biweekly payment on the condominium?
    2. List the given data:
       Price of the residence = $105,000.00,
       n = 12 for monthly P&I and 26 for biweekly escrow.
       (The P & I portion of the monthly payment will be halved later.)
       Rate = 8.5%,   Down payment rate = 5%,
       Time = 30 years,   Taxes = $2,000.00,
       Insurance = $250.00,   PMI = 0.48% of amount financed annually.
    3. Identify the unknowns as the amortization of the loan and the escrow payment.

Step 2: Devise a Plan.
    1. Calculate the down payment: Down payment = (0.05)(Price of residence).
    2. Determine the amount financed:
       Amount financed = price of residence − down payment.
    3. Select the amortization formula to calculate the monthly payment on the loan, (Principal & Interest):

$$P = \dfrac{\left(\dfrac{AR}{n}\right)\left(1+\dfrac{R}{n}\right)^{(nT)}}{\left(1+\dfrac{R}{n}\right)^{(nT)} - 1}.$$

4. Convert the given data into the appropriate units.
5. Find the intermediate values that make the formula easier to manipulate.

   Determine $\dfrac{AR}{n}$, $\left(1+\dfrac{R}{n}\right)^{(nT)}$, $\left(1+\dfrac{R}{n}\right)^{(nT)} - 1$ and store these values in your calculator's memory without rekeying the data.
6. Substitute the appropriate values into the formula and use the order of operations to simplify.
7. Divide the amortization result by 2.
8. Find the amount of money placed in escrow.
   a. PMI = 0.0048 (Amount financed).
   b. Escrow = $\dfrac{\text{Taxes} + \text{Insurance} + \text{PMI}}{26}$.
9. Determine the biweekly payment for the residence:
   Payment = P & I **plus** Escrow.

Step 3: Execute the Plan.
1. Down Payment = 0.05($105,000.00) = $5,250.00.
2. Amount financed:
   A = $105,000.00 − $5,250.00 = $99,750.00.
3. Monthly P&I Payment: $P = \dfrac{\left(\dfrac{AR}{n}\right)\left(1+\dfrac{R}{n}\right)^{(nT)}}{\left(1+\dfrac{R}{n}\right)^{(nT)} - 1}$.
4. Rate: 8.5% = 0.085 = R
5. $\dfrac{AR}{n} = \dfrac{(\$99,750)(.085)}{12} = \$706.5625$.

   (Store in calculator's memory 1 without rekeying the data.)

   $\left(1+\dfrac{R}{n}\right)^{(nT)} = \left(1+\dfrac{0.085}{12}\right)^{(12\times 30)} \approx 12.69249879$.

   (Store in calculator's memory 2 without rekeying the data.)

   $\left(1+\dfrac{R}{n}\right)^{(nT)} - 1 = \left(1+\dfrac{0.085}{12}\right)^{(12\times 30)} - 1 \approx 11.69249879$.

   (Store in calculator's memory 3 without rekeying the data.)
6. Substitute, recall the appropriate stored values from memory and simplify. (Note: The denominator is one less than the second factor of the numerator.)

   $P = \dfrac{(\$706.5625)(12.69249879)}{11.69249879} \approx \$766.9911999$.

7. $\dfrac{P}{2} = \dfrac{\$766.9911999}{2} \approx \$383.50$

8a. PMI = 0.0048($99750) = $478.80.

b. Escrow = $\dfrac{\text{Taxes} + \text{Insurance} + \text{PMI}}{26} =$
$\dfrac{\$2,000 + \$250 + \$478.80}{26} \approx \$104.95$.

9. Total bi-weekly payment: $383.50 + $104.95 = $488.45.
The actual bi-weekly payment is $488.45.

Step 4: Check Your Work.
1. Check the numbers and your calculations.
2. Is $488.45 a reasonable bi-weekly payment for this condominium?
3. Was the question answered?

Interest–only mortgage

Another way to finance a home is to apply for an interest–only mortgage. The borrower only repays the interest on the loan for a fixed period. The loan balance does not change during this time. At the end of this period, the interest–only loan converts to a fixed rate mortgage for the remainder of the term of the mortgage. For example, in a 30 year interest–only mortgage with a 10 year interest–only period, the loan balance remains unchanged for the first 10 years and the balance is amortized over the remaining 20 years.

Example 3: Find the two monthly payments on a $105,000 condominium that requires 5% down with the remainder of the balance financed at an interest–only fixed rate of 8.5% for 10 years and then the loan is amortized over the remaining 20 years. The property taxes are $2,000 annually, the homeowners insurance is $250 annually, and the private mortgage insurance (PMI) is 0.48%.

(c) 2006 JupiterImages Corp

Solution:
Step 1: Picture and understand the problem.
1. Question: What are the two monthly payments on the condominium?
   a. Monthly payment, first 10 years → Interest–only and escrow.
   b. Monthly payment, remaining 20 years → amortization and escrow.
2. List the given data:
   Price of the residence = $105,000.00,   n = 12 pay periods annually,
   Rate = 8.5%,   Down payment rate = 5%,
   Time = 10 years (interest–only), 20 years (amortized),
   Taxes = $2,000.00 annually,   Insurance = $250 annually,
   PMI = 0.48% of amount financed annually.

Step 2: Devise a Plan.
1. Calculate the down payment: Down payment = (0.05)(Price of residence).
2. Determine the amount financed:
   Amount financed = price of residence − down payment.
3. For the first 10 years.
   a. Interest–only = $\dfrac{\text{rate}}{12}$(amount financed).
   b. PMI = 0.0048 (Amount financed).
   c. Escrow = $\dfrac{\text{Taxes} + \text{Insurance} + \text{PMI}}{12}$.
   d. Monthly payment A = Interest–only + Escrow.
4. For the remaining 20 years.
   a. Select the amortization formula to calculate the monthly payment on the loan, (Principal & Interest):
   $$P = \dfrac{\left(\dfrac{AR}{n}\right)\left(1+\dfrac{R}{n}\right)^{(nT)}}{\left(1+\dfrac{R}{n}\right)^{(nT)} - 1}.$$
5. Convert the given data into the appropriate units.
6. Find the intermediate values that make the formula easier to manipulate.
   Determine $\dfrac{AR}{n}$, $\left(1+\dfrac{R}{n}\right)^{(nT)}$, $\left(1+\dfrac{R}{n}\right)^{(nT)} - 1$ and store these values in your calculator's memory without rekeying the data.
7. Substitute the appropriate values into the formula and simplify.
8. Find the amount of money placed in escrow.
   a. PMI = 0.0048 (Amount financed).
   b. Escrow = $\dfrac{\text{Taxes} + \text{Insurance} + \text{PMI}}{12}$.
9. Determine the monthly payment for the residence:
   Payment = P & I **plus** Escrow.

Step 3: Execute the Plan.
1. Down Payment = 0.05($105,000.00) = $5,250.00.
2. Amount financed:
   A = $105,000.00 − $5,250.00 = $99,750.00.
3. For the first 10 years.
   a. Interest–only = $\dfrac{0.085}{12}$($99,750) ≈ $706.56.
   b. PMI = 0.0048 ($99,750.00) = $478.80.

c. Escrow = $\dfrac{\text{Taxes} + \text{Insurance} + \text{PMI}}{12} =$

$\dfrac{\$2{,}000 + \$250 + \$478.80}{12} = \$227.40.$

d. Monthly payment A = $706.56 + $227.40 = $933.96.

4. For the remaining 20 years

Monthly Payment: $P = \dfrac{\left(\dfrac{AR}{n}\right)\left(1+\dfrac{R}{n}\right)^{(nT)}}{\left(1+\dfrac{R}{n}\right)^{(nT)} - 1}.$

5. Rate: 8.5% = 0.085 = R,   n = 12 pay periods annually,   T = 20 years.

6. $\dfrac{AR}{n} = \dfrac{(\$99{,}750)(.085)}{12} = \$706.5625.$

(Store in calculator's memory 1 without rekeying the data.)

$\left(1+\dfrac{R}{n}\right)^{(nT)} = \left(1+\dfrac{0.085}{12}\right)^{(12 \times 20)} \approx 5.441242569.$

(Store in calculator's memory 2 without rekeying the data.)

$\left(1+\dfrac{R}{n}\right)^{(nT)} - 1 = \left(1+\dfrac{0.085}{12}\right)^{(12 \times 20)} - 1 \approx 4.441242569.$

(Store in calculator's memory 3 without rekeying the data.)

7. Substitute, recall the appropriate stored values from memory and simplify. (Note: The denominator is one less than the second factor of the numerator.)

$P = \dfrac{(\$706.5625)(5.441242569)}{4.441242569} \approx \$865.6536753 \approx \$865.65.$

8a. PMI = 0.0048($99750) = $478.80.

b. Escrow = $\dfrac{\text{Taxes} + \text{Insurance} + \text{PMI}}{12} =$

$\dfrac{\$2{,}000 + \$250 + \$478.80}{12} = \$227.40.$

9. Total Monthly Payment: $865.65 + $227.40 = $1,093.05.

Step 4: Check Your Work.
1. Check the numbers and your calculations.
2. Is $933.96 a reasonable monthly payment for the first 10 years?
3. Is $1,093.05 a reasonable monthly payment for the next 20 years?
4. Was the question answered?

170

For the next three examples, consider the purchase of single family residence, using different purchase plans.

Example 4: Find the monthly payment on a $140,990 single family home that requires 5% down with the remainder of the balance financed at 8.125% for 30 years. The annual property taxes are $3,000.00, the homeowners insurance is $300 per year and the private mortgage insurance (PMI) is 0.36% of the amount financed.

Solution:
Step 1: Picture and understand the problem.
  1. Question: What is the total monthly payment?
  2. List the given data: Price of the residence = $140,990.00
     Rate = 8.125%, Down payment rate = 5%,
     n = 12 pay periods annually, Time = 30 years,
     Taxes = $3,000.00 annually, Insurance = $300.00 annually,
     PMI = 0.36% of amount financed annually.
  3. Identify the unknowns as the amortization of the loan and the escrow payment.

Step 2: Devise a Plan.
  1. Calculate the down payment:
     Down payment = (0.05)(Price of residence).
  2. Determine the amount financed:
     Amount financed = price of residence – down payment.
  3. Select the amortization formula to calculate the monthly payment on the loan, (Principal & Interest):

     $$P = \frac{\left(\dfrac{AR}{n}\right)\left(1+\dfrac{R}{n}\right)^{(nT)}}{\left(1+\dfrac{R}{n}\right)^{(nT)} - 1}.$$

  4. Convert the given data into the appropriate units.
  5. Find the intermediate values that will make the formula easier to manipulate.
     Determine $\dfrac{AR}{n}$, $\left(1+\dfrac{R}{n}\right)^{(nT)}$, $\left(1+\dfrac{R}{n}\right)^{(nT)} - 1$ and store these values in your calculator's memory without rekeying the data.
  6. Substitute the appropriate values into the formula and use the order of operations to simplify.

7. Find the amount of money placed in escrow.
   a. PMI = 0.0036 (Amount financed).
   b. Escrow = $\dfrac{\text{Taxes} + \text{Insurance} + \text{PMI}}{12}$.
8. Determine the monthly payment for the residence:
   Payment = P & I **plus** Escrow.

Step 3: Execute the Plan.
1. Down Payment = 0.05(140,990.00) = $7,049.50.
2. Amount financed: A = $140,990.00 − $7,049.50 = $133,940.50.
3. Monthly Payment: $P = \dfrac{\left(\dfrac{AR}{n}\right)\left(1+\dfrac{R}{n}\right)^{(nT)}}{\left(1+\dfrac{R}{n}\right)^{(nT)} - 1}$.
4. Rate: 8.125% = 0.08125 = R.
5. $\dfrac{AR}{n} = \left(\dfrac{(\$133{,}940.50)(0.08125)}{12}\right) \approx \$906.8888021$.

   (Store in calculator's memory 1 without rekeying the data.)

   $\left(1+\dfrac{R}{n}\right)^{(n\times T)} = \left(1+\dfrac{0.08125}{12}\right)^{(12\times 30)} \approx 11.3507646$.

   (Store in calculator's memory 2 without rekeying the data.)

   $\left(1+\dfrac{R}{n}\right)^{(nT)} - 1 = \left(1+\dfrac{0.08125}{12}\right)^{(12\times 30)} - 1 \approx 10.3507646$.

   (Store in calculator's memory 3 without rekeying the data.)
6. Substitute, recall the appropriate stored values from memory and simplify. (Note: The denominator is one less than the second factor of the numerator.)

   $P = \dfrac{(\$906.8888021)(11.3507646)}{10.3507646} \approx \$994.50$.
7. Find the amount of money placed in escrow.
   a. PMI = 0.0036 ($133,940.50) ≈ $482.19.
   b. Escrow = $\dfrac{\text{Taxes} + \text{Insurance} + \text{PMI}}{12} = \dfrac{\$3{,}000 + \$300 + \$482.19}{12} \approx \$315.18$.
8. Total Monthly Payment: $994.50 + $315.18 = $1,309.68.

Step 4: Check Your Work.
1. Check the numbers and your calculations.
2. Does $1,309.68 monthly seem reasonable for this house?
3. Was the question answered?

Example 5: Find the bi-weekly payment on a $140,990 single family home that requires 5% down with the remainder of the balance financed at 8.125% for 30 years. The annual property taxes are $3,000.00, the homeowners insurance is $300 per year and the private mortgage insurance is 0.36% of the amount financed.

Solution:
Step 1: Picture and understand the problem.
1. Question: What is the total biweekly payment?
2. List the given data: Price of the residence = $140,990.00,
Rate = 8.125%, Down payment rate = 5%,
n = 12 for monthly P&I and 26 for biweekly escrow.
Time = 30 years, Taxes = $3,000.00 annually,
Insurance = $300.00 annually, PMI = 0.36% of amount financed.
3. Identify the unknowns as the amortization of the loan and the escrow payment.

Step 2: Devise a Plan.
1. Calculate the down payment:
Down payment = (0.05)(Price of residence).
2. Determine the amount financed:
Amount financed = price of residence − down payment.
3. Select the amortization formula to calculate the monthly payment on the loan, (Principal & Interest):

$$P = \frac{\left(\dfrac{AR}{n}\right)\left(1+\dfrac{R}{n}\right)^{(nT)}}{\left(1+\dfrac{R}{n}\right)^{(nT)} - 1}.$$

4. Convert the given data into the appropriate units.
5. Find the intermediate values that will make the formula easier to manipulate.
Determine $\dfrac{AR}{n}$, $\left(1+\dfrac{R}{n}\right)^{(nT)}$, $\left(1+\dfrac{R}{n}\right)^{(nT)} - 1$ and store in memory without rekeying the data..

6. Substitute the appropriate values into the formula and simplify.
7. Divide the monthly payment by 2.
8. Find the amount of money placed in escrow:
   a. PMI = 0.0036 (Amount financed).
   b. Escrow = $\dfrac{\text{Taxes} + \text{Insurance} + \text{PMI}}{26}$.
9. Determine the biweekly payment for the residence:
   Payment = P & I **plus** Escrow.

Step 3: Execute the Plan.
1. Down Payment = 0.05(140,990.00) = $7,049.50.
2. Amount financed: A = $140,990.00 − $7,049.50 = $133,940.50.
3. Biweekly Payment: $P = \dfrac{\left(\dfrac{AR}{n}\right)\left(1+\dfrac{R}{n}\right)^{(nT)}}{\left(1+\dfrac{R}{n}\right)^{(nT)} - 1}$.
4. Rate: 8.125% = 0.08125 = R.
5. $\dfrac{AR}{n} = \left(\dfrac{(\$133{,}940.50)(0.08125)}{12}\right) \approx \$906.8888021$.

   (Store in calculator's memory 1 without rekeying the data.)

   $\left(1+\dfrac{R}{n}\right)^{(n\times T)} = \left(1+\dfrac{0.08125}{12}\right)^{(12\times 30)} \approx 11.3507646$.

   (Store in calculator's memory 2 without rekeying the data.)

   $\left(1+\dfrac{R}{n}\right)^{(nT)} - 1 = \left(1+\dfrac{0.08125}{12}\right)^{(12\times 30)} - 1 \approx 10.3507646$.

   (Store in calculator's memory 3 without rekeying the data.)
6. Substitute, recall the appropriate stored values from memory and simplify. (Note: The denominator is one less than the second factor of the numerator.)

   $P = \dfrac{(\$906.8888021)(11.3507646)}{10.3507646} \approx \$994.50$.
7. $\dfrac{P}{2} = \dfrac{\$994.50}{2} = \$497.25$.
8. Find the amount of money placed in escrow.
   a. PMI = 0.0036 ($133,940.50) ≈ $482.19.
   b. Escrow = $\dfrac{\text{Taxes} + \text{Insurance} + \text{PMI}}{26} =$
   $\dfrac{\$3{,}000 + \$300 + 482.19}{26} \approx \$145.47$.

9. Total bi-weekly payment: $497.25+ $145.47 = $642.72.
   The actual biweekly payment is $642.72.

Step 4: Check Your Work.
   1. Check the numbers and your calculations.
   2. Does $642.72 seem a reasonable bi-weekly payment for this house?
   3. Was the question answered?

Example 6: Find the two monthly payments on a $140,990 single family home that requires 5% down with the remainder of the balance financed at an interest–only rate of 8.125% for 7 years and then the loan is amortized over the remaining 23 years. The annual property taxes are $3,000.00, the homeowners insurance is $300 per year and the private mortgage insurance is 0.36% of the amount financed.

Solution:
Step 1: Picture and understand the problem.
   1. Question: What are the two monthly payments on the townhouse?
      a. Monthly payment A (first 7 years) = Interest–only and escrow.
      b. Monthly payment B (remaining 23 years) = amortization and escrow.
   2. List the given data:
      Price of the residence = $140,990.00,
      Rate = 8.125%,   Down payment rate = 5%,
      n = 12 pay periods annually,   Time = 30 years,
      Taxes = $3,000.00 annually,   Insurance = $300.00 annually,
      PMI = 0.36% of amount financed annually.

Step 2: Devise a Plan.
   1. Calculate the down payment: Down payment = (0.05)(Price of residence).
   2. Determine the amount financed:
      Amount financed = price of residence − down payment.
   3. For the first 7 years.
      a. Interest–only = $\dfrac{\text{rate}}{12}$(amount financed).
      b. PMI = 0.0036 (Amount financed).
      c. Escrow = $\dfrac{\text{Taxes} + \text{Insurance} + \text{PMI}}{12}$.
      d. Monthly payment A = Interest–only + Escrow.

4. For the remaining 23 years.
   a. Select the amortization formula to calculate the monthly payment on the loan, (Principal & Interest):
   $$P = \frac{\left(\frac{AR}{n}\right)\left(1+\frac{R}{n}\right)^{(nT)}}{\left(1+\frac{R}{n}\right)^{(nT)} - 1}.$$
5. Convert the given data into the appropriate units.
6. Find the intermediate values that make the formula easier to manipulate.
   Determine $\frac{AR}{n}$, $\left(1+\frac{R}{n}\right)^{(nT)}$, $\left(1+\frac{R}{n}\right)^{(nT)} - 1$ and store these values in your calculator's memory without rekeying the data.
7. Substitute the appropriate values into the formula and simplify.
8. Find the amount of money placed in escrow.
   a. PMI = 0.0036 (Amount financed).
   b. Escrow = $\frac{\text{Taxes} + \text{Insurance} + \text{PMI}}{12}$.
9. Determine the monthly payment for the residence:
   Payment = P & I plus Escrow.

Step 3: Execute the Plan.
1. Down Payment = 0.05(140,990.00) = $7,049.50.
2. Amount financed: A = $140,990.00 − $7,049.50 = $133,940.50.
3. For the first 7 years.
   a. Interest–only = $\frac{0.08125}{12}(\$133,940.50) \approx \$906.89$.
   b. PMI = 0.0036 ($133,940.50) ≈ $482.19.
   c. Escrow = $\frac{\text{Taxes} + \text{Insurance} + \text{PMI}}{12} =$
   $\frac{\$3,000 + \$300 + 482.19}{12} \approx \$315.18$.
   d. Monthly payment A = $906.89 + $315.18 = $1,222.07.
4. For the remaining 23 years
   Monthly Payment: $P = \frac{\left(\frac{AR}{n}\right)\left(1+\frac{R}{n}\right)^{(nT)}}{\left(1+\frac{R}{n}\right)^{(nT)} - 1}.$
5. Rate: 8.125% = 0.08125 = R

176

        n = 12 pay periods annually   T = 23 years.

6. $\dfrac{AR}{n} = \dfrac{(\$133{,}940.50)(.08125)}{12} \approx \$906.8854167$

(Store in calculator's memory 1 without rekeying the data.)

$\left(1 + \dfrac{R}{n}\right)^{(nT)} = \left(1 + \dfrac{0.08125}{12}\right)^{(12 \times 23)} \approx 6.439506556.$

(Store in calculator's memory 2 without rekeying the data.)

$\left(1 + \dfrac{R}{n}\right)^{(nT)} - 1 = \left(1 + \dfrac{0.08125}{12}\right)^{(12 \times 23)} - 1 \approx 5.439506556.$

(Store in calculator's memory 3 without rekeying the data.)

7. Substitute, recall the appropriate stored values from memory and simplify. (Note: The denominator is one less than the second factor of the numerator.)

$P = \dfrac{(\$906.8854167)(6.439506556)}{5.439506556} \approx \$1{,}073.61.$

8. Find the amount of money placed in escrow.
   a. PMI = 0.0036 ($133,940.50) ≈ $482.19.
   b. Escrow = $\dfrac{\text{Taxes} + \text{Insurance} + \text{PMI}}{12} =$

$\dfrac{\$3{,}000 + \$300 + 482.19}{12} \approx \$315.18.$

9. Total Monthly Payment: $1,073.61 + $315.18 = $1,388.79.

Step 4: Check Your Work.
   1. Check the numbers and your calculations.
   2. Is $1,073.61 a reasonable monthly payment for the first 7 years?
   3. Is $1,388.79 a reasonable monthly payment for the next 23 years?
   4. Was the question answered?

Total Cost of the Mortgage

Comparing the total cost of the two mortgage options from Examples 4 and 6. **Only the P&I portion of the actual monthly payment is applied to the loan.**

Ex. 4: 30 year fixed rate; 360 payments of $994.50   360($994.50) = $358,020
Ex. 6: Interest-Only; 84 payments of $906.89   and   276 payments of $1,073.61
                      84($906.89) = $76,178.76   and   276($1,073.61) = $293,316.36
                              $76,178.76 + $293,316.36 = $372,495.12

While the Interest-Only is the most expensive option, all three mortgages require paying a large amount of interest for the original $133,940.50 borrowed. Strategies for saving on the interest payments are presented in the next section.

Nightmare Mortgages

DANGER!!!!!
(c) 2006 Jupiter Images Corp.

Option Adjustable Rate Mortgage (ARM)
The option adjustable rate mortgage may be the riskiest home loan product ever created. Under the option adjustable rate mortgage, the home buyer has the choice of paying less than the P & I portion of the monthly payment.

What the home buyer doesn't realize is that the difference between the payment and the amount due is **ADDED TO THE REMAINING BALANCE!** When the adjustable period runs out, the mortgage converts to a fixed mortgage at a rate higher than the adjustable period. The remaining balance may be larger than the original amount of money borrowed!

Many home buyers could only afford making the minimum or less payments during the adjustable period. When the mortgage resets to a fixed mortgage they cannot afford to make the monthly payments and have to default on their loans. **The banks have to foreclose on the property and the home buyer has to declare bankruptcy!**

⚠ A Good Credit Rating Works For You. ⚠

The higher your credit score, the less you pay for a mortgage. The table displays the monthly P&I payment for a $200,000 mortgage for different credit ratings.

Credit Score	APR	Monthly P&I
720 - 850	5.793%	$1,173
700 - 719	5.918%	$1,189
675 - 699	6.456%	$1,258
620 - 674	7.606%	$1,413
560 - 619	8.531%	$1,542
500 - 559	9.289%	$1,651

To improve your credit score
1. **Pay your bills on time.** Do not make late payments.
2. **Reduce your credit card balances.** Do not max. out your credit limits.
3. **Limit your credit applications.** Do not apply for credit cards from all the stores you shop. Each time you apply for credit, the lender checks your credit rating. Too many inquires will lower your rating.
4. **Establish a credit history.** Use a single credit card or gasoline card to establish a credit history. Using a credit card wisely will help you get credit for your first major credit purchase (car).

? Cognitive Problems ?

1. What is the purpose of the escrow account?
2. Why would a lending institution require the borrower to make payments into an escrow account?
3. What other charges may a lending institution include in an escrow account to guarantee its payment?
4. Compare the monthly and bi-weekly payments from Examples 1 and 2. Make a conjecture about your comparison.

(c) 2006 Jupiter Images Corp.

Exercise 3.1

1. A $96,000 townhouse requires 5% down and the remainder is financed at 7.9% for 30 years. The property taxes are $2,100 annually, the insurance is $230 per year and the private mortgage insurance is 0.32% of the amount financed. Find the actual monthly payment.

(c) 2006 JupiterImages Corp.

2. A $324,800 house requires 10% down and the remainder is financed at 7.75% for 30 years. The property taxes are $4,870 annually, the insurance is $409 per year and the private mortgage insurance is 0.6% of the amount financed. Find the actual monthly payment.

3. A $249,000 lakefront condominium requires 10% down and the remainder is financed at 8.4% for 30 years. The property taxes are $6,200 annually, the insurance is $550 per year and the private mortgage insurance is 0.56% of the amount financed. Find the actual monthly payment.

(c) 2006 JupiterImages Corp

4. A $494,800 house requires 5% down and the remainder is financed at 8.125% for 30 years. The property taxes are $10,840 annually, the insurance is $1,086 per year and the private mortgage insurance is 0.64% of the amount financed. Find the actual monthly payment.

(c) 2006 JupiterImages Corp.

5. A $96,000 townhouse requires 5% down and the remainder is financed at 7.9% for 30 years. The property taxes are $2,100 annually, the insurance is $230 per year and the private mortgage insurance is 0.32% of the amount financed. Find the actual bi-weekly payment.

(c) 2006 JupiterImages Corp

6. A $324,800 house requires 10% down and the remainder is financed at 7.75% for 30 years. The property taxes are $4,870 annually, the insurance is $409 per year and the private mortgage insurance is 0.6% of the amount financed. Find the actual bi-weekly payment.

7. A $249,000 lakefront condominium requires 10% down and the remainder is financed at 8.4% for 30 years. The property taxes are $6,200 annually, the insurance is $550 per year and the private mortgage insurance is 0.56% of the amount financed. Find the actual bi-weekly payment.

(c) 2006 JupiterImages corp

8. A $494,800 house requires 5% down and the remainder is financed at 8.125% for 30 years. The property taxes are $10,840 annually, the insurance is $1,086 per year and the private mortgage insurance is 0.64% of the amount financed. Find the actual bi-weekly payment.

(c) 2006 JupiterImages Corp.

9. Find the two monthly payments on a $96,000 townhouse that requires 5% down with the remainder of the balance financed at an interest–only rate of 7.9% for 5 years and then the loan is amortized over the remaining 25 years. The property taxes are $2,100 annually, the insurance is $230 per year and the private mortgage insurance is 0.32% of the amount financed.

(c) 2006 JupiterImages Corp

10. Find the two monthly payments on a $324,800 house requires 10% down and the remainder of the balance financed at an interest–only of 7.75% for 10 years and then the loan is amortized over the remaining 20 years. The property taxes are $4,870 annually, the insurance is $409 per year and the private mortgage insurance is 0.6% of the amount financed.

11. Find the two monthly payments on a $249,000 lakefront condominium requires 10% down and the remainder of the balance financed at an interest–only of 8.4% for 7 years and then the loan is amortized over the remaining 23 years. The property taxes are $6,200 annually, the insurance is $550 per year and the private mortgage insurance is 0.56% of the amount financed.

(c) 2006 JupiterImages Corp

12. Find the two monthly payments on a $494,800 house requires 5% down and the remainder is financed at an interest–only of 8.125% for 10 years and then the loan is amortized over the remaining 20 years. The property taxes are $10,840 annually, the insurance is $1,086 per year and the private mortgage insurance is 0.64% of the amount financed.

(c) 2006 JupiterImages Corp.

3.2 Strategies to Save on Interest Charges

Objective: Paying extra on the monthly payment and refinancing can save thousands of dollars on interest charges.

At the start of the loan, a major portion of the periodic payment is applied to interest. Let's examine the first three payments for the condominium of Example 1 from the previous section. The $994.39 monthly payment is comprised of two parts. The Principal and Interest (P & I) portion, $766.99, and the escrow portion, $227.40. The P & I portion of the monthly payment applies to the $99,750.00 loan at 8.5%. These payments are determined in the same manner as the car payments and are listed in the table below. An explanation of the first three payments follows.

Payment #	P & I	Interest	Principal	Balance
---	----------	-------	----------	$99,750.00
1	$766.99	$706.56	$ 60.43	$99,689.57
2	$766.99	$706.13	$ 60.86	$99,628.71
3	$766.99	$705.70	$ 61.29	$99,567.42

The first month (payment #1) The initial balance is $99,750.00.	
$99,750(0.085)\left(\dfrac{1}{12}\right) = \$706.56$	Interest = PRT. P is the current loan balance, $99,750.00 R = 0.085 and T is one month $\left(\dfrac{1}{12}\right)$.
$766.99 - $706.56 = $60.43	Payment minus interest is principal. The Principal is applied to the loan.
$99,750.00 - $60.43 = $99,689.57	Previous loan balance minus the principal equals the new loan balance.

The second month (payment #2) This process is repeated, using the new balance ($99,689.57).	
$99,689.57(0.085)\left(\dfrac{1}{12}\right) = \$706.13$	Interest = PRT. P is the current loan balance, $99,689.57 R = 0.085 and T is one month $\left(\dfrac{1}{12}\right)$.
$766.99 - $706.13 = $60.86	Payment minus interest is principal. The Principal is applied to the loan.
$99,689.57 - $60.86 = $99,628.71	Previous loan balance minus the principal equals the new loan balance.

The third month (payment #3) This process is repeated, using the new balance ($99,628.71).	
$\$99{,}628.71(0.085)\left(\dfrac{1}{12}\right) = \$705.70$	Interest = PRT. P is the current loan balance, $99628.71, R = 0.085 and T is one month $\left(\dfrac{1}{12}\right)$.
$\$766.99 - \$705.70 = \$61.29$	Payment minus interest is principal. The Principal is applied to the loan.
$\$99{,}628.71 - \$61.29 = \$99{,}567.42$	Previous loan balance minus the principal equals the new loan balance.

For the term of this loan,
 a. The money repaid is 360 payments × $766.99 = $276,116.40.
 b. The initial loan was $99,750.
 c. The interest charge is $276,116.40 − $99,750 = $176,366.40.

The interest charge is considerably more money than the amount initially borrowed! Following are strategies to decrease the interest charge.

Finance Strategies to Save on Interest Charges

Strategy #1: Paying an extra amount each month on the principal.
This strategy is recommended by the financial advisors at the Chicago Tribune. To save on interest charges pay an extra $50 per payment on the principal. The following table shows the effect of the extra $50 per month on the first three payments of the mortgage of Example 1. Notice that the extra $50 goes directly toward the principal, reducing your balance by an extra $50 and lowering your interest for the following month. This results in repaying the 30 year mortgage in less than 30 years. The first three monthly payments with an extra $50 per month ($766.99 + $50 = $816.99) are shown in the following table.

Payment #	P & I + $50 extra	Interest	Principal	Balance
---	----------	-------	----------	$99,750.00
1	$766.99 + $50.00 = $816.99	$706.56	$110.43	$99,639.57
2	$766.99 + $50.00 = $816.99	$705.78	$111.21	$99,528.36
3	$766.99 + $50.00 = $816.99	$704.99	$112.00	$99,416.36

First three monthly payments at the regular payment ($766.99)

Payment #	P & I	Interest	Principal	Balance
---	----------	-------	----------	$99,750.00
1	$766.99	$706.56	$ 60.43	$99,689.57
2	$766.99	$706.13	$ 60.86	$99,628.71
3	$766.99	$705.70	$ 61.29	$99,567.42

Note: Using a financial calculator, paying an extra $50 per month on the principal for a $99,750 mortgage at 8.5%, results in repaying the 30 year mortgage in 284 payments (23 years and 8 months). **This reduces your repayment period by 6 years and 4 months!**

The following bar graphs demonstrate the savings in accrued interest for the condominium from Example 1 based on a 30 year $99,750 mortgage with a fixed rate of 8.5%. Compare the interest paid for the two monthly payment plans. Option A is the regular payment of $766.99. Option B pays an extra $50 per month on the principal, ($766.99 + $50.00 = $816.99).

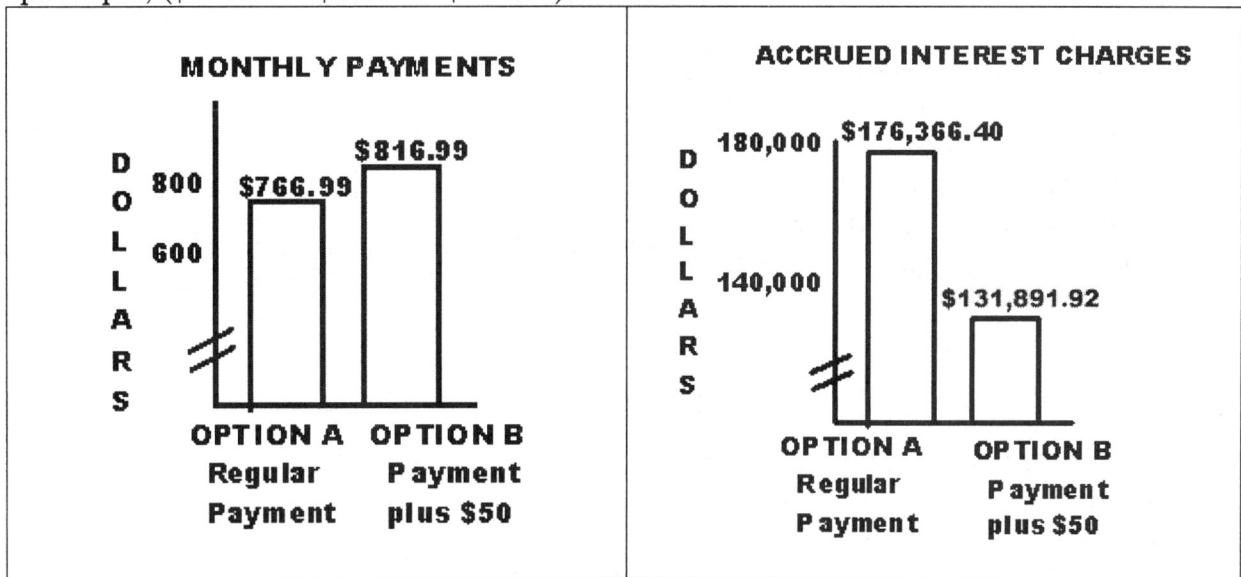

Option A:
1. The amount of money spent repaying the loan is 360 ($766.99) = $276,116.40.
2. Interest paid is Total Repayments − Amount Borrowed = Interest charged.
   $276,116.40 − $99,750 = $176,366.40

Option B (Payment + $50.00 extra on the principal):
1. Using a financial program, the amount of money spent repaying the loan is 283 payments of $816.99 and a final payment of $433.75.
   The amount of money spent repaying the loan is
   283($816.99) + $413.75 = $231,208.17 + $433.75 = $231,641.92.

2. Interest paid is Total Repayments − Amount Borrowed = Interest charged.
   $231,641.92 − $99,750 = $131,891.92

This is a savings of $176,366.40 − $131,891.92 = $44,474.48 in interest charges.

183

(c) 2006 Jupiter Images Corp

Comment: Your extra payment does not have to be $50. It can be any amount. Whatever extra amount you apply toward the principal lowers the remaining balance by that amount. These extra payments will pay off a 30 year mortgage before the 30 year time period and yield a savings on interest fees. The following table shows the time and money saved by paying an extra amount on your principal. The mortgage is for $99,750 at 8.5% for 30 years.

	Dollars saved in interest payments	Repayment period reduced by
Extra $25 applied monthly to the principal	$26,248.05	3 years, 8 months
Extra $50 applied monthly to the principal	$44,474.48	6 years, 4 months
Extra $75 applied monthly to the principal	$58,135.31	8 years, 5 months
Extra $100 applied monthly to the principal	$68,877.57	10 years

Bi-weekly mortgage

The bi-weekly plan requires $26 \times \frac{1}{2}$ monthly payments = 13 monthly payments per year. With the additional monthly payment, the homebuyer will automatically pay off a 30 year loan in less than 30 years. The formula for finding the number of payments it takes to repay a bi-weekly mortgage is

Number of Payments to Repay a Bi-weekly Mortgage

$$x = \frac{\log\left(1 + \frac{L(R)}{26F}\right)}{\log\left(1 + \frac{R}{26}\right)}$$

Where: x is the number of payments to repay the loan
L = the amount of the loan (mortgage).
R = Interest Rate as a decimal
F = the first payment's principal (interest not included)
log = a built in function of your calculator.

Note: Your instructor will demonstrate how to use the "log" key.

From the previous section, a $99,750 loan (mortgage) at 8.5% for 30 years had a bi-weekly payment of $383.50. The P&I portion of the first payment was distributed in the following manner. Using the simple interest formula I = PRT,

$$\text{Interest} = (\$99{,}750)(0.085)\left(\frac{1}{26}\right) \approx \$326.11.$$

Principal = Payment − Interest = $383.50 − $326.11 = $57.39.

Using the formula for number of payments produces:

$$x = \frac{\log\left(1 + \frac{(\$99{,}750)(0.085)}{(\$57.39)(26)}\right)}{\log\left(1 + \frac{0.085}{26}\right)} = \frac{\log(6.682275122)}{\log(1.0032692321)} = \frac{0.8249243524}{0.0014174931} \approx 581.960053.$$

For any fraction of a payment, bump up the payment number to the next number. It will take 582 payments to repay this mortgage.

26 payments for 30 years = 780. 780 − 582 = 198 payments saved.

$$\frac{198 \text{ payments}}{26 \text{ payments annually}} = 7.615384615 \approx 7 \text{ years, } 7 \text{ months}$$

Strategy #2: Refinancing your mortgage at a lower rate for a shorter time.
When interest rates are falling, it is financially sound to refinance. Peruse the article.

Homeowners Joining Rush To Refinance
By Lisa Holton, STAFF WRITER
Falling interest rates have sent a growing number of homeowners rushing to banks to refinance their mortgages. With most lenders waiving the hefty fees, refinancing is attracting a much broader range of borrowers. Experts say that if mortgage rates fall at least one percentage point below what you now pay, you should consider refinancing. Most lenders aren't charging points, fees based on a percentage of the amount you borrow, to refinance loans. Banks are also seeing homeowners going from 30-year mortgages to 15-year mortgages It can save you 15-20 percent over the life of the loan.

Reprinted with special permission from the Chicago Sun-Times, Inc.

The following graph displays the fluctuations in the interest rates of 30 year fixed rate mortgages. Observe the wide variation in rates from 1994 to 2006.

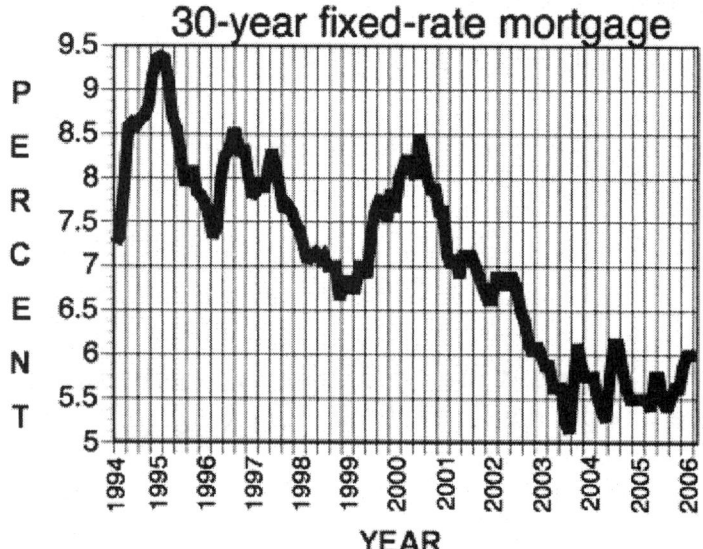

Refinancing allows the homeowner to save money on interest payments. These savings are demonstrated in the following examples.

Example 1:

After 7 years, 2 months, the remaining balance for the condominium in Example 1 is $92,626.77. The current 30 year mortgage has a P&I of $766.99 and an interest rate of 8.5%. If the homeowner refinances the remaining balance at 7% for 15 years, how much money will be saved?

(c) 2006 JupiterImages Corp

Solution:

Step 1: Picture and understand the problem.
1. Question: How much money is saved by refinancing?
2. List the given data:
   Current mortgage: P&I = $766.99, Rate = 8.5%, Term = 30 years, n = 12 pay periods annually,
   Time: Payments were made for 7 years and 2 months.
   New mortgage: Amount refinanced = $92,626.77, Rate = 7%, Time = 15 years.
3. Identify the unknown as the savings (difference) between the payments left on the current mortgage and the total refinanced payment.

Step 2: Devise a Plan
1. Determine the number of payments remaining in the current mortgage.
2. Calculate the expense of keeping the current mortgage.
3. Determine the number of payments in the refinanced mortgage.

4. Use the amortization formula to calculate the monthly payment on the refinanced loan (Principal and Interest): $P = \dfrac{\left(\dfrac{AR}{n}\right)\left(1+\dfrac{R}{n}\right)^{(nT)}}{\left(1+\dfrac{R}{n}\right)^{(nT)} - 1}$.

5. Convert the given data into the appropriate units.
6. Find the intermediate values that make the formula easier to manipulate.

Determine $\dfrac{AR}{n}$, $\left(1+\dfrac{R}{n}\right)^{(nT)}$, $\left(1+\dfrac{R}{n}\right)^{(nT)} - 1$ and store these values in your calculator's memory.

7. Substitute the values into the formula and simplify.
8. Calculate the total cost of the refinanced mortgage.
9. Find the money saved (difference) between the current mortgage and the refinanced mortgage.

Step 3: Execute the Plan.

1. 7 years 2 months = $(7 \; \cancel{\text{years}})\left(\dfrac{12 \text{ months}}{1 \; \cancel{\text{year}}}\right)$ + 2 months =

    84 + 2 months = 86 payments made.

    30 years = $(30 \; \cancel{\text{years}})\left(\dfrac{12 \text{ months}}{1 \; \cancel{\text{year}}}\right)$ = 360 payments.

    360 − 84 = 274 payments left in the current loan.

2. 274 payments of $766.99.
    274 ($766.99) = **$210,155.26 repays the current loan.**

3. 15 years = $(15 \; \cancel{\text{years}})\left(\dfrac{12 \text{ months}}{1 \; \cancel{\text{year}}}\right)$ = 180 payments.

4. Monthly payment $P = \dfrac{\left(\dfrac{AR}{n}\right)\left(1+\dfrac{R}{n}\right)^{(n \times T)}}{\left(1+\dfrac{R}{n}\right)^{(n \times T)} - 1}$.

5. Rate: 7% = 0.07 = R.

6. $\dfrac{AR}{n} = \dfrac{(\$92,626.77)(0.07)}{12}$ = $540.322825.

    (Store in calculator's memory 1.)

$$\left(1+\frac{R}{n}\right)^{(nT)} = \left(1+\frac{0.07}{12}\right)^{(12\times15)} = 2.848946731$$

(Store in calculator's memory 2.)

$$\left(1+\frac{R}{n}\right)^{(nT)} - 1 = \left(1+\frac{0.07}{12}\right)^{(12\times15)} - 1 = 1.848946731$$

(Store in calculator's memory 3.)

7. Substitute, recall the appropriate stored values from memory and simplify. (Note: The denominator is one less than the second factor of the numerator.)

$$P = \frac{(\$540.322825)(2.848946731)}{1.848946731} \approx \$832.5555951.$$

The new monthly payment (P&I) is $832.56.
8. 180 payments of $832.56.
180 ($832.56) = **$149,860.80 repays the refinanced mortgage.**
9. Current Loan Repayment − Refinanced Loan Repayment = Savings
Note: Step 2 − Step 9.
$210,155.26 − $149,860.80 = $60,294.46.
$60,294.46 is saved by refinancing.

Step 4: Check Your Work.
 1. Check the numbers and your calculations.
 2. Is the answer reasonable?
 3. Was the question answered?

Most banks will display the remaining balance of the mortgage on the monthly statement. However, if the remaining balance is unknown, it can be calculated using the remaining balance for an amortized loan formula presented in the previous chapter.

REMAINING BALANCE FOR A SIMPLE INTEREST LOAN USING THE AMORTIZATION METHOD

$$B = P\left(\frac{\left(1+\frac{R}{n}\right)^{x} - 1}{\left(\frac{R}{n}\right)\left(1+\frac{R}{n}\right)^{x}}\right)$$

where: B is the remaining balance on the loan.
 P is the fixed periodic payment.
 R is the annual interest rate as a decimal.
 n is the number of compound periods in a year.
 x is the number of remaining payments.

This scenario is presented in Example 2.
Example 2:
After 7 years and 9 months, the homeowner in example 4 from the previous section reads the real estate section of the newspaper and finds a 15 year mortgage charges 7%. The current 30 year mortgage has a P&I of $994.50 and an interest rate of 8.125%. If the homeowner refinances the remaining balance how much money will be saved?

Solution:
Step 1: Picture and understand the problem.
    1. Question: How much money is saved by refinancing?
    2. List the given data:
       Current mortgage: P&I = $994.50, Rate = 8.125%, Term = 30 years,
       Time: Payments made for 7 years and 9 months,
       n = 12 pay periods annually.
       New mortgage: Rate = 7%, Time = 15 years,
       n = 12 pay periods annually.
    3. Identify the unknown as the possible savings (difference) between the payments left on the current mortgage and the refinanced payment plan.

Step 2: Devise a Plan
    1. Determine the number of payments remaining in the current mortgage.
    2. Calculate the expense of keeping the current mortgage.
    3. Select the formula that calculates the remaining balance

$$B = P\left(\frac{\left(1+\frac{R}{n}\right)^x - 1}{\left(\frac{R}{n}\right)\left(1+\frac{R}{n}\right)^x}\right).$$

    4. Convert the given data into the appropriate units.
    5. Find the intermediate values that will make the formula easier to manipulate.
       Determine $\left(\frac{R}{n}\right)$, $\left(1+\frac{R}{n}\right)^x$, $\left(1+\frac{R}{n}\right)^x - 1$ and store these values in your calculator's memory.
    6. Substitute the values into the formula and use the order of operations to simplify.
    7. Determine the number of payments in the refinanced mortgage.

8. Select the amortization formula to calculate the monthly payment on the refinanced loan (Principal and Interest): $P = \dfrac{\left(\dfrac{AR}{n}\right)\left(1+\dfrac{R}{n}\right)^{(nT)}}{\left(1+\dfrac{R}{n}\right)^{(nT)} - 1}$.

9. Convert the given data into the appropriate units.
10. Find the intermediate values that will make the formula easier to manipulate.

    Determine $\dfrac{AR}{n}$, $\left(1+\dfrac{R}{n}\right)^{(nT)}$, $\left(1+\dfrac{R}{n}\right)^{(nT)} - 1$ and store these values in your calculator's memory.

11. Substitute the values into the formula and simplify.
12. Calculate the expense of the refinanced mortgage.
13. Find the money saved (difference) between the current mortgage and the refinanced mortgage.

Step 3: Execute the Plan.

1. 7 years 9 months = $(7 \text{ years})\left(\dfrac{12 \text{ months}}{1 \text{ year}}\right) + 9 \text{ months} = (84 + 9)$ months

   93 payments made.

   30 years = $(30 \text{ years})\left(\dfrac{12 \text{ months}}{1 \text{ year}}\right) = 360$ payments.

   360 − 93 = 267 payments left in the current loan, x = 267.

2. 267 payments of $994.50.
   267 ($994.50) = **$265,531.50 repays the current loan.**

3. Use the formula

   $B = P\left(\dfrac{\left(1+\dfrac{R}{n}\right)^x - 1}{\left(\dfrac{R}{n}\right)\left(1+\dfrac{R}{n}\right)^x}\right)$.

4. R = 8.125%, = 0.08125.

5. $\left(\dfrac{R}{n}\right) = \dfrac{0.08125}{12} \approx 0.0067708333$.

   (Store in calculator's memory 1.)

   $\left(1+\dfrac{R}{n}\right)^x = \left(1+\dfrac{0.08125}{12}\right)^{267} \approx 6.060060526$.

   (Store in calculator's memory 2.)

$$\left(1+\frac{R}{n}\right)^x - 1 = \left(1+\frac{0.08125}{12}\right)^{267} - 1 \approx 5.060060526.$$

(Store in calculator's memory 3.)

6. Substitute, recall the appropriate stored values from memory and use the order of operations to simplify.

$$B = 994.50\left(\frac{(5.060060526)}{(0.0067708333)(6.060060526)}\right) \approx \$122,642.6183.$$

The remaining balance is $122,642.62.

7. 15 years = $(15 \text{ years})\left(\frac{12 \text{ months}}{1 \text{ year}}\right)$ = 180 payments.

8. Monthly payment $P = \dfrac{\left(\dfrac{AR}{n}\right)\left(1+\dfrac{R}{n}\right)^{(nT)}}{\left(1+\dfrac{R}{n}\right)^{(nT)} - 1}.$

9. Rate: 7% = 0.07 = R.

10. $\dfrac{AR}{n} = \dfrac{(\$122,642.62)(0.07)}{12} \approx \$715.4152833.$

    Note: "A" is the remaining balance calculated in step 6.
    (Store in calculator's memory 1.)

$$\left(1+\frac{R}{n}\right)^{(nT)} = \left(1+\frac{0.07}{12}\right)^{(12\times 15)} \approx 2.848946731.$$

(Store in calculator's memory 2.)

$$\left(1+\frac{R}{n}\right)^{(nT)} - 1 = \left(1+\frac{0.07}{12}\right)^{(12\times 15)} - 1 \approx = 1.848946731.$$

(Store in calculator's memory 3.)

11. Substitute, recall the appropriate stored values from memory and simplify. (Note: The denominator is one less than the second factor of the numerator.)

$$P = \frac{(\$715.4152833)(2.848946731)}{1.848946731}$$

$P \approx \$1,102.346541$

The new monthly payment (P&I) is $1,102.35.

12. 180 payments of $1,102.35
    180 ($1,102.35) = **$198,423.00 repays the refinanced mortgage**.

13. $265,531.50 − $198,423.00 = $67,108.50
    (Note: Step 2 minus step 12.)
    $67,108.50 is saved by refinancing.

192

Step 4: Check Your Work.
  1. Check the numbers and your calculations.
  2. Is the answer reasonable?
  3. Was the question answered?

Remaining Balance for a Bi-Weekly Mortgage
Since you are making an extra payment every year by using the bi-weekly payment plan, it does not take 30 years to repay a 30 year mortgage. A different formula is required to calculate the remaining balance for a bi-weekly mortgage.

Remaining Balance for a Bi-Weekly Mortgage

$$B = L - F\left(\frac{\left(1 + \frac{R}{26}\right)^m - 1}{\frac{R}{26}}\right)$$

Where: B is the remaining balance on the loan.
L is the amount of the loan (mortgage).
F is the first month's principal.
R is the interest rate as a decimal.
m is the number of payments made.

Example 3:
After 2 years, 6 months, the homeowners in example 5 of the previous section find a 15 year bi-weekly mortgage at 7%. The current 30 year bi-weekly mortgage was for $133,940.50. The P&I is $497.25 and the interest rate is 8.125%. If the owner refinances the remaining balance how much money will be saved?

Solution:
Step 1: Picture and understand the problem.
  1. Question: How much money is saved by refinancing?
  2. List the given data:
     Current mortgage: Loan = $133,940.50, P&I = $497.25, R = 8.125%, Term = 30 years, Time: Payments made for 2 years, 6 months.
     n = 26 pay periods annually.
     New mortgage: Rate = 7%, Time = 15 years, n = 26 periods annually.
  3. Identify the unknown as the savings (difference) between the payments left on the current mortgage and the refinanced payment plan.

Step 2: Devise a Plan.
  1. Find the principal on the first bi-weekly payment.

2. Use this formula to determine the number of payments required to repay the current bi-weekly mortgage.

$$x = \frac{\log\left(1 + \frac{L(R)}{26F}\right)}{\log\left(1 + \frac{R}{26}\right)}$$

3. Determine the number of payments made.
4. Determine the number of payments remaining in the current mortgage.
5. Calculate the expense of keeping the current mortgage.
6. Use this formula to calculate the remaining balance on the current bi-weekly mortgage

$$B = L - F\left(\frac{\left(1 + \frac{R}{26}\right)^m - 1}{\frac{R}{26}}\right).$$

7. Use the amortization formula to calculate the monthly payment on the refinanced loan (Principal and Interest): $P = \dfrac{\left(\dfrac{AR}{n}\right)\left(1 + \dfrac{R}{n}\right)^{(nT)}}{\left(1 + \dfrac{R}{n}\right)^{(nT)} - 1}.$

8. Divide the monthly payment by 2 to find the bi-weekly payment.
9. Find the principal on the first bi-weekly payment.
10. Use this formula to determine the number of payments required to repay the new bi-weekly mortgage.

$$x = \frac{\log\left(1 + \frac{L(R)}{26F}\right)}{\log\left(1 + \frac{R}{26}\right)}.$$

11. Calculate the expense of the refinanced mortgage.
12. Find the money saved (difference) between the current mortgage and the refinanced mortgage.

Step 3: Execute the Plan.
1. Find the principal on the first bi-weekly payment.
   The interest on the first payment is I = PRT.

$$I = (\$133{,}940.50)(0.08125)\left(\frac{1}{26}\right) \approx \$418.56$$

Principal = Payment − Interest = $497.25 − $418.56 = $78.69.

2. Determine the number of payments required to repay the current mortgage. $x = \dfrac{\log\left(1 + \dfrac{\$133{,}940.50(0.08125)}{26(\$78.69)}\right)}{\log\left(1 + \dfrac{0.08125}{26}\right)} \approx 590.86$.

   Bumping up to the next payment, x = 591. It takes 591 bi-weekly payments to repay this loan.

3. Determine the number of payments made.

   2 years, 6 months = $2 + \dfrac{6}{12}$ years

   $\left(2 + \dfrac{6}{12} \cancel{\text{years}}\right)\left(\dfrac{26 \text{ payments}}{1 \cancel{\text{year}}}\right) = 65$ payments

4. Determine the number of payments remaining in the current mortgage.
   591 − 65 = 526 payments remaining.

5. Calculate the expense of keeping the current mortgage.
   526($497.25) = **$261,553.50 to repay the current mortgage**.

6. Calculate the remaining balance on the current mortgage using

   $B = \$133{,}940.50 - \$78.69\left(\dfrac{\left(1 + \dfrac{0.08125}{26}\right)^{65} - 1}{\dfrac{0.08125}{26}}\right) \approx \$128{,}278.91$.

7. Use the amortization formula to calculate the monthly payment on the refinanced loan (Principal and Interest):

   $P = \dfrac{\left(\dfrac{\$128{,}278.91(0.07)}{12}\right)\left(1 + \dfrac{0.07}{12}\right)^{(12 \times 15)}}{\left(1 + \dfrac{0.07}{12}\right)^{(12 \times 15)} - 1} \approx \$1{,}153.01$.

8. Divide the monthly payment by 2 to find the bi-weekly payment.
   $\dfrac{\$1{,}153.01}{2} \approx \$576.51$.

9. Find the principal on the first bi-weekly payment.
   The interest on the first payment is I = PRT.

   $I = (\$128{,}278.91)(0.07)\left(\dfrac{1}{26}\right) \approx \$345.37$

   Principal = Payment − Interest = $576.51 − $345.37 = $231.14.

10. Determine the number of payments required to repay the new mortgage. Use $x = \dfrac{\log\left(1 + \dfrac{\$128,278.91(0.07)}{26(\$231.14)}\right)}{\log\left(1 + \dfrac{0.07}{26}\right)} \approx 339.9286267$.

Bumping up to the next payment, x = 340. It takes 340 bi-weekly payments to repay this loan.

11. Calculate the expense of the refinanced mortgage.
    340($576.51) = **$196,013.40 to repay the refinanced mortgage.**

12. Find the money saved (difference) between the current mortgage and the refinanced mortgage.
    $261,553.50 - $196,013.40 = $65,540.10.
    $65,540.10 is saved by refinancing.

Step 4: Check Your Work.
1. Check the numbers and your calculations.
2. Is the answer reasonable?
3. Was the question answered?

Example 4:
After 9 years 4 months, the homeowner in example 6 of the previous section reads the real estate section of the newspaper and finds a 15 year fixed rate mortgage of 7% monthly. The current 30 year mortgage has an interest–only period of 7 years. The mortgage then converts to a 23 year fixed rate mortgage with the balance amortized at an interest rate of 8.125% monthly. The P&I portion of the amortized mortgage is $1,073.61. If the homeowner refinances the remaining balance how much money will be saved? Note: The homeowner paid for 2 years 4 months on his 23 year fixed rate mortgage.

Solution:
Step 1: Picture and understand the problem.
1. Question: How much money is saved by refinancing?
2. List the given data:
   Current mortgage: P&I = $1,073.61,   Rate = 8.125%,
   Term = 30 years: 7 years interest–only and 23 years amortized,
   Time: Payments made for 9 years and 4 months,
   n = 12 pay periods annually.
   New mortgage: Rate = 7%,   Time = 15 years,   n = 12 pay periods.

3. Identify the unknown as the savings (difference) between the payments left on the current mortgage and the refinanced payment plan.

Step 2: Devise a Plan.
1. Determine the number of payments remaining in the current mortgage. 9 years 4 months is 2 years 4 months into the amortized portion of the loan. (The interest–only portion of the loan was the first 7 years.)
2. Calculate the expense of keeping the current mortgage.
3. Select the remaining balance formula $B = P\left(\dfrac{\left(1+\dfrac{R}{n}\right)^x - 1}{\left(\dfrac{R}{n}\right)\left(1+\dfrac{R}{n}\right)^x}\right)$.
4. Convert the current rate into the appropriate units.
5. Find the intermediate values that will make the formula easier to manipulate.

   Determine $\left(\dfrac{R}{n}\right)$, $\left(1+\dfrac{R}{n}\right)^x$, $\left(1+\dfrac{R}{n}\right)^x - 1$ and store these values in your calculator's memory.
6. Substitute the values into the formula and use the order of operations to simplify.
7. Determine the number of payments in the refinanced mortgage.
8. Select the amortization formula to calculate the monthly payment on the refinanced loan (Principal and Interest): $P = \dfrac{\left(\dfrac{AR}{n}\right)\left(1+\dfrac{R}{n}\right)^{(nT)}}{\left(1+\dfrac{R}{n}\right)^{(nT)} - 1}$.
9. Convert the new rate into the appropriate units.
10. Find the intermediate values that will make the formula easier to manipulate.

    Determine $\dfrac{AR}{n}$, $\left(1+\dfrac{R}{n}\right)^{(nT)}$, $\left(1+\dfrac{R}{n}\right)^{(nT)} - 1$ and store these values in your calculator's memory.
11. Substitute the values into the formula and simplify.
12. Calculate the expense of the refinanced mortgage.
13. Find the money saved (difference) between the current mortgage and the refinanced mortgage.

Step 3: Execute the Plan.

1. 2 years 4 months = $(2 \text{ years})\left(\dfrac{12 \text{ months}}{1 \text{ year}}\right) + 4 \text{ months} =$

   $(24 + 4)$ months = 28 payments made.

   23 years = $(23 \text{ years})\left(\dfrac{12 \text{ months}}{1 \text{ year}}\right) = 276$ payments.

   $276 - 28 = 248$ payments left in the current loan, $x = 248$.

2. 248 payments of $1,073.61.
   248 ($1,073.61) = **$266,255.28 repays the current loan.**

3. Calculate the remaining balance using the formula

   $$B = P\left(\dfrac{\left(1+\dfrac{R}{n}\right)^x - 1}{\left(\dfrac{R}{n}\right)\left(1+\dfrac{R}{n}\right)^x}\right).$$

4. $R = 8.125\%, = 0.08125$.

5. $\left(\dfrac{R}{n}\right) = \dfrac{0.08125}{12} \approx 0.0067708333$.

   (Store in calculator's memory 1.)

   $\left(1+\dfrac{R}{n}\right)^x = \left(1+\dfrac{0.08125}{12}\right)^{248} \approx 5.330833038$

   (Store in calculator's memory 2.)

   $\left(1+\dfrac{R}{n}\right)^x - 1 = \left(1+\dfrac{0.08125}{12}\right)^{248} - 1 \approx 4.330833038$

   (Store in calculator's memory 3.)

6. Substitute, recall the appropriate stored values from memory and use the order of operations to simplify.

   $B = \$1,073.61\left(\dfrac{(4.330833038)}{(0.0067708333)(5.330833038)}\right) \approx \$128,819.2555$

   The remaining balance is $128,819.26.

7. 15 years = $(15 \text{ years})\left(\dfrac{12 \text{ months}}{1 \text{ year}}\right) = 180$ payments.

8. Monthly payment $P = \dfrac{\left(\dfrac{AR}{n}\right)\left(1+\dfrac{R}{n}\right)^{(nT)}}{\left(1+\dfrac{R}{n}\right)^{(nT)} - 1}$.

9. Rate: 7% = 0.07 = R.

10. $\dfrac{AR}{n} = \dfrac{(\$128{,}819.26)(0.07)}{12} \approx \$751.44568433$.

   Note: "A" is the remaining balance calculated in Step 6.
   (Store in calculator's memory 1.)

   $\left(1+\dfrac{R}{n}\right)^{(nT)} = \left(1+\dfrac{0.07}{12}\right)^{(12\times 15)} \approx 2.848946731$

   (Store in calculator's memory 2.)

   $\left(1+\dfrac{R}{n}\right)^{(nT)} - 1 = \left(1+\dfrac{0.07}{12}\right)^{(12\times 15)} - 1 \approx\ = 1.848946731$

   (Store in calculator's memory 3.)

11. Substitute, recall the appropriate stored values from memory and simplify. (Note: The denominator is one less than the second factor of the numerator.)

    $P = \dfrac{(\$751.4456833)(2.848946731)}{1.848946731} \approx \$1{,}157.863927$.

    The new monthly payment (P&I) is $1,157.86.

12. 180 payments of $1,157.86.
    180 ($1,157.86) = **$208414.80 repays the refinanced mortgage**.

13. $266,255.28 − $208,414.80 = $57,840.48.   Note: Step 2 minus Step 12.
    $57,840.48.is saved by refinancing.

Step 4: Check Your Work.
    1. Check the numbers and your calculations.
    2. Is the answer reasonable?
    3. Was the question answered?

❓ Cognitive Problems ❓

1. What is the advantage of refinancing a mortgage at a lower rate for a shorter time even though the monthly payments may be larger than the current payment?
2. Compare the monthly, bi-weekly and interest–only mortgage payments of examples 1, 2, and 3 and explain your observations.

(c) 2006 Jupiter Images Corp

Exercise 3.2

1. A $91,2000 mortgage is financed at 7.9% for 30 years. The P&I portion of the monthly payment is $662.85. After 10 years, 9 months the homeowner is considering refinancing the mortgage.
   A. Calculate the remaining balance for the home mortgage.
   B. Calculate the new monthly payment using the remaining balance from Part A at 6.6% for 15 years?
   C. How much money, if any, can be saved by refinancing?

(c) 2006 Jupiter Images Corp

2. A $292,320 mortgage is financed at 7.75% for 30 years. The P&I portion of the monthly payment is $2,094.22. After 8 years the owner is considering refinancing.
   A. Calculate the balance for the mortgage.
   B. Calculate the new monthly payment using the balance from Part A at 6.3% for 10 years?
   C. How much money can be saved by refinancing?

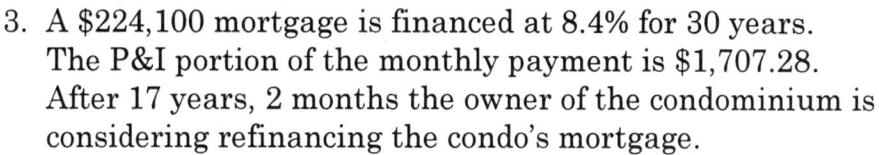

3. A $224,100 mortgage is financed at 8.4% for 30 years. The P&I portion of the monthly payment is $1,707.28. After 17 years, 2 months the owner of the condominium is considering refinancing the condo's mortgage.
   A. Calculate the remaining balance for the condominium's mortgage.
   B. Calculate the new monthly payment using the remaining balance from Part A at 7.1% for 10 years?
   C. How much money, if any, can be saved by refinancing?

(c) 2006 Jupiter Images Corp

4. A $470,060 mortgage is financed at 8.125% for 30 years. The P&I portion of the monthly payment is $3,490.18. After 11 years, 5 months the owner is considering refinancing.
   A. Calculate the balance for the mortgage.
   B. Calculate the new monthly payment using the balance from Part A at 7% for 15 years?
   C. How much money can be saved by refinancing?

(c) 2006 Jupiter Images Corp.

5. A $91,200 mortgage is financed at 7.9% for 30 years. The P&I portion of the bi-weekly payment is $331.42. The first month's principal is $54.31. It will take 597 payments to repay the mortgage. After 7 years (182 payments), the owner is considering refinancing.
   A. Calculate the balance for the home mortgage.
   B. Calculate the new bi-weekly payment using the remaining balance from Part A at 6.6% for 15 years?
   C. How much money, if any, can be saved by refinancing?

(c) 2006 Jupiter Images Corp

6. A $292,320 mortgage is financed at 7.75% for 30 years. The P&I portion of the bi-weekly payment is $1,047.11. The first month's principal is $175.80. It takes 600 payments to repay the mortgage. After 15 years (390 payments) the owner wants to refinance.
   A. Calculate the balance for the home mortgage.
   B. Calculate the new bi-weekly payment using the balance from Part A at 6.3% for 10 years?
   C. How much money can be saved by refinancing?

7. A $224,100 mortgage is financed at 8.4% for 30 years. The P&I portion of the bi-weekly payment is $853.64. The first month's principal is $129.62. It takes 585 payments to repay the mortgage. After 12 years 8 months (330 payments) the owner wants to refinance.
   A. Calculate the balance for the home mortgage.
   B. Calculate the new bi-weekly payment using the balance from Part A at 7.2% for 10 years?
   C. How much money, if any, can be saved by refinancing?

(c) 2006 Jupiter Images Corp

8. A $470,060 mortgage is financed at 8.125% for 30 years. The P&I portion of the bi-weekly payment is $1,745.09. The first month's principal is $276.15. It takes 591 payments to repay the mortgage. After 7 years 9 months (202 payments) the owner is considering refinancing the mortgage.
   A. Calculate the balance for the home mortgage.
   B. Calculate the new bi-weekly payment using the balance from Part A at 7% for 15 years?
   C. How much money can be saved by refinancing?

(c) 2006 Jupiter Images Corp.

9. A $91,200 mortgage is financed at an interest–only rate of 7.9% for 5 years. The loan is amortized over the remaining 25 years at the same rate. The P&I portion of the monthly payment is $697.87. After 7 years 8 months the owner wants to refinance.
   A. Calculate the remaining balance for the mortgage.
   B. Calculate the new monthly payment using the remaining balance from Part A at 6.3% for 15 years?
   C. How much money, if any, can be saved by refinancing?

(c) 2006 Jupiter Images Corp

10. A $292,320 mortgage is financed at an interest–only of 7.75% for 10 years. The loan is amortized over the remaining 20 years at the same rate. The P&I portion of the monthly payment is $2,399.80. After 11 years 3 months the owner wants to refinance.
    A. Calculate the balance for the mortgage.
    B. Calculate the new monthly payment using the balance from Part A at 6.2% for 15 years?
    C. How much money can be saved by refinancing?

11. A $224,100 mortgage is financed at an interest–only of 8.4% for 7 years. The loan is amortized over the remaining 23 years at the same rate. The P&I portion of the monthly payment is $1,568.70 for the interest–only period and $1,836.53 for the amortized period. After 5 years the owner is considering refinancing.
    A. Calculate the remaining balance for the mortgage.
    B. Calculate the new monthly payment using the remaining balance from Part A at 7.1% for 15 years?
    C. How much money can be saved by refinancing?

(c) 2006 Jupiter Images Corp

12. A $470,060 mortgage is financed at an interest–only of 8.125% for 10 years. Then the loan is amortized over the remaining 20 years. The P&I portion of the monthly payment is $3,182.70 for the interest–only period and $3,968.42 for the amortized period. After 6 years 7 months the owner wants to refinance.
    A. Calculate the balance for the mortgage.
    B. Calculate the new monthly payment using the balance from Part A at 7% for 15 years?
    C. How much money can be saved by refinancing?

(c) 2006 Jupiter Images Corp.

3.3 CLOSING COSTS
Objective: The hidden costs of purchasing a home are revealed.

The purchase of a home is the "American dream" for many people. This dream can quickly become a nightmare for the unprepared home buyer. The down payment is only one factor involved in applying for a mortgage. Another factor is the annual income of the borrower. The following table is taken from a federal publication, *"Opening the Door to a Home of Your Own"*, published by the Fannie Mae Foundation. The Fannie Mae Foundation is America's largest supplier of conventional home mortgages. The chart displays the size of the mortgage you might qualify for based on your annual income and the interest rate currently being quoted for a 30 year fixed mortgage. (Note: All dollar figures are in thousands.)

Rates	\$15	\$20	\$25	\$30	\$35	\$40	\$45	\$50	\$55	**\$60**	\$65	\$70
5.5%	\$55	\$73	\$92	\$110	\$128	\$147	\$165	\$184	\$202	**\$220**	\$239	\$257
6.0%	\$52	\$70	\$87	\$104	\$122	\$139	\$156	\$174	\$191	**\$209**	\$226	\$243
6.6%	\$49	\$66	\$82	\$99	\$115	\$132	\$148	\$165	\$181	**\$198**	\$214	\$231
7.0%	\$47	\$63	\$78	\$94	\$110	\$125	\$141	\$157	\$172	**\$188**	\$204	\$219
7.5%	\$46	\$60	\$75	\$89	\$104	\$119	\$134	\$149	\$164	**\$179**	\$194	\$209
8.0%	**\$45**	**\$57**	**\$71**	**\$85**	**\$99**	**\$114**	**\$128**	**\$142**	**\$156**	**\$170**	**\$185**	**\$199**
8.5%	\$41	\$54	\$68	\$81	\$95	\$108	\$122	\$135	\$149	**\$163**	\$176	\$190
9.0%	\$39	\$52	\$65	\$78	\$91	\$104	\$117	\$129	\$142	**\$155**	\$168	\$181
9.5%	\$37	\$50	\$62	\$74	\$87	\$99	\$111	\$124	\$136	**\$149**	\$161	\$173

Annual Income (Thousands of Dollars)

If the borrower's annual income is \$60,000 and the current interest rate for a 30 year fixed mortgage is 8.0%, then \$170,000 is the maximum mortgage that the borrower can receive.

Other factors involved in applying for a mortgage are various one-time fees, called "closing costs". These closing costs are fees the borrower must pay to the bank, lawyer, title company, county, state, and insurance company. After the homebuyer pays these fees, he can legally move into the residence. An example of a closing cost document follows.

ITEMIZATION OF SETTLEMENT CHARGES	
Down Payment	$
Loan Application Fee	$
Points (pts)	$
Title Search	$
Title Insurance	$
Recording Fee	$
Document Preparation Fee	$
Credit Search	$
Appraisal Fee	$
Survey	$
Closing Fee	$
Interest (max. 1 month)	$
Taxes (max. 1 month)	$
Insurance	$
PMI	$
Legal (attorney)	$
Total Amount Due on Closing	$

Here is a brief explanation of these fees:

1. *Down Payment* is the amount of money (a percentage of the selling price of the property) required by the lender as one of the conditions of the loan.
2. *Loan Application Fee* is a fee charged by the lender to take your application. It may or may not be refundable if the lender declines the application.
3. *Points* are an upfront cash payment based on a percentage of the loan. For example, a $100,000 mortgage with one point, means the buyer will pay an upfront cash payment of 1 percent of the $100,000 or $1,000. Points are usually associated with lower interest rates and are negotiable.
4. *Title Search Fee* pays for the research for liens against the property that could affect the sale of the property.
5. *Title Insurance* is the premium for an insurance policy. This protects the buyer on matters pertaining to the title on the land. The title company pays all legal fees in the defense of the title.
6. *Recording Fee* pays the recording of the deed with the county.
7. *Document Preparation Fee* is charged for the preparation of all the required documentation in the sale of the property.
8. *Credit Search Fee* pays a credit service bureau to provide the lender with a report detailing the buyer's credit history and rating.
9. *Appraisal Fee* pays for an appraiser to research and assess the market value of the property.

204

10. **Survey Fee** pays a surveyor to survey the property and prepare a "plot plan." This certifies that all structures and other improvements on the property do not violate any zoning laws and do not encroach on anyone else's property.
11. **Closing Fee** pays for finalizing the closing and pertaining to the issuance of the title insurance policy (document preparation, notarization, recording and maintenance).
12. **Interest Charge** is the interest payment due from the closing date to the first mortgage payment.
13. **Tax Charge** is the property tax payment due from the closing date to the first mortgage payment.
14. **Insurance** is the premium for a hazard (fire, storm, etc.) insurance policy for the residence.
15. **Private Mortgage Insurance** is required if the down payment is less than 20%. Mortgage insurance protects the mortgage lender against financial loss if the borrower defaults.
16. **Legal Fees** are the attorney's fees for the handling of the closing.

(c) 2006 JupiterImages Corp

Consider the closing costs on the $105,000 condominium from Example 1 in the previous section. It requires 5% down, 0 points, a $300 application fee, a fixed rate of 8.5% for 30 years. The property taxes are $2,000 annually, the homeowners insurance is $250 annually, and the private mortgage insurance (PMI) is 0.48%.

The closing costs calculations are:
a. The down payment on the home is 0.05·($105,000) = $5,250.00.
b. The amount financed is $105,000 − $5,250.00 = $99,750.00.
c. The loan application fee is $300.
d. Zero points means 0% of the amount financed which is $0.00.
e. The interest for one month is $\dfrac{0.085}{12}(\$99{,}750.00) \approx \$706.56$.
f. The tax for one month is $\dfrac{\$2{,}000}{12} \approx \$166.67$.
g. The insurance premium is $250.
h. Private Mortgage Insurance is 0.0048($99,750) = $478.80.
i. The title search, title insurance, recording fee, documentation preparation fee, credit search, appraisal, survey, closing fee and legal fee are **standard**. These standard fees will remain the same for problems in this section.

Following is a listing of the more common closing costs:

ITEMIZATION OF SETTLEMENT CHARGES	
Down Payment	$ 5,250.00
Loan Application Fee	$ 300.00
Points (0 pts.)	$ ----------
Title Search	$ 250.00
Title Insurance	$ 400.00
Recording Fee	$ 50.00
Document Preparation Fee	$ 125.00
Credit Search	$ 50.00
Appraisal Fee	$ 350.00
Survey	$ 275.00
Closing Fee	$ 50.00
Interest (max. 1 month)	$ 706.56
Taxes (max. 1 month)	$ 166.67
Insurance	$ 250.00
PMI	$ 478.80
Legal (Attorney)	$ 500.00
Total Amount Due on Closing	$ 9,202.03

This means that in addition to the $5,250.00, down payment, the buyer needs an additional $3,952.03 in closing costs. On closing day $ 9,202.03 is due.

Points

Points are optional and are purchased by the borrower to lower the interest rate on a mortgage. If you have the available funds, the Chicago Tribune financial advisor recommends, purchasing points as a strategy to lower the interest payments over the life of the loan. The scenario is a $100,000 mortgage loan. One point equals 1 percent of the loan amount. Compare the following mortgage rates. The "No points" rate is 7.875%. The "3 points" rate is 7.5%.

Mortgage	Rate	Points
30 year fixed	7.875%	0
30 year fixed	7.5%	3

Paying 3 points (0.03 × $100,000 = $3,000) will not necessarily lower your interest rate by 0.375%. These numbers are negotiable. Payment amounts do not include taxes and insurance. These savings are shown in the following diagram.

206

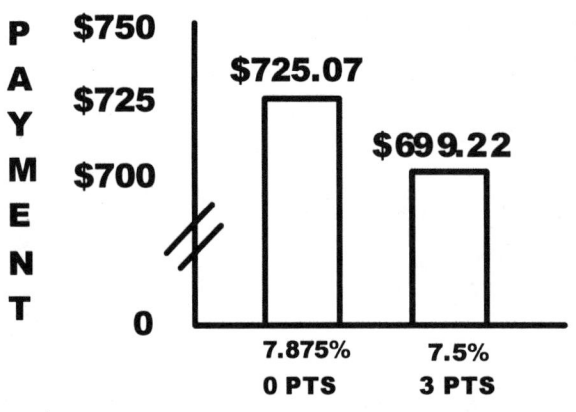

MONTHLY PAYMENTS WITHOUT VS WITH POINTS

Paying points can lower your loan's interest rate and monthly payment, but it is an added up-front cost. The above graph shows a $25.85 per month difference in the two monthly payments. This amounts to 360 × $25.85 = $9,306.00 over the 30 years of the mortgage. Subtracting the $3,000 spent in purchasing points, there is a savings of $9,306.00 − $3,000 = $6,306 over the term of the loan.

Example 1:
Determine the closing costs on the $105,000 condominium from Example 1 in the previous section. It requires 5% down, a $300 loan application fee, **2 points for a negotiated lower interest rate of 8% for 30 years.** The property taxes are $2,000 annually, the homeowners insurance is $250 annually, and the private mortgage insurance (PMI) is 0.48%.

(c) 2006 JupiterImages Corp

Solution:
Step 1: Picture and understand the problem.
1. Question: How much money is needed to close on the condominium?
2. Determine the down payment and closing costs.
3. The title search, title insurance, recording, documentation, credit search, appraisal, survey, closing and legal fees are standard.
4. The application fee, points, interest, taxes, and insurance are not standard and have to be calculated.

Step 2: Develop a plan.
1. Use the given standard fees.

2. Calculate the nonstandard fees: down payment, points, interest, and taxes.
3. Fill in the insurance premium.
4. Calculate the total closing costs.

Step 3: Execute the plan.
1. Fill in the standard fees.

ITEMIZATION OF SETTLEMENT CHARGES	
Down Payment	$
Loan Application Fee	$
Points (pts.)	$
Title Search	$ 250.00
Title Insurance	$ 400.00
Recording Fee	$ 50.00
Document Preparation Fee	$ 125.00
Credit Search	$ 50.00
Appraisal Fee	$ 350.00
Survey	$ 275.00
Closing Fee	$ 50.00
Interest (max. 1 month)	$
Taxes (max. 1 month)	$
Insurance	$
PMI	$
Legal (Attorney)	$ 500.00
Total Amount Due on Closing	$

2. Calculate the nonstandard fees:
Down payment = (0.05)($105,000.00) = $5,250.00.
Amount financed = $105,000 − $5,250 = $99,750.
Loan Application Fee is $300.
Two points are two percent of the amount financed:
  (0.02)($99,750) = $1,995.
Interest: 1 month's worth of interest on the amount financed.
$\frac{0.08}{12}$($99,750) = $665.00, One month's interest is $665.00.

Taxes: 1 month's worth of taxes: $\frac{\$2,000}{12} \approx \$166.6666667 = \$166.67$.

Insurance is $250.00.
Private Mortgage Insurance is 0.0048($99,750) = $478.80.

Fill in the chart with these nonstandard fees:

ITEMIZATION OF SETTLEMENT CHARGES	
Down Payment	$ 5,250.00
Loan Application Fee	$ 300.00
Points (2 pts.)	$ 1,995.00
Title Search	$ 250.00
Title Insurance	$ 400.00
Recording Fee	$ 50.00
Document Preparation Fee	$ 125.00
Credit Search	$ 50.00
Appraisal Fee	$ 350.00
Survey	$ 275.00
Closing Fee	$ 50.00
Interest (max. 1 month)	$ 665.00
Taxes (max. 1 month)	$ 166.67
Insurance	$ 250.00
PMI	$ 478.80
Legal (Attorney)	$ 500.00
Total Amount Due on Closing	$ 11,155.47

This means that in addition to the $5,250.00, down payment, the buyer needs an additional $5,905.47 in closing costs. On closing day $11,155.47 is due.

Step 4: Check your work.
1. Check the numbers and your calculations.
2. Does $11,155.47 seem reasonable to close on this home?
3. Was the total closing cost calculated?

Example 2:

Determine the closing costs on the $140,990 house from Example 4 in the previous section. It requires 5% down, 0 points, a $300 application fee, a fixed rate of 8.125% for 30 years. The annual property taxes are $3,000.00, the homeowners insurance is $300 per year and the private mortgage insurance is 0.36% of the amount financed.

Solution:
Step 1: Picture and understand the problem.
1. Question: How much money is needed to close on the house?
2. Determine the down payment and closing costs.

3. The title search, title insurance, recording, documentation, credit search, appraisal, survey, closing and legal fees are standard.
4. The application fee, points, interest, taxes, and insurance are not standard and have to be calculated.

Step 2: Develop a plan.
1. Fill in the standard fees.
2. Calculate the down payment, points, interest, and taxes.
3. Fill in the insurance premium.
4. Calculate the total closing costs.

Step 3: Execute the plan.
1. Fill in standard fees.

ITEMIZATION OF SETTLEMENT CHARGES	
Down Payment	$
Loan Application Fee	$
Points (pts.)	$
Title Search	$ 250.00
Title Insurance	$ 400.00
Recording Fee	$ 50.00
Document Preparation Fee	$ 125.00
Credit Search	$ 50.00
Appraisal Fee	$ 350.00
Survey	$ 275.00
Closing Fee	$ 50.00
Interest (max. 1 month)	$
Taxes (max. 1 month)	$
Insurance	$
PMI	$
Legal (Attorney)	$ 500.00
Total Amount Due on Closing	$

2. Down payment = (0.05)($140,990) = $7,049.50.
Amount financed = $140,990 − $7,049.50 = $133,940.50.
Loan Application Fee is $300.
Interest: 1 month's worth of interest on the amount financed.

$$\frac{0.08125}{12}(\$133{,}940.50) \approx \$906.89.$$

Taxes: 1 month's worth of taxes. $\dfrac{\$3{,}000}{12} = \$250.00$.

One month's of taxes is $250.00.
Insurance is $300.00.
PMI = 0.0036 ($133,940.50) ≈ $482.19.

210

Fill in the chart with these nonstandard fees:

ITEMIZATION OF SETTLEMENT CHARGES	
Down Payment	$ 7,049.50
Loan Application Fee	$ 300.00
Points (pts.)	$
Title Search	$ 250.00
Title Insurance	$ 400.00
Recording Fee	$ 50.00
Document Preparation Fee	$ 125.00
Credit Search	$ 50.00
Appraisal Fee	$ 350.00
Survey	$ 275.00
Closing Fee	$ 50.00
Interest (max. 1 month)	$ 906.89
Taxes (max. 1 month)	$ 250.00
Insurance	$ 300.00
PMI	$ 482.19
Legal (Attorney)	$ 500.00
Total Amount Due on Closing	$11,338.58

In addition to the $7,049.50, down payment, the buyer needs an additional $4,289.08 in closing costs. On closing day $11,338.58 is due.

Step 4: Check your work.
 1. Check the numbers and your calculations.
 2. Does $11,338.58 seem reasonable to close on this home?
 3. Was the total cost of closing calculated?

Example 3:
Determine the closing costs on the $140,990 house from Example 4 in the previous section. It requires 5% down, a $300 application fee, **2.5 points for a negotiated lower fixed interest rate of 7.5% for 30 years.** The annual property taxes are $3,000.00, the homeowners insurance is $300 per year and the private mortgage insurance is 0.36% of the amount financed.

Solution:
Step 1: Picture and understand the problem.
 1. Question: How much money is needed to close on the house?
 2. Determine the down payment and closing costs.

3. The title search, title insurance, recording, documentation, credit search, appraisal, survey, closing and legal fees are standard.
4. The application fee, points, interest, taxes, and insurance are not standard and have to be calculated.

Step 2: Develop a plan.
1. Fill in the standard fees.
2. Calculate the down payment, points, interest, and taxes.
3. Fill in the insurance premium.
4. Calculate the total closing costs.

Step 3: Execute the plan.
1. Fill in standard fees.

ITEMIZATION OF SETTLEMENT CHARGES	
Down Payment	$
Loan Application Fee	$
Points (pts.)	$
Title Search	$ 250.00
Title Insurance	$ 400.00
Recording Fee	$ 50.00
Document Preparation Fee	$ 125.00
Credit Search	$ 50.00
Appraisal Fee	$ 350.00
Survey	$ 275.00
Closing Fee	$ 50.00
Interest (max. 1 month)	$
Taxes (max. 1 month)	$
Insurance	$
PMI	$
Legal (Attorney)	$ 500.00
Total Amount Due on Closing	$

2. Down payment = (0.05)($140,990) = $7,049.50.
Amount financed = $140,990 − $7,049.50 = $133,940.50.
Loan Application Fee is $300.
Two and a half points are two and a half percent of the amount financed: (0.025)($133,940.50) = $3,348.51.
Interest: 1 month's worth of interest on the amount financed.
$\frac{0.075}{12}$ ($133,940.50) ≈ $837.128125.
One month's interest is $837.13.
Taxes: 1 month's worth of taxes. $\frac{\$3,000}{12} = \$250.00$.

One month's of taxes is $250.00.
Insurance is $300.00.
PMI = 0.0036 ($133,940.50) ≈ $482.19.
Fill in the chart with the nonstandard fees:

ITEMIZATION OF SETTLEMENT CHARGES	
Down Payment	**$ 7,049.50**
Loan Application Fee	**$ 300.00**
Points (pts.)	**$ 3,348.51**
Title Search	$ 250.00
Title Insurance	$ 400.00
Recording Fee	$ 50.00
Document Preparation Fee	$ 125.00
Credit Search	$ 50.00
Appraisal Fee	$ 350.00
Survey	$ 275.00
Closing Fee	$ 50.00
Interest (max. 1 month)	**$ 837.13**
Taxes (max. 1 month)	**$ 250.00**
Insurance	**$ 300.00**
PMI	**$ 482.19**
Legal (Attorney)	$ 500.00
Total Amount Due on Closing	**$14,617.33**

In addition to the $7,049.50, down payment, the buyer needs an additional $7,567,83 in closing costs. On closing day $14,617.33 is due.

Step 4: Check your work.
1. Check the numbers and your calculations.
2. Does $14,617.33 seem reasonable to close on this home?
3. Was the total cost of closing calculated?

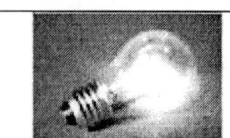
(c) 2006 JupiterImages Corp.

Optional Student Project

"Purchasing a Residence" can be found on page 255.

? Cognitive Problems ?

1. Explain the difference between the standard and nonstandard closing fees.
2. Why does the lender require the buyer to purchase homeowner's insurance?

3. What is the purpose of the Title search? Give an example of a lien.
4. Why does the lender want an appraisal of the property?

Exercise 3.3.

(c) 2006 JupiterImages Corp.

1. Determine the closing costs on a $96,000 townhouse. It requires 5% down, $300 application fee, **3 points for a negotiated fixed lower fixed interest rate of 7.5% for 30 years.** The property taxes are $2,100 annually, the insurance is $230 per year and the private mortgage insurance is 0.32% of the amount financed.

(c) 2006 JupiterImages Corp.

ITEMIZATION OF SETTLEMENT CHARGES	
Down Payment	$
Loan Application Fee	$
Points (pts.)	$
Title Search	$ 250.00
Title Insurance	$ 400.00
Recording Fee	$ 50.00
Document Preparation Fee	$ 125.00
Credit Search	$ 50.00
Appraisal Fee	$ 350.00
Survey	$ 275.00
Closing Fee	$ 50.00
Interest (max. 1 month)	$
Taxes (max. 1 month)	$
Insurance	$
PMI	$
Legal (Attorney)	$ 500.00
Total Amount Due on Closing	$

2. Determine the closing costs on the same $96,000 townhouse. It requires 5% down, 0 points, a $300 application fee, and a fixed rate of 7.9% for 30 years. The property taxes are $2,100 annually, the insurance is $230 per year and the private mortgage insurance is 0.32% of the amount financed.

ITEMIZATION OF SETTLEMENT CHARGES	
Down Payment	$
Loan Application Fee	$
Points (pts.)	$
Title Search	$ 250.00
Title Insurance	$ 400.00
Recording Fee	$ 50.00
Document Preparation Fee	$ 125.00
Credit Search	$ 50.00
Appraisal Fee	$ 350.00
Survey	$ 275.00
Closing Fee	$ 50.00
Interest (max. 1 month)	$
Taxes (max. 1 month)	$
Insurance	$
PMI	$
Legal (Attorney)	$ 500.00
Total Amount Due on Closing	$

3. Determine the closing costs on a $324,800 house. It requires 10% down, 0 points, a $325 application fee, and a fixed rate of 7.75% for 30 years. The property taxes are $4,870 annually, the insurance is $409 per year and the private mortgage insurance is 0.6% of the amount financed.

ITEMIZATION OF SETTLEMENT CHARGES	
Down Payment	$
Loan Application Fee	$
Points (pts.)	$
Title Search	$ 250.00
Title Insurance	$ 400.00
Recording Fee	$ 50.00
Document Preparation Fee	$ 125.00
Credit Search	$ 50.00
Appraisal Fee	$ 350.00
Survey	$ 275.00
Closing Fee	$ 50.00
Interest (max. 1 month)	$
Taxes (max. 1 month)	$
Insurance	$
PMI	$

215

Legal (Attorney)	$ 500.00
Total Amount Due on Closing	$

4. Determine the closing costs on the same $324,800 house. It requires 5% down, a $325 application fee, and **1.5 points for a negotiated lower fixed interest rate of 7.4 % for 30 years**. The property taxes are $4,870 annually, the insurance is $409 per year and the private mortgage insurance is 0.6%.

ITEMIZATION OF SETTLEMENT CHARGES	
Down Payment	$
Loan Application Fee	$
Points (pts.)	$
Title Search	$ 250.00
Title Insurance	$ 400.00
Recording Fee	$ 50.00
Document Preparation Fee	$ 125.00
Credit Search	$ 50.00
Appraisal Fee	$ 350.00
Survey	$ 275.00
Closing Fee	$ 50.00
Interest (max. 1 month)	$
Taxes (max. 1 month)	$
Insurance	$
PMI	$
Legal (Attorney)	$ 500.00
Total Amount Due on Closing	$

5. Determine the closing costs on a $249,000 lakefront condominium. It requires 10% down, 0 points, a $500 application fee, and a fixed rate of 8.4% for 30 years. The property taxes are $6,200 annually, the insurance is $550 per year and the private mortgage insurance is 0.56% of the amount financed.

(c) 2006 JupiterImages Corp

ITEMIZATION OF SETTLEMENT CHARGES	
Down Payment	$
Loan Application Fee	$
Points (pts.)	$
Title Search	$ 250.00
Title Insurance	$ 400.00
Recording Fee	$ 50.00

Document Preparation Fee	$	125.00
Credit Search	$	50.00
Appraisal Fee	$	350.00
Survey	$	275.00
Closing Fee	$	50.00
Interest (max. 1 month)	$	
Taxes (max. 1 month)	$	
Insurance	$	
PMI	$	
Legal (Attorney)	$	500.00
Total Amount Due on Closing	$	

6. Determine the closing costs on the same $249,000 lakefront condominium. It requires 10% down, a $500 application fee, and **2.5 points for a negotiated lower fixed rate of 8% for 30 years**. The property taxes are $6,200 annually, the insurance is $550 per year and the private mortgage insurance is 0.56%.

ITEMIZATION OF SETTLEMENT CHARGES		
Down Payment	$	
Loan Application Fee	$	
Points (pts.)	$	
Title Search	$	250.00
Title Insurance	$	400.00
Recording Fee	$	50.00
Document Preparation Fee	$	125.00
Credit Search	$	50.00
Appraisal Fee	$	350.00
Survey	$	275.00
Closing Fee	$	50.00
Interest (max. 1 month)	$	
Taxes (max. 1 month)	$	
Insurance	$	
PMI	$	
Legal (Attorney)	$	500.00
Total Amount Due on Closing	$	

7. Determine the closing costs on a $494,800 house. It requires 5% down, a $300 application fee, and 2.5 points for a negotiated lower fixed interest rate of 6.8% for 30 years. The property taxes are $10,840 annually, the insurance is $1,086 per year and the private mortgage insurance is 0.64% of the amount financed.

(c) 2006 JupiterImages Corp.

ITEMIZATION OF SETTLEMENT CHARGES	
Down Payment	$
Loan Application Fee	$
Points (pts.)	$
Title Search	$ 250.00
Title Insurance	$ 400.00
Recording Fee	$ 50.00
Document Preparation Fee	$ 125.00
Credit Search	$ 50.00
Appraisal Fee	$ 350.00
Survey	$ 275.00
Closing Fee	$ 50.00
Interest (max. 1 month)	$
Taxes (max. 1 month)	$
Insurance	$
PMI	$
Legal (Attorney)	$ 500.00
Total Amount Due on Closing	$

8. Determine the closing costs on the same $494,800 house. It requires 5% down, 0 points, a $300 application fee, and a fixed rate of 7.125% for 30 years. The property taxes are $10,840 annually, the insurance is $1,086 per year and the private mortgage insurance is 0.64% of the amount financed.

ITEMIZATION OF SETTLEMENT CHARGES	
Loan Expenses	
Down Payment	$
Loan Application Fee	$
Points (pts.)	$
Title Search	$ 250.00
Title Insurance	$ 400.00
Recording Fee	$ 50.00
Document Preparation Fee	$ 125.00
Credit Search	$ 50.00
Appraisal Fee	$ 350.00
Survey	$ 275.00
Closing Fee	$ 50.00
Interest (max. 1 month)	$
Taxes (max. 1 month)	$
Insurance	$
PMI	$
Legal (Attorney)	$ 500.00
Total Amount Due on Closing	$

3.4 ANNUITIES THAT GROW
Objective: Identify and calculate the sinking fund and future value formulas.

Annuities
Thinking about saving for college, retirement, dream car or home? One way to achieve these goals is to establish an annuity. An annuity is an account that has periodic transactions. These transactions can be deposits or withdrawals during predetermined intervals. In this section ordinary annuities involving multiple deposits will be developed. An ordinary annuity pays interest at the end of each compound period. With each deposit and interest payment, the annuity grows and the account gets larger.

Sinking Fund: In a specific time frame, a periodic payment reaches a fixed amount of money.

How much money must be set aside at regular intervals (periods) to achieve a fixed goal (amount)? When a known amount of money is needed at a future date, the periodic deposit can be determined by using the sinking fund formula. The sinking fund is a growth annuity. A growth annuity is an account whose balance increases over time. The formula for the sinking fund is

SINKING FUND

$$P = \dfrac{\dfrac{AR}{n}}{\left(1+\dfrac{R}{n}\right)^{(nT)} - 1}$$

where: A is the future value of the account.
P is the fixed periodic deposit.
R is the annual interest rate as a decimal.
n is the number of compound periods in a year.
T is the time in years.

For example, a school board determines that the district will need $3,000,000 in 5 years for an addition. The building fund money is deposited annually into an account that pays 6% interest compounded annually. How much money must be transferred annually into the building fund in order to reach the $3,000,000 goal?

Use the sinking fund formula to solve this problem: $P = \dfrac{\dfrac{AR}{n}}{\left(1+\dfrac{R}{n}\right)^{(nT)} - 1}$.

A = $3,000,000, Rate = 6% = 0.06 = R, n = 1 compound periods annually, T = 5 years.

$\dfrac{AR}{n} = \dfrac{\$3,000,000(0.06)}{1} = \$180,000$ **(Store in your calculator's memory.)**

$\left(1+\dfrac{R}{n}\right)^{(nT)} - 1 = \left(1+\dfrac{0.06}{1}\right)^{(1 \times 5)} - 1 = (1.06)^5 - 1 \approx 0.3382255776$ **(Store in your calculator's memory.)**

Recall the appropriate stored values from memory and simplify.

$P = \dfrac{\$180,000}{0.3382255776} \approx \$532,189.2013$

The annual deposit is $532,189.20.

The following table demonstrates how 5 annual deposits of $532,189.20 will grow to the desired $3,000,000 in 5 years when earning 6% interest compounded annually.

Deposit Number	Deposit	Interest Calculated On	Interest	Balance
1	$532,189.20			$532,189.20
After year 1		$532,189.20	$31,931.35	$564,120.55
2	$532,189.20			$1,096,309.75
After year 2		$1,096,309.75	$65,778.59	$1,162,088.34
3	$532,189.20			$1,694,277.54
After year 3		$1,694,277.54	$101,656.65	$1,795,934.19
4	$532,189.20			$2,328,123.39
After year 4		$2,328,123.39	$139,687.40	$2,467,810.79
5	$532,189.20			**$2,999,999.99**

220

After the fifth deposit of $532,189.20, the building fund has grown to $2,999,999.99. Shy by a mere one cent of the desired goal, the sinking fund formula works.

Example 1: A couple needs $10,697.67 in 18 months for the down payment and closing costs on their condominium. How much money should they deposit monthly into an account that earns 4.5% interest compounded monthly in order to achieve their goal?

(c) 2006 JupiterImages Corp

Solution:
  Step 1: Understand and picture the problem.
   1. Question: What is the amount of money that needs to be periodically deposited?
   2. List the given data: A = $10,697.67,   Time = 18 months,
       n = 12 periods annually,
       Rate = 4.5% compounded monthly.
   3. Identify the "sinking fund" formula as the formula needed to determine the amount of each periodic deposit to reach $10,697.67 in 18 months.

  Step 2: Devise a plan.
   1. Use the appropriate formula.
   2. Convert the given data into the appropriate units.
   3. Perform the intermediate calculations that make the formula easier to manipulate. Determine : $\dfrac{AR}{n}$ and $\left(1+\dfrac{R}{n}\right)^{(nT)} - 1$ and store these values in your calculator's memory.
   4. Substitute the appropriate values into the formula and simplify.

  Step 3: Execute the plan.
   1. The formula is: $P = \dfrac{\dfrac{AR}{n}}{\left(1+\dfrac{R}{n}\right)^{(nT)} - 1}$.
   2. Convert the data: Rate = 4.5% = 0.045 = R,

   $$T = 18 \text{ months} = \left(\dfrac{\cancel{18}^{3} \cancel{\text{months}}}{}\right)\left(\dfrac{1 \text{ year}}{\cancel{12}_{2} \cancel{\text{months}}}\right) = \dfrac{3}{2} \text{ years} = 1.5 \text{ years}.$$

3. Perform the intermediate calculation:
$$\frac{AR}{n} = \frac{\$10{,}697.67(0.045)}{12} \approx \$40.1162625$$
(Store in calculator's memory 1.)
$$\left(1+\frac{R}{n}\right)^{(nT)} - 1 = \left(1+\frac{0.045}{12}\right)^{\left(12\left(\frac{18}{12}\right)\right)} - 1 = (1.00375)^{18} - 1 \approx 0.0696952053$$
(Store in calculator's memory 2.)
4. Recall the appropriate stored values from memory and simplify.
$$P = \frac{\$40.1162625}{0.0696952053} \approx \$575.5957291$$
The periodic deposit is $575.60

Step 4: Check your work.
1. Check the numbers and your calculations.
2. Is it reasonable for 18 deposits of $575.60 totaling $10,360.80 to grow to $10,697.67 in 18 months?
3. Was the question answered?

Example 2: Due to housing developments, the planning committee of a village board of trustees has determined that the village will need to build another fire station in 6 years. The village will need $4,000,000 in 6 years for this project. How much money must be deposited every quarter into an account that earns 6.3% interest compounded quarterly.

Solution:
Step 1: Understand and picture the problem.
1. Question: What is the amount of money that needs to be periodically deposited?
2. List the given data: A = $4,000,000, Time = 6 years,
    Rate = 6.3% compounded monthly,
    n = 4 periods annually.
3. Identify the "sinking fund" formula as the formula needed to determine the amount of each periodic deposit to reach $4,000,000 in 6 years.

Step 2: Devise a plan.
1. Use the appropriate formula.
2. Convert the given data into the appropriate units.

3. Perform the intermediate calculations that make the formula easier to manipulate. Determine: $\dfrac{AR}{n}$ and $\left(1+\dfrac{R}{n}\right)^{(nT)} - 1$ and store these values in your calculator's memory.
4. Substitute the appropriate stored values into the formula and simplify.

Step 3: Execute the plan.

1. The formula is: $P = \dfrac{\dfrac{AR}{n}}{\left(1+\dfrac{R}{n}\right)^{(nT)} - 1}$.

2. Convert the data: Rate = 6.3% = 0.063 = R, T = 6 years.
3. Perform the intermediate calculation:
$\dfrac{AR}{n} = \dfrac{\$4,000,000(0.063)}{4} = \$63,000.00$.
(Store in calculator's memory 1.)
$\left(1+\dfrac{R}{n}\right)^{(nT)} - 1 = \left(1+\dfrac{0.063}{4}\right)^{(4\times 6)} - 1 = (1.01575)^{24} - 1 \approx 0.4550701918$.
(Store in calculator's memory 2.)
4. Recall the appropriate stored values from memory and simplify.
$P = \dfrac{\$63,000}{0.4550701918} \approx \$138,440.1816$.
The periodic (quarterly) deposit is $138,440.18.

Step 4: Check your work.
1. Check the numbers and your calculations.
2. Is it reasonable for 24 deposits of $138,440.18 totaling $3,322,564.32 to grow to $4,000,000.00 in 6 years?
3. Was the question answered?

Future Value of an Annuity: Determine the balance in the account after making periodic deposits for a predetermined time period.
Multiple deposits distinguish an annuity from the compound interest problems in Chapter 2, where only a single deposit was made. The following table shows the growth of monthly payments of $100 paid into an ordinary 6 months annuity that earns 6% interest compounded monthly. (Note the interest is paid at the end of the month. Each deposit remains in the account for a month before it collects interest.)

Deposit Number	Deposit	Interest Calculated On	Interest	Balance
1	$100.00			$100.00
After 1 month		$100.00	$0.50	$100.50
2	$100.00			$200.50
After 2 months		$200.50	1.00	$201.50
3	$100.00			$301.50
After 3 months		$301.50	$1.51	$303.01
4	$100.00			$403.01
After 4 months		$403.01	$2.02	$405.03
5	$100.00			$505.03
After 5 months		$505.03	$2.53	$507.56
6	$100.00			**$607.56**

At the end of 6 months, the account has $607.56. The formula which calculate the future vale directly follows:

FUTURE VALUE OF AN ANNUITY

$$A = P \left(\frac{\left(1 + \frac{R}{n}\right)^{(nT)} - 1}{\frac{R}{n}} \right)$$

where: A is the future value of the account.
P is the fixed periodic deposit.
R is the interest rate as a decimal.
n is the number of compound periods in a year.
T is the time in years.

Example 3: A person deposits $100 into an annuity every month. The account earns 6% interest compounded monthly. If these deposits continue for 6 months, what is the future value of the account?
Solution:
  Step 1: Understand and picture the problem.
    1. Question: Find the future value of the annuity.
    2. List the given data: P = $100,   Rate = 6% compounded monthly,
        Time = 6 months, n = 12 periods annually.
    3. Identify the "future value of an annuity" as the unknown.

224

Step 2: Devise a plan.
1. Use the appropriate formula.
2. Convert the given data into the appropriate units.
3. Perform the intermediate calculations that make the formula easy to manipulate. Determine: $\dfrac{R}{n}$ and $\left(1+\dfrac{R}{n}\right)^{(nT)} - 1$ and store these values in your calculator's memory.
4. Substitute the appropriate values into the formula and simplify.

Step 3: Execute the plan.
1. The formula is: $A = P\left[\dfrac{\left(1+\dfrac{R}{n}\right)^{(nT)} - 1}{\dfrac{R}{n}}\right]$.

2. Convert the data: Rate = 6% = 0.06 = R,

$T = 6 \text{ months} = \left(6 \text{ months}\right)\left(\dfrac{1 \text{ year}}{12 \text{ months}}\right) = \dfrac{1}{2} \text{ year} = 0.5 \text{ year}.$

3. Perform the intermediate calculations:
$\dfrac{R}{n} = \dfrac{0.06}{12} = 0.005$ (Store in calculator's memory 1.)

$\left(1+\dfrac{R}{n}\right)^{(nT)} - 1 = \left(1+\dfrac{0.06}{12}\right)^{(12 \times 0.5)} - 1 = (1.005)^6 - 1 = 0.0303775094$

(Store in calculator's memory 2.)

4. Recall the appropriate stored values from memory and simplify.
$\$100\left(\dfrac{0.03037755094}{0.005}\right) \approx \$607.5501879.$

The future value of the annuity is $607.55.

Step 4: Check your work.
1. Check the numbers and your calculations.
2. This number matches the balance in the above annuity table within one cent. A rounding error of one cent for each month is acceptable.
3. Was the question answered?

Future Value of an Annuity

In an advertisement for an individual retirement account (IRA) a bank publishes the following table. This plan requires quarterly deposits of $500 (A maximum of $2,000 a year is allowed by law). According to this chart, if you started this plan at the age of 20 and continued making quarterly deposits until you were 65, you would have contributed $90,000 to the account. At 14% interest compounded quarterly, this money would have grown to $6,970,690.

"RETIRE WITH A MILLION" IRA PLAN

	IRA GROWTH			
Age	Total Deposit at Age 65	8% Approx. Value at Age 65	11% Approx. Value at Age 65	14% Approx. Value at Age 65
20	$90,000	$858,021	$2,382,678	$6,970,690
25	$80,888	$569,248	$1,377,319	$3,496,125
30	$70,000	$374,912	$792,954	$1,749,927
35	$60,000	$244,129	$453,291	$872,347
40	$50,000	$156,116	$255,862	$431,306
45	$40,000	$96,886	$141,106	$209,653
50	$30,000	$57,026	$74,405	$98,258
55	$20,000	$30,201	$35,635	$42,275
60	$10,000	$12,149	$13,099	$14,140
Compounded quarterly with deposits of $500 each quarter.				

Example 4: A person deposits $500 into an annuity every quarter. The account earns quarterly interest at 14%. If these deposits continue for 45 years, what is the future value of the account?

Solution:

Step 1: Understand and picture the problem.
1. Question: What is the future value of the account?
2. List the given data: P = $500, Time = 45 years, Rate = 14% quarterly, n = 4 periods annually.
3. Identify the "future value of an annuity" as the unknown.

Step 2: Devise a plan.
1. Use the appropriate formula.
2. Convert the given data into the appropriate units.

3. Perform the intermediate calculations that make the formula easier to manipulate. Determine: $\dfrac{R}{n}$, $\left(1+\dfrac{R}{n}\right)^{(nT)} - 1$ and store these values in your calculator's memory.
4. Substitute the appropriate values into the formula and simplify.

Step 3: Execute the plan.

1. The formula is: $A = P\left(\dfrac{\left(1+\dfrac{R}{n}\right)^{(nT)} - 1}{\dfrac{R}{n}}\right)$.

2. Convert the data: Rate = 14% = 0.14 = R.
3. Perform the intermediate calculations:

$\dfrac{R}{n} = \dfrac{0.14}{4} = 0.035$ **(Store in your calculator's memory 1.)**

$\left(1+\dfrac{R}{n}\right)^{(nT)} - 1 = \left(1+\dfrac{0.14}{4}\right)^{(4\times 45)} - 1 = (1.035)^{180} - 1 = 487.948325.$

(Store in your calculator's memory 2.)

4. Recall the appropriate stored values from memory and simplify.

$\$500\left(\dfrac{487.948325}{0.035}\right) \approx \$6{,}970{,}690.358.$

The future value of the annuity is $6,970,690.36.

Step 4: Check your work.
 1. Check the numbers and your calculations.
 2. The result is reasonable. It matches the figure in the table.
 3. Was the question answered?

Advertising

It's the purpose of advertising to get your attention, present the product in a favorable manner and stimulate your interest in their product. This is usually accomplished through the use of a "hook". The "hook" is something in the advertisement that gets your attention. An account growing from $90,000 to $6,970,690 in 45 years is the "hook". However, while this is true for 14%, there is no bank that will guarantee you a fixed rate of 14% for 45 years

Continuous Compound Interest

Another bank advertises an individual retirement account (IRA), but they offer **continuous compound interest**. In their table annual deposits of $2,000 for 45 years grows to $8,321,564.

SEE HOW MUCH YOUR MONEY GROWS

Age	Total Deposit at Age 65	10% Approx. Value at Age 65	12% Approx. Value at Age 65	14% Approx. Value at Age 65
		Compounded continuously at assumed rates		
20	$90,000	$1,870,843	$3,898,253	$8,321,564
30	$70,000	$674,959	$1,161,771	$2,040,538
40	$50,000	$235,018	$337,559	$491,656
50	$30,000	$73,173	$89,311	$109,707
60	$10,000	$13,633	$14,540	$15,519

Let's check this advertisement. The formula for the future value of an annuity using continuous compound interest is:

FUTURE VALUE OF A CONTINUOUS COMPOUNDING ANNUITY

$$A = \frac{P(e^R)(e^{(RT)} - 1)}{e^R - 1}$$

where: A is the future value of the account.
P is the fixed periodic deposit.
e is a constant, use e^x key on your calculator
(Note: e is approximately 2.7182818)
R is the annual rate as a decimal.
T is the time in years.

Example 5: A person deposits $2,000 annually into an IRA account that pays 14% interest compounded continuously. After making deposits for 45 years, how much money is in the account?

Solution:
Step 1: Understand and picture the problem.
1. Question: What is the future value of the account?
2. List the given data: P = $2,000,   T = 45 years,
Rate = 14% continuous compound interest.
3. Identify the "future value of an continuous compound annuity" as the unknown.

Step 2: Devise a plan.
1. Use the appropriate formula.
2. Convert the given data into the appropriate units.
3. Perform the intermediate calculations that make the formula easier to manipulate. Determine e^R, $e^R - 1$, $e^{RT} - 1$ and store these values in your calculator's memory.
4. Substitute the appropriate values into the formula and simplify.

Step 3: Execute the plan.
1. The formula is: $A = \dfrac{P(e^R)(e^{(RT)} - 1)}{e^R - 1}$.
2. Convert the data: Rate = 14% = 0.14 = R.
3. Perform the intermediate calculations:
$e^R = e^{0.14} \approx 1.150273799$, $\quad e^R - 1 \approx 0.150273799$,
$e^{RT} - 1 = e^{(0.14 \times 45)} - 1 = e^{6.3} - 1 \approx 543.5719101$.
(Store these values in your calculator's memory.)
4. Recall the appropriate stored values from memory and simplify.
$A = \dfrac{\$2{,}000(1.150273799)(543.5719101)}{0.150273799} \approx \$8{,}321{,}564.115$
The future value of the annuity is $8,321,564.12.

Step 4: Check your work
1. Check the numbers and your calculations.
2. The result is reasonable. It matches the figure in the table.
3. Was the question answered?

More Advertising

It's the purpose of advertising to get your attention, present the product in a favorable manner and stimulate your interest in their product. This is usually accomplished through the use of a "hook". The "hook" is something in the advertisement that gets your attention. An account growing from $90,000 to $8,321,564 in 45 years is the "hook". However, while this is true for 14%, there is no bank that will guarantee you a fixed rate of 14% for 45 years.

? Cognitive Problems ?

1. Explain the difference between a single deposit earning compound interest and an annuity earning the same compound interest.

2. Define an ordinary annuity.
3. Explain the difference between a sinking fund and the future value of an annuity.
4. Explain the difference between a sinking fund and amortization.

Exercise 3.4

(c) 2006 JupiterImages Corp.

1. A couple needs $9,000.00 in 18 months for the down payment and closing costs on their first townhouse. How much money should they deposit monthly into an account that earns 6.5% interest compounded monthly?

(c) 2006 JupiterImages Corp.

2. A couple needs $49,000 in 27 months for the down payment and closing costs on their home. How much money should they deposit monthly into an account that earns 7.2% interest compounded monthly?

3. A person needs $38,000 in 18 months for the down payment and closing costs on her lakefront condominium. How much money should she deposit monthly into an account that earns 6.8% interest compounded monthly?

(c) 2006 JupiterImages Corp

4. A couple needs $37,000.00 in 3 years for the down payment and closing costs on their first home. How much money should they deposit monthly into an account that earns 6.5% interest compounded monthly?

(c) 2006 JupiterImages Corp.

5. Andre' needs $50,000 in 6 years to purchase his dream car without a loan. How much money should he deposit monthly into an account that earns 7.9% interest compounded monthly?

(c) 2006 JupiterImages Corp.

6. A couple needs $6,000 in 3 years for their dream vacation. How much money should they deposit monthly into an account that earns 6% interest compounded monthly?

7. A steel mill needs $12,000,000 in 6 years to renovate a furnace. How much money does it deposit every quarter into an account that earns 6.8% interest compounded quarterly?

8. A city needs $8,000,000 in 7 years to build a new library. How much money must it deposit every 3 months into an account that earns 4.8% interest compounded quarterly?

9. A small airline needs $50,000,000 in 10 years to replace its oldest airplane. How much money must it deposit every 6 months into an account that earns 7.2% interest compounded semi-annually?

10. A park district needs $20,000,000 in 5 years to build a new aquatic and health center. How much money must it deposit every year into an account that earns 6.4% interest compounded annually?

11. As is the case with many advertisements, the IRA in Example 4, is presented in a favorable way. Try the same problem with a more modest rate of 5%. A person deposits $500 into an annuity every quarter for 45 years. The account earns 5% interest compounded quarterly. What is the future value of this account?

12. A person deposits $500 into an annuity every quarter for 30 years. The account earns 6.7% interest compounded quarterly. What is the future value of this account

13. As is the case with many advertisements, the IRA in Example 5, is presented in a favorable way. Try the same problem with a more modest rate of 5%. A person deposits $2,000 annually for 45 years into an annuity. The account earns 5% interest compounded continuously. What is the future value of this account?

14. A person deposits $2,000 annually for 25 years into an annuity. The account earns 6.6% interest compounded continuously. What is the future value of this account?

15. A person deducts $300 a month from his paycheck and deposits it into a savings account. He continues this deduction for 38 years. The account pays 5.4% interest compounded monthly. What is the future value of this account?

16. A person deposits $3,600 a year into a savings account. She continues these annual deposits for 38 years. The account pays 5.4% interest compounded continuously. What is the future value of this account?

17. At the end of each year, a person employed part-time deposits $2,000 into an individual retirement account (IRA). She continues this savings plan for 22 years. The retirement account pays 6.4% interest compounded annually. What is the future value of this retirement account?

18. At the end of each year, a person employed part-time deposits $2,000 into an individual retirement account (IRA). She continues this savings plan for 22 years. The retirement account pays 6.4% interest compounded continuously. What is the future value of this retirement account?

19. A school district deposits $200,000 four times a year into a building fund for 10 years. This account pays 6.4% interest compounded quarterly. What is the future value of this account?

20. A village deposits $75,000 twice a year into a building fund for a future water treatment sight. The village has been making these deposits for 8 years. The account earns 5.8% interest compounded semi-annually. What is the future value of this account?

3.5 ANNUITIES THAT DECAY
Objective: Identify and calculate the present value and amortization formulas.

In this section ordinary annuities involving multiple withdrawals will be developed. An ordinary annuity pays interest at the end of each compound period. With each interest payment and withdrawal, the annuity decays and the account becomes smaller.

Present Value of an Annuity: Find the lump sum of money needed for future withdrawals.

Law cases that result in cash settlements and state lotteries use the present value of an annuity. Present value of an annuity calculates the amount of money that must be in the account for the distribution of the monetary award (future periodic withdrawals of a fixed amount of money for a predetermined period of time). Rather than growing, this account decays with each withdrawal. The present value formula is:

(c) 2006 JupiterImages Corp.

PRESENT VALUE OF AN ANNUITY

$$A = P \left(\frac{\left(1 + \frac{R}{n}\right)^{(nT)} - 1}{\left(\frac{R}{n}\right)\left(1 + \frac{R}{n}\right)^{(nT)}} \right)$$

where: A is the present value of the account.
P is the fixed periodic withdrawal.
R is the annual interest rate as a decimal.
n is the number of compound periods in a year.
T is the time in years.

The following example demonstrates the purpose of the present value of an annuity.

Example 1: A student wants to withdraw $8,000 from an account every year for 4 years to pay for college tuition. How much money must be in an account that pays 5% interest compounded annually, in order to meet the student's future tuition needs?

Solution:
Step 1: Understand and picture the problem.
1. Question: Find the amount of money needed in the account.

2. List the given data: P = $8,000, Rate = 5% compounded annually, n = 1 pay period annually, Time = 4 years.
3. Identify the "present value of an annuity" as the unknown.

Step 2: Devise a plan.
1. Use the appropriate formula.
2. Convert the given data into the appropriate units.
3. Perform the intermediate calculations that make the formula easier to manipulate. Determine: $\frac{R}{n}$, $\left(1+\frac{R}{n}\right)^{(nT)}$, and $\left(1+\frac{R}{n}\right)^{(nT)}-1$ store these values in your calculator's memory.
4. Substitute the appropriate values into the formula and use the order of operations to simplify.

Step 3: Execute the plan.
1. The formula is: $A = P\left(\dfrac{\left(1+\frac{R}{n}\right)^{(nT)} - 1}{\left(\frac{R}{n}\right)\left(1+\frac{R}{n}\right)^{(nT)}}\right).$

2. Convert the data: Rate = 5% = 0.05 = R.
3. Perform the intermediate calculations: $\dfrac{R}{n} = \dfrac{0.05}{1} = 0.05.$

(Store in calculator's memory 1.)

$\left(1+\dfrac{R}{n}\right)^{(nT)} = \left(1+\dfrac{0.05}{1}\right)^{(1\times 4)} = (1.05)^4 \approx 1.21550625.$

(Store in calculator's memory 2.)

$\left(1+\dfrac{R}{n}\right)^{(nT)} - 1 = \left(1+\dfrac{0.05}{1}\right)^{(1\times 4)} - 1 = (1.05)^4 - 1 \approx 0.21550625.$

(Store in calculator's memory 3.)

4. Substitute, recall the appropriate stored values from memory and simplify: (Note: The numerator is one less than the second factor of the denominator.)

$A = \$8,000\left(\dfrac{0.21550625}{(0.05)(1.21550625)}\right) \approx \$28,367.60403.$

The present value is $28,367.60.

Step 4: Check your work.
1. Check the numbers and your calculations.

2. Is it reasonable for $28,367.60 to produce 4 payments of $8,000 totaling $32,000 in 4 years? The following table verifies these results.
3. Was the question answered?

The following chart shows the disbursement of four payments of $8,000 from the account starting with $28,367.60 at 5% interest compounded annually. This account is an ordinary annuity, the interest is paid at the end of the each period. Thus, the money remains in the account for a year, before the first withdrawal is made.

Withdrawal Number	Withdrawal	Interest Calculated on	Interest 5%	Balance
End of year 1		$28,367.60	$1,418.38	$29,785.98
1	$8,000.00			$21,785.98
End of year 2		$21,785.98	$1,089.30	$22,875.28
2	$8,000.00			$14,875.28
End of year 3		$14,875.28	$743.76	$15,619.04
3	$8,000.00			$7,619.04
End of year 4		$7,619.04	$380.95	$7,999.99
4	$8,000.00			−$0.01

Note: This difference of one cent is trivial. This is a rounding error caused by the rounding done in the intermediate steps of the problem.

For the first year:
a. The $28,367.60 is deposited and remains in the account untouched for a year.
b. At the end of the first year, $1,418.38 is paid in interest.
I = PRT = $28,367.60 (0.05) (1) = $1,418.38.
The account grows to $29,785.98. $28,367.60 + $1,418.38 = $29,785.98.
c. Next, the first withdrawal of $8,000 is taken out. The account decays to $21,785.98. This amount of money is used to calculate the interest for the second year.

Withdrawal Number	Withdrawal	Interest Calculated on	Interest 5%	Balance
End of year 1		$28,367.60	$1,418.38	$29,785.98
1	$8,000.00			$21,785.98

For the second year:
   a. At the end of the second year, $1,089.30 is paid in interest. The account grows to $22,875.28.
   b. Next, the second withdrawal of $8,000 is taken out. The account decays to $14,875.28. This amount is used to calculate the interest for the third year.

Withdrawal Number	Withdrawal	Interest Calculated on	Interest 5%	Balance
End of year 2		$21,785.98	$1,089.30	$22,875.28
2	$8,000.00			$14,875.28

For the third year:
   a. At the end of the third year, $743.76 is paid in interest. The account grows to $15,619.04.
   b. Next, the third withdrawal of $8,000 is taken out. The account decays to $7,619.04. This amount is used to calculate the interest for the fourth year.

Withdrawal Number	Withdrawal	Interest Calculated on	Interest 5%	Balance
End of year 3		$14,875.28	$743.76	$15,619.04
3	$8,000.00			$7,619.04

For the fourth year:
   a. At the end of the fourth year, $380.95 is paid in interest. The account grows to $7,999.99.
   b. Next, the fourth withdrawal of $8,000 is taken ouy. The account decays to −$0.01. The present value of an annuity formula works.

Withdrawal Number	Withdrawal	Interest Calculated on	Interest 5%	Balance
End of year 4		$7,619.05	$380.95	$8,000.00
4	$8,000.00			−$0.01

Note: This difference of one cent is trivial. This is a rounding error caused by rounding that was done in the intermediate steps of the problem.

Example 2: A very lucky gambler wins the $3,000,000 lottery. The winnings are paid in monthly installments of $12,500 over a period of 20 years. How much money must the lottery commission deposit into a 6% compounded monthly account to cover these payments? This is the lump sum awarded, present value, instead of taking an annuity.

Solution:
Step 1: Understand and picture the problem.
1. Question: Find the amount of money needed to open the account.
2. List the given data: Rate = 6% compounded monthly, P = $12,500.
   n = 12 pay periods annually, Time = 20 years.
3. Identify the "present value of an annuity" as the unknown.

Step 2: Devise a plan.
1. Use the appropriate formula.
2. Convert the given data into the appropriate units.
3. Calculate the periodic (monthly) payment.
4. Perform the intermediate calculations that will make the formula easier to manipulate.
   Determine: $\frac{R}{n}$, $\left(1+\frac{R}{n}\right)^{(nT)}$, and $\left(1+\frac{R}{n}\right)^{(nT)} - 1$
   and store these values in your calculator's memory.
5. Substitute the appropriate values into the formula and use the order of operations to simplify.

Step 3: Execute the plan.
1. The formula is: $A = P\left(\dfrac{\left(1+\frac{R}{n}\right)^{(nT)} - 1}{\left(\frac{R}{n}\right)\left(1+\frac{R}{n}\right)^{(nT)}}\right)$.

2. Convert the data: Rate = 6% = 0.06 = R.
3. The periodic payment is $\dfrac{\$3,000,000}{\frac{12 \text{ months}}{\text{year}}(20 \text{ years})} = \$12,500$.

4. Perform the intermediate calculations: $\dfrac{R}{n} = \dfrac{0.06}{12} = 0.005$.

   (Store in calculator's memory 1.)

   $\left(1+\frac{R}{n}\right)^{(nT)} = \left(1+\frac{0.06}{12}\right)^{(12\times 20)} = (1.005)^{240} \approx 3.310204476$.

   (Store in calculator's memory 2.)

   $\left(1+\frac{R}{n}\right)^{(nT)} - 1 = \left(1+\frac{0.06}{12}\right)^{(12\times 20)} - 1 \approx 2.310204476$.

   (Store in calculator's memory 3.)

5. Substitute, recall the appropriate stored values from memory and simplify: (Note: The numerator is one less than the second factor of the denominator.)

$$A = \$12{,}500 \left(\frac{(2.310204476)}{(0.005)(3.310204476)} \right) \approx \$1{,}744{,}759.646.$$

The present value is $1,744,759.65.

Step 4: Check your work.
1. Check the numbers and your calculations.
2. Is it result reasonable?
3. Was the question answered?

Amortization Revisited

The amortization formula, another decaying annuity formula, can be used to calculate the periodic withdrawal from an account with a fixed rate. Consider the savings account used in Example 1. The account initially has $28,367.60 and pays 5% interest, compounded annually. How much money can be withdrawn every year for four years, before the account decays to zero? The amortization formula will determine the amount withdrawn.

AMORTIZATION OF A LOAN

$$P = \frac{\left(\dfrac{AR}{n}\right)\left(1+\dfrac{R}{n}\right)^{nT}}{\left(1+\dfrac{R}{n}\right)^{nT} - 1}$$

where: A is the amount of the loan.
P is the fixed periodic payment.
R is the annual interest rate.
n is the number of compound periods in a year.
T is the number of years.

Example 3: An account has $28,367.60 and pays 5% interest compounded annually. How much can be withdrawn every year, if its takes four years for the account to decay to zero?

Solution:
Step 1: Understand and picture the problem.
1. Question: Find the amount of the annual withdrawal.
2. List the given data: A = $28,367.60, n = 1 pay period annually, Time = 4 years, Rate = 5% compounded annually.
3. Identify the amortization formula as the unknown.

Step 2: Devise a plan.
1. Use the appropriate formula.
2. Convert the given data into the appropriate units.
3. Perform the intermediate calculations that will make the formula easier to manipulate.

Determine: $\dfrac{R}{n}$, $\left(1+\dfrac{R}{n}\right)^{(nT)}$, and $\left(1+\dfrac{R}{n}\right)^{(nT)} - 1$.

and store these values in your calculator's memory.
4. Substitute the appropriate values into the formula and use the order of operations to simplify.

Step 3: Execute the plan.

1. The formula is: $P = \dfrac{\left(\dfrac{AR}{n}\right)\left(1+\dfrac{R}{n}\right)^{nT}}{\left(1+\dfrac{R}{n}\right)^{nT} - 1}$.

2. Convert the data: Rate = 5% = 0.05 = R.
3. Perform the intermediate calculations:

$\dfrac{AR}{n} = \dfrac{\$28,367.60(0.05)}{1} = \$1,418.38.$

(Store in calculator's memory 1.)

$\left(1+\dfrac{R}{n}\right)^{nT} = \left(1+\dfrac{0.05}{1}\right)^{(1\times 4)} = (1.05)^4 = 1.21550625.$

(Store in calculator's memory 2.)

$\left(1+\dfrac{R}{n}\right)^{(nT)} - 1 = \left(1+\dfrac{0.05}{1}\right)^{(1\times 4)} - 1 = (1.05)^4 - 1 = 0.21550625.$

(Store in calculator's memory 3.)

4. Substitute, recall the appropriate stored values from memory and simplify: (Note: The denominator is one less than the second factor of the numerator.)

$P = \dfrac{\$1,418.38(1.21550625)}{0.21550625} \approx \$7,999.998863.$

The annual withdrawal is $8,000.00.

Step 4: Check your work
1. Check the numbers and calculations.
2. The result of $8,000 is reasonable and agrees with the information given in Example 1.
3. Was the question answered?

Note: Due to interest being deposited into the account as the withdrawals are made, the account will payout more than its initial value. The account pays out 4 payments of $8,000 or $32,000. This is more than the account initial balance of $28,367.60.

? Cognitive Problems ?

1. Explain the difference between the present value of an annuity and amortization.
2. Explain the difference between the future value and present value of an annuity.
3. Explain the difference between the sinking fund and amortization.

Exercise 3.5

(c) 2006 JupiterImages Corp.

1. A college trust fund is established as the first place prize in a national talent search competition. The winner gets $20,000 every year for 4 years to pay college expenses. How much money must be in an account that earns 6% interest compounded annually, in order to meet this need?

2. A fund is established as the first place prize in a lottery, the winner gets $12,500 every quarter for 20 years (a $1,000,000 prize). How much money must first be in the account that earns 6.4 % interest compounded quarterly, in order to meet this need?

3. A plumber wants to start his own business. He figures he needs $3,000 a month for 2 years before his business becomes established. How much money must be in an account that earns 4.3% interest, compounded monthly, before he starts out on his own.

4. The grand prize in a multi-state lottery is $45,000,000. The winner has a choice of taking a monthly payment of $150,000 for 25 years or a lump sum of money. If the money is placed in an account that earns 5.2% interest compounded monthly, how much is the lump sum payment?

5. An artist wants to withdraw $2,000 every month for 3 years as she establishes her private studio. How much money must be in the account that pays 3.8% interest compounded monthly, in order to meet her future needs?

6. As settlement in a court case, a person is awarded a lump sum of $500,000. The money is deposited into an account that earns 6.2 % interest compounded monthly.
   A. How much money can this person withdraw from the account every month for 6 years?
   B. How much money will the person actually receive in the 6 years?

7. A person inherits the lump sum of $100,000 from a life insurance policy on a relative. The money is deposited into an account that earns 5.8% interest compounded quarterly.
   A. How much money can this person withdraw every quarter for 10 years?
   B. How much money will the person actually receive in the 10 years?

8. A person receives a lump sum lottery prize of $1,000,000. The money is deposited into an account that earns 4.9 % interest compounded monthly.
   A. How much money can this person withdraw every month for 20 years?
   B. How much money will the person actually receive in the 20 years?

9. A federally funded program receives a lump sum of $350,000. The money is deposited into an account that earns 5.2 % interest compounded annually.
   A. How much money can the program withdraw every year (their annual budget) for 4 years.
   B. How much money will the person actually receive in the 4 years?

10. A federally funded program receives a lump sum of $750,000. The money is deposited into an account that earns 6% interest compounded annually.
    A. How much money can the program withdraw every year (their annual budget) for 3 years.
    B. How much money will the person actually receive in the 3 years?

3.6 Annuities Revisited

Objective: Compare present value of an annuity, future value of an annuity, sinking fund and amortization formulas

In addition to saving for a home, college education, car, or vacation, annuities are used for retirement planning. Social Security is no longer a reliable retirement plan. The following political cartoon demonstrates how the Social Security system operates. It indicates that an individual retirement account (IRA) or retirement plan is a necessity.

Copyright © 1997 by The Record. Reprinted by permission of Jimmy Margulies/The Record

Commentary on the cartoon: The Social Security system spends money taken from current workers to pay the benefits of today's retirees. In the near future, there may not be enough people working to pay all the retirees at today's rate. There is a strong probability the Social Security system will go bankrupt. Social Security's options are increase the money taken from the workers paychecks or decrease the amount of money paid to retirees. Neither option is popular. An alternative is to establish your own retirement fund. The four annuity formulas: present value, future value, sinking fund and amortization describe different retirement plans. Before discussing these plans, lets review the previously studied annuity formulas.

ANNUITIES THAT GROW — The balance of the account increases over time. (deposits)	ANNUITIES THAT DECAY — The balance of the account decreases over time. (withdrawals)
FUTURE VALUE OF AN ANNUITY (Interest, not Continuous) $$A = P\left(\frac{\left(1+\frac{R}{n}\right)^{(nT)} - 1}{\frac{R}{n}}\right)$$ where: A is the future value of the account. P is the fixed periodic deposit. R is the interest rate as a decimal. n is the compound periods in a year. T is the time in years. **Continuous Compound Interest** $$A = \frac{P(e^R)(e^{(RT)} - 1)}{e^R - 1}$$ where: A is the future value of the account. P is the fixed periodic deposit. e is a constant, use e^x key on the calculator R is the annual rate as a decimal. T is the time in years. **SINKING FUND** $$P = \frac{\frac{AR}{n}}{\left(1+\frac{R}{n}\right)^{(nT)} - 1}$$ where: A is the future value of the account. P is the fixed periodic deposit. R is the interest rate as a decimal. n is the compound periods in a year. T is the time in years	**PRESENT VALUE OF AN ANNUITY** $$A = P\left(\frac{\left(1+\frac{R}{n}\right)^{(nT)} - 1}{\left(\frac{R}{n}\right)\left(1+\frac{R}{n}\right)^{(nT)}}\right)$$ (how much lump sum to have.) where: A is the present value of the account. P is the fixed periodic withdrawal. R is the interest rate as a decimal. n is the compound periods in a year. T is the time in years. **AMORTIZATION FORMULA** $$P = \frac{\left(\frac{AR}{n}\right)\left(1+\frac{R}{n}\right)^{nT}}{\left(1+\frac{R}{n}\right)^{nT} - 1}$$ (know what lump sum is.) where: A is the amount of the loan. P is the fixed periodic payment. R is the interest rate as a decimal. n is the compound periods in a year. T is the number of years

Notice that the FUTURE VALUE and PRESENT VALUE formulas are both solved for A, the total amount of money. A similar relationship exists for the SINKING FUND and AMORTIZATION formulas. They are both solved for P, the periodic transaction. The only way to distinguish between the formulas is to determine if the account is growing or decaying. The FUTURE VALUE and SINKING FUND are *growing* accounts. These accounts build funds for the future. The PRESENT VALUE and AMORTIZATION are *decaying* accounts. These accounts supply the funds for future projects and "zero out" in the future. Study the following examples to understand these relationships.

Example 1: Planning ahead for retirement, a person wants to withdraw $1,200 every month, for 22 years, to supplement his pension check. How much money must be in an account that pays 6.5% interest compounded monthly, in order to meet these needs?

Solution:
Step 1: Understand and picture the problem.
1. Question: Identify the problem as solving for *"A"*, the total amount of money in the account. Since the retiree will be receiving monthly checks from this account for 22 years, the account's balance is *decaying*. The account will "zero out" after 22 years. Thus, it is a *present value* problem. Find the present value of the account.
2. List the given data: P = $1,200,   Rate = 6.5% compounded monthly, n = 12 pay periods annually,   Time = 22 years.

Step 2: Devise a plan.
1. Use the appropriate formula.
2. Convert the given data into the appropriate units.
3. Perform the intermediate calculations that will make the formula easier to manipulate.

Determine: $\dfrac{R}{n}$, $\left(1+\dfrac{R}{n}\right)^{(nT)}$, and $\left(1+\dfrac{R}{n}\right)^{nT} - 1$ and store these values in your calculator's memory.

4. Substitute the appropriate values into the formula and use the order of operations to simplify.

Step 3: Execute the plan.

1. The formula is: $A = P \left[\dfrac{\left(1+\dfrac{R}{n}\right)^{(nT)} - 1}{\left(\dfrac{R}{n}\right)\left(1+\dfrac{R}{n}\right)^{(nT)}} \right]$.

2. Convert the data: Rate = 6.5% = 0.065 = R.

3. Perform the intermediate calculations: $\dfrac{R}{n} = \dfrac{0.065}{12} \approx 0.00541666667$.

   (Store in calculator's memory 1.)

   $\left(1+\dfrac{R}{n}\right)^{(nT)} = \left(1+\dfrac{0.065}{12}\right)^{(12\times 22)} = (1.00541666667)^{264} \approx 4.162604717$.

   (Store in calculator's memory 2.)

   $\left(1+\dfrac{R}{n}\right)^{nT} - 1 = \left(1+\dfrac{0.065}{12}\right)^{(12\times 22)} - 1 \approx 3.162604717$.

   (Store in calculator's memory 3.)

4. Substitute, recall the appropriate stored values from memory and simplify: (Note: The numerator is one less than the second factor of the denominator.)

   $A = \$1,200 \left(\dfrac{3.162604717}{(0.00541666667)(4.162604717)}\right) \approx \$168{,}317.3472$.

   The present value is $168,317.35.

Step 4: Check your work.
    1. Check the numbers and your calculations.
    2. Is the answer reasonable?
    3. Was the question answered?

Let's look at how this amount of money is saved for retirement.

Example 2: A person needs $168,317.35 in 36 years in order to retire. How much money should be deposited every month into an account that earns 6.2% interest compounded monthly in order to achieve this goal?

Solution:
Step 1: Understand and picture the problem.
    1. Question: Identify the problem as solving for *"P"*, the periodic transaction for the account. Since the person will be making monthly deposits into this account for 36 years, the account's balance is ***growing***. After 36 years of making monthly deposits, the account will have "grown" to $168.317.35. Thus, it is a ***sinking fund*** problem. Find the amount of money that needs to be periodically deposited.
    2. List the given data: A = $168,317.35,   n = 12 pay periods annually,
                               Rate = 6.2% compounded monthly,
                               Time = 36 years.

Step 2: Devise a plan.
    1. Use the appropriate formula.
    2. Convert the given data into the appropriate units.

3. Perform the intermediate calculations that make the formula easier to manipulate.

   Determine: $\dfrac{AR}{n}$ and $\left(1+\dfrac{R}{n}\right)^{(nT)} - 1$ and store these values in your calculator's memory.

4. Substitute the appropriate values into the formula and use the order of operations to simplify.

Step 3: Execute the plan.

1. The formula is: $P = \dfrac{\dfrac{AR}{n}}{\left(1+\dfrac{R}{n}\right)^{(nT)} - 1}$.

2. Convert the data: Rate = 6.2% = 0.062 = R.

3. Perform the intermediate calculations:

   $\dfrac{AR}{n} = \dfrac{\$168{,}317.35(0.062)}{12} \approx \$869.6396417$.

   (Store in calculator's memory 1.)

   $\left(1+\dfrac{R}{n}\right)^{(nT)} - 1 = \left(1+\dfrac{0.062}{12}\right)^{(12 \times 36)} - 1 \approx (1.0051666667)^{432} - 1 \approx 8.265091949$.

   (Store in calculator's memory 2.)

4. Substitute, recall the appropriate stored values from memory and simplify.

   $P = \dfrac{\$869.6396417}{8.265091949} \approx \$105.2183868$. The periodic deposit is $105.22.

Step 4: Check your work
   1. Check the numbers and your calculations.
   2. Is the answer reasonable?
   3. Was the question answered?

Summary of Examples 1 and 2

The person makes 12 payments a year for 36 years equaling 432 monthly deposits of $105.22, that totals to $432 \times \$105.22 = \$45{,}455.04$.

Due to compound interest, this amount grows to $168,317.35 over the 36 years. Then from this account, the retiree receives $1,200 every month for 22 years, which is

$$\dfrac{\$1{,}200}{\cancel{\text{month}}}\left(\dfrac{12 \ \cancel{\text{months}}}{\cancel{\text{year}}}\right)(22 \ \cancel{\text{years}}) = \$316{,}800$$

Due to compound interest, the retiree contributes $45,455.04 and receives $316,800.

	What if the person in Examples 1 and 2 waited until middle age to start saving for retirement? Instead of saving for 36 years, they saved for 18 years. This is the scenario for Example 3.

Example 3: A person needs $168,317.35 in 18 years in order to retire. How much money should be deposited every month into an account that earns 6.2% interest compounded monthly in order to achieve this goal?

Solution:

Step 1: Understand and picture the problem.
1. Question: Identify the problem as solving for **"P"**, the periodic transaction for the account. Since the person will be making monthly deposits into this account for 18 years, the account's balance is **growing**. After 18 years of making monthly deposits, the account will have "grown" to $168.317.35. Thus, it is a ***sinking fund*** problem. Find the amount of money that needs to be periodically deposited.
2. List the given data: A = $168,317.35, n = 12 pay periods annually,
Rate = 6.2% compounded monthly,
Time = 18 years.

Step 2: Devise a plan.
1. Use the appropriate formula.
2. Convert the given data into the appropriate units.
3. Perform the intermediate calculations that make the formula easier to manipulate.

Determine: $\dfrac{AR}{n}$ and $\left(1+\dfrac{R}{n}\right)^{(nT)} - 1$ and store these values in your calculator's memory.

4. Substitute the appropriate values into the formula and simplify.

Step 3: Execute the plan.

1. The formula is: $P = \dfrac{\dfrac{AR}{n}}{\left(1+\dfrac{R}{n}\right)^{(nT)} - 1}$.

2. Convert the data: : Rate = 6.2% = 0.062 = R.
3. Perform the intermediate calculations:

$\dfrac{AR}{n} = \dfrac{\$168{,}317.35(0.062)}{12} \approx \$869.6396417.$

(Store in calculator's memory 1.)

$$\left(1+\frac{R}{n}\right)^{(nT)} - 1 = \left(1+\frac{0.062}{12}\right)^{(12\times18)} - 1 \approx 2.043861355.$$

(Store in calculator's memory 2.)

4. Substitute, recall the appropriate stored values from memory and simplify

$$P = \frac{\$869.6396417}{2.0438161355} \approx \$425.4885682. \quad \text{The periodic deposit is } \$425.49.$$

Step 4: Check your work.
1. Check the numbers and your calculations.
2. Is the answer reasonable?
3. Was the question answered?

Summary of Examples 2 and 3
Notice the difference between saving for 18 years and 36 years. The periodic payment for 18 years is $425.49. The periodic payment for 36 years is $105.22. **The monthly payments are four times greater while the time period is one-half.**

Consider another retirement possibility.

Example 4: A person deposits $125 every month for 40 years into an account. The account earns 5.7% interest compounded monthly. How much money will be in the account when this person retires?

Solution:
Step 1: Understand and picture the problem.
1. Question: Identify the problem as solving for **"A"**, the total amount of money in the account. Since the person is depositing $125 every month for 40 years, the account's balance is *growing*. Thus, it is a *future value* problem. Find the future value of the annuity.
2. List the given data: P = $125, Rate = 5.7% compounded monthly, Time = 40 years, n = 12 pay periods annually.

Step 2: Devise a plan.
1. Use the appropriate formula.
2. Convert the given data into the appropriate units.
3. Perform the intermediate calculations that make the formula easier to manipulate. Determine: $\frac{R}{n}$ and $\left(1+\frac{R}{n}\right)^{(nT)} - 1$ and store these values in your calculator's memory.
4. Substitute the appropriate values into the formula and simplify.

247

Step 3: Execute the plan.

1. The formula is: $A = P\left(\dfrac{\left(1+\dfrac{R}{n}\right)^{(nT)} - 1}{\dfrac{R}{n}}\right)$.

2. Convert the data: Rate = 5.7% = 0.057 = R.
3. Perform the intermediate calculations:

$\dfrac{R}{n} = \dfrac{0.057}{12} = 0.00475.$ **(Store in calculator's memory 1.)**

$\left(1+\dfrac{R}{n}\right)^{(nT)} - 1 = \left(1+\dfrac{0.057}{12}\right)^{(12\times 40)} - 1 = (1.00475)^{480} - 1 \approx 8.724048914.$

(Store in calculator's memory 2.)

4. Substitute, recall the appropriate stored values and simplify.

$A = \$125\left(\dfrac{8.724048914}{0.00475}\right) \approx \$229{,}580.2346.$

The future value of the annuity is $229,580.23.

Step 4: Check your work.
1. Check the numbers and calculations.
2. Is the answer reasonable?
3. Was the question answered?

Using $229,580.23 as the amount of money in the account for retirement, what amount of money can be withdrawn every month from the account that earns 6.2% interest compounded monthly? The goal is to receive checks every month for 24 years.

Example 5: An account earning 6.2% interest compounded monthly has a balance of $229,580.23. How much money can be withdrawn every month for 24 years?

Solution:

Step 1: Understand and picture the problem.
1. Question: Identify the problem as solving for **"P"**, the periodic transaction for the account. Since the person will be making monthly withdrawals from this account for 24 years, the account's balance is *decaying*. The account will be "zeroed out" in 24 years. Thus, it is an *amortization* problem. Find the monthly withdrawal.
2. List the given data: A = $229,580.23, n = 12 pay periods annually, Time = 24 years, Rate = 6.2% compounded monthly.

Step 2: Devise a plan.
1. Use the appropriate formula.
2. Convert the given data into the appropriate units.

3. Perform the intermediate calculations that make the formula easier to manipulate.

   Determine: $\dfrac{AR}{n}$, $\left(1+\dfrac{R}{n}\right)^{(nT)}$, and $\left(1+\dfrac{R}{n}\right)^{(nT)} - 1$ and store these values in your calculator's memory.

4. Substitute the appropriate values into the formula and simplify.

Step 3: Execute the plan.

1. The formula is: $P = \dfrac{\left(\dfrac{AR}{n}\right)\left(1+\dfrac{R}{n}\right)^{nT}}{\left(1+\dfrac{R}{n}\right)^{nT} - 1}$.

2. Convert the data: Rate = 6.2% = 0.062 = R.

3. Perform the intermediate calculations.

   $\dfrac{AR}{n} = \dfrac{\$229{,}580.23(0.062)}{12} \approx \$1{,}186.164522$.

   (Store in calculator's memory 1.)

   $\left(1+\dfrac{R}{n}\right)^{nT} = \left(1+\dfrac{0.062}{12}\right)^{(12 \times 24)} \approx (1.051666667)^{288} \approx 4.411298936$.

   (Store in calculator's memory 2.)

   $\left(1+\dfrac{R}{n}\right)^{(nT)} - 1 = \left(1+\dfrac{0.062}{12}\right)^{(12 \times 24)} - 1 \approx 3.411298936$.

   (Store in calculator's memory 3.)

4. Substitute, recall the appropriate stored values from memory and simplify: (Note: The denominator is one less than the second factor of the numerator.)

   $P = \dfrac{(\$1{,}186.164522)(4.411298936)}{3.411298936} \approx \$1{,}533.880903$.

   The monthly withdrawal is $1,533.88.

Step 4: Check your work
   1. Check the numbers and calculations.
   2. Is the answer reasonable?
   3. Was the question answered?

Summary of Examples 4 and 5

The person makes 12 payments per year for 40 years equaling 480 monthly deposits of $125 that totals to $480 \times \$125 = \$60{,}000$. Due to compound interest, this amount grows to $229,580.23 over 40 years. Then from this account the retiree withdraws $1,533.88 every month for 24 years, which is

$$\frac{\$1,533.88}{\text{month}} \left(\frac{12 \text{ month}}{\text{year}} \right) (24 \text{ years}) = \$441,757.44$$

Due to compound interest, the retiree contributes $60,000 and receives $441,757.44.

Retirement Options

The $229,580.23 will yield a $1,533.88 monthly check for 24 years. If the retiree is worried about outliving their retirement income, they can purchase a lifetime annuity from an insurance company. A lifetime annuity will provide a monthly check for the retiree's life. However, this monthly check will be less than the $1,533.88. At age 60, the monthly check may be $1,483.84 for life. To include your spouse, the monthly check may be $1,233.81 for the life of the surviving spouse. However, in order to take advantage of any of these options, you must have saved $229,580.23 at the time of your retirement. **You must start early!**

 (c) 2006 JupiterImages Corp.	Don't wait till middle age to start an IRA or join a 401 plan! START EARLY!
Time is the major factor when it comes to retirement programs. Compare the results of the following exercise problems 1 & 3, 2 & 4, 6 & 8, and 7 & 9. The only difference in these problems is the *time* factor. Notice the magnitude of money available when a person started saving early.	

	While the IRA provides funds in the future. There is an immediate income tax advantage to opening an IRA account. Go to the Appendix C, located in the back of the textbook and compare the income tax returns. The first return is for a full time student, working part time with no benefits, and no IRA account. The second return is for that same student, but with an IRA account.

	IRA'S? IRA's are mentioned throughout this section. What is an IRA? While there are several types of Individual Retirement Accounts (IRA), the traditional IRA is described.

The Traditional IRA

The Individual Retirement Account (IRA) brings together two tremendously powerful forces, both of which benefit you: 1) compound interest, and 2) tax savings. An IRA is a personal retirement savings plan. While an IRA is similar to a savings account, its primary function is to provide the account holder with money when they retire. Money may be withdrawn from an IRA at any time, but on withdrawal it will be taxed. Withdrawals from a traditional IRA prior to age 59.5 will result in a 10% excise tax as well as an ordinary income tax.

There are several exceptions to the 10% penalty for IRA withdrawals prior to age 59.5. The early withdrawal penalty does not apply for distributions that:

1. Occur because of the IRA owner's disability.
2. Are a series of "substantially equal periodic payments" made over the life expectancy of the IRA owner.
3. Are used to pay for unreimbursed medical expenses that exceed 7.5% of adjusted gross income (AGI).
4. Are used to pay medical insurance premiums after the IRA owner has received unemployment compensation for more than 12 weeks.
5. Are used to pay the costs of a first-time home purchase (subject to a lifetime limit of $10,000).
6. Are used to pay for the qualified expenses of higher education for the IRA owner and/or eligible family members.
7. Are used to pay back taxes because of an Internal Revenue Service levy placed against the IRA.

David Wolpe, Chicago Sun-Times, Financial Section, "Motley Fool"

(c) 2006 JupiterImages Corp.

Optional Student Project

"A Personal Retirement Plan" can be found on page 257.

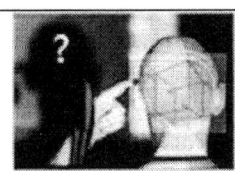
(c) 2006 JupiterImages Corp.

Chapter Review

Chapter review problems can be found on page 258.

? **Cognitive Problems** ?

1. Explain the difference between an annuity that grows and an annuity that decays.
2. Which annuities grow?
3. Which annuities decay?
4. What is the difference between the future value of an annuity and the present value of an annuity.
5. Explain the difference between a sinking fund account and the amortization formula.
6. Explain the difference between a sinking fund account and the present value of an annuity.
7. Explain the difference between the future value of an annuity and the amortization formula.

Exercise 3.6

(c) 2006 JupiterImages Corp.

1. A retiree wants to withdraw $1,500 every month, for 24 years, to supplement their Social Security check.
   a. How much money must be in an account that pays 6.2% interest compounded monthly, in order to meet their needs?
   b. How much money must be deposited into an account that pays 6% interest compounded monthly in order to achieve this amount? The monthly deposits will be made for 40 years.
   c. How much money was actually deposited into the account?
   d. How much money will actually be paid out from this account?

2. A retiree wants to withdraw $1,000 every month, for 21 years, to supplement their Social Security check.
   a. How much money must be in an account that pays 7% interest compounded monthly, in order to meet their needs?
   b. How much money must be deposited into an account that pays 6.6% interest compounded monthly in order to achieve this amount? The monthly deposlits will be made for 45 years.
   c. How much money was actually deposited into the account?
   d. How much money will actually be paid out from this account?

3. A retiree wants to withdraw $1,500 every month, for 24 years, to supplement their Social Security check.
   a. How much money must be in an account that pays 6.2% interest compounded monthly, in order to meet their needs?
   b. How much money must be deposited into an account that pays 6% interest compounded monthly in order to achieve this amount? The monthly deposits will be made for 24 years.
   c. How much money was actually deposited into the account?
   d. How much money will actually be paid out from this account?

4. A retiree wants to withdraw $1,000 every month, for 21 years, to supplement their Social Security check.
   a. How much money must be in an account that pays 7% interest compounded monthly, in order to meet their needs?
   b. How much money must be deposited every month for 25 years into an account that pays 6.6% interest compounded monthly in order to achieve this amount?
   c. How much money was actually deposited into the account?
   d. How much money will actually be paid out from this account?

5. A retiree wants to withdraw $2,000 every month, for 24 years, to supplement their Social Security check.
   a. How much money must be in an account that pays 6.2% interest compounded monthly, in order to meet their needs?
   b. How much money must be deposited into an account that pays 6% interest compounded monthly in order to achieve this amount? The monthly deposits will be made for 40 years.
   c. How much money was actually deposited into the account?
   d. How much money will actually be paid out from this account?

6. A person deposits $200 every month, for 42 years, into an account that earns 5.9% interest compounded monthly.
   a. How much money will be in an account when the person retires?
   b. How much money can be withdrawn every month for 26 years from this account? The account earns 6.7% interest compounded monthly.
   c. How much money was actually deposited into the account?
   d. How much money will actually be paid out from this account?

7. A person deposits $175 every month, for 40 years, into an account that earns 4.8% interest compounded monthly.
   a. How much money will be in an account when the person retires?
   b. How much money can be withdrawn every month for 20 years from this account? The account earns 5.8% interest compounded monthly.

c. How much money was actually deposited into the account?
d. How much money will actually be paid out from this account?

8. A person deposits $200 every month, for 22 years, into an account that earns 5.9% interest compounded monthly.
   a. How much money will be in an account when the person retires?
   b. How much money can be withdrawn every month for 26 years from this account? The account earns 6.7% interest compounded monthly.
   c. How much money was actually deposited into the account?
   d. How much money will actually be paid out from this account?

9. A person deposits $175 every month, for 21 years, into an account that earns 4.8% interest compounded monthly.
   a. How much money will be in an account when the person retires?
   b. How much money can be withdrawn every month for 20 years from this account? The account earns 5.8% interest compounded monthly.
   c. How much money was actually deposited into the account?
   d. How much money will actually be paid out from this account?

10. A person deposits $300 every month, for 45 years, into an account that earns 6% interest compounded monthly.
    a. How much money will be in an account when the person retires?
    b. How much money can be withdrawn every month for 26 years from this account? The account earns 5.8% interest compounded monthly.
    c. How much money was actually deposited into the account?
    d. How much money will actually be paid out from this account?

255

PROJECT: PURCHASING A RESIDENCE
Objective: Calculate the money needed to purchase a residence.

Choose one of the following residences and perform the calculations to purchase it.

 (c) 2006 JupiterImages Corp.	 (c) 2006 JupiterImages Corp.	 (c) 2006 JupiterImages Corp.
Residence A: A family wants to purchase a town house. It sells for $183,000 cash or with a 5% down payment and 1.5 points, the remainder of the balance can be financed at a fixed rate of 7.9% for 30 years. The mortgage application fee is $350, the annual property taxes are $4,824.96, the annual insurance premium is $420 and the PMI is 0.3%.	**Residence B:** A family wants to purchase a condominium. It sells for $349,000 cash or with a 10% down payment and 2.0 points, the remainder of the balance can be financed at a fixed rate of 7.5% for 30 years. The mortgage application fee is $350, the annual property taxes are $8,256.36, the annual insurance premium is $1,100 and the PMI is 0.36%.	**Residence C:** A family wants to purchase a house. It sells for $2,100,000 cash or with a 10% down payment and 2.5 points, the remainder of the balance can be financed at a fixed rate of 7.0% for 30 years. The mortgage application fee is $350, the annual property taxes are $12,248.96, the annual insurance premium is $2,400 and the PMI is 0.49%.

I. Calculate the closing costs for this residence.

ITEMIZATION OF SETTLEMENT CHARGES	
Loan Expenses	
Down Payment	$
Loan Application Fee	$
Points (pts.)	$
Title Search	$ 250.00
Title Insurance	$ 400.00

Recording Fee	$ 50.00
Document Preparation Fee	$ 125.00
Credit Search	$ 50.00
Appraisal Fee	$ 350.00
Survey	$ 275.00
Closing Fee	$ 50.00
Interest (max. 1 month)	$
Taxes (max. 1 month)	$
Insurance	$
PMI	$
Legal (Attorney)	$ 500.00
Total Amount Due on Closing	$

II. Finance the remainder of the mortgage according to the terms described by the residence above. What is the actual monthly payment (P&I and escrow) on this residence given that:
  A. The payments are monthly.
  B. The payments are bi-weekly.
  C. The first ten years are interest only. The remaining twenty years are amortized at the same rate. Find the:
    1. First ten years monthly payment.
    2. Remaining twenty years monthly payment.

III. Find the remaining balance on the mortgage after 8 years and 11 months for :
  A. The monthly payment option.
  B. The bi-weekly payment option.
  C. Interest only option.

IV. Find the new P&I portion of the refinanced mortgage at 6% for 15 years for:
  A. The monthly payment option. The new payments are monthly.
  B. The bi-weekly payment option. The new payments are bi-weekly.
  C. The interest only option. The new payments are monthly. *which Balance do you use?*

V. How much money, if any, would the family save if they refinance?
  A. The monthly payment option.
  B. The bi-weekly payment option.
  C. The interest only option.

PROJECT: PERSONAL RETIREMENT PLANS

Objective: Calculate the money needed to achieve the financial goals of these retirees.

I. Charlotte contributes to a 401k plan at work.
   A. She wants to receive $1,800 a month for 26 years. This money will be in an account that pays 5.2% interest compounded monthly. How much money must be in this account in order for Charlotte to achieve her goal?
   B. Using the result from Part A, how much money must Charlotte deposit monthly into a 5% compounded monthly account to achieve this amount in 40 years?
   C. How much money did Charlotte actually deposit into the account?
   D. How much money will Charlotte actually receive from the account?

II. What if Charlotte waited?
    Using the result from Problem I, Part A, how much money must Charlotte deposit monthly into a 5.2% compounded monthly account to achieve this amount in 20 years?

III. André establishes a retirement account. He deposits $250 every month for 42 years into an account that earns 5.2% interest compounded monthly.
   A. When André retires, how much money is in the account?
   B. Using the result from Part A, how much money can André withdraw every month, for 28 years? The account earns 6.1% interest compounded monthly.
   C. How much money did André actually deposit into the account?
   D. How much money will André actually receive from the account?

IV. What if André waited? André establishes a retirement account. He deposits $250 every month for 21 years into an account that earns 5.2% interest compounded monthly.
   A. When André retires, how much money is in the account?
   B. Using the result from Part A, how much money can André withdraw every month, for 28 years? The account earns 6.1% interest compounded monthly.
   C. How much money did André actually deposit into the account?
   D. How much money will André actually receive from the account?

CHAPTER 3 REVIEW

Section 3.1: Actual Monthly Payment and Refinancing

1. A $284,900 house requires 10% down and the remainder is financed at 7.2% for 30 years. The property taxes are $5,200, the insurance is $490 per year and the private mortgage insurance (PMI) is 0.46%. Find the actual monthly payment.

2. A $125,600 condominium requires 5% down and the remainder is financed at 8.4% for 30 years. The property taxes are $4,200, the insurance is $350 per year and the private mortgage insurance (PMI) is 0.38%. Find the bi-weekly payment.

3. Find the two monthly payments on a $200,600 townhouse that requires 5% down with the remainder of the balance financed at an interest–only fixed rate of 8.4% for 7 years and then the loan is amortized over the remaining 23 years. The property taxes are $4,800, the insurance is $400 per year and the private mortgage insurance (PMI) is 0.42%.

3.2 Strategies to Save on Interest Charges

4. After 12 years, 2 months the homeowner in problem 1 is considering refinancing the mortgage. The P&I portion of the monthly payment is $1,740.48.
   A. Calculate the remaining balance for the home mortgage.
   B. Calculate the new monthly payment using the remaining balance from Part A, at 6.125% for 15 years?
   C. How much money, if any, can be saved by refinancing?

5. After 7 years, 6 months (195 payments), the owner of the condominium in problem 2 wants to refinance the original 30 year, 8.4%, $119,320 mortgage. The P&I portion of the bi-weekly payment is $454.51. The first month's principal is $69.01. It takes 585 payments to repay the original mortgage.
   A. Calculate the remaining balance for the condominium's mortgage.
   B. Calculate the new bi-weekly payment using the remaining balance from Part A, at 7.2% for 15 years?
   C. How much money, if any, can be saved by refinancing?

6. After 9 years, 7 months, the owner of the townhouse in problem 3 is considering refinancing the townhome's mortgage. The P&I portion of the monthly $1,561.75.
   A. Calculate the remaining balance for the townhome's mortgage.
   B. Calculate the new monthly payment using the remaining balance from Part A, at 7% for 15 years?
   C. How much money, if any, can be saved by refinancing?

Section 3.3: Closing Costs

7. Determine the closing costs on the $284,900 house in problem 1. It requires 10% down, 1.25 points, a $300 application fee, and a fixed rate of 7.2% for 30 years. The property taxes are $5,200, the insurance is $490 per year and the private mortgage insurance (PMI) is 0.46%.

ITEMIZATION OF SETTLEMENT CHARGES	
Down Payment	$
Loan Application Fee	$
Points (pts.)	$
Title Search	$ 250.00
Title Insurance	$ 400.00
Recording Fee	$ 50.00
Document Preparation Fee	$ 125.00
Credit Search	$ 50.00
Appraisal Fee	$ 350.00
Survey	$ 275.00
Closing Fee	$ 50.00
Interest (max. 1 month)	$
Taxes (max. 1 month)	$
Insurance	$
PMI	$
Legal (Attorney)	$ 500.00
Total Amount Due on Closing	$

8. Determine the closing costs on the $125,600 condominium in problem 2. It requires 5% down, 0.75 points, a $300 application fee, and a fixed rate of 8.4% for 30 years. The property taxes are $4,200, the insurance is $350 per year and the private mortgage insurance (PMI) is 0.38%.

ITEMIZATION OF SETTLEMENT CHARGES	
Down Payment	$
Loan Application Fee	$
Points (pts.)	$
Title Search	$ 250.00
Title Insurance	$ 400.00
Recording Fee	$ 50.00
Document Preparation Fee	$ 125.00
Credit Search	$ 50.00
Appraisal Fee	$ 350.00
Survey	$ 275.00
Closing Fee	$ 50.00
Interest (max. 1 month)	$

Taxes (max. 1 month)	$
Insurance	$
PMI	
Legal (Attorney)	$ 500.00
Total Amount Due on Closing	$

9. Determine the closing costs on the $200,600 townhouse in problem 3. It requires 5% down with the remainder of the balance financed at an interest–only fixed rate of 8.4% for 7 years and then the loan is amortized over the remaining 23 years. It has zero points and the loan application fee is $350. The property taxes are $4,800, the insurance is $400 per year and the private mortgage insurance (PMI) is 0.42%.

ITEMIZATION OF SETTLEMENT CHARGES	
Down Payment	$
Loan Application Fee	$
Points (pts.)	$
Title Search	$ 250.00
Title Insurance	$ 400.00
Recording Fee	$ 50.00
Document Preparation Fee	$ 125.00
Credit Search	$ 50.00
Appraisal Fee	$ 350.00
Survey	$ 275.00
Closing Fee	$ 50.00
Interest (max. 1 month)	$
PMI	$
Legal (Attorney)	$ 500.00
Total Amount Due on Closing	$

Section 3.3: Annuities that Grow

10. A couple desires $42,780.00 in 2 years for the down payment and closing costs on the house in problem 1. How much money should they deposit monthly into an account that earns 5.75% interest compounded monthly?

11. A couple desires $9,420.00 in 18 months for the down payment and closing costs on the townhouse in problem 2. How much money should they deposit monthly into an account that earns 6.5% interest compounded monthly?

12. A person deposits $450 into an annuity every quarter for 45 years. The account earns 5.7% interest compounded quarterly. What is the future value of this account?

13. A person deposits $2,200 annually for 45 years into an annuity. The account earns 6.3% interest compounded continuously. What is the future value of this account?

Section 3.4: Annuities that Decay
14. A fund is established as the first place prize in a lottery, the winner gets $6,666.67 every month for 25 years (a $2,000,000 prize). How much money must first be in the account that earns 6.4 % interest compounded monthly, in order to meet this need?

15. An electrician wants to start his own business. He figures he needs $4,200 a month for 2 years before his business becomes established. How much money must be in an account that earns 4.3% interest, compounded monthly, before he starts out on his own.

16. As a settlement in a court case, a person is awarded a lump sum of $790,000. How much money can this person withdraw every month for 10 years. The money is deposited into an account that earns 6.2 % interest compounded monthly.

17. A person receives a lump sum lottery prize of $1,240,000. How much money can this person withdraw every month for 20 years. The money is deposited into an account that earns 4.8 % interest compounded monthly.

Section 3.5: Annuities
18. A retiree wants to withdraw $2,200 every month, for 24 years, to supplement their Social Security check.
    a. How much money must be in an account that pays 6.2% interest compounded monthly, in order to meet their needs?
    b. How much money must be deposited into an account that pays 6% interest compounded monthly in order to achieve this amount? The monthly deposits will be made for 37 years.
    c. How much money was actually deposited into the account?
    d. How much money will actually be paid out from this account?

19. A retiree wants to withdraw $3,100 every month, for 24 years, to supplement their Social Security check.
    a. How much money must be in an account that pays 6.2% interest compounded monthly, in order to meet their needs?
    b. How much money must be deposited into an account that pays 6% interest compounded monthly in order to achieve this amount? The monthly deposits will be made for 23 years.

c. How much money was actually deposited into the account?
d. How much money will actually be paid out from this account?

20. A person deposits $222 every month, for 38 years, into an account that earns 5.3% interest compounded monthly.
    a. How much money will be in an account when the person retires?
    b. How much money can be withdrawn every month for 26 years from this account? The account earns 6.7% interest compounded monthly.
    c. How much money was actually deposited into the account?
    d. How much money will actually be paid out from this account?

21. A person deposits $222 every month, for 23 years, into an account that earns 5.3% interest compounded monthly.
    a. How much money will be in an account when the person retires?
    b. How much money can be withdrawn every month for 26 years from this account? The account earns 6.7% interest compounded monthly.
    c. How much money was actually deposited into the account?
    d. How much money will actually be paid out from this account?

Chapter 4: Geometry and Home Renovation

Chapter Overview
In this chapter, the fundamental shapes of triangles, rectangles, and circles will be studied. These figures will be combined to form other shapes. The perimeter, area and in some cases the volume of these figures will be studied. These geometric concepts will be applied to home construction, remodeling or renovation.

4.1 Perimeter and Circumference
Objective: Investigate properties for fundamental shapes: triangles, rectangles and circles. Calculate the perimeter of these shapes.

Perimeter
- **One-dimensional** measurements are made in one direction.
- The distance on the outside of a shape. Analogous to erecting a fence.
- Units: inches, feet, meters, centimeters, etc.

These units express the distance from one end of an object to the other end. Perimeter is the distance required to outline the object. How long is the rope? How tall is that building? How far is it from Chicago to New York?

Perimeter measures the distance around the shape. The amount of fencing needed to enclose a yard or the distance covered by a jogger who runs around a golf course are perimeter problems.

Perimeter and Rectangles

The rectangle is a four-sided figure comprised of four line segments (sides) and four right angles. The opposite sides of a rectangle are the same length. Consider the following rectangle ABCD. Only two sides are labeled because the opposite sides of a rectangle are the same length. Thus, $m\overline{AD} = m\overline{BC}$ measures 5 feet. Similarly $m\overline{AB} = m\overline{CD}$ measures 12 ft. To find the perimeter of a rectangle, calculate the sum of the four sides of the figure.

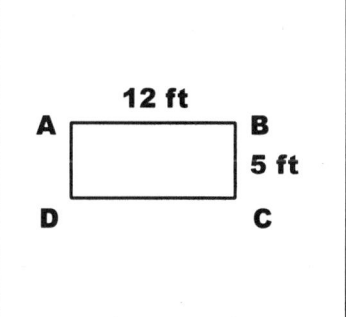

Example 1: Find the perimeter of the following rectangle ABCD.

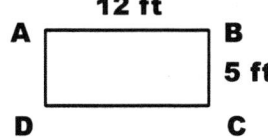

264

Solution:
Step 1: Understand and Picture the Problem.
    1. Question: Find the perimeter of the rectangle.
    2. Copy the figure and list the dimensions.

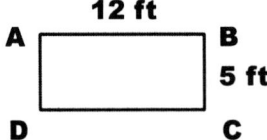

Step 2: Develop a Plan.
    1. Use the property that the opposite sides of a rectangle are the same length to determine the unlabeled sides.
    2. Find the sum of the four sides.

Step 3: Execute the plan.
    1. Side $m\overline{AB} = 12$ ft so its opposite side $m\overline{CD} = 12$ ft.
       Side $m\overline{AD} = 5$ ft so its opposite side $m\overline{BC} = 5$ ft.
    2. 12 ft + 5 ft + 12 ft + 5 ft = 34 ft. The perimeter is 34 feet.

Step 4: Check your work.
    1. Check the numbers and your calculations.
    2. Is the answer reasonable?
    3. Was the question answered?

Squares

A special case of the rectangle is the square. All sides of a square are the same length. Consider the following rectangle, MATH, which is a square.

Only one side needs to be labeled because all sides of a square are the same length. Thus, $m\overline{MA} = m\overline{AT} = m\overline{TH} = m\overline{HM}$ and each side measures 6 meters.

Example 2: Find the perimeter of the following square MATH

Solution:
Step 1: Understand and Picture the Problem.
    1. Question: Find the perimeter of the square.
    2. Copy the figure and list the dimensions.

Step 2: Develop a Plan.
    1. Use the property that all sides of a square are the same length to determine the unlabeled sides.
    2. Find the sum of the four sides.

Step 3: Execute the plan.
    1. Sides, $m\overline{MA} = 6$ m, $m\overline{AT} = 6$ m, $m\overline{TH} = 6$ m and $m\overline{HM} = 6$ m.
    2. 6 m + 6 m + 6 m + 6 m = 24 m. The perimeter is 24 meters.

Step 4: Check your work.
    1. Check the numbers and your calculations.
    2. Is the answer reasonable?
    3. Was the question answered?

DIMENSIONAL ANALYSIS

Dimensional analysis uses definitions, (such as, 1 foot = 12 inches), as "factors of one" to convert the given units into the desired units. Dimensional analysis is the process of multiplying by 1 (the product remains the same) to convert the units of a variable into the desired units. For example, the perimeter of the rectangle in Example 1 is 34 feet. This measurement can be converted to inches. The following steps demonstrate the conversion of feet to inches.
1. 12 inches = 1 foot
2. This definition can be written as $\dfrac{12 \text{ inches}}{1 \text{ foot}}$ or $\dfrac{1 \text{ foot}}{12 \text{ inches}}$.
3. Since the objective is to convert feet to inches, the term inches must be in the numerator (upper portion of the fraction). Use $\dfrac{12 \text{ inches}}{1 \text{ foot}}$.
4. Multiplication produces, $\dfrac{34 \text{ feet}}{1} \cdot \dfrac{12 \text{ inches}}{1 \text{ foot}} = \dfrac{34 \cancel{\text{ feet}}}{1} \cdot \dfrac{12 \text{ inches}}{1 \cancel{\text{ foot}}} = 408 \text{ inches}$.

Note: The fraction $\dfrac{\text{feet}}{\text{foot}}$ is a factor of 1 and can be removed from the problem. Multiplication by 1 does not change the value of the product.

Example 3: Given 3.28 feet is one meter, convert the perimeter from Example 1 to meters.
Solution:
  Step 1: Understand and Picture the Problem.
    Question: Rewrite the number of feet as the number of meters required to enclose the shape.

  Step 2: Develop a Plan.
    Use dimensional analysis. Since distance is one dimensional, **one factor** of $\frac{1 \text{ meter}}{3.28 \text{ feet}}$, is required to convert feet into meters.
    Note: "meters", the desired unit, must be in the numerator.

  Step 3: Execute the plan.
    $$P = \frac{34 \text{ feet}}{1} \cdot \frac{1 \text{ meter}}{3.28 \text{ feet}} = \frac{34 \; \cancel{\text{feet}}}{1} \cdot \frac{1 \text{ meter}}{3.28 \; \cancel{\text{feet}}} = \frac{34 \text{ meters}}{3.28} \approx 10.37 \text{ meters}.$$
    Note: A fraction of a meter of fencing cannot be purchased. 11 meters of fencing would be required to enclose the figure. Any fraction is bumped up to the next whole number.

  Step 4: Check your work.
    1. Check the numbers and calculations.
    2. Is the answer reasonable?
    3. Was the question answered?

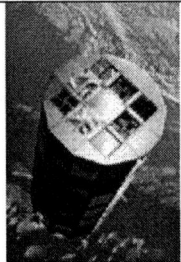
(c) 2006 JupiterImages Corp.

Failure to convert from "English" units to "Metric" units can have disastrous consequences. Peruse the article.

Metric Mishap Caused Loss of NASA Orbitor
(CNN) -- NASA lost a $125 million Mars orbiter because a Lockheed Martin engineering team used English units of measurement while the agency's team used the more conventional metric system for a key spacecraft operation.

The units mismatch prevented navigation information from transferring between the Mars Climate Orbiter spacecraft team at Lockheed Martin and the flight team at NASA's Jet Propulsion Laboratory.

The navigation mishap killed the mission on a day when engineers had expected to celebrate the craft's entry into Mars' orbit. After a 286-day journey, the probe fired its engine on September 23 to push itself into orbit.

The engine fired but the spacecraft came about 100 km closer than planned and beneath the level at which it could function properly. The spacecraft's propulsion system overheated and was disabled.

That probably stopped the engine from completing its burn, so the Climate Orbiter likely plowed through the atmosphere, continued out beyond Mars and now could be orbiting the sun, he said.

Perimeter and Triangles
A triangle is a three-sided figure comprised of three line segments and three vertices. The vertices are the endpoints of the line segments, where the line segments intersect. In the following triangle, the line segments are $\overline{AB}$, $\overline{BC}$ and $\overline{AC}$. The vertices are the endpoint of the line segments, namely A, B and C. To find the perimeter of a triangle, add the lengths of the three sides of the figure in the specified units of measurement.

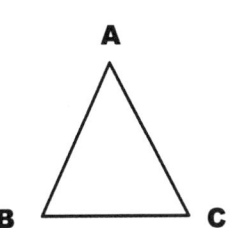

Example 4: Find the perimeter of the following triangle.

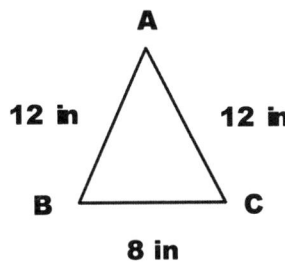

Solution:
  Step 1: Understand and Picture the Problem.
    1. Question: Find the perimeter, the distance around the triangle.
    2. Copy the figure and list the dimensions.

  Step 2: Develop a Plan.
    Add the lengths of the three sides to find the perimeter.

  Step 3: Execute the plan.
    P = 12 ft + 12 ft + 8 ft = 32 ft.

Step 4: Check your work.
1. Check the numbers and your calculations.
2. Is the answer reasonable?
3. Was the question answered?

In most perimeter problems the lengths of the three sides must be given to find the perimeter. However, there is one type of triangle that only requires knowing two sides. The length of the third side of that triangle can be calculated by a procedure named the **Pythagorean Theorem**. This special triangle is called a right triangle.

Right Triangles

In a right triangle, the reference point is the right angle (90°). A right angle is the union of two perpendicular line segments. For example a horizontal line segment and a vertical line segment meet at a common endpoint (vertex) to form a right angle. The following figures are examples of right angles. Notice that the right angle can be rotated.

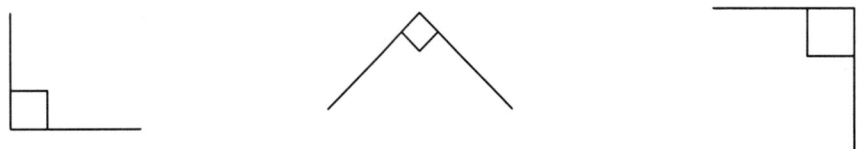

The square in the corner, ⌐, inserted at the vertex where the two line segments intersect, denotes the right angle (90°). Only one of the triangle's angles can be a right angle. The right angle is located at vertex C, formed by the horizontal line segment $\overline{BC}$ and the vertical line segment $\overline{AC}$

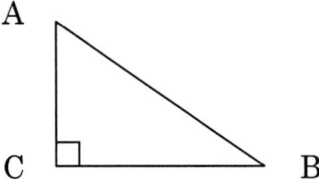

Other examples of right triangles are △ DEF, △ FUN and △ GEM.

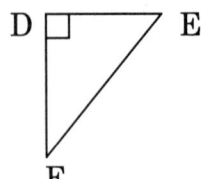  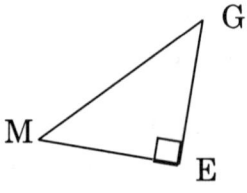

Pythagorean Theorem
The Pythagorean Theorem was discovered by an ancient Greek named Pythagorus around 500 BC. He found the relationship between the lengths of the sides of a right triangle. A right triangle is a triangle that has one right angle (90 degrees).

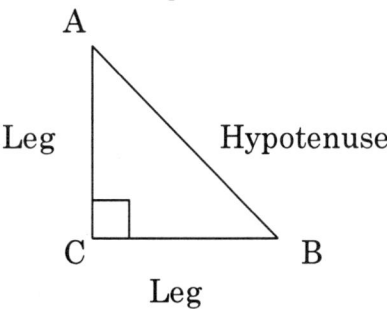

The two sides that make up the right angle $\left(\overline{AC} \text{ and } \overline{BC}\right)$ are called "legs." The third side $\left(\overline{AB}\right)$ is always the longest side and is named the "hypotenuse". The Pythagorean Theorem states

> **Pythagorean Theorem (units)**
> $$a^2 + b^2 = c^2$$
> where: a is the length of one leg.
> b is the length of the other leg.
> c is the length of the hypotenuse.

Given the length of any two sides of a right triangle, the Pythagorean Theorem allows you to calculate the length of the unknown side. To find the missing side, use this translation of the Pythagorean Theorem.

> **Pythagorean Theorem**
> 1. Square the lengths of the two known sides. Write the larger value first.
> 2. Make a decision. If you are solving for the
>   a. Hypotenuse, then **ADD**.
>   b. Leg, then **SUBTRACT**.
> 3. Square root the result.

Use the Pythagorean Theorem to calculate the length of the third side of any right triangle.

Example 5: Find the perimeter of the right triangle

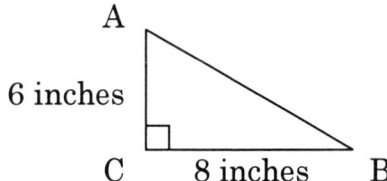

269

270

Solution:
Step 1: Understand and Picture the Problem.
    1. Question: Find the perimeter of the triangle
    2. Copy the figure and list the dimensions.
    3. Identify the triangle as a right triangle.

Step 2: Develop a Plan.
    1. Identify the known sides as either legs or hypotenuse.
    2. Use the Pythagorean Theorem to solve for the length of the third side.
    3. Add the three values to find the perimeter.

Step 3: Execute the plan.
    1. Side $\overline{AC}$ is a leg and measures 6 inches.
       Side $\overline{BC}$ is a leg and measures 8 inches.

2. Step	Action
Square the lengths of the two sides that you know. Write the larger value first.	$8^2 = 64,\quad 6^2 = 36$ 64, 36
Since you are solving for the **HYPOTENUSE**, **ADD** the numbers.	$64 + 36 = 100$
Square root the result.	$\sqrt{100} = 10$

    The length of the third side (hypotenuse) is 10 inches.
    3. Perimeter = 6 in + 8 in + 10 in = 24 in.

Step 4: Check your work.
    1. Check the numbers and your calculations.
    2. Is the answer reasonable?
    3. Was the question answered?

Example 6: Find the perimeter of the right triangle

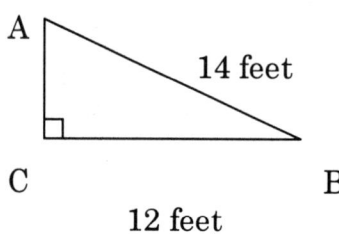

271

Solution:
Step 1: Understand and Picture the Problem.
　　1. Question: Find the perimeter of the triangle.
　　2. Copy the figure and list the dimensions.
　　3. Identify the triangle as a right triangle.

Step 2: Develop a Plan.
　　1. Identify the known sides as either legs or hypotenuse.
　　2. Use the Pythagorean Theorem to find the length of the third side.
　　3. Add the lengths of the three sides to find the perimeter.

Step 3: Execute the plan.
　　1. Side $\overline{AB}$ is the hypotenuse and measures 14 ft.
　　　Side $\overline{BC}$ is a leg and measures 12 ft.

2.

Step	Action
Square the lengths of the two sides that you know. Write the larger value first.	$14^2 = 196$,　$12^2 = 144$ 196,　144
Since you are solving for a **LEG**, **SUBTRACT**.	$196 - 144 = 52$
Square root the result.	$\sqrt{52} \approx 7.2$

　　　The length of the third side (leg) is approximately 7.2 ft.
　　　Note: The third side of a right triangle is not always a perfect square. It is possible to have a solution that is not a whole number.
　　3. Perimeter = 12 ft + 14 ft + 7.2 ft = 33.2 ft.
　　　Note: If fencing was being purchased to enclose this triangular area, then 34 feet of fencing would be required. A fraction of a foot of fencing cannot be purchased, so any fraction is bumped up to the next whole number.

Step 4: Check your work.
　　1. Check the numbers and your calculations.
　　2. Is the answer reasonable?
　　3. Was the question answered?

Example 7: Find the perimeter of the right triangle.

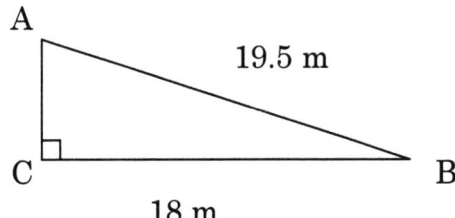

Solution:
Step 1: Understand and Picture the Problem.
1. Question: Find the perimeter of the triangle.
2. Copy the figure and list the dimensions.
3. Identify the triangle as a right triangle.

Step 2: Develop a Plan.
1. Identify the known sides as either legs or hypotenuse.
2. Use the Pythagorean Theorem to find the length of the third side.
3. Add the three sides to find the perimeter.

Step 3: Execute the plan.
1. Side $\overline{AB}$ is the hypotenuse and measures 19.5 meters.
   Side $\overline{BC}$ is a leg and measures 18 meters.

2. Step	Action
Square the lengths of the two sides that you know. Write the larger value first.	$19.5^2 = 380.25$, $18^2 = 324$ 380.25, 324
Since you are solving for a **LEG**, **SUBTRACT**.	$380.25 - 324 = 56.25$
Square root the result.	$\sqrt{56.25} = 7.5$

The length of the third side (leg) is 7.5 meters.
3. Perimeter = 19.5 m + 18 m + 7.5 m = 45 meters.

Step 4: Check your work.
1. Check the numbers and your calculations,
2. Is the answer reasonable?
3. Was the question answered?

Example 8: Use the definition, 1 meter (m) equals 100 centimeters (cm), to convert the perimeter from Example 7 to centimeters.

Solution:
Step 1: Understand and Picture the Problem.
Question: Rewrite the number of meters as the number of centimeters required to enclose the outside of the shape.

Step 2: Develop a Plan.
Use dimensional analysis. Since distance is one dimensional, **one factor** of $\dfrac{100 \text{ cm}}{1 \text{ meter}}$, is required to convert meters to centimeters.

Note: "centimeters", the desired unit, must be in the numerator.

Step 3: Execute the plan.
$$P = \frac{7.5 \text{ meters}}{1} \cdot \frac{100 \text{ centimeters}}{1 \text{ meter}} = \frac{7.5 \; \cancel{\text{meters}}}{1} \cdot \frac{100 \text{ centimeters}}{1 \; \cancel{\text{meter}}} = 750 \text{ cm}.$$

Step 4: Check your work.
    1. Check the numbers and calculations.
    2. Is the answer reasonable?
    3. Was the question answered?

A Right Triangular Building

This is a triangular building in Chicago, across the street from the Lincoln Park Zoo. It is bordered by three streets, Lincoln Park West, Clark and Dickens. You are looking down Clark Street in this picture. The footprint (foundation) of this building is a right triangle. On the corner of Dickens and Lincoln Park West is the R.J. Grunt's restaurant. It is the first restaurant in the "Lettuce Entertain You" family of restaurants.

(c) 2006 JupiterImages Corp.

Historical Perspective
Pythagoras of Samos (about 580 – 500 BC)
Greece

Pythagoras was not the first person to discover the 3-4-5 right triangle. The Chinese knew this theorem. It is attributed to Tschou-Gun who lived in 1100 BC. The theorem was also known to the Babylonians more than a thousand years before Pythagoras. Pythagoras and his wife Theana, also a philosopher-mathematician, founded a school devoted to the study of mathematics. This led to the proof (geometric demonstration) of the theorem that bears his name:

The Pythagorean Theorem: $a^2 + b^2 = c^2$.

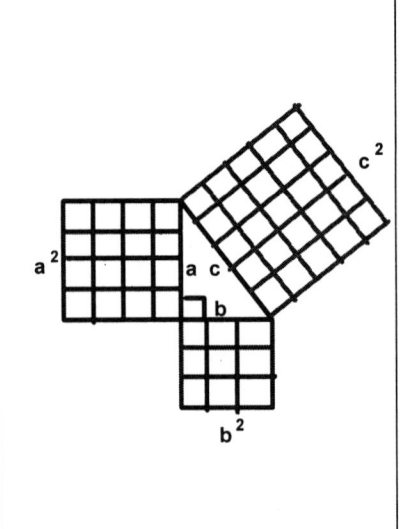

View the figure on the left. The legs of the right triangle are a = 4, b = 3, and the hypotenuse c = 5.
1. Squaring a = 4 x 4 = 16. The geometric representation of squaring 4 is a square with 4 rows and 4 columns producing 16 squares, with each square having a side length of 1 unit.
2. Squaring b = 3 x 3 = 9. The geometric representation of squaring 3 is a square with 3 rows and 3 columns producing 9 squares, with each square having a side length of 1 unit.
3. Summing the unit squares produces 25 squares (16 + 9 = 25). This equals the square produced by squaring the hypotenuse c = 5. Squaring 5 produces a larger square with 25 squares, with each square having a side length of 1 unit.

For triangles, rectangles and squares, the perimeter is found by adding the lengths of all the sides. How do you find the perimeter of a circle?

Circumference and Circles

Imagine a 20 foot diameter circular swimming pool in your yard. The diameter of a circle is a line segment that has its endpoints on the circle and passes through the center $\left(\overline{AB}\right)$.

First prepare the ground for the pool. A stake is driven into the ground (C, center of the circle). Attach the stake to a 10 a feet string ($\overline{CD}$, radius). At the other end of the string attach a can of spray paint to mark the ground, the points of the circle. With the string pulled taut, spray around the stake. A 20 foot circle is marked on the ground. The circumference of the circle is the sprayed portion of the circle on the ground.

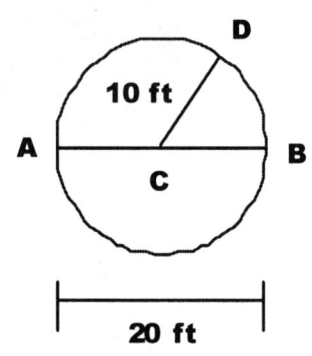

Definition: A circle is the set of all points that are a fixed distance (radius) from a fixed center point.

Vocabulary for a Circle
Center: The fixed point in the center of the circle.
Radius: The distance from the center to any point on the circle.
Diameter: The distance from one point on the circle to the another point on the circle going through the center. A diameter equals two radii.
Circumference: Distance around the outside of the circle.

Since a circle is not constructed of straight line segments, its circumference (perimeter) is not found by adding the lengths of its sides. The circumference of a circle is found by a formula. This formula is based on the constant named *pi* (π.).

(c) 2006 JupiterImages Corp.

Explore your calculator. Press the appropriate keys to display the value of π (pi).

The formula for circumference is

Circumference of a Circle
$C = \pi d = 2\pi r$
Where: C is the Circumference.
d is the diameter.
r is the radius.
π is a constant. (Use the π key on your calculator.)

Example 9: Find the circumference of circle with diameter 16 feet. Round your answer to tenths.

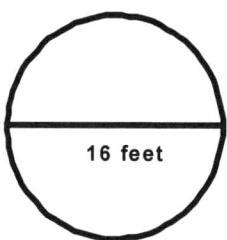
16 feet

Solution:
Step 1: Understand and Picture the Problem.
1. Question: Find the distance around a circle (circumference.)
2. Copy the figure.
3. The diameter of the circle is 16 feet.

Step 2: Develop a Plan.
Use the circumference formula, $C = \pi d$.

Step 3: Execute the plan.

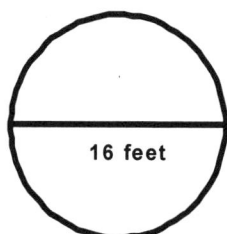
16 feet

Substitute the values into the formula. Use the π key on your calculator.
C = π(16 feet) ≈ 50.26548246 feet ≈ 50.3 feet (rounded to tenths.)

Step 4: Check your work.
    1. Check the numbers and your calculations.
    2. Is the answer reasonable?
    3. Was the question answered?

Example 10: Find the circumference of circle with radius 8 feet. Round to tenths.

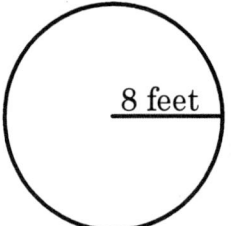

Solution:
Step 1: Understand and Picture the Problem.
    1. Question: Find the distance called the circumference.
    2. Copy the figure.
    3. The radius of the circle is 8 feet.

Step 2: Develop a Plan.
    Use the circumference formula, C = 2πr.

Step 3: Execute the plan.

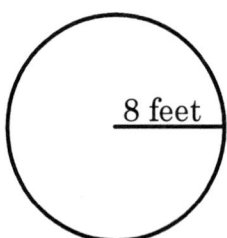

Substitute the values into the formula. Use the π key on your calculator.
C = 2π(8 feet) ≈ 50.26548246 feet ≈ 50.3 feet.
(The circumference is rounded to tenths.)

Step 4: Check your work.
    1. Check the numbers and your calculations.
    2. Is the answer reasonable?
    3. Was the question answered?

277

Sectors, Partial Circles
It is possible to work with sectors, partial circles. To find the length of a sector, one needs to know the following property of a circle.

Degrees
There are 360 degrees in a full circle.

 (c) 2006 JupiterImages Corp.	**Historical Perspective** **The Sumerians (2900 – 1800 BC)** **What is now Saudi Arabia.**
The Sumerians developed a number system based on the unit 60. They divided the hour into 60 minutes and the circle into 360 degrees, as we still do today. They also developed basic algebra and geometry.	

Example 11: Find the perimeter (rounded to tenths) of the following sector.

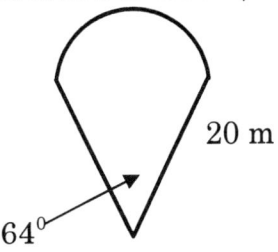

Solution:

Step 1: Understand and Picture the Problem.
    1. Question: Find the distance around the sector.
    2. Copy the figure.
    3. The sector has a radius of 20 meters and an angle of 64 degrees.

Step 2: Develop a Plan.
    1. Use the circumference formula $C = 2\pi r$.
    2. Find the fractional portion of the circumference corresponding to the sector.
    3. Add the lengths of the two radii to the result of part 2.

Step 3: Execute the plan.

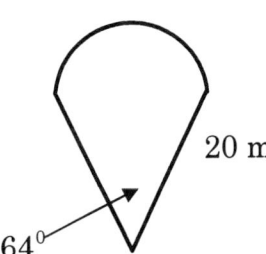

    1. Substitute the appropriate values into the formula, $C = 2\pi(20 \text{ feet})$
    2. Use the π key on your calculator.

278

    3. Out of 360 degrees, the sector is 64 degrees. The sector is $\dfrac{64}{360}$ of a circle.

       The sector (arc) is $\dfrac{64}{360} 2\pi (20 \text{ meters}) \approx 22.340214$ meters.

    4. 22.34021443 m + 20 m + 20 m ≈ 62.340214 meters.
       The perimeter of the sector is approximately 62.3 meters.

Step 4: Check your work.
    1. Check the numbers and your calculations.
    2. Is the answer reasonable?
    3. Was the question answered?

Example 12: Find the perimeter (rounded to tenths) of the following sector.

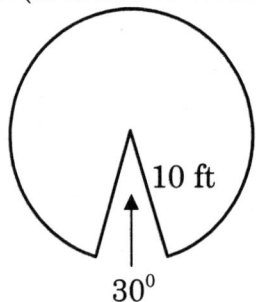

Solution:
Step 1: Understand and Picture the Problem.
    1. Question: Find the distance around the sector.
    2. Copy the figure.
    3. The radius of the sector is 10 feet.
    4. The angle outside the sector measures 30 degrees.

Step 2: Develop a Plan.
    1. Use the circumference formula C = 2πr.
    2. Find the fractional portion of the circumference matching the sector.
    3. Add the lengths of the two radii to the result of part 2.

Step 3: Execute the plan.

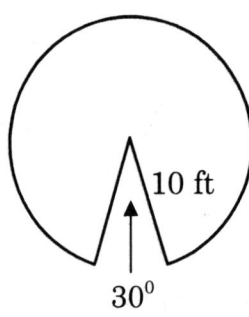

1. Substitute the appropriate values into the formula, C = 2π(10 feet).
2. Use the π key on your calculator.
3. Out of 360 degrees, the sector is 330 degrees. The sector is $\frac{330}{360}$ of a circle. Thus the arc is $\frac{330}{360}$ 2π(10 feet) ≈ 57.59586532 feet.
4. 57.59586532 ft + 10 ft + 10 ft ≈ 77.59586532 feet. The perimeter is 77.6 feet.

Step 4: Check your work.
1. Check the numbers and your calculations.
2. Is the answer reasonable?
3. Was the question answered?

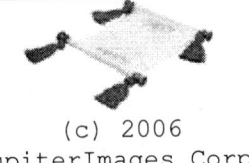 (c) 2006 JupiterImages Corp.	**Historical Perspective** **Archimedes of Syracuse (287 – 212 BC) Greece** **Aryabhata I (476 - 550) India**

The first person to calculate pi with reasonable accuracy was Archimedes, around 250 BC. He considered regular polygons inscribed in and circumscribed around a circle. Since the area of the circle is between the areas of the inscribed and circumscribed polygons, he estimated the value of pi between $3\frac{10}{71}$ and $3\frac{1}{7}$.
Aryabhata gave a surprisingly accurate approximation for π. He wrote in the *Aryabhatiya*: *Add four to one hundred, multiply by eight and then add sixty-two thousand. The result is approximately the circumference of a circle of diameter twenty thousand.* Performing the math: 4 + 100 = 104, 8(104) = 832, 832 + 62,000 = 62,832. Next, Circumference = π(d)   62,832 = π(20,000) Solving for π produces: $\pi = \frac{62,832}{20,000} = 3.1416$, correct to 4 places.

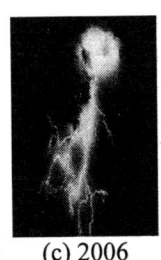 (c) 2006 JupiterImages Corp.	**Light Year** A light year is the distance light travels in a year. The speed of light is approximately $\frac{186,300 \text{ miles}}{\text{second}}$. Using dimensional analysis the units of time (miles per second) are converted to (miles per year). $\frac{186,300 \text{ mi}}{\text{sec}}\left(\frac{60 \text{ sec}}{\text{min}}\right)\left(\frac{60 \text{ min}}{\text{hr}}\right)\left(\frac{24 \text{ hr}}{\text{day}}\right)\left(\frac{365.25 \text{ day}}{\text{year}}\right) \approx 5.87918088 \times 10^{12}$ $5.87918088 \times 10^{12} \approx 5,879,180,880,000$ miles per year.

|  (c) 2006 JupiterImages Corp. | **Optional Student Project**

Project: "National Debt" can be found on page 338. |

   Cognitive Problems

1. Given the lengths of two sides of a triangle, can the third side be found of any triangle? Explain.
2. In a right triangle, what is the difference between a leg and hypotenuse?
3. Can a triangle have more than one right angle? Try drawing one.
4. Is a square a rectangle? Is a rectangle a square? Explain.
5. What is the difference between the radius and diameter of a circle?
6. Are the units of measurement necessary when computing the perimeter? Explain.
7. Explore the value of π (pi). Measure the circumference and diameters of several circles and compare the ratio of $\dfrac{\text{Circumference}}{\text{Diameter}}$.

   **Exercise 4.1**

(c) 2006 JupiterImages Corp.

I. Find the perimeter (circumference) for the following shapes. Round your answers to tenths.

1.   2.   3.   4.

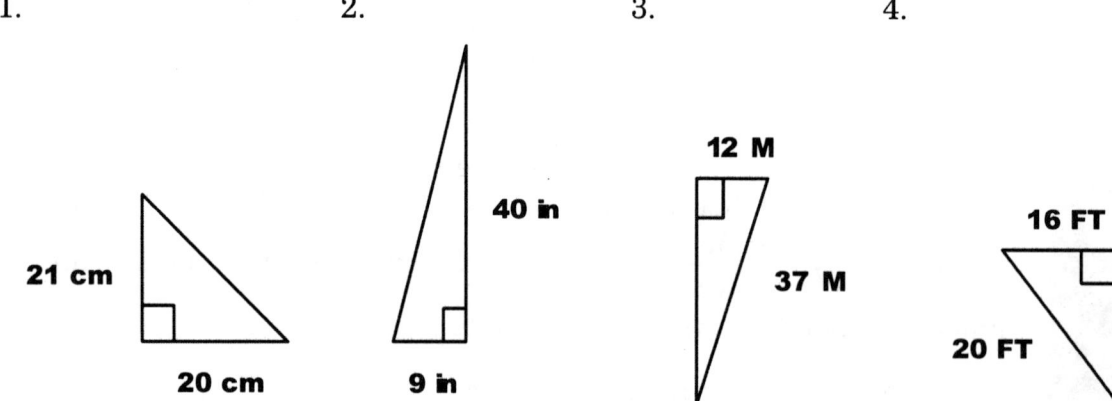

5.

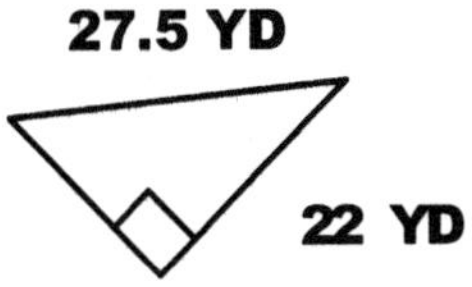

27.5 YD
22 YD

6.

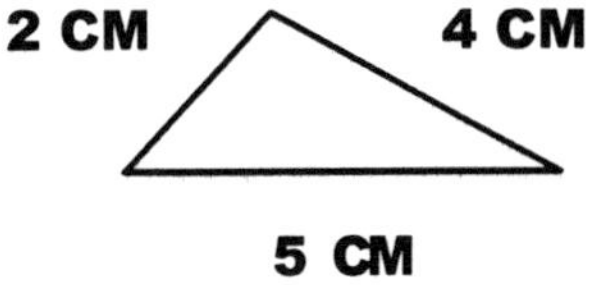

2 CM   4 CM
5 CM

7.

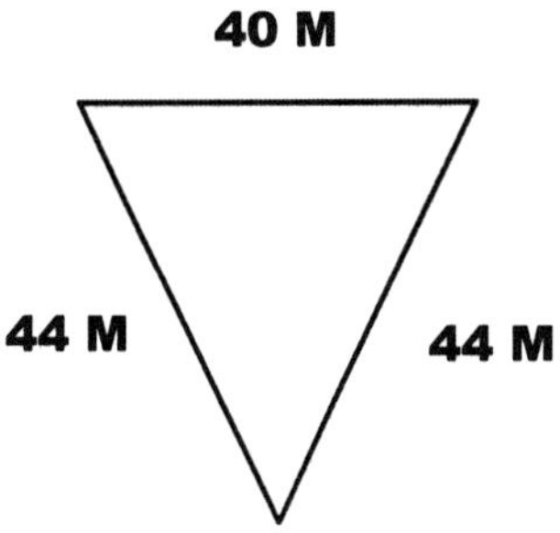

40 M
44 M   44 M

8.

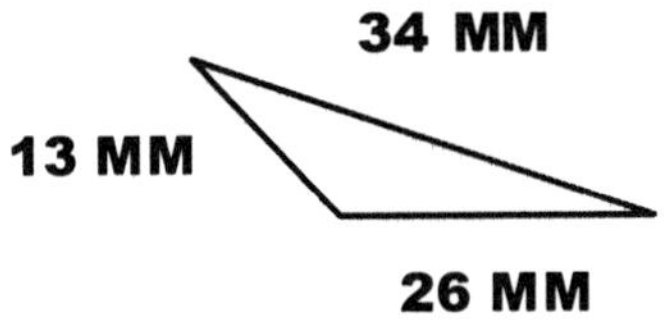

34 MM
13 MM
26 MM

9.

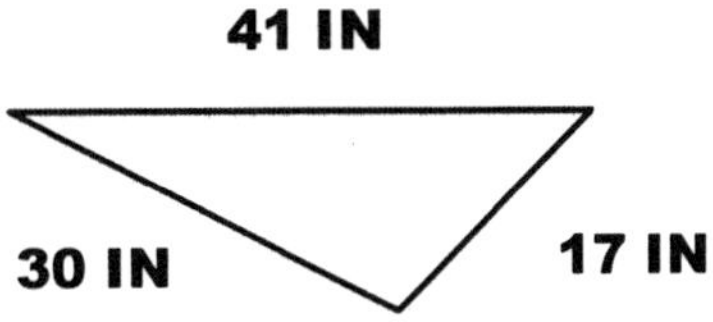

41 IN
30 IN   17 IN

10.

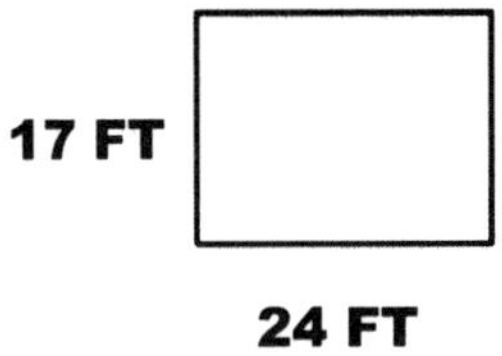

17 FT
24 FT

11.

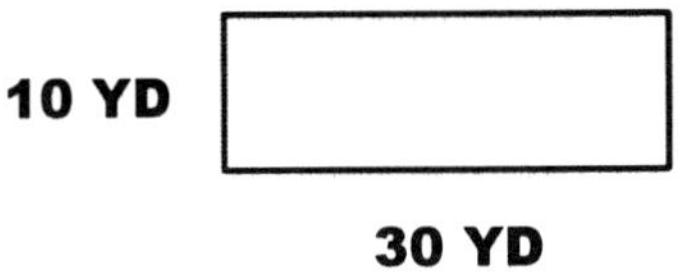

10 YD
30 YD

12.

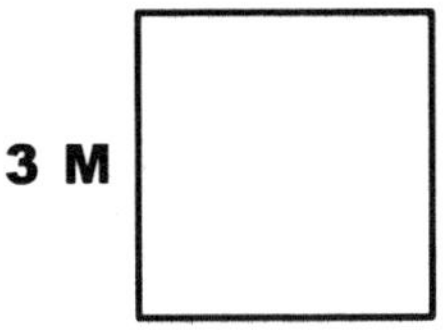

3 M

13.

48 MM

14.

32 FT

15.

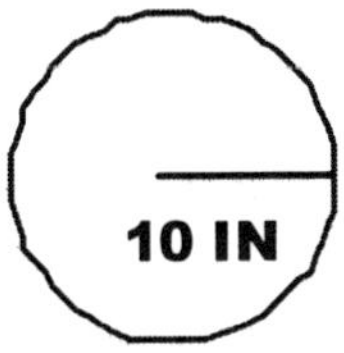

10 IN

16.

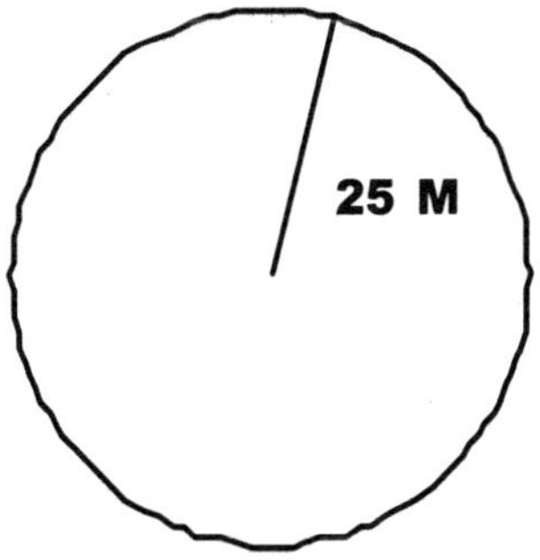

25 M

282

17.   18.   19.   20.

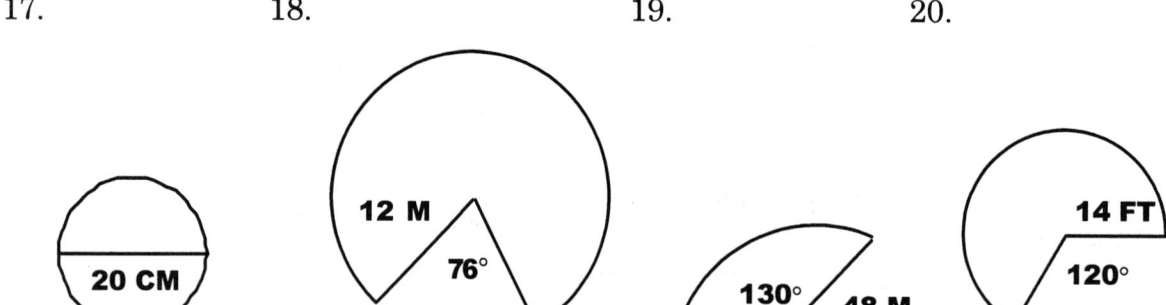

II. Use Dimensional Analysis to convert the perimeters from Problem I to the specified dimensions. Round to thousandths.

Dimensions
1,000 millimeters (mm) = 1 meter
100 centimeters (cm) = 1 meter
3.28 ft ≈ 1 meter
1.09 yd ≈ 1 meter
3 feet (ft) = 1 yard (yd)
36 inches (in) = 1 yard (yd)
12 inches (in) = 1 foot (ft)
1 inch ≈ 25.4 mm
1 inch ≈ 2.54 cm

1. Convert the perimeter from Exercise 1 to meters.
2. Convert the perimeter from Exercise 2 to feet.
3. Convert the perimeter from Exercise 3 to centimeters.
4. Convert the perimeter from Exercise 4 to yards.
5. Convert the perimeter from Exercise 5 to feet.
6. Convert the perimeter from Exercise 6 to meters.
7. Convert the perimeter from Exercise 7 to centimeters.
8. Convert the perimeter from Exercise 8 to meters.
9. Convert the perimeter from Exercise 9 to yards.
10. Convert the perimeter from Exercise 10 to yards.
11. Convert the perimeter from Exercise 11 to feet.
12. Convert the perimeter from Exercise 12 to centimeters.
13. Convert the perimeter from Exercise 13 to inches.
14. Convert the perimeter from Exercise 14 to inches.
15. Convert the perimeter from Exercise 15 to yards.
16. Convert the perimeter from Exercise 16 to centimeters.
17. Convert the perimeter from Exercise 17 to meters.
18. Convert the perimeter from Exercise 18 to centimeters.
19. Convert the perimeter from Exercise 19 to feet.
20. Convert the perimeter from Exercise 20 to yards.

4.2 AREA
Objective: Calculate the area of rectangles, triangles and circles.

Perimeter
- **One-dimensional** measurements are made in **one** direction.
- The distance around a shape. Analogous to erecting a fence.
- Units: inches, feet, meters, centimeters, etc.

Area
- **Two-dimensional** measurements are made in **two** directions.
- Cover the inside of a shape. Analogous to laying tile on a floor.
- Units: square inches, ft^2, square meters, cm^2, etc.

A square inch is a square whose sides are each one inch. A square foot is a square whose sides are each one foot. A square yard is a square whose sides are each one yard.

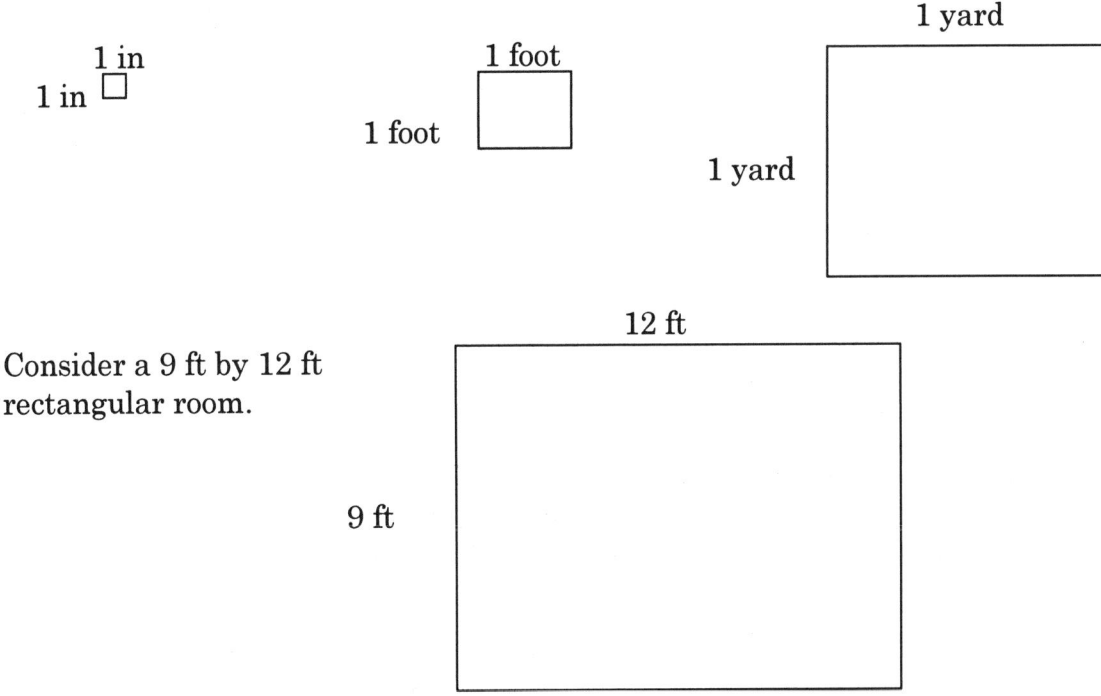

Consider a 9 ft by 12 ft rectangular room.

Since the room's dimensions are given in feet, the area will be square feet. Area is the number of squares required to cover the floor of this room. The two dimensional aspect of area means that not only will the length (one dimension) of the room be covered, but also the width (the second dimension) of the room will be covered. Using square feet, a square unit measures one foot on each side.

It takes 12 tiles to cover one row in this room.

12 ft

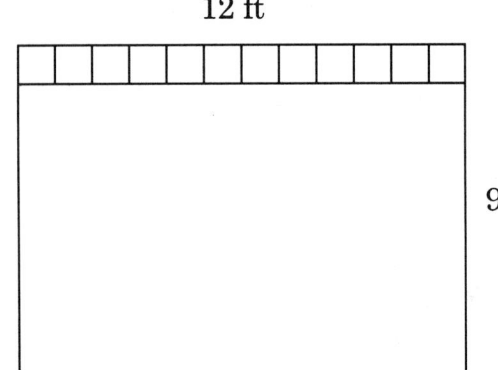

9 ft

It takes 24 tiles (2 rows of 12 tiles) to cover two rows in this room.

12 ft

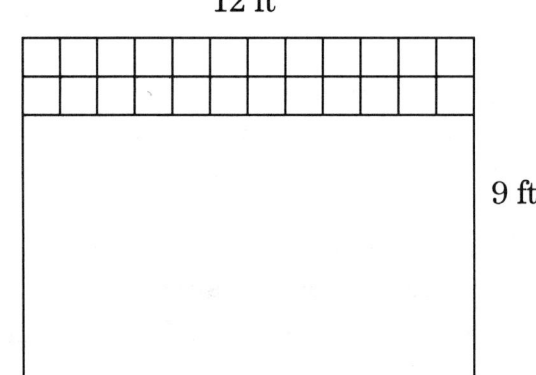

9 ft

Continuing to add rows of tile, it takes 108 tiles (9 rows of 12 tiles) to completely cover the inside of this room. Therefore, the area of the room is 108 square feet.

12 ft

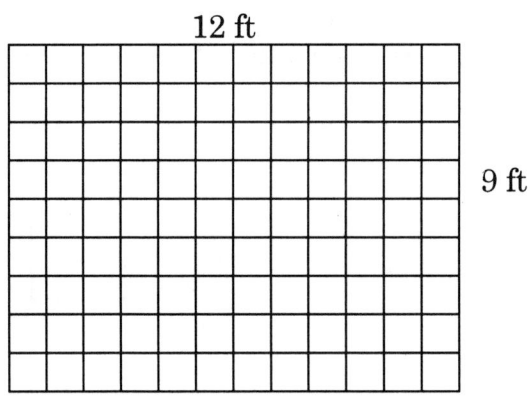

9 ft

The area of a rectangle is represented by the following formula:

Area of a rectangular region (square units)
Area = (length)(width)

The area of the 9 ft by 12 ft rectangular room is (9 ft)(12 ft) = 108 sq ft. Consider the same 9 ft by 12 ft rectangular room but this time use square yards (carpeting). The square in this situation has side lengths of 1 yard. Since the room's dimensions are given in feet, they need to be converted to yards by using the definition of a yard; 1 yard equals 3 feet. This is written as $\dfrac{1 \text{ yard}}{3 \text{ feet}}$ or $\dfrac{3 \text{ feet}}{1 \text{ yard}}$ depending on whether the conversion is from feet to yards or from yards to feet.

A. To convert 12 feet to yards: $(12 \text{ feet})\left(\dfrac{1 \text{ yard}}{3 \text{ feet}}\right) = \left(\cancel{12} \cancel{\text{ feet}}\right)\left(\dfrac{1}{\cancel{3}} \dfrac{\text{yard}}{\cancel{\text{feet}}}\right) = 4 \text{ yards}$.

(with 12 reduced to 4 and 3 reduced to 1)

Note: The sought for unit of measure need to be in the numerator.

B. To convert 4 yards to feet: $(4 \text{ yards})\left(\dfrac{3 \text{ feet}}{1 \text{ yard}}\right) = (4 \cancel{\text{ yards}})\left(\dfrac{3 \text{ feet}}{1 \cancel{\text{ yard}}}\right) = 12$ feet.

Note: The sought for unit of measure need to be in the numerator.

The calculation of area in square yards requires finding the number of squares, one yard on a side, necessary to cover the inside of the figure. Notice that **both** measurements (length and width) must be converted.

Since 12 feet equals 4 yards, it takes 4 tiles cover one row in this room.

Since 9 feet equals 3 yards, it takes 3 rows of 4 tiles or 12 tiles to cover the inside of the room. The area is 12 square yards.

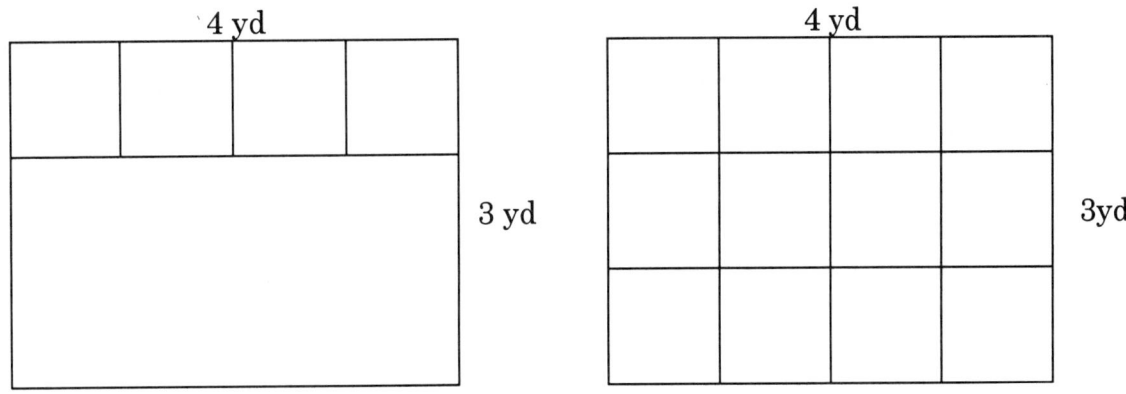

Dimensional Analysis is an alternate approach to calculating the square yards needed to cover the inside of this room.

DIMENSIONAL ANALYSIS

Dimensional analysis uses definitions, (such as, 1 yard = 3 feet), and "factors of one" to convert the given units into the desired units. The area of the room is 108 square feet. This area can be converted to square yards. The following steps demonstrate the conversion of 108 square feet into square yards. (Note: square feet equals ft^2, which equals *feet **times** feet*.)

108 square feet = 108 ft^2 = 108 (feet)(feet).

$$\dfrac{108(\text{feet})(\text{feet})}{1} \cdot \dfrac{1 \text{ yard}}{3 \text{ feet}} \cdot \dfrac{1 \text{ yard}}{3 \text{ feet}} =$$

$$\dfrac{108(\cancel{\text{feet}})(\cancel{\text{feet}})}{1} \cdot \dfrac{1 \text{ yard}}{3 \cancel{\text{ feet}}} \cdot \dfrac{1 \text{ yard}}{3 \cancel{\text{ feet}}} = \dfrac{108(\text{yard})(\text{yard})}{9} = 12 \text{ square yards}.$$

Notice that **two factors** of $\frac{1 \text{ yard}}{3 \text{ feet}}$ are necessary to convert square feet to square yards. **Both the length and the width (two dimensions) have to be converted** into yards. The square yard unit is a square that is 1 yard on each side. Since 1 yard equals 3 feet, 1 square yard = (1 yard)(1 yard) = (3 feet)(3 feet) = 9(feet)(feet). A square yard is 9 square feet.

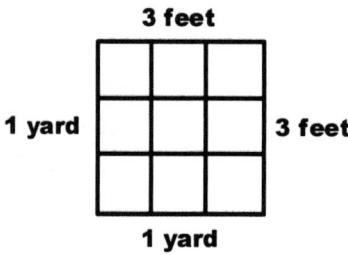

It takes 9 square feet (not 3 square feet) to make a square one yard. Notice that one square foot is smaller than one square yard. Further study reveals that the square foot is $\frac{1}{9}$ the size of the square yard.

Example 1: Find the area (sq. ft.) for the figure below.

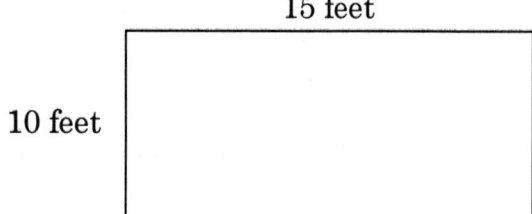

Solution:
Step 1: Understand and Picture the Problem.
  1. Problem: Find the area. Determine the number of square feet required to cover the inside of the shape.
  2. Copy the figure. The dimensions of the rectangle are 10 feet by 15 ft.

Step 2: Develop a Plan.
  Use the formula: Area = (length)(width)

Step 3: Execute the plan.

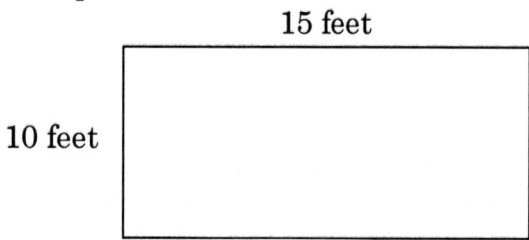

Area = (15 ft)(10 ft) = 150 square feet. The area of the figure is 150 sq ft.

Step 4: Check your work.
1. Check the numbers and your calculations.
2. Is the answer reasonable?
3. Was the question answered?

Example 2: Convert the area from Example 1 to square meters.
Solution:
Step 1: Understand and Picture the Problem.
1. Question: Determine the number of square meters required to cover the inside of the shape in Example 1.
2. The area of the shape in Example 1 is 150 square feet.

Step 2: Develop a Plan.
Use dimensional analysis to determine the number of square meters required to cover the inside of the shape. Since 1 meter ≈ 3.28 feet, use **two factors** of $\dfrac{1 \text{ meter}}{3.28 \text{ feet}}$ to convert square feet to square meters.

Step 3: Execute the plan.
$$A = \dfrac{150(\text{feet})(\text{feet})}{1} \cdot \dfrac{1 \text{ meter}}{3.28 \text{ feet}} \cdot \dfrac{1 \text{ meter}}{3.28 \text{ feet}}$$

$$A = \dfrac{150(\cancel{\text{feet}})(\cancel{\text{feet}})}{1} \cdot \dfrac{1 \text{ meter}}{3.28 \cancel{\text{ feet}}} \cdot \dfrac{1 \text{ meter}}{3.28 \cancel{\text{ feet}}} = \dfrac{150(\text{meter})(\text{meter})}{10.7584}$$

A ≈ 13.94259369 sq m. The area is approximately 13.9 square meters.

Step 4: Check your work.
1. Check the numbers and calculations.
2. Is the answer reasonable?
3. Was the question answered?

Example 3: A square ceramic tile measures 4 inches on a side. There are 108 tiles per carton and a carton costs $23.76. What will it cost to cover the floor in Example 1? (**Note:** You cannot purchase a fraction of a tile or a fraction of a carton.)
Solution:
Step 1: Understand and Picture the Problem.
1. Question: How many 4" by 4" tiles are needed to cover 150 square feet?

2. List the given information:
   Each tile is 4 inches by 4 inches.
   The area of the room in Example 1 is 150 sq ft.
3. 12 inches equal 1 foot.

Step 2: Develop a Plan.
1. The area must be converted from square feet to square inches to match the dimensions of the tile. Use Dimensional Analysis to convert the total area of square feet to square inches. Use two factors of $\frac{12 \text{ inches}}{1 \text{ foot}}$ for the conversion.
2. Find the area of a single tile.
3. Divide the area of the floor (square inches) by the area of a single tile (square inches). Since you cannot purchase a fraction of a tile, bump up the number of tiles to the next whole number.
4. Divide the number of tiles required to cover the floor by the number of tiles per carton. Since you cannot purchase a fraction of a carton, bump up the number of cartons to the next whole number.

Step 3: Execute the plan.
1. $A = \frac{150(\text{feet})(\text{feet})}{1} \cdot \frac{12 \text{ inches}}{1 \text{ foot}} \cdot \frac{12 \text{ inches}}{1 \text{ foot}}$

   $A = \frac{150(\cancel{\text{feet}})(\cancel{\text{feet}})}{1} \cdot \frac{12 \text{ inches}}{1 \cancel{\text{foot}}} \cdot \frac{12 \text{ inches}}{1 \cancel{\text{foot}}} = 21{,}600$ square inches.

2. The area of a tile is (4 in)(4 in) = 16 sq in.
3. The number of 4 inch square ceramic tiles needed to cover the floor of Example 1 is 21,600 sq in ÷ 16 sq in per tile.

   $(21{,}600 \text{ in in})\left(\frac{1 \text{ tile}}{16 \text{ in in}}\right).$

   $\left(\cancel{21{,}600}^{1{,}350} \cancel{\text{in}} \cancel{\text{in}}\right)\left(\frac{1 \text{ tile}}{\cancel{16} \cancel{\text{in}} \cancel{\text{in}}}\right) = 1{,}350$ tiles.

4. The number of cartons needed is 1,350 tiles ÷ 108 tiles per carton.

   $(1{,}350 \text{ tiles})\left(\frac{1 \text{ carton}}{108 \text{ tiles}}\right)$

   $(1{,}350 \cancel{\text{tiles}})\left(\frac{1 \text{ carton}}{108 \cancel{\text{tiles}}}\right) = \frac{1{,}350}{108}$ cartons = 12.5 cartons.

   Since a fraction of a carton cannot be purchased, 13 cartons of tiles are needed.

5. At $23.76 per carton, 13 cartons would cost:
$$\frac{\$23.76}{\text{carton}}(13 \text{ carton}) = \frac{\$23.76}{\cancel{\text{carton}}}(13 \cancel{\text{ cartons}}) = \$308.88.$$

Step 4: Check your work.
1. Check the numbers and calculations.
2. Is the answer reasonable?
3. Was the question answered?

Example 4: Find the area (sq. ft.) for the figure below

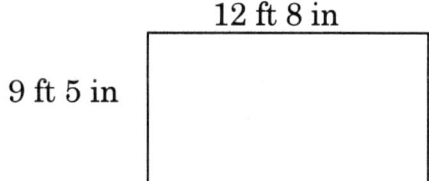
12 ft 8 in

9 ft 5 in

Solution:
Step 1: Understand and Picture the Problem.
1. Question: Find the area. Determine the number of square feet required to cover the inside of the shape.
2. Copy the figure and list the dimensions.

Step 2: Develop a Plan.
1. Area = (length)(width)
2. Use fractions to represent inches as a fractional part of a foot.

Step 3: Execute the plan.

12 ft 8 in

9 ft 5 in

$$\text{Area} = \left(12 + \frac{8}{12}\right)\left(9 + \frac{5}{12}\right) \approx 119.2777778 \text{ square feet}.$$

Note: The parenthesis keys on your calculator must be used.
The area of the figure is approximately 119.3 square feet.

Step 4: Check your work.
1. Check the numbers and your calculations.
2. Is the answer reasonable?
3. Was the question answered?

TRIANGLES

Rectangles can be partitioned to form triangles. Returning to the 9 ft by 12 ft room whose area is 108 sq ft. Cut the room in half diagonally.

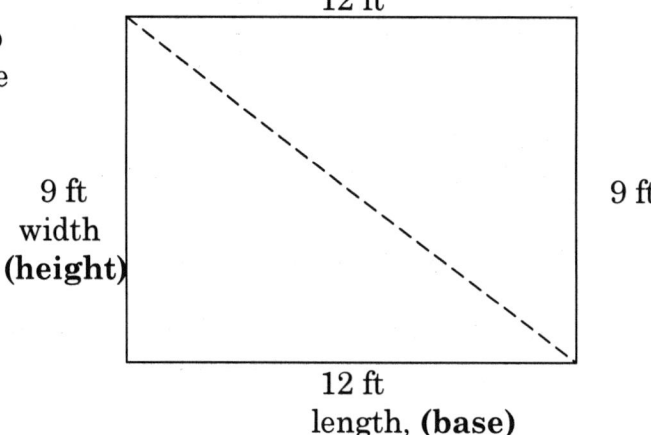

The width of the room is the **height** of the triangle and the length of the room is the **base** of the triangle. The height and the base of a triangle must always be at right angles to each other. The area of each triangle above is 54 sq. ft., half of the rectangle's area. By dividing a rectangle in half, the area of each triangle is calculated. The following formula allows the calculation of the area for all triangles.

Area of a Triangle (square units)

$$\text{Area} = 0.5(\text{base})(\text{height}) = \frac{(\text{base})(\text{height})}{2} = \frac{bh}{2}$$

The height (sometimes called an altitude) of the triangle forms a right angle with the base and extends to the highest point of the triangle. The base is always a side of the triangle, while the height may fall inside or outside the triangle. The following examples demonstrate triangles with a base of 6 meters and an altitude (height) of 5 meters.

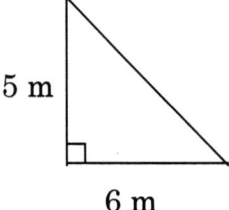 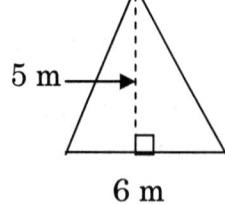 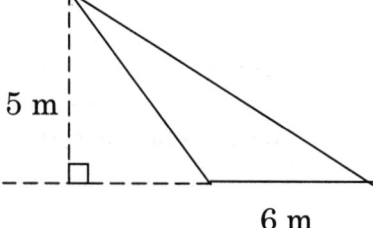

Example 5 Find the area (sq in) of the triangle.

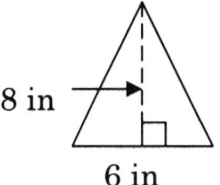

Solution:
Step 1: Understand and picture the problem.
1. Question: Find the area of the triangle.
2. Copy the figure and list the dimensions.

Step 2: Devise a plan.
1. Find the dimensions of the base and height.
2. Substitute these values into the formula and simplify.

Step 3: Execute the plan.

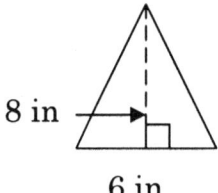

1. Base = 6 inches, Height = 8 inches.
2. Area = 0.5(base)(height) = 0.5 (6 in)(8 in) = 24 sq in.

The area of the triangle is 24 square inches.

Step 4: Check your work.
1. Check the numbers and your calculations.
2. Is the answer reasonable?
3. Was the question answered?

Example 6: Find the area (sq ft) of the triangle.

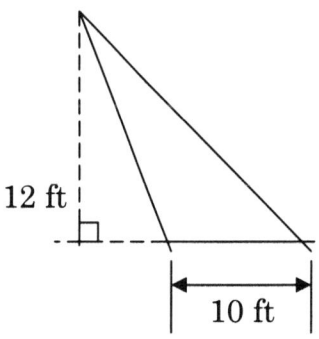

Solution:
Step 1: Understand and picture the problem.
1. Question: What is the area of a triangle?
2. Copy the figure and list the dimensions.

Step 2: Devise a plan.
1. Find the base and height dimensions.
2. Substitute these values into the formula and simplify.

Step 3: Execute the plan.

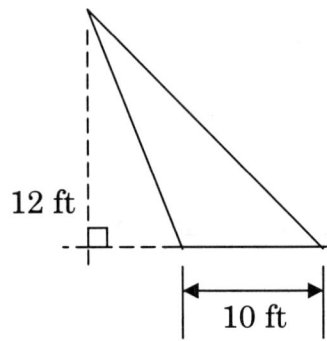

1. Base = 10 ft, Height = 12 ft.
2. Area = $0.5(\text{base})(\text{height}) = 0.5\,(10\text{ ft})(12\text{ ft}) = 60$ sq ft.

The area of the triangle is 60 square feet.

Step 4: Check your work.
1. Check the numbers and calculations.
2. Is the answer reasonable?
3. Was the question answered?

CIRCLES
In the previous section, the Circumference of a circle was defined as

> **Circumference of a Circle (units)**
> Circumference = $\pi d = 2\pi r$

In order to find the area of a circle, a circle is cut into slices and arranged in the following manner.

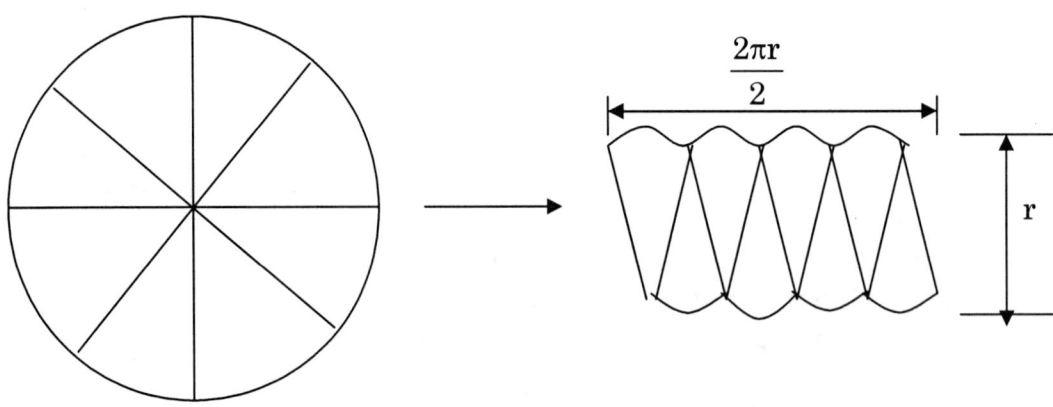

The slices form a rectangle. The formula for the "Area of a Rectangle = Length times Width". The width of the rectangle is represented by the radius of the circle. The length of the rectangle is represented by half of the circumference of a circle. Substituting these values into the rectangle area formula produces:

$$\text{Area} = (\text{Length})(\text{Width})$$
$$\text{Area} = \frac{2\pi r}{2}(r)$$
$$\text{Area} = \frac{\cancel{2}\pi r}{\cancel{2}}(r) = \pi r^2$$

The area of a circle is defined by the formula:

> **Area of a Circle (square units)**
> Area = πr^2
> Where: r is the radius
> π is a constant
> (Use the π key on your calculator.)

Example 7: Find the area of a circle with an 8 inch radius.

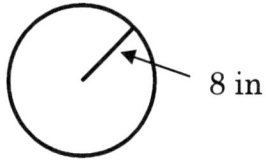

Solution:
Step 1: Understand and picture the problem.
1. Question: Find the area of a circle with an 8 inch radius.
2. Identify the parts of a circle. The radius is 8 inches.

Step 2: Devise a plan.
1. Copy the figure.
2. Use the formula; Area = πr^2.
3. Substitute the value for the radius and use the π key of your calculator and simplify.

Step 3: Execute the plan.

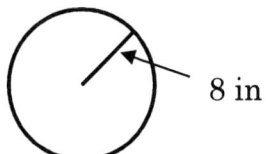

294

Area = πr² = π(8)² ≈ 201.06193 sq in.

The area of the circle is approximately 201.1 square inches.

Step 4: Check your work.
1. Check the numbers and your calculations.
2. Is the answer reasonable?
3. Was the question answered?

Example 8: Find the area of a circle with a 10 foot diameter.

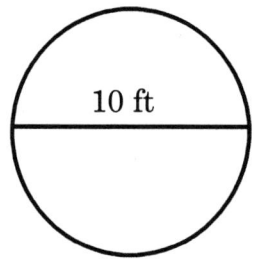

Solution:
Step 1: Understand and picture the problem.
1. Problem: Find the area of a circle with a 10 foot diameter.
2. Identify the parts of a circle. The diameter is 10 feet.

Step 2: Devise a plan.
1. Copy the figure.
2. Convert the diameter to the radius: 1 diameter = 2 radii.
   0.5 diameter = 1 radius.
3. Use the formula; Area = πr² .
4. Substitute the value for the radius and use the π key of your calculator and simplify.

Step 3: Execute the plan.

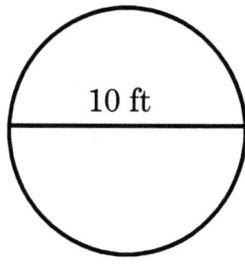

Radius = 0.5 x 10 feet = 5 ft.
Area = πr² = π (5 ft.)(5 ft.) ≈ 78.539816 sq ft.
The area of the circle is approximately 78.5 square feet.

Step 4: Check your work.
    1. Check the numbers and your calculations.
    2. Is the answer reasonable?
    3. Was the question answered?

Example 9: Find the area (sq ft rounded to tenths) for the following sector:

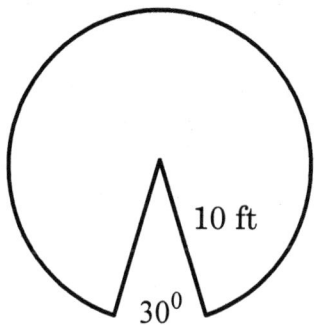

Solution:
Step 1: Understand and Picture the Problem.
    1. Question: Find the area of the sector, fractional portion of the circle.
    2. The radius of the sector is 10 feet.
    3. The angle outside the sector is 30 degrees.

Step 2: Develop a Plan.
    1. Copy the figure.
    2. Use the Area formula, Area $= \pi r^2$.
    3. Find the fractional portion of the circle corresponding to the sector.

Step 3: Execute the plan.

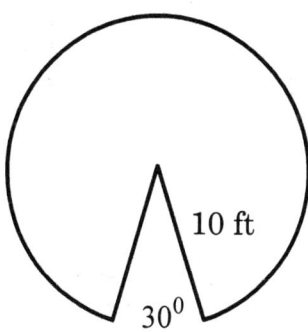

Area $= \pi r^2$.
Out of 360 degrees, the sector has 330 degrees.
The sector is $\dfrac{330}{360}$ of a circle.
The area of the sector is $\dfrac{330}{360} \pi (10 \text{ feet})^2 \approx 287.9793266$ sq ft.
Thus the area of the sector is approximately 288.0 square feet.

296

Step 4: Check your work.
   1. Check the numbers and your calculations.
   2. Is the answer reasonable?
   3. Was the question answered?

? **Cognitive Problems** **?**

1. Explain the difference between perimeter and area.
2. Explain the significance of the term "square" in an area problem.
3. Give at least one example of a "real life" problem that requires area.
4. Explain why one square foot equals 144 square inches.
5. Explain why area is measured in square units.

(c) 2006 JupiterImages Corp.

Exercise 4.2

I. Find the area for the following figures. Bump up the area to the next whole number.

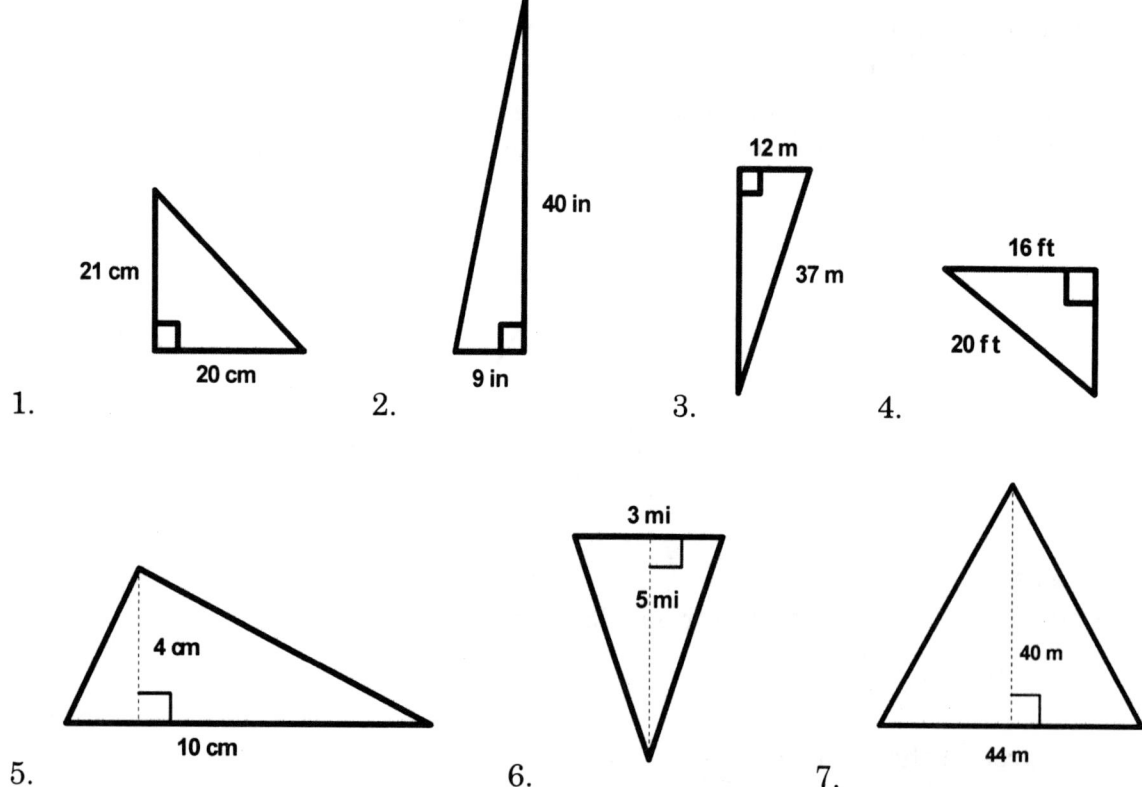

1.   2.   3.   4.

5.   6.   7.

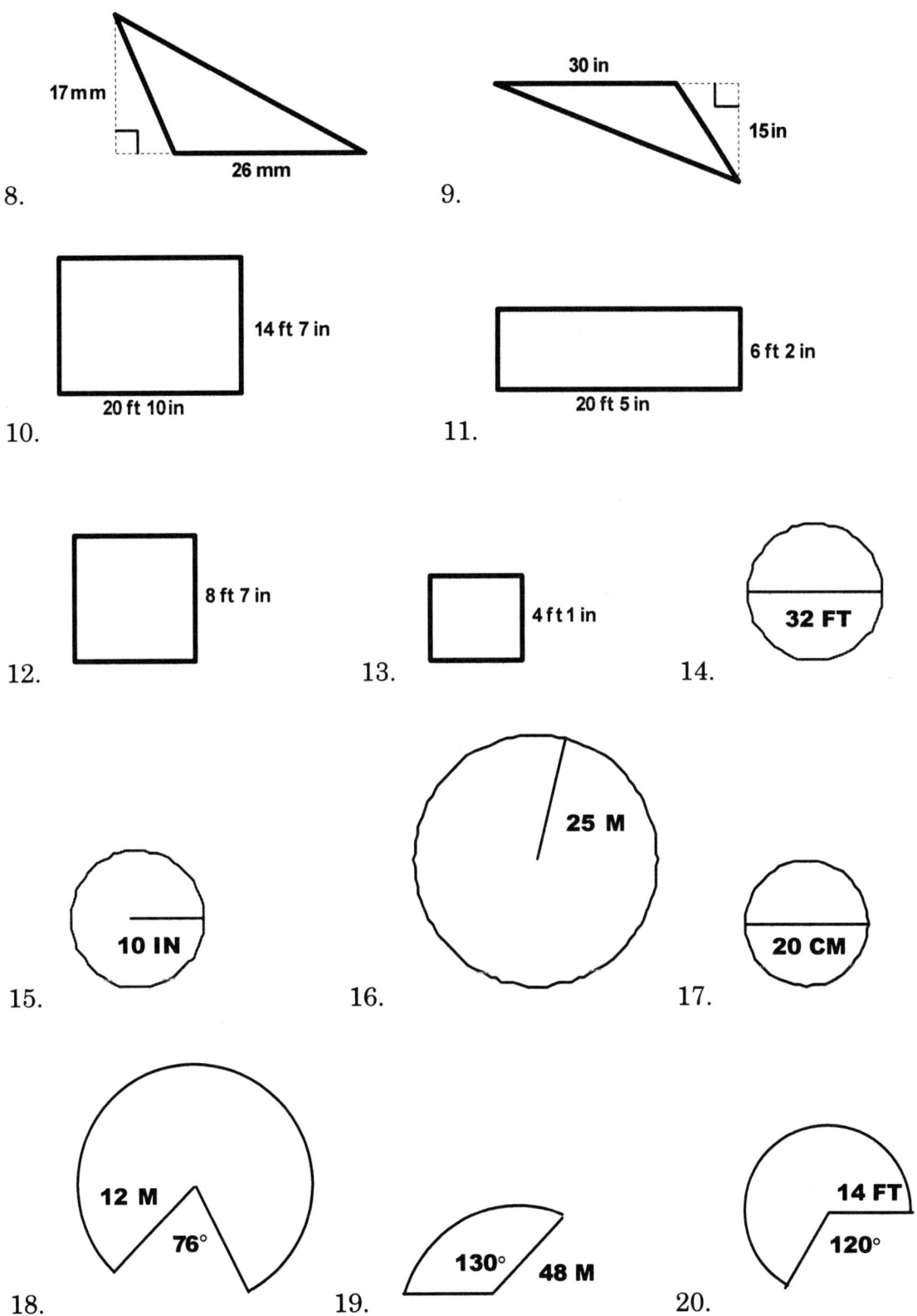

II. Use Dimensional Analysis to convert the areas from Problem I to the specified dimensions. Bump up the areas to the next whole number.

Dimensions
1,000 millimeters (mm) = 1 meter
100 centimeters (cm) = 1 meter
3.28 ft ≈ 1 meter
1.09 yd ≈ 1 meter
1 inch ≈ 25.4 mm
1 inch ≈ 2.54 cm
1 mile = 5,280 feet
3 feet (ft) = 1 yard (yd)
36 inches (in) = 1 yard (yd)
12 inches (in) = 1 foot (ft)

1. Convert the area from Exercise 1 to square inches.
2. Convert the area from Exercise 2 to square feet.
3. Convert the area from Exercise 3 to square yards.
4. Convert the area from Exercise 4 to square yards.
5. Convert the area from Exercise 5 to square inches.
6. Convert the area from Exercise 6 to square feet.
7. Convert the area from Exercise 7 to square centimeters.
8. Convert the area from Exercise 8 to square inches.
9. Convert the area from Exercise 9 to square millimeters.
10. Convert the area from Exercise 10 to square yards.
11. Convert the area from Exercise 11 to square yards.
12. Convert the area from Exercise 12 to square inches.
13. Convert the area from Exercise 13 to square inches.
14. Convert the area from Exercise 14 to square inches.
15. Convert the area from Exercise 15 to square feet.
16. Convert the area from Exercise 16 to square yards.
17. Convert the area from Exercise 17 to square meters.
18. Convert the area from Exercise 18 to square centimeters.
19. Convert the area from Exercise 19 to square millimeters.
20. Convert the area from Exercise 20 to square yards.

4.3 PERIMETER AND AREA OF IRREGULAR SHAPES

Objective: Calculate the perimeter and area of shapes made from rectangles, triangles and circles.

What strategy would you take if you have to calculate the area of a shape that is not a regular shape (rectangle, triangle, or circle)? This problem can be resolved by drawing horizontal or vertical lines to partition the region into rectangles, triangles or circles (sectors). For example, find the area (square feet) of the following room.

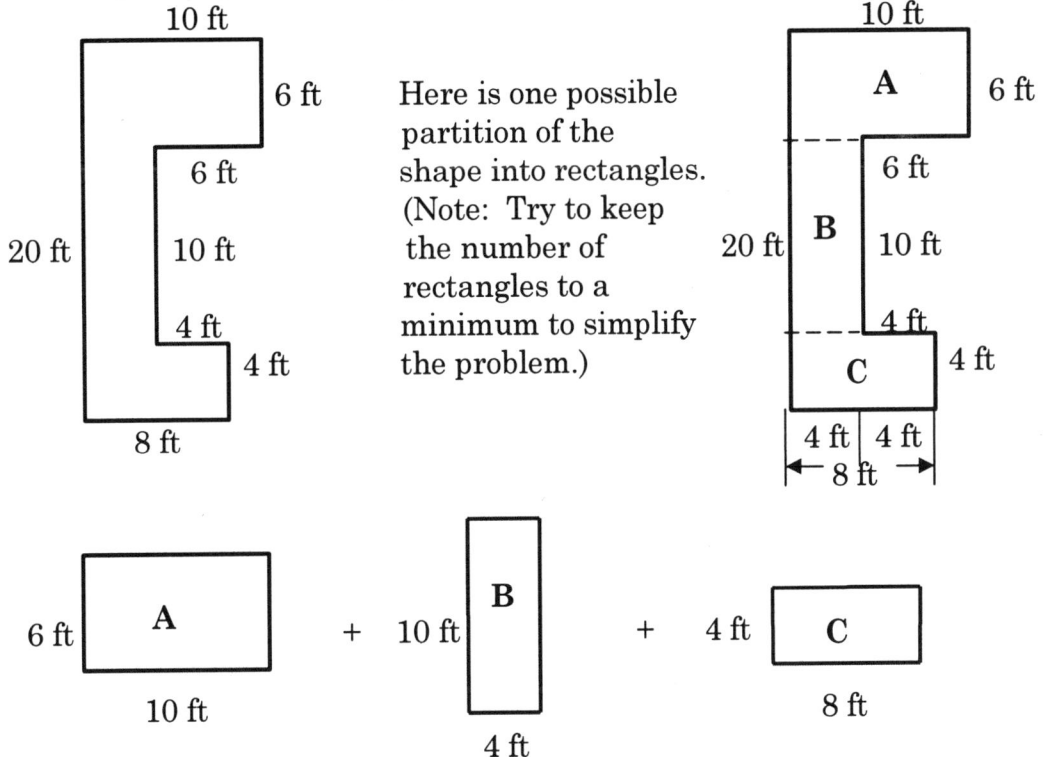

Here is one possible partition of the shape into rectangles. (Note: Try to keep the number of rectangles to a minimum to simplify the problem.)

Then the area is the sum of the area of the three partitions.
The area of: **Section A** is (10 ft)(6 ft) =   60 sq. ft
              Section B is (10 ft)(4 ft) =   40 sq. ft
              Section C is (8 ft)(4 ft) = + 32 sq. ft
                                                132 sq. ft

The area of the figure is 132 square feet.
(Note: Try partitioning this figure using vertical lines and compare the results. Check your work if the area is different.)

The perimeter of an irregular shape is the sum of the sides. For the above figure, add the lengths of the sides to find the perimeter.

300

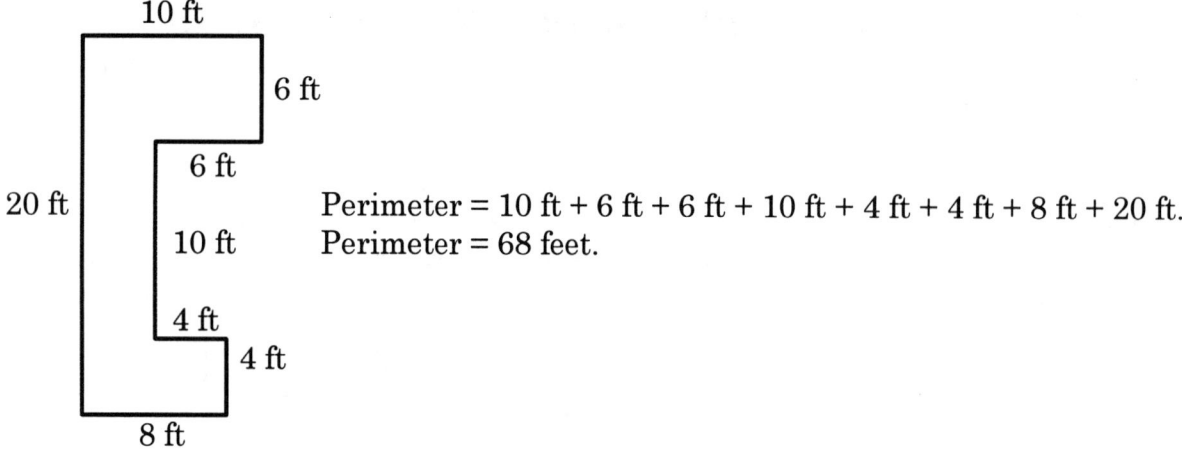

Perimeter = 10 ft + 6 ft + 6 ft + 10 ft + 4 ft + 4 ft + 8 ft + 20 ft.
Perimeter = 68 feet.

Example 1: Find the area (sq. ft) for the figure below.

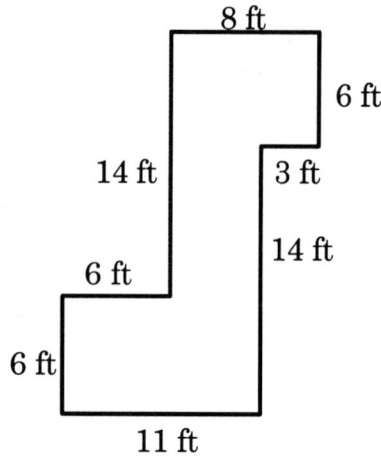

Solution:
  Step 1: Understand and Picture the Problem.
       1. Question: Find the area.
       2. Copy the figure and list the dimensions.

  Step 2: Develop a Plan.
       1. Partition the figure into rectangles using horizontal and/or vertical lines.
       2. Find the area of each rectangular section, using the area formula, Area = (length)(width).
       3. Sum the section's areas to get the total area of the figure.

Step 3: Execute the plan.

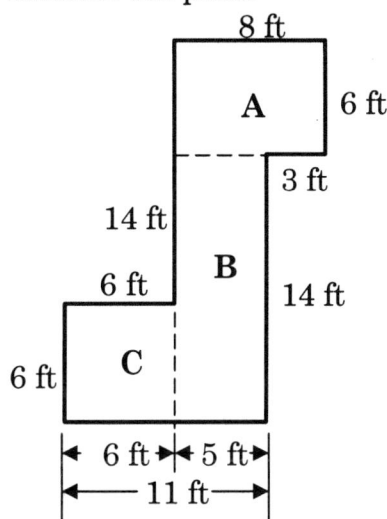

The area of
Section **A** is (8 ft)(6 ft) = 48 sq ft
Section **B** is (14 ft)(5 ft) = 70 sq ft
Section **C** is (6 ft)(6 ft) = + 36 sq ft
                                154 sq ft

The area of the figure is 154 sq ft.

Step 4: Check your work.
    1. Check the numbers and your calculations.
    2. Is the answer reasonable?
    3. Was the question answered?

Example 2: Find the perimeter for the figure in Example 1.
Solution:
Step 1: Understand and Picture the Problem.
    1. Question: Find the distance around the figure.
    2. Copy the figure and list the dimensions.

Step 2: Develop a Plan.
    Find the sum of the sides of the figure.

Step 3: Execute the Plan.

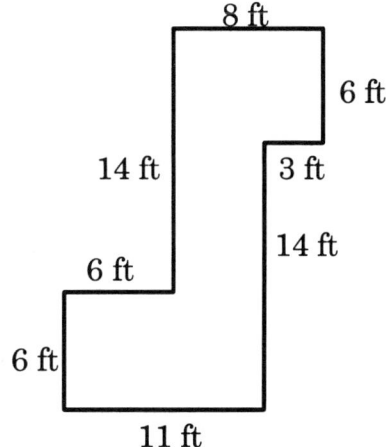

Perimeter = 8 ft + 6 ft + 3 ft + 14 ft + 11 ft + 6 ft + 6 ft + 14 ft.
Perimeter = 68 feet.

Step 4: Check Your Work.
1. Check the numbers and your calculations.
2. Is the answer reasonable?
3. Was the question answered?

DIMENSIONAL ANALYSIS (Revisited)
Suppose the area of the figure in Example 1 is needed in square yards. For instance, carpet is sold by the square yard, not the square foot. *"Dimensional Analysis"* is a technique used to convert square feet to square yards. Dimensional analysis uses definitions, (in this case, 1 yard = 3 feet), and "factors of one" to convert the units of one problem into the desired units. Since the area of Example 1 is 154 sq ft and **square feet** means **feet** *times* **feet**, the following dimensional analysis problem demonstrates the conversion of square feet into square yards.

$$154 \text{ ft}^2 = \frac{154(\text{ft})(\text{ft})}{1} \times \frac{1 \text{ yd}}{3 \text{ ft}} \times \frac{1 \text{ yd}}{3 \text{ ft}}.$$

$$154 \text{ ft}^2 = \frac{154(\cancel{\text{ft}})(\cancel{\text{ft}})}{1} \times \frac{1 \text{ yd}}{3 \cancel{\text{ft}}} \times \frac{1 \text{ yd}}{3 \cancel{\text{ft}}} = \frac{154 \text{ yd}^2}{9} \approx 17.111 \text{ yd}^2.$$

The area of the figure in Example 1 is approximately 18 square yards.
(Note: In real applications take any decimal remainder up to the next whole number. You cannot purchase a fraction of a square yard of carpeting and you do not want to end up short.) Notice in the dimensional analysis problem that two factors of $\frac{1 \text{ yd}}{3 \text{ ft}}$ are necessary for converting square feet into square yards. Both the length and the width (two dimensions) have to be converted into yards.

A square yard is a square that is 1 yard on each side. One square yard is comprised of 9 square feet. Using dimensional analysis:

$$1 \text{ yd}^2 = \frac{1(\text{yd})(\text{yd})}{1} \times \frac{3 \text{ ft}}{1 \text{ yd}} \times \frac{3 \text{ ft}}{1 \text{ yd}}.$$

$$1 \text{ yd}^2 = \frac{1(\cancel{\text{yd}})(\cancel{\text{yd}})}{1} \times \frac{3 \text{ ft}}{1 \cancel{\text{yd}}} \times \frac{3 \text{ ft}}{1 \cancel{\text{yd}}} = 9 \text{ ft}^2.$$

Compare the following figures:

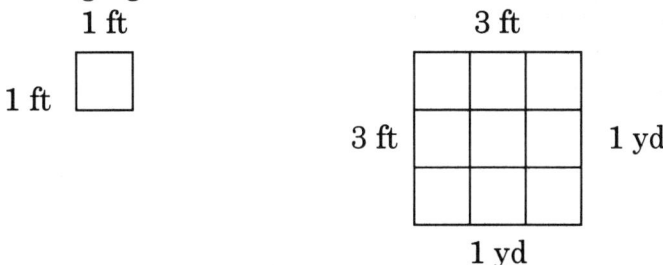

It takes 9 square feet to make 1 square yard.

Example 3: A square ceramic tile measures 4 inches on a side. Using dimensional analysis, how many tiles will it take to cover a floor with the dimensions of Example 1?

Solution:
Step 1: Understand and Picture the Problem.
1. Question: How many 4" by 4" tiles are needed to cover 154 square feet?
2. Each tile is 4 inches by 4 inches.
3. There are 12 inches in 1 foot.
4. The area of the room in Example 1 is 154 sq ft.

Step 2: Develop a Plan.
1. The area must be converted from square feet to square inches to keep the units the same. Using Dimensional Analysis, convert the total area from square feet to square inches.
2. Find the area of a single tile.
3. Divide the area of the floor (square inches) by the area of a single tile (square inches). Since you cannot purchase a fraction of a tile, bump up the number of tiles to the next whole number, if necessary.

Step 3: Execute the plan.
1. $154 \text{ ft}^2 = \dfrac{154(\text{ft})(\text{ft})}{1} \times \dfrac{12 \text{ in}}{1 \text{ ft}} \times \dfrac{12 \text{ in}}{1 \text{ ft}}$.

    $154 \text{ ft}^2 = \dfrac{154(\cancel{\text{ft}})(\cancel{\text{ft}})}{1} \times \dfrac{12 \text{ in}}{1 \cancel{\text{ft}}} \times \dfrac{12 \text{ in}}{1 \cancel{\text{ft}}} = 22{,}176 \text{ in}^2$.

2. The area of a tile is (4 in)(4 in) = 16 sq in per tile.
3. The number of 4 inch square ceramic tiles needed to cover the floor of Example 1 is: $\dfrac{22{,}176 \text{ in}^2}{1} \div \dfrac{16 \text{ in}^2}{1 \text{ tile}}$.

    $\dfrac{22{,}176 \ \cancel{\text{in}^2}}{1} \times \dfrac{1 \text{ tile}}{16 \ \cancel{\text{in}^2}} = \dfrac{22{,}176}{16} = 1{,}386 \text{ tiles}$.

304

Step 4: Check your work.
1. Check the numbers and your calculations.
2. Is the answer reasonable?
3. Was the question answered?

Example 4: Find the area (sq ft) for the figure below.

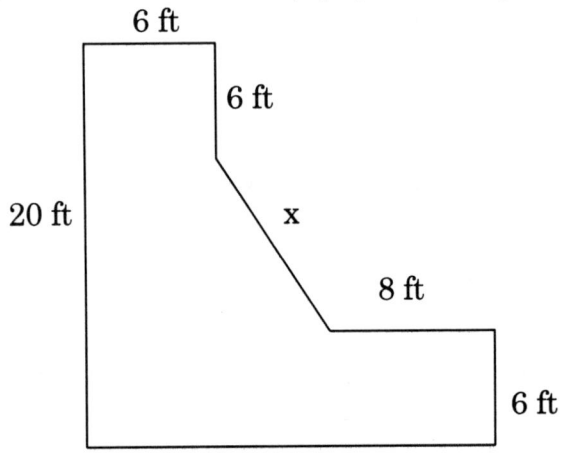

Solution:
Step 1: Understand and Picture the Problem.
1. Question: Find the area (sq ft) of the figure.
2. Copy the figure and list the dimensions.

Step 2: Develop a Plan.
1. Partition the figure into rectangles and triangles.
2. Find the area of each section and the sum of these areas.

Step 3: Execute the Plan

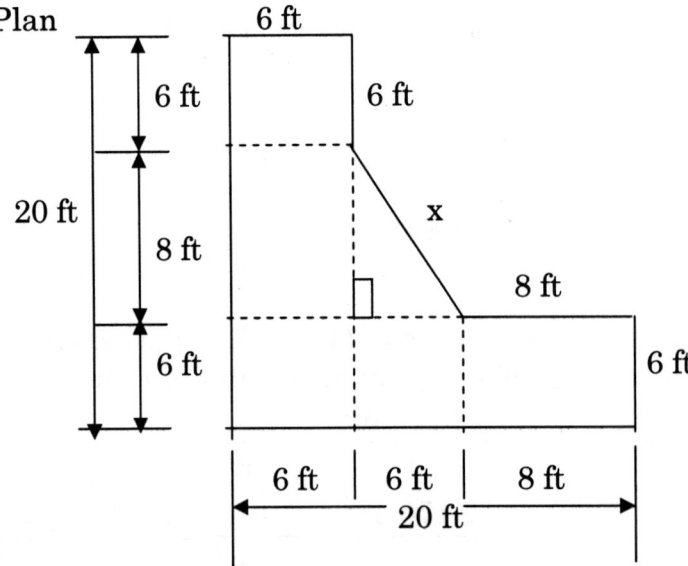

After finding specific lengths, the partitioned figure can be simplified to the figure below

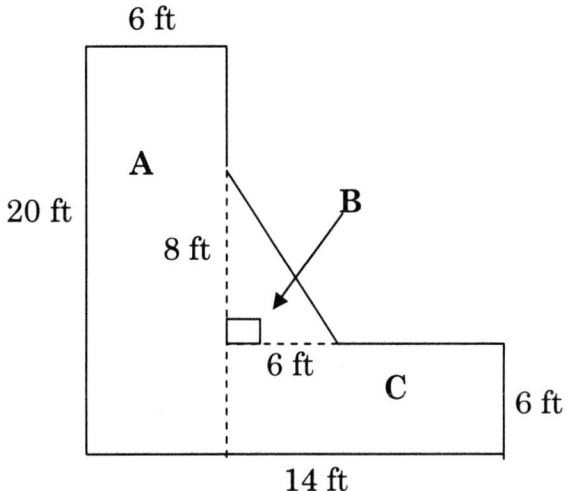

The area of **Section A:** Rectangle A = lw   (6 ft)(20 ft)   =   120 sq ft
                Section B: Triangle A = 0.5bh 0.5(8 ft)(6 ft) =   24 sq ft
                Section C: Rectangle A = lw   (6 ft)(14 ft)   = + 84 sq ft
                                                                    228 sq ft

The area is 228 square feet.

Step 4: Check your work.
　　　1. Check the numbers and your calculations.
　　　2. Is the answer reasonable?
　　　3. Was the question answered?

Example 5: Find the perimeter for the figure in Example 4.
Solution:
Step 1: Understand and Picture the Problem.
　　　1. Perimeter is the distance around the outside of the figure.
　　　2. One side of the figure is unknown.
　　　3. Question: Find the perimeter for the figure in Example 4.

Step 2: Develop a Plan.
　　　1. Use the partitioning from Example 4.
　　　2. Use the Pythagorean Theorem to solve for the third side (hypotenuse) of the right triangle.
　　　3. Add the sides of the entire figure to calculate its perimeter.

Step 3: Execute the Plan.

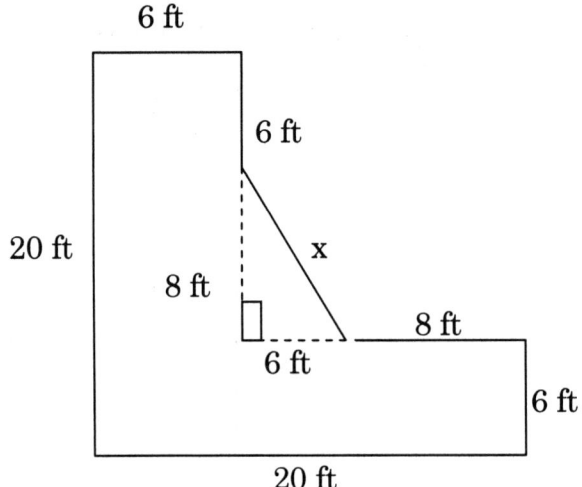

Find the missing side:

Step	Action
Square the lengths of the two sides that you know and put the larger value first.	$8^2 = 64$, $6^2 = 36$ 64, 36
Since you are solving for the hypotenuse, ADD	$64 + 36 = 100$
Square root the answer.	$\sqrt{100} = 10$

The length of the third side of the triangle is 10 ft.
P = 6 ft + 6 ft + 10 ft + 8 ft + 6 ft + 20 ft + 20 ft = 76 ft.
The perimeter of the figure is 76 feet.

Step 4: Check Your Work.
1. Check the numbers and your calculations.
2. Is the answer reasonable?
3. Was the question answered?

Example 6: Find the area for the following shape

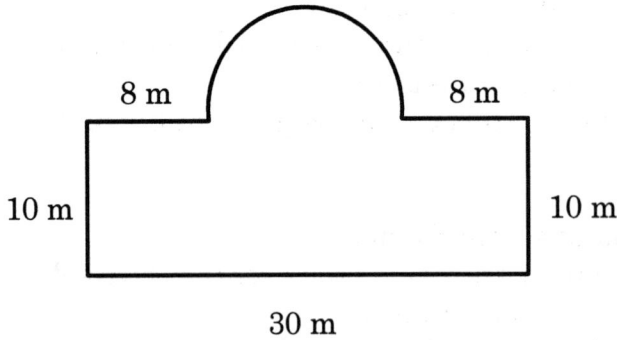

Solution:
Step 1: Understand and Picture the Problem.
1. Question: Find the area of the figure.
2. Copy the figure and list the dimensions.

Step 2: Develop a plan.
1. Partition the figure into a rectangle and circle (semi-circle) by drawing horizontal and vertical lines.
2. Use the appropriate formulas, to find the area of each section.
3. Sum the areas to get the total area of the figure.
   If necessary, bump up the area to the next whole number.

Step 3: Execute the Plan.
1. One possible partition is:

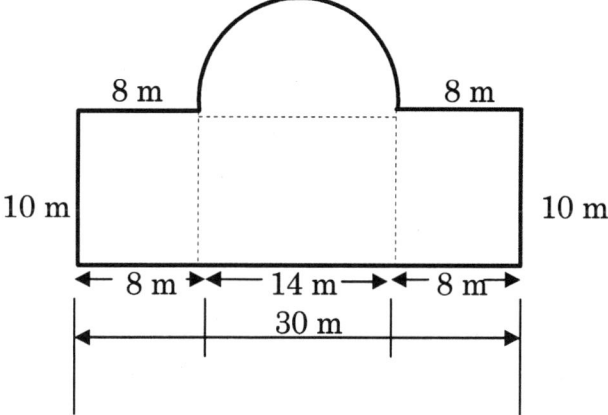

The partition can be simplified to the following figure.

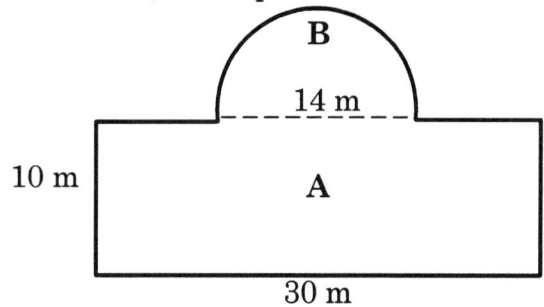

2. The area of **Section A** (rectangle) is:
Area = (Length)(Width) = (10 m)(30 m) = 300 sq m

The area of **Section B** (Semi-circle) is one-half the area of a circle. Radius is one-half the diameter. The radius is $\frac{14}{2}$ m = 7 m.

The area of **Section B** is:
Area = $\frac{\pi r^2}{2} = \frac{\pi(7)^2}{2} \approx$ 76.96902001 sq m. (Use the π key.)

3. The area of the figure is: 300 sq m
                              + 76.96902001 sq m
                                376.96902991 sq m
The area of the figure is approximately 377.0 sq m.

307

308

Step 4: Check your work.
1. Check your numbers and your calculations.
2. Is the answer reasonable?
3. Was the question answered?

Example 7: Find the perimeter for the figure in Example 6.
Solution:
Step 1: Understand and Picture the Problem.
1. Question: Find the perimeter, the distance around the outside of the figure.
2. One side of the figure is unknown.
3. Copy the figure and its dimensions.

Step 2: Develop a Plan.
1. Use the partitioning from Example 6.
2. Use the Circumference formula to calculate the length of the semi-circle.
3. Add the sides of the figure to calculate its perimeter. If necessary, bump up the perimeter to the next whole number.

Step 3: Execute the Plan.
1.
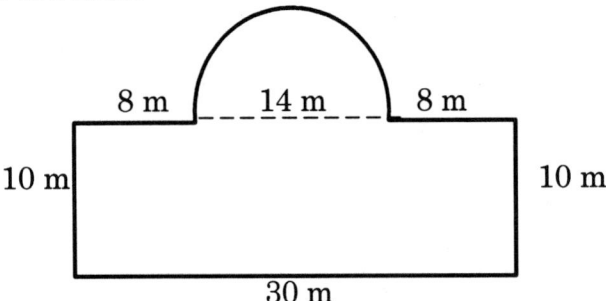

2. The length of the semi-circle is one-half the circumference of circle. The circumference of a circle is $C = \pi d$. The length of the semi-circle is $\dfrac{\pi d}{2} = \dfrac{\pi(14)}{2} \approx 21.99114858$ m.
(Use the π key and store in memory.)
3. The perimeter of the figure is:
P = 8 m + 21.99114858 m + 8 m + 10 m + 30 m + 10 m = 87.99114858 m.
P ≈ 88 m.

Step 4: Check Your Work.
1. Check your numbers and your calculations.
2. Is the answer reasonable?
3. Was the question answered?

? Cognitive Problems ?

1. Is there usually more than one way to partition a figure?
2. Will the figure have the same area if it is partitioned differently? Explain.
3. Explain the difference between one and two dimensional measurements.
4. Explain the difference between one square yard and one square foot.

(c) 2006 JupiterImages Corp.

Exercise 4.3

I. Find the perimeter and area. Bump answers up to the next whole number.

1.
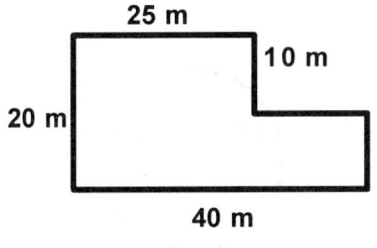

2.
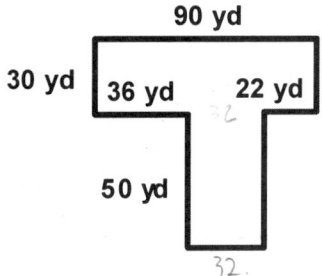

3.
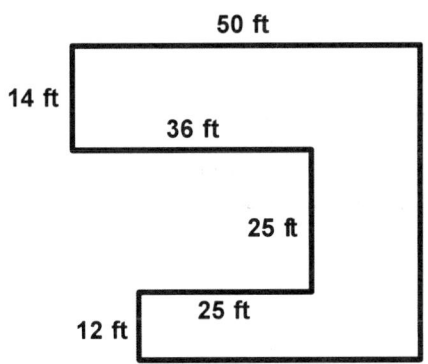

4.
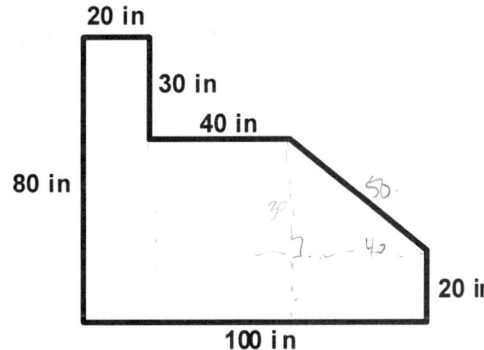

5.
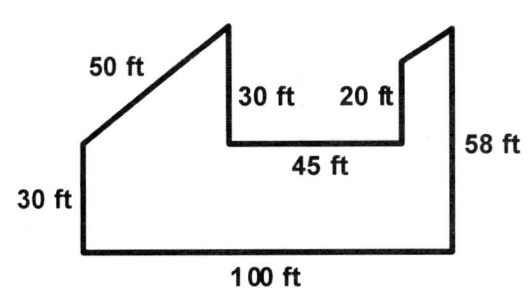

6.
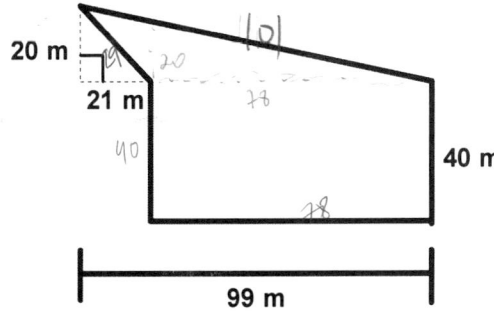

309

310

7.

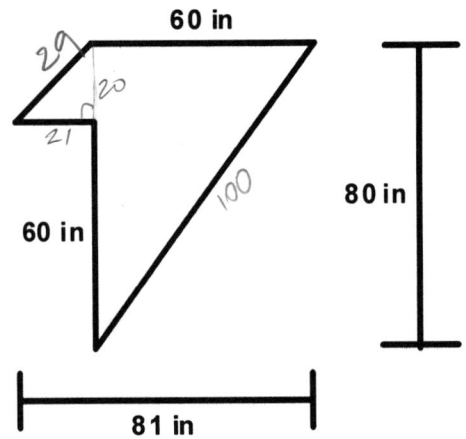

8.

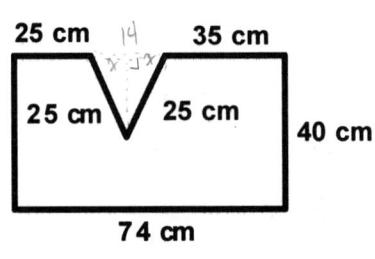

9.

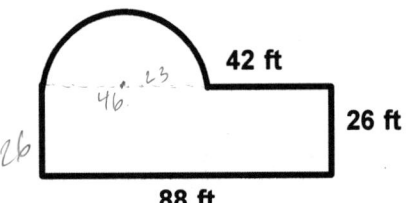

10.

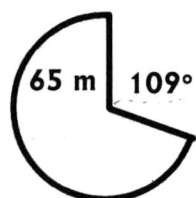

11.

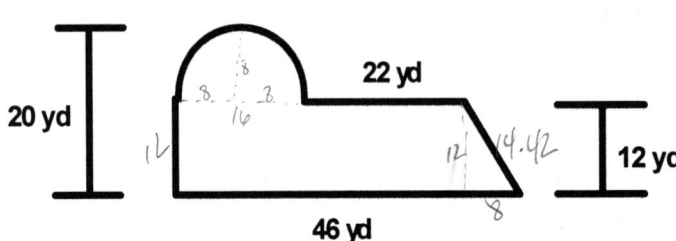

12.

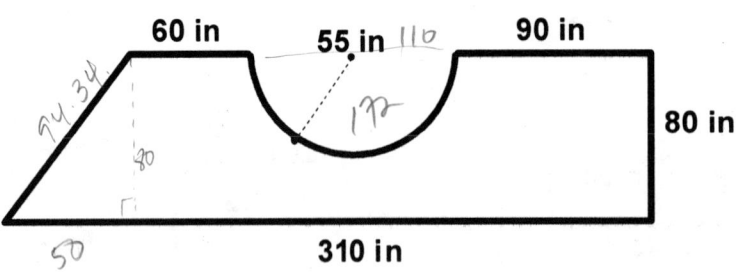

II. Use Dimensional Analysis to convert the perimeters and areas from Problem I to the specified dimensions. Bump answers up to the next whole number.

> **Dimensions**
>
> 1,000 millimeters (mm) = 1 meter
> 100 centimeters (cm) = 1 meter
> 3.28 ft ≈ 1 meter
> 1.09 yd ≈ 1 meter
> 1 inch ≈ 25.4 mm
> 1 inch ≈ 2.54 cm
> 1 mile = 5,280 feet
> 3 feet (ft) = 1 yard (yd)
> 36 inches (in) = 1 yard (yd)
> 12 inches (in) = 1 foot (ft)

1. Convert the perimeter from Exercise 1 to yards.
   Convert the area from Exercise 1 to square yards.
2. Convert the perimeter from Exercise 2 to feet.
   Convert the area from Exercise 2 to square feet.
3. Convert the perimeter from Exercise 3 to yards.
   Convert the area from Exercise 3 to square yards.
4. Convert the perimeter from Exercise 4 to feet.
   Convert the area from Exercise 4 to square feet.
5. Convert the perimeter from Exercise 5 to yards.
   Convert the area from Exercise 5 to square yards.
6. Convert the perimeter from Exercise 6 to centimeters.
   Convert the area from Exercise 6 to square centimeters.
7. Convert the perimeter from Exercise 7 to yards.
   Convert the area from Exercise 7 to square yards.
8. Convert the perimeter from Exercise 8 to inches.
   Convert the area from Exercise 8 to square inches.
9. Convert the perimeter from Exercise 9 to yards.
   Convert the area from Exercise 9 to square yards.
10. Convert the perimeter from Exercise 10 to yards.
    Convert the area from Exercise 10 to square yards.
11. Convert the perimeter from Exercise 11 to feet.
    Convert the area from Exercise 11 to square feet.
12. Convert the perimeter from Exercise 12 to yards.
    Convert the area from Exercise 12 to square yards.

4.4 VOLUME
Objective: Compute the volume of regular solids.
Sugar cubes, dice, and children's alphabet playing blocks are examples of cubes. How many cubes would it take to fill a room by orderly stacking them in rows and layers? Volume is the number of cubes it takes to fill the inside of a three-dimensional shape.

Perimeter
- **One-dimensional** measurements are made in one direction.
- The distance around a shape is analogous to erecting a fence.
- Units: inches, feet, meters, centimeters, etc.

Area
- **Two-dimensional** measurements are made in **two** directions.
- Covering the inside of a shape is analogous to laying tile on a floor.
- Units: square inches, ft^2, square meters, cm^2, etc.

Volume
- **Three-dimensional** measurements are made in **three** directions.
- The number of cubes it takes to fill a room from floor to ceiling is analogous
- to filling a small rectangular box with bouillon cubes.
- Units: cubic inches, ft^3, cubic meters, cc (cubic centimeters), etc.

The three-dimensional aspect of volume means that three separate measurements are made. The length of the shape (one dimension) and the width of the shape (the second dimension) are covered with cubes forming one layer. The height of the shape (the third dimension) is then filled using layers of cubes. Similar to a square, the sides of a cube are all the same length.

Consider a 3 ft. by 4 ft. by 5 ft. rectangular shipping crate. Since the crate's dimensions are in feet, the volume is measured in cubic feet.

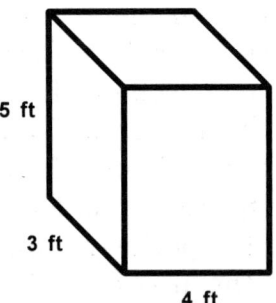

A cubic foot is a cube whose sides (length, width and height) are each one foot long.

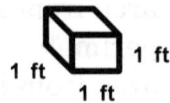

The calculation of volume is done in two stages. First the floor of the crate is covered with cubes by finding the area of the floor Area = 3 ft x 4 ft = 12 sq ft
It takes 12 cubes to cover the floor and make one layer of cubes.

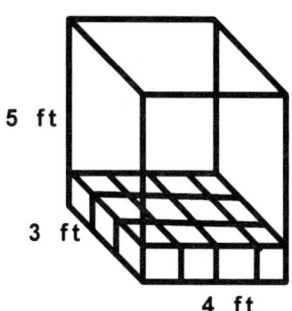

Then the cubes are layered filling the shape.

Volume = $\left(\dfrac{12 \text{ cubes}}{\text{layer}}\right)(5 \text{ layers})$.

Volume = $\left(\dfrac{12 \text{ cubes}}{\cancel{\text{layer}}}\right)(5 \cancel{\text{layers}})$ = 60 cubes.

Volume = 60 cu ft or 60 ft³.
The volume of the shipping crate is 60 cubic feet.

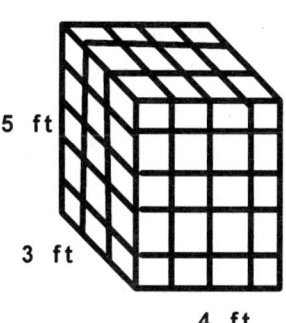

The formula for volume of a rectangular shipping crate is:

Volume of a rectangular solid (cubic units)
Volume = (Area of the base) × (height)
Volume = Length × Width × Height

Regular Solids
The volume of a rectangular crate can be generalized to define the volume of regular solids. A regular solid is a figure whose opposite sides (floor and ceiling, front and back, left region and right region) have the same dimensions. Either one of these sides may be referred to as the base. The distance between the regular bases is called the height (altitude) or depth (thickness). The volume formula for a regular solid becomes:

Volume of a regular solid (cubic units)
Volume = (Area of regular side) × (Distance between bases)
Volume = (Area of the base) × (Height)
Volume = (Area of the base) × (Depth)

Consider the two circular solids (cylinders, concrete columns). The first figure has a diameter of 10 feet and a height (altitude) of 12 feet. The second figure has a diameter of 10 feet and a depth (thickness) of 12 feet.

314

In both figures consider the circle as the base. The formula for the area of a circle is
$A = \pi r^2$. The volume of the first circular solid (cylinder) is
Volume = (Area of the base) × (Height)
Volume = $\left(\pi\left(5^2 \text{ ft}^2\right)\right) \times (12 \text{ ft}) \approx 942.4777961$ ft³. (Use the calculator's π button.)

Bumping up to the next whole number, the volume is 943 cubic feet.

The volume of the second circular solid (cylinder) is
Volume = (Area of the base) × (Height)
Volume = $\left(\pi\left(5^2 \text{ ft}^2\right)\right) \times (12 \text{ ft}) \approx 942.4777961$ ft³. (Use the calculator's π button.)

Bumping up to the next whole number, the volume is 943 cubic feet.
The two circular solids have the same dimensions and the same volumes.

Example 1:
A volume of one cubic foot holds approximately 7.48 gallons. Use dimensional analysis to convert 943 cu ft to gallons.
Solution:
 Step 1: Understand and Picture the Problem.
  1. Question: Convert 943 cubic feet to gallons.
  2. There are 7.48 gallons in a cubic foot.

 Step 2: Develop a Plan.
  Use Dimensional Analysis to convert cubic feet to gallons. The term gallons must be in the numerator and cubic feet must be in the denominator.

 Step 3: Execute the plan.
$$\left(\frac{943 \text{ ft}^3}{1}\right)\left(\frac{7.48 \text{ gallons}}{1 \text{ ft}^3}\right) = \left(\frac{943 \cancel{\text{ ft}^3}}{1}\right)\left(\frac{7.48 \text{ gallons}}{1 \cancel{\text{ ft}^3}}\right) = 7,053.64 \text{ gallons}$$

Bumping up to the next whole number, a container with a volume of 943 cubic feet will hold approximately 7,054 gallons.

Step 4: Check your work.
1. Check the numbers and your calculations.
2. Is the answer reasonable?
3. Was the question answered?

Consider the figure in Example 1 from the previous section (Perimeter and Area of Irregular Shapes). Suppose this is a three-dimension figure with the following dimensions.

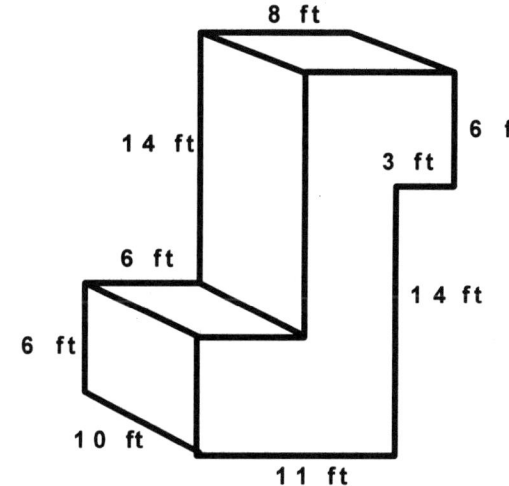

Since the area of the base is 154 sq ft, the volume of this figure is:
Volume = (Area of regular side) × (Distance between bases)
Volume = (154 sq ft) × (10 ft) = 1540 cu ft.
The volume of the figure is 1,540 cubic feet or 1,540 ft³.

Example 2: Find the volume of the solid with the bases of Example 4 from the previous section (Perimeter and Area of Irregular Shapes). The distance between bases is 8 feet 5 inches.
Solution:

Step 1: Understand and picture the problem.
1. Question: Find the volume of the solid.
2. Copy the figure and list its dimensions.
3. Example 4 gives the area of the base as 228 sq ft.

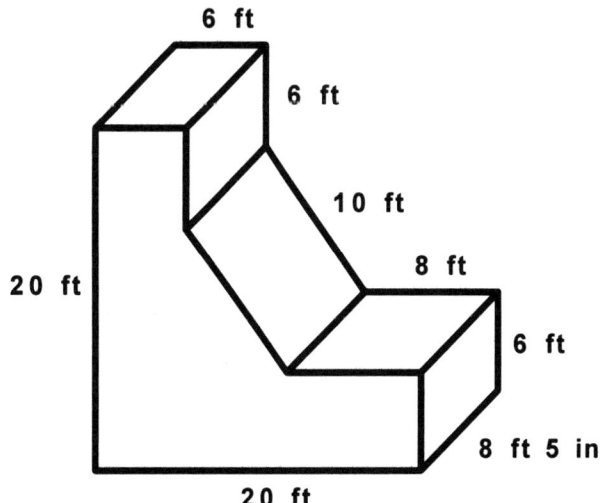

316

Step 2: Develop a Plan.
1. Use the Area from Example 4.
2. Use the volume formula:
Volume = (Area of base) × (Distance between bases).

Step 3: Execute the plan.
1. Area = 228 sq ft.
2. Volume = (Area of base) × (Distance between bases)
Volume = (228 sq ft) × $\left(8 + \dfrac{5}{12}\right)$ ft = 1,919 cu ft.

(Note: Use the parenthesis keys on your calculator.)
The volume of the figure is 1,919 cubic feet.

Step 4: Check your work.
1. Check the numbers and your calculations.
2. Is the answer reasonable?
3. Was the question answered?

Example 3: Find the volume of the solid with the bases of Example 6 of the previous section (Perimeter and Area of Irregular Shapes). The distance between bases is 12 meters.
Solution:

Step 1: Understand and picture the problem.
1. Question: Find the volume of the figure.
2. Copy the figure and list its dimensions.
3. Example 6 gives the area of the base as approximately 377.0 sq m.

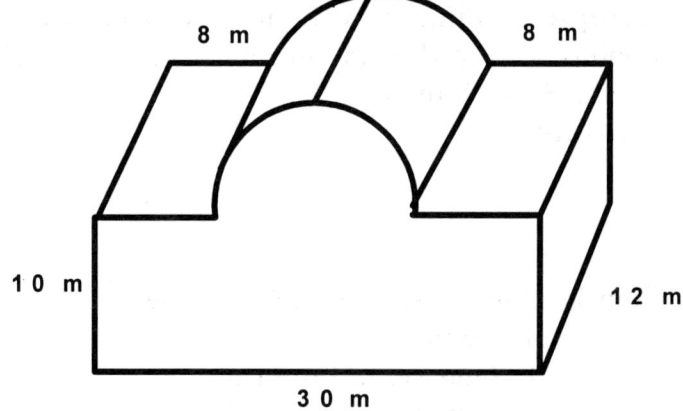

Step 2: Develop a Plan.
1. Use the Area from Example 6.
2. Use the volume formula:
Volume = (Area of the base) × (Distance between bases).

Step 3: Execute the plan.
1. Area ≈ 377 sq m.
2. Volume = (Area of the base) × (Distance between bases).
   Volume = (377 sq m) × (12 m) = 4,524 m³.
   The volume of the figure is approximately 4,524 cubic meters.

Step 4: Check your work.
1. Check the numbers and your calculations.
2. Is the answer reasonable?
3. Was the question answered?

DIMENSIONAL ANALYSIS (Revisited)
Dimensional analysis can also be used to convert the volume of any figure to different cubic units. Dimensional analysis uses definitions, (such as; 1 yard = 3 feet), and "factors of one" to convert the units of one problem into the desired units. Since the volume of Example 2 is 1,919 cu ft and **cubic feet** means **(feet)(feet)(feet)**, the following demonstrates the conversion of cubic feet into cubic yards using dimensional analysis.

$$1{,}919 \text{ ft}^3 = \frac{1{,}919(\text{ft})(\text{ft})(\text{ft})}{1} \times \frac{1 \text{ yd}}{3 \text{ ft}} \times \frac{1 \text{ yd}}{3 \text{ ft}} \times \frac{1 \text{ yd}}{3 \text{ ft}}.$$

$$1{,}919 \text{ ft}^3 = \frac{1{,}919(\cancel{\text{ft}})(\cancel{\text{ft}})(\cancel{\text{ft}})}{1} \times \frac{1 \text{ yd}}{3 \cancel{\text{ft}}} \times \frac{1 \text{ yd}}{3 \cancel{\text{ft}}} \times \frac{1 \text{ yd}}{3 \cancel{\text{ft}}} = \frac{1{,}919 \text{ yd}^3}{27} \approx 71.07407407 \text{ cu yd.}$$

Bumping up to the next whole number, the volume of the figure is approximately 72 cubic yards.

Notice that **three factors** of $\frac{1 \text{ yd}}{3 \text{ ft}}$ are necessary for converting cubic feet into cubic yards. The length, width and height (three dimensions) have to be converted into yards. The cubic yard unit is a cube that is 1 yard on each side. The cubic yard is also comprised of 27 cubic feet. Using dimensional analysis:

$$1 \text{ cu yd} = \frac{1(\text{yd})(\text{yd})(\text{yd})}{1} \left(\frac{3 \text{ ft}}{1 \text{ yd}}\right)\left(\frac{3 \text{ ft}}{1 \text{ yd}}\right)\left(\frac{3 \text{ ft}}{1 \text{ yd}}\right).$$

$$1 \text{ cu yd} = \frac{1(\cancel{\text{yd}})(\cancel{\text{yd}})(\cancel{\text{yd}})}{1}\left(\frac{3 \text{ ft}}{1 \cancel{\text{yd}}}\right)\left(\frac{3 \text{ ft}}{1 \cancel{\text{yd}}}\right)\left(\frac{3 \text{ ft}}{1 \cancel{\text{yd}}}\right) = 27 \text{ cu ft.}$$

Example 4: Use dimensional analysis to convert 1,919 cu ft to cubic inches.
Solution:
Step 1: Understand and Picture the Problem.
1. Question: Convert 1,919 cubic feet to cubic inches.
2. There are 12 inches in 1 foot.

Step 2: Develop a Plan.
Use Dimensional Analysis to convert cubic feet to cubic inches. Use three factors of $\dfrac{12 \text{ in}}{1 \text{ ft}}$ to perform the conversion.

Step 3: Execute the plan.
$$V = 1{,}919 \text{ ft}^3 = \dfrac{1{,}919(\text{ft})(\text{ft})(\text{ft})}{1}\left(\dfrac{12 \text{ in}}{1 \text{ ft}}\right)\left(\dfrac{12 \text{ in}}{1 \text{ ft}}\right)\left(\dfrac{12 \text{ in}}{1 \text{ ft}}\right).$$
$$V = 1{,}919 \text{ ft}^3 = \dfrac{1{,}919(\cancel{\text{ft}})(\cancel{\text{ft}})(\cancel{\text{ft}})}{1}\left(\dfrac{12 \text{ in}}{1 \cancel{\text{ft}}}\right)\left(\dfrac{12 \text{ in}}{1 \cancel{\text{ft}}}\right)\left(\dfrac{12 \text{ in}}{1 \cancel{\text{ft}}}\right).$$
V = 3,316,032 cu in.
The volume of the figure is 3,316,032 cubic inches.

Step 4: Check your work.
1. Check the numbers and your calculations.
2. Is the answer reasonable?
3. Was the question answered?

Automotive conversions

The automotive industry used to measure engines in cubic inches. Today liters are used to measure the engine size.

Note:
1 liter = 1,000 milliliters (ml) = 1,000 cubic centimeters (cc).

(c) 2006 JupiterImages Corp.

Example 5: Given 1 liter equals 1,000 cubic centimeters (cc) and 1 inch equals 2.54 centimeters, use dimensional analysis to convert 3.2 liters to cubic inches.
Solution:
Step 1: Understand and Picture the Problem.
1. Question: Convert 3.2 liters to cubic inches.
2. There are 1,000 cc per liter.
3. There are 2.54 centimeters in 1 inch.
Step 2: Develop a Plan.
1. Use Dimensional Analysis to convert liters to cubic centimeters. Use one factor of $\dfrac{1{,}000 \text{ cc}}{1 \text{ liter}}$ to perform the conversion.

2. Use Dimensional Analysis to convert cubic centimeters to cubic inches. Use three factors of $\frac{1 \text{ in}}{2.54 \text{ cm}}$ to perform the conversion.

Step 3: Execute the plan.

$$V = 3.2 \text{ liters} = \frac{3.2 \text{ liters}}{1}\left(\frac{1,000 \text{ cc}}{1 \text{ liter}}\right) = \frac{3.2 \cancel{\text{liters}}}{1}\left(\frac{1000 \text{ cc}}{1 \cancel{\text{liter}}}\right) = 3,200 \text{ cc}$$

$$V = 3,200 \text{ cc} = \frac{3,200(\cancel{\text{cm}})(\cancel{\text{cm}})(\cancel{\text{cm}})}{1}\left(\frac{1 \text{ in}}{2.54 \cancel{\text{cm}}}\right)\left(\frac{1 \text{ in}}{2.54 \cancel{\text{cm}}}\right)\left(\frac{1 \text{ in}}{2.54 \cancel{\text{cm}}}\right)$$

$V \approx 195.3$ cu in.

Bumping up to the next whole number, the volume of the engine is approximately 196 cubic inches.

Step 4: Check your work.
1. Check the numbers and your calculations.
2. Is the answer reasonable?
3. Was the question answered?

?   **Cognitive Problems**   **?**

1. Explain why the unit of volume is a "cubic" measurement.
2. Explain why volume unit conversions must be done with three factors of one.
3. Give an example of a "real life" volume measurement.

Exercise 4.4

(c) 2006 JupiterImages Corp.

I. Find the volume of these three-dimensional solids based on the figures in Exercise 4.3. Bump up volumes to the next whole number.

1.
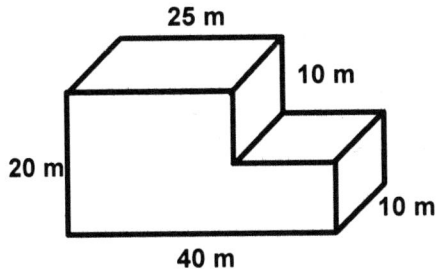

2.
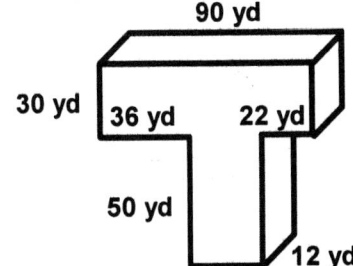

320

3.

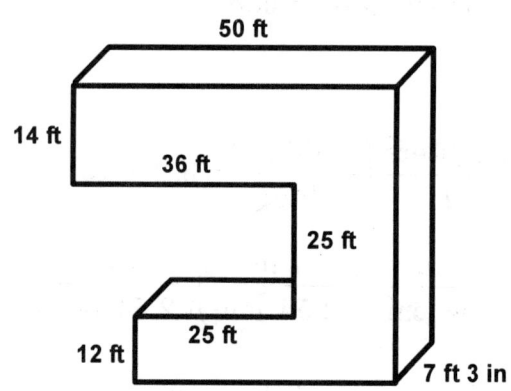

4.

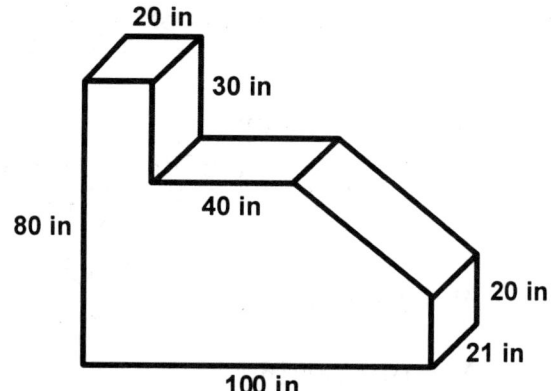

5.

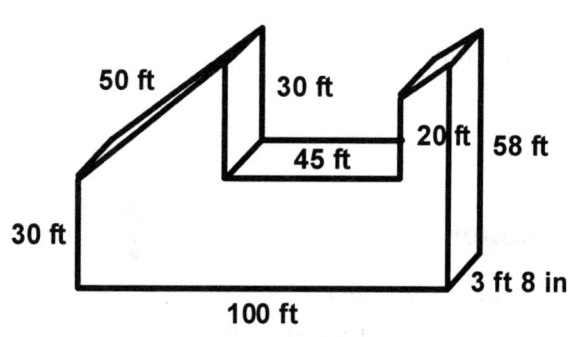

6.

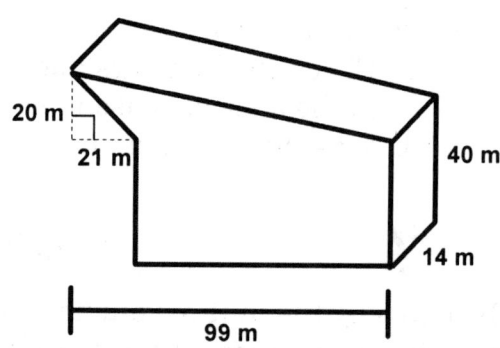

7.

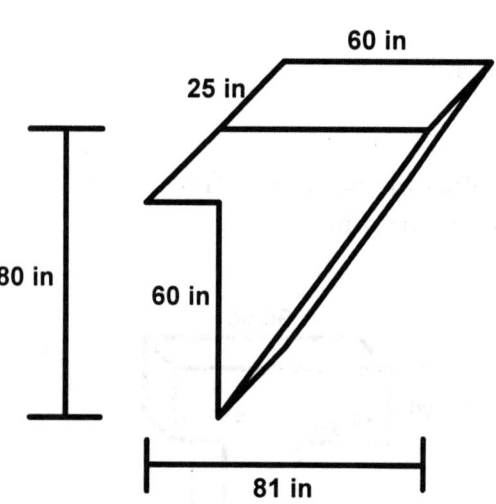

8.

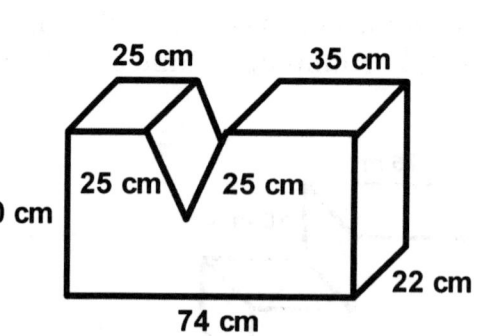

9.

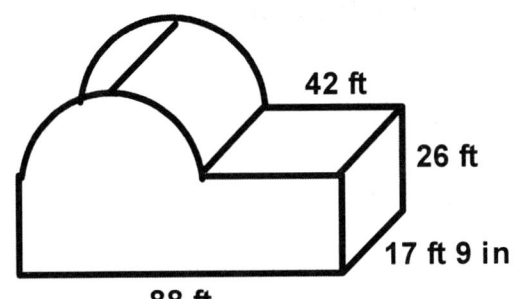

10.

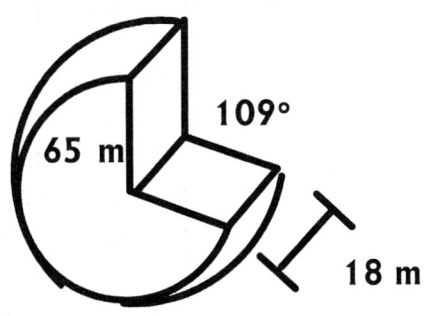

11.

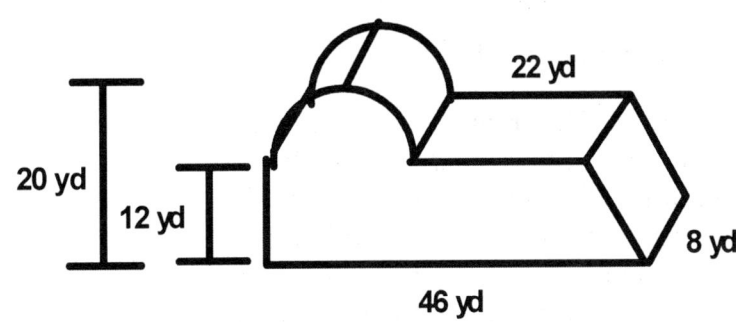

12.

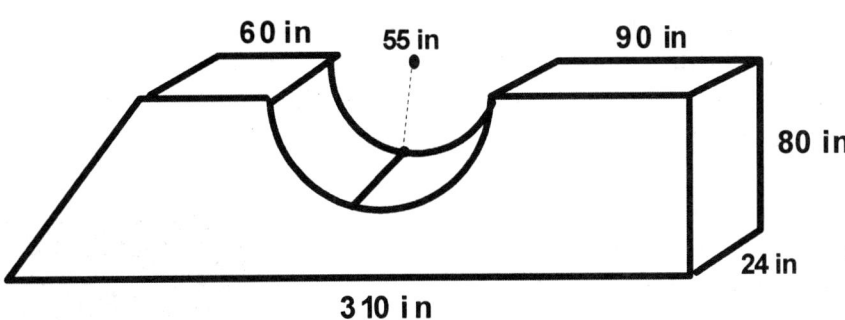

II. Use Dimensional Analysis to convert the volumes from Problem I to the specified dimension. Bump up volumes to the next whole number.

Dimensions
1,000 millimeters (mm) = 1 meter
100 centimeters (cm) = 1 meter
3.28 ft ≈ 1 meter
1.09 yd ≈ 1 meter
1 inch ≈ 25.4 mm
1 inch ≈ 2.54 cm
1 mile = 5,280 feet
3 feet (ft) = 1 yard (yd)
36 inches (in) = 1 yard (yd)
12 inches (in) = 1 foot (ft)
1 cu ft ≈ 7.48 gallons

1. Convert the volume of Exercise 1 to cubic feet.
2. Convert the volume of Exercise 2 to cubic feet.
3. Convert the volume of Exercise 3 to gallons.
4. Convert the volume of Exercise 4 to cubic feet.
5. Convert the volume of Exercise 5 to cubic yards.
6. Convert the volume of Exercise 6 to cubic yards.
7. Convert the volume of Exercise 7 to cubic feet.
8. Convert the volume of Exercise 8 to cubic inches.
9. Convert the volume of Exercise 9 to cubic yards.
10. Convert the volume of Exercise 10 to cubic yards.
11. Convert the volume of Exercise 11 to cubic feet.
12. Convert the volume of Exercise 12 to cubic yards.

III. Automotive Conversions.
Round answer to the tenths.
Note: 1 liter = 1,000 ml = 1,000 cc.
1. Convert a 327 cubic inch engine to liters.
2. Convert a 409 cubic inch engine to liters.
3. Convert a 454 cubic inch engine to liters.
4. Convert a 2.2 liter engine to cubic inches.
5. Convert a 3.8 liter engine to cubic inches.
6. Convert a 5.7 liter engine to cubic inches.

(c) 2006 JupiterImages Corp.

4.5 DO-IT-YOURSELF

Objective: Determine the quantity needed and the cost of materials for "do-it-yourself" projects.

Note: A fraction of a tile, gallon of paint, or bolt of wallpaper cannot be purchased. Another tile, gallon of paint, bolt of wallpaper must be purchased to cover this fraction. For this reason the answer to perimeter, area, and volume problems will be bumped up to the next whole number. If a shape has to be partitioned, the intermediate perimeters and areas will be rounded to tenths. One concern in renovation projects is the quantity of material needed. Another is the material cost. Use the following floor plan for Examples 1 through 8.

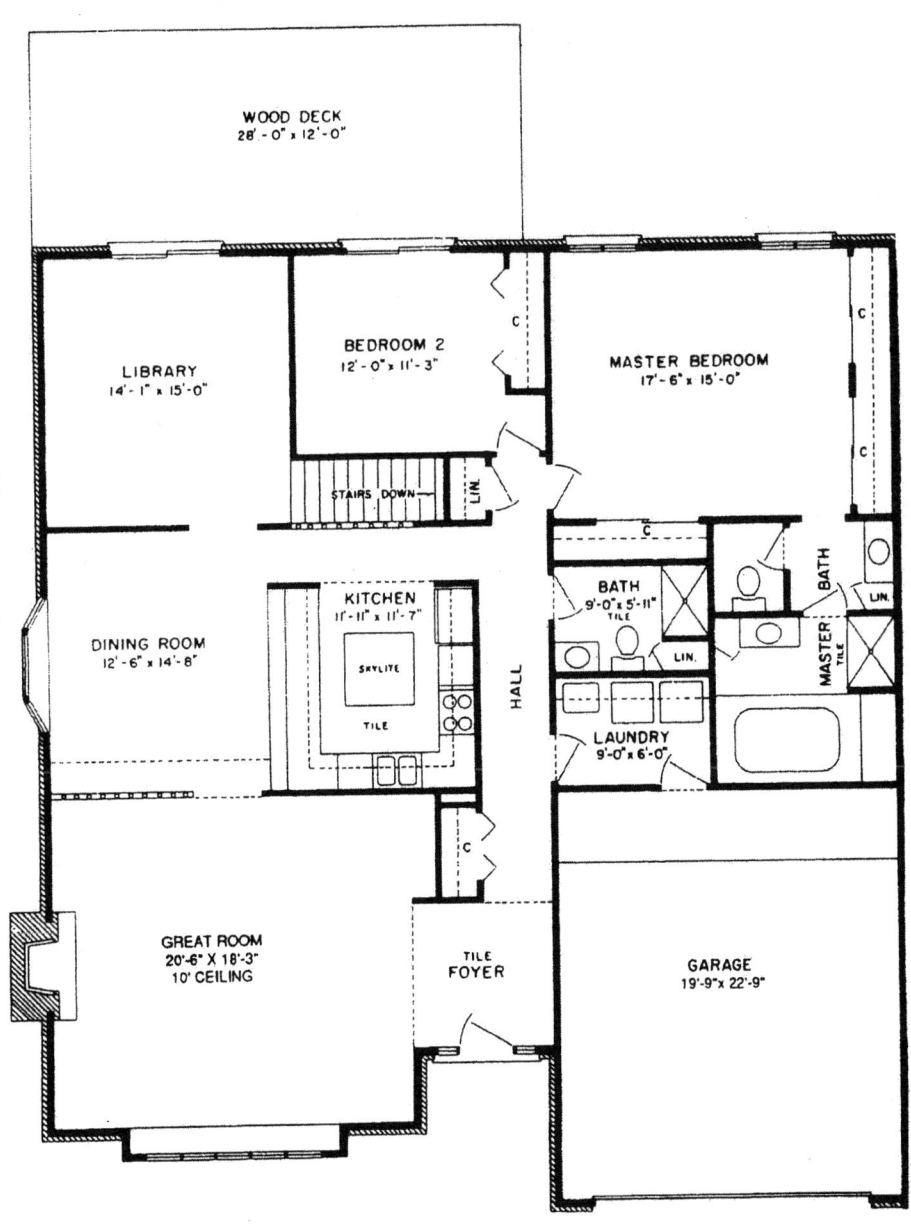

Example 1: A 12 inch by 12 inch tile (1 sq ft) costs $0.79. What would it cost to tile the KITCHEN?
Solution:
  Step 1: Understand and picture the problem.
   1. Question: Find the cost of tiling the kitchen.
   2. Draw a diagram of the room.
   3. List the given data and label it on the diagram.
      Length = 11 ft 11 in,   Width = 11 ft 7 in,
      Cost is $0.79 a sq ft.

  Step 2: Establish a plan.
   1. To find the area of the kitchen, use the formula for the area of a rectangle. Area = (Length) x (Width)
   2. Substitute the room's dimensions into the formula to calculate the area.
   3. Multiply the area of the room by the cost per square foot to find the cost of tiling the kitchen.

  Step 3: Execute the plan.
   1. Area = (Length)(Width).
   2. Use the parenthesis key on your calculator and substitute the kitchen's dimensions for length and width in the following manner.
      $$\text{Area} = \left(11+\frac{11}{12}\right) \times \left(11+\frac{7}{12}\right) \approx 138.0347222 \text{ sq ft}$$
      Bump up the area of the kitchen to 139 sq ft (Note: You cannot purchase a fraction of a tile.)
   3. $\text{Cost} = \left(\dfrac{139(\text{ft})(\text{ft})}{1}\right) \times \left(\dfrac{\$0.79}{(\text{ft})(\text{ft})}\right)$ (Note: Area is square feet.)

      $\text{Cost} = \left(\dfrac{139(\cancel{\text{ft}})(\cancel{\text{ft}})}{1}\right) \times \left(\dfrac{\$0.79}{(\cancel{\text{ft}})(\cancel{\text{ft}})}\right) = \$109.81$

      It costs $109.81 to tile the kitchen floor.

  Step 4: Check your work.
   1. Check the numbers and your calculations.
   2. Is the answer reasonable?
   2. Was the question answered?

Example 2: A gallon of paint costs $21.97 and covers 350 sq ft. What does it cost to purchase the paint needed to cover the 4 walls of the DINING ROOM? (The standard height of a wall is 8 feet.)

Solution:

Step 1: Understand and Picture the problem.
1. Question: Calculate the cost of purchasing the paint needed to cover the four walls of the dining room.
2. Draw a simple diagram of the walls to be painted.
3. List the given data and label it on each diagram:
   Length = 12 ft 6 in,   Width = 14 ft 8 in,   Height = 8 ft.
   Since the floor is 12 ft 6 in by 14 ft 8 in, two walls have these dimensions for its base. The height of the walls are 8 feet.

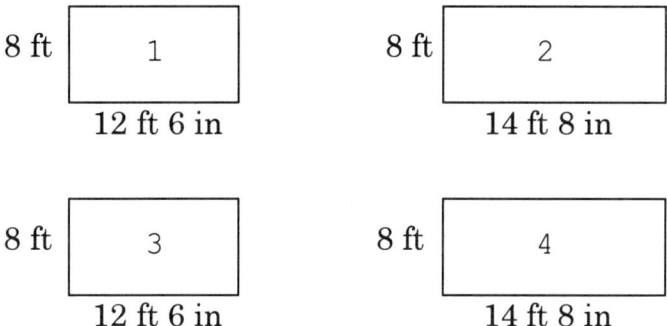

3. Paint costs $\frac{\$21.97}{\text{gallon}}$ and covers $\frac{350 \text{ sq ft}}{\text{gallon}}$.

Step 2: Establish a plan.
1. Use the formula for the area of a rectangle:
   Area = (Length) x (Width) = (Base) x (Height).
2. Substitute the dining room's dimensions into the area formula to calculate the area of each wall.
3. Add the areas of the four walls to get the total area.
4. Calculate how many gallons are needed by dividing the total area by
   $$\frac{350(\text{ft})(\text{ft})}{1 \text{ gallon}}$$
5. Multiply the number of gallons needed by $\frac{\$21.97}{\text{gallon}}$.

Step 3: Execute the plan.
1. Area = (Length) x (Width) = (Base) x (Height).
2. Using the parenthesis key on your calculator, substitute the dining room's dimensions into the formula in the following manner.

$$\text{Wall 1} = \left(12 + \frac{6}{12}\right) \times (8) = 100 \text{ sq ft.}$$

$$\text{Wall 2} = \left(14 + \frac{8}{12}\right) \times (8) \approx 117.3 \text{ sq ft.}$$

$$\text{Wall 3} = \left(12 + \frac{6}{12}\right) \times (8) = 100 \text{ sq ft.}$$

$$\text{Wall 4} = \left(14 + \frac{8}{12}\right) \times (8) \approx 117.3 \text{ sq ft.}$$

3. Total area = 100 + 117.3 + 100 + 117.3 = 434.6 sq ft.
   Bump the area up to 435 square feet.

4. Gallons needed: $\dfrac{435(\text{ft})(\text{ft})}{1} \times \dfrac{1 \text{ gallon}}{350(\text{ft})(\text{ft})} =$

   $\dfrac{435(\cancel{\text{ft}})(\cancel{\text{ft}})}{1} \times \dfrac{1 \text{ gallon}}{350(\cancel{\text{ft}})(\cancel{\text{ft}})} \approx 1.2$ gallons.

   You cannot purchase a fraction of a gallon. Bump up to the next whole number. 2 gallons of paint are needed to paint the dining room.

5. Cost = $\dfrac{2 \text{ gallons}}{1} \times \dfrac{\$21.97}{\text{gallon}}$.

   Cost = $\dfrac{2 \cancel{\text{gallons}}}{1} \times \dfrac{\$21.97}{\cancel{\text{gallon}}} = \$43.94.$

   It costs $43.94 to paint the four walls of the dining room.

Step 4: Check your work.
   1. Check the numbers and your calculations.
   2. Is the answer reasonable?
   3. Was the question answered?

Example 3: Using the data from Example 2, is there enough paint leftover to paint the ceiling?
Solution:
Step 1: Understand and Picture the problem.
   1. Question: Is enough paint remaining to paint the ceiling?
   2. Review Example 2 and transfer the pertinent information.
      The total area of the walls is 435 sq feet.
      The two gallons of paint purchased will cover 700 sq feet.

$$2 \;\cancel{\text{gallons}} \left(\frac{350 \text{ ft}^2}{\cancel{\text{gallon}}} \right) = 700 \text{ square feet.}$$

There is 700 sq ft − 435 sq ft = 265 sq ft of paint remaining.

3. Sketch the ceiling and label its dimensions on the diagram:
   Length = 12 ft 6 in
   Width = 14 ft 8 in

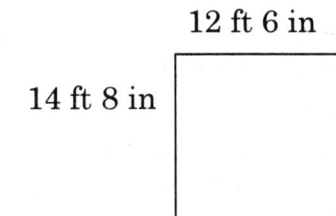

Step 2: Establish a plan.
1. Use the equation, Area = (Length) x (Width) = (Base) x (Height)
2. Substitute the dining room's dimensions into the area formula to calculate the area of the ceiling.
3. If the area of the ceiling is less than 265 sq ft, than there is enough paint remaining to paint the ceiling.

Step 3: Execute the plan.
1. Area = (Length) x (Width)
2. Using the parenthesis key on your calculator, substitute the dining room's dimensions into the formula in the following manner.

   Area of ceiling = $\left(12 + \frac{6}{12}\right) \times \left(14 + \frac{8}{12}\right) \approx 183.3333333$ sq ft.

   Bumping up the area of the ceiling to 184 sq ft.
3. Since the area of the ceiling, 184 sq ft, is less than the 265 sq ft of paint remaining, there is enough paint left over to paint the ceiling.

Step 4: Check your work.
1. Check the numbers and calculations.
2. Is the answer reasonable?
3. Was the question answered?

Example 4: What is the cost of purchasing acoustical ceiling panels for the LIBRARY. The ceiling panels are 2 ft by 2 ft and cost $2.79 each.
Solution:
Step 1: Understand and Picture the problem.
1. Question: Determine the cost of purchasing ceiling panels for the library.
2. Sketch the library and label its dimensions on the diagram.

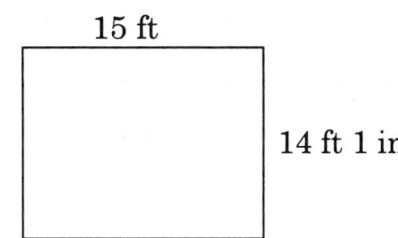

Length = 15 ft, Width = 14 ft 1 in.
3. List the given data: Cost is $2.79 per tile (2 ft by 2 ft).

Step 2: Establish a plan.
1. Use the formula for the area of a rectangle:
Area = (Length) x (Width) = (Base) x (Height).
2. Substitute the room's dimensions into the formula and calculate the area.
3. Calculate the area of a ceiling tile.
4. Calculate the number of ceiling tiles by dividing the area of the ceiling by the area of one ceiling tile.
5. Calculate total cost of the ceiling tiles by multiplying the number of tiles by the cost per tile.

Step 3: Execute the plan.
1. Area = (Length) x (Width) = (Base) x (Height).
2. Using the parenthesis key on your calculator, substitute the library's dimensions into the formula in the following manner.

$$\text{Area} = (15) \times \left(14 + \frac{1}{12}\right) = 211.25.$$

Bumping up the area of the library's ceiling to 212 sq ft.
3. Area of a ceiling tile = (2 ft)(2 ft) = 4 sq ft.
4. Number of tiles needed is:

$$\frac{\text{Area of Ceiling}}{\text{Area of a Tile}} = \frac{212(\text{ft})(\text{ft})}{4(\text{ft})(\text{ft})} = \frac{\overset{53}{\cancel{212}}(\cancel{\text{ft}})(\cancel{\text{ft}})}{\underset{1}{\cancel{4}}(\cancel{\text{ft}})(\cancel{\text{ft}})} = 53 \text{ tiles.}$$

5. Cost = $53 \text{ tiles}\left(\frac{\$2.79}{\text{tile}}\right) = 53 \,\cancel{\text{tiles}} \left(\frac{\$2.79}{\cancel{\text{tile}}}\right) = \$147.87.$

The cost of ceiling tiles for the library is $147.87.

Step 4: Check your work.
1. Check the numbers and calculations.
2. Is the answer reasonable?
3. Was the question answered?

EXAMPLE 5: Antique oak ceiling molding costs $25.95 per 8 foot section. What would it cost in material to install this molding in the GREAT ROOM?

Solution:
  Step 1: Understand and Picture the problem.
    1. Question: Determine the cost of
       antique oak ceiling molding for
       the great room.
    2. Sketch the ceiling of the great
       room and label its dimensions
       on the diagram.
       Length = 20 ft 6 in,   Width = 18 ft 3 in.
    3. List the given data: Cost of molding is $25.95 per 8 foot section.

  20 ft 6 in

  18 ft 3 in

  Step 2: Establish a plan.
    1. Use the formula for the perimeter of a rectangle:
       Perimeter = Find the sum of the four sides.
    2. Substitute the room's dimensions into the formula and calculate the
       perimeter.
    3. Find the number of sections of molding needed by dividing the
       perimeter by 8.
    4. Calculate cost of the ceiling molding by multiplying the number of
       sections by the cost per section.

  Step 3: Execute the plan.
    1. Perimeter = The sum of the four sides.
    2. Substitute the great room's dimensions into the formula in the
       following manner.
       $$\text{Perimeter} = \left(20 + \frac{6}{12}\right) + \left(18 + \frac{3}{12}\right) + \left(20 + \frac{6}{12}\right) + \left(18 + \frac{3}{12}\right) = 77.5 \text{ ft.}$$
       Bump up the perimeter to 78 ft.
    3. $\frac{78 \text{ ft}}{1}\left(\frac{1 \text{ section}}{8 \text{ ft}}\right) = \frac{78 \cancel{\text{ ft}}}{1}\left(\frac{1 \text{ section}}{8 \cancel{\text{ ft}}}\right) = 9.75$ sections

       Bumping up the number of sections to 10 sections.
       (Note: You cannot purchase a fraction of a section of molding.)
    4. Cost = $10 \text{ sections} \times \left(\frac{\$25.95}{\text{section}}\right) = 10 \cancel{\text{ sections}} \times \left(\frac{\$25.95}{\cancel{\text{section}}}\right) = \$259.50$

  Step 4: Check your work.
    1. Check the numbers and calculations.
    2. Is the answer reasonable?
    3. Was the question answered?

EXAMPLE 6: Oak paneling cost $5.89 per sheet. A sheet of paneling is 8 feet tall and 4 feet wide. What is the cost of paneling for the walls of the LIBRARY.
Solution:
Step 1: Understand and Picture the problem.
1. Question: Determine the cost of oak paneling for the library.
2. Sketch the library and label its dimensions on the diagram.
Width = 14 ft 1 in,   Length = 15 ft.
3. List the given data:
Cost of one 4 ft by 8 ft panel is $5.89.

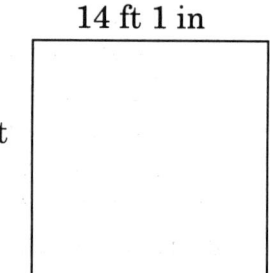

Step 2: Establish a plan.
1. This may be considered a perimeter problem. Since both the panel and walls of the room are 8 feet tall, the height may be ignored. View this as a one dimensional problem. Place four foot sections, the width of a panel, along the base of the room.
2. Use the formula for the perimeter of a rectangle:
Perimeter = The sum of the four sides.
3. Substitute the dimensions into the formula and find the perimeter.
4. Divide the perimeter by 4 feet to determine the number of panels.
5. Calculate cost by multiplying the number of panels by the panel cost.

Step 3: Execute the plan.
1. Use the concept of perimeter to solve the problem.
2. Perimeter = The sum of the four sides.
3. Substitute the library's dimensions into the formula manner.

$$\text{Perimeter} = \left(14 + \frac{1}{12}\right) + (15) + \left(14 + \frac{1}{12}\right) + (15) = 58.16666667.$$

Bump the perimeter up to 59 ft.

4. The number of panels needed is

$$59 \text{ ft}\left(\frac{\text{panel}}{4 \text{ ft}}\right) = 59 \; \cancel{\text{ft}} \left(\frac{\text{panel}}{4 \cancel{\text{ft}}}\right) = 14.75 \text{ panels.}$$ Bumping up the library requires 15 panels.

Note: You cannot purchase a fraction of a panel, so the result was bumped up to the next whole number.

5. Cost = $15 \text{ panels} \times \left(\frac{\$5.89}{\text{panel}}\right) = 15 \; \cancel{\text{panels}} \times \left(\frac{\$5.89}{\cancel{\text{panel}}}\right) = \$88.35.$

The cost of paneling for the library is $88.35.

Step 4: Check your work.
   1. Check the numbers and calculations.
   2. Is the answer reasonable?
   3. Was the question answered?

Example 7: What is the cost of materials (concrete) to replace the present wood deck with a cement patio? Concrete costs $71.75 a "*yard*". (In this context, a cubic yard is understood for the term "*yard*".) The concrete patio has a depth of 4 inches.
Solution:
   Step 1: Understand and Picture the problem.
      1. Sketch the patio and label its dimensions.
         Length = 28 ft, Width = 12 ft,
         Depth = 4 in.

         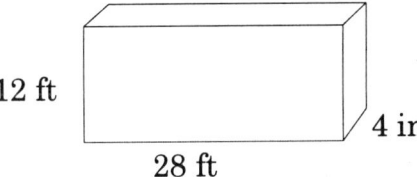

      2. List the given data:
         Cost is $71.75 per cubic yard.
      3. Problem: Find the cost of cement for the patio.

   Step 2: Establish a plan.
      1. Substitute the patio's dimensions into the formula for area of a rectangle: Area = (Length) x (Width), and the formula for the volume of a rectangular solid: Volume = (Area of the base) x (Depth)
      2. Calculate the volume of the deck in cubic feet.
      3. Use dimensional analysis to convert cubic feet into cubic yards.
      4. Calculate total cost by multiplying the volume by the cost per yard (cubic yard).

   Step 3: Execute the plan.
      1. Area = (Length) x (Width) .
         Volume = (Area of the base) x (Depth).
      2. Using the parenthesis key on your calculator, substitute the patio's dimensions into the calculator in the following manner.
      3. Area = (28 ft) x (12 ft) = 336 sq ft.
         Volume = 336 sq ft x $\left(\dfrac{4}{12} \text{ ft}\right)$ = 112 cu ft.

         Note: Units need to be the same, 4 inches = $\dfrac{4}{12}$ ft.

      4. Volume = $\dfrac{112(\text{ft})(\text{ft})(\text{ft})}{1} \times \dfrac{1 \text{ yd}}{3 \text{ ft}} \times \dfrac{1 \text{ yd}}{3 \text{ ft}} \times \dfrac{1 \text{ yd}}{3 \text{ ft}}$ (Volume is cubic feet.)

         Volume = $\dfrac{112(\cancel{\text{ft}})(\cancel{\text{ft}})(\cancel{\text{ft}})}{1} \times \dfrac{1 \text{ yd}}{3 \cancel{\text{ft}}} \times \dfrac{1 \text{ yd}}{3 \cancel{\text{ft}}} \times \dfrac{1 \text{ yd}}{3 \cancel{\text{ft}}} \approx 4.1481481$ cu yds.

332

Since you cannot purchase a fraction of a cubic yard, 5 "*yards*" of concrete need to be ordered.

5. $\text{Cost} = 5(\text{yd})(\text{yd})(\text{yd})\left(\dfrac{\$71.75}{(\text{yd})(\text{yd})(\text{yd})}\right).$

$\text{Cost} = 5(\cancel{\text{yd}})(\cancel{\text{yd}})(\cancel{\text{yd}})\left(\dfrac{\$71.75}{(\cancel{\text{yd}})(\cancel{\text{yd}})(\cancel{\text{yd}})}\right) = \$358.75.$

Concrete costs $358.75 for the patio.

Step 4: Check your work.
1. Check the numbers and calculations.
2. Is the answer reasonable?
3. Was the question answered?

Example 8: Use the diagram to find the cost of concrete to pour the foundation for this home. Concrete costs $71.75 a "*yard*". (A cubic yard is understood for the term "*yard*".) The foundation has a height of 7 ft 9 inches and a thickness of 10 inches. The floor is not included, only the concrete for the walls is required.

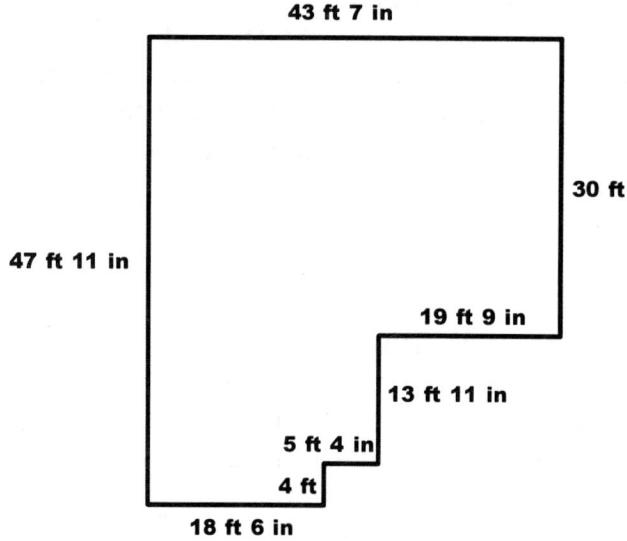

Solution:
Step 1: Understand and Picture the problem.
1. Blueprint of foundation is given.
Cost is $71.75 per cubic yard.

Height is 7 ft 9 inches.
Thickness is 10 inches.
2. Problem: Find the cost of pouring the foundation.

Step 2: Establish a plan.
1. This may be considered a perimeter problem. Substitute the foundation's perimeter as the length, 7 ft 9 inches as the width and 10 inches as the depth into the formula for volume of a rectangle. Volume = (Length) x (Width) x (Depth).
2. Calculate the volume in cubic feet.
3. Use dimensional analysis to convert cubic feet into cubic yards.
4. Calculate total cost by multiplying the volume by the cost per yard (cubic yard).

Step 3: Execute the plan.
1. Using the parenthesis key on your calculator, substitute the dimensions into the calculator in the following manner.

$$\text{Perimeter} = \left(43 + \frac{7}{12}\text{ft}\right) + 30\text{ ft} + \left(19 + \frac{9}{12}\text{ft}\right) + \left(13 + \frac{11}{12}\text{ft}\right) + \left(5 + \frac{4}{12}\text{ft}\right)$$
$$+ 4\text{ ft} + \left(18 + \frac{6}{12}\text{ft}\right) + \left(47 + \frac{11}{12}\text{ft}\right).$$

Perimeter = 183 feet.

2. $\text{Volume} = (183\text{ ft}) \times \left(7 + \frac{9}{12}\text{ft}\right) \times \left(\frac{10}{12}\text{ft}\right) = 1{,}181.875$ cubic feet.

Bump the volume up to 1,182 cubic feet.

3. $\text{Volume} = \frac{1{,}182(\text{ft})(\text{ft})(\text{ft})}{1} \times \frac{1\text{ yd}}{3\text{ ft}} \times \frac{1\text{ yd}}{3\text{ ft}} \times \frac{1\text{ yd}}{3\text{ ft}}.$

$\text{Volume} = \frac{1{,}182(\cancel{\text{ft}})(\cancel{\text{ft}})(\cancel{\text{ft}})}{1} \times \frac{1\text{ yd}}{3\cancel{\text{ft}}} \times \frac{1\text{ yd}}{3\cancel{\text{ft}}} \times \frac{1\text{ yd}}{3\cancel{\text{ft}}} \approx 43.77777778$ cu yds.

Since you cannot purchase a fraction of a cubic yard, 44 "*yards*" of concrete need to be ordered.

4. $\text{Cost} = \left(\frac{44(\text{yd})(\text{yd})(\text{yd})}{1}\right) \times \left(\frac{\$71.75}{(\text{yd})(\text{yd})(\text{yd})}\right).$

$\text{Cost} = \left(\frac{44(\cancel{\text{yd}})(\cancel{\text{yd}})(\cancel{\text{yd}})}{1}\right) \times \left(\frac{\$71.75}{(\cancel{\text{yd}})(\cancel{\text{yd}})(\cancel{\text{yd}})}\right) = \$3{,}157.00.$

It costs $3,157.00 for the concrete for the foundation.

Step 4: Check your work.
1. Check the numbers and calculations.
2. Is the answer reasonable?
3. Was the question answered?

 (c) 2006 JupiterImages Corp.	**Optional Student Project** Project: Sweat Equity can be found on page 339.
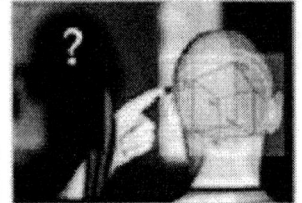	**Chapter Review** Chapter review problems can be found on page 343.

? Cognitive Problems ?

1. Explain the difference between perimeter, area, and volume.
2. Explain why a paneling problem can be viewed as a perimeter problem.
3. Explain why the concrete patio problem is a volume problem.
4. Explain why the carpeting problem is an area problem.
5. Give an example of another perimeter problem that can be applied to this house.
6. Give an example of another area problem that can be applied to this house.
7. Give an example of another volume problem that can be applied to this house.

(c) 2006 JupiterImages Corp.

Exercise 4.5

Use the floor plan on the next page for problems 1 through 8. All walls are 8 ft tall. Perimeter, area, and volume problems will be bumped up to the next whole number. If a shape has to be partitioned, the intermediate perimeters and areas will be rounded to tenths.

1. What is the cost of installing $15\frac{1}{4}$ inch by $15\frac{1}{4}$ inch ceramic squares in the dining room? A carton of tile costs $37.39 and covers 12.94 sq feet.

2. What is the cost of installing a wood floor in the living room. The cost of wood planks is $49.82 per carton? A carton covers 18.8 square feet.

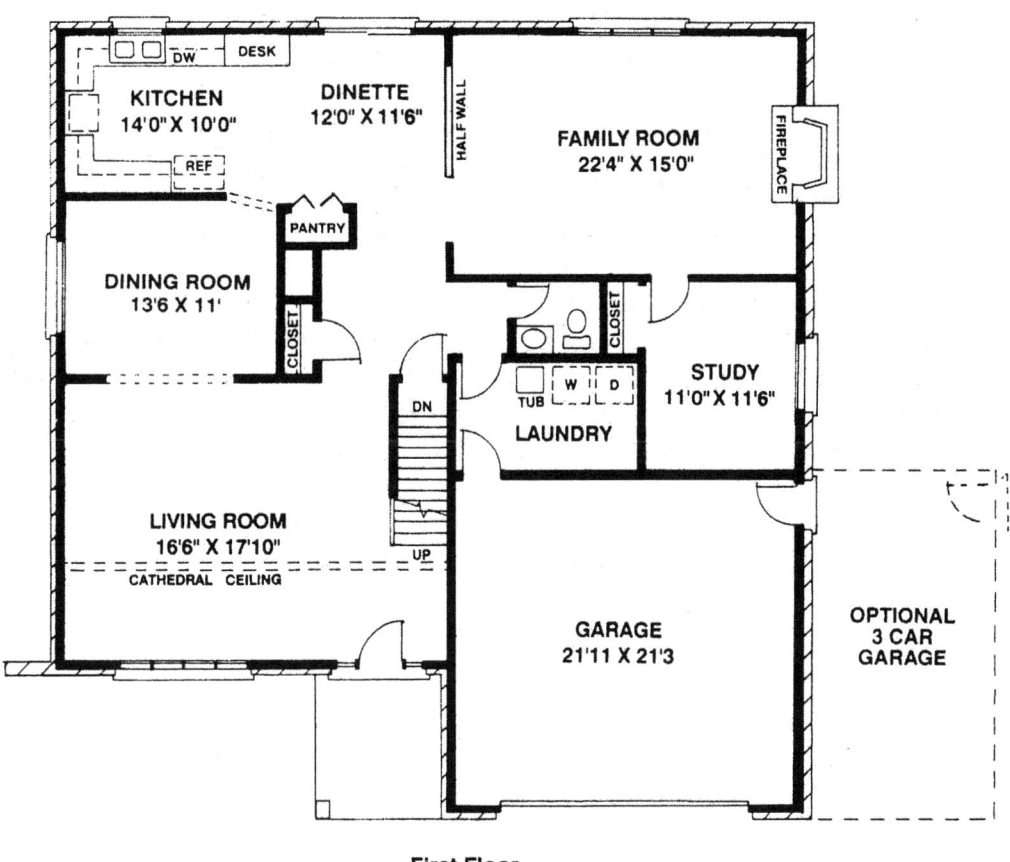

First Floor

3. What is the cost of wallpapering the family room? A double bolt of wallpaper cover 56.37 square feet and costs $14.96.

4. What is the cost of painting the living room, ceiling included? A gallon of paint covers 350 sq. ft. and costs $17.98 per gallon.

5. What is the cost of installing vinyl flooring in the kitchen? The cost of vinyl flooring is $9.66 per square yard.

6. What is the cost of installing a border around the ceiling the dining room? A roll of border covers 15 ft. and costs $12.47

7. What is the cost of installing wood paneling in the study? The panels are 8 ft tall and 4 ft wide and cost $19.99 per panel.

8. Use the chart to determine what size ceiling fan is needed for the living room.

Blade Size	Volume
29 inch fan	Up to 400 cu ft
36 inch fan	Up to 600 cu ft
42 inch fan	Up to 800 cu ft
50 inch or 52 inch fan	Up to 3,200 cu ft
54 inch fan	Over 3,200 cu ft

9. What is the cost of installing sod in the yard? The outside dimensions of the home measures 32 ft by 64 ft. A 30 ft by 20 ft rectangular region (driveway), an 18 ft (diameter) swimming pool (circular) and a 24 ft by 32 ft rectangular region (garden) have no sod. A roll of sod measures 6 feet (Length) and 18 inches (Width) and costs $1.19 per roll. (Note: The house is on a cul-de-sac.)

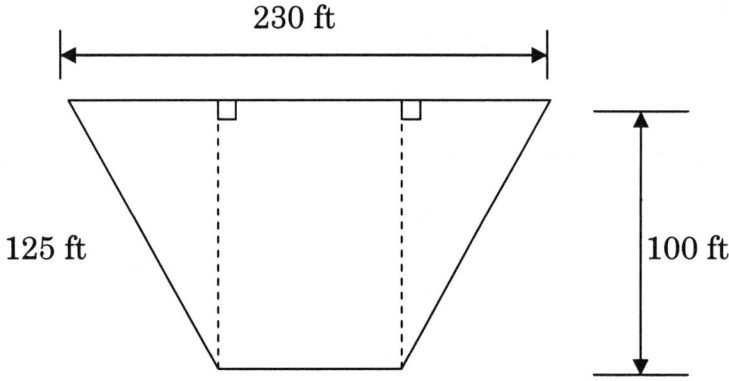

10. Concrete costs $94.75 a "*yard*". (In this context, a cubic yard is understood for the term "*yard*".) Find the cost of pouring the concrete floor for a 3 car garage. The dimensions are 32 feet wide by 21 feet 3 inches long and 6 inches thick.

11. What is the cost of material to shingle the roof? The roof sections are rectangular and there are:
    2 sections measuring 38 ft 10 inches by 15 feet 3 inches
    2 sections measuring 26 ft 6 inches by 14 ft 5 in.
    2 sections measuring 16 ft by 17 ft. 11 inches.

    A bundle of shingles cost $12.79 and covers $33\frac{1}{3}$ square feet.

12. Use the diagram to find the cost of materials (concrete) to pour the foundation for this home (not including the floor). Concrete costs $94.75 a "*yard*". The concrete foundation has a height of 7 ft 9 inches and a thickness of 10 inches

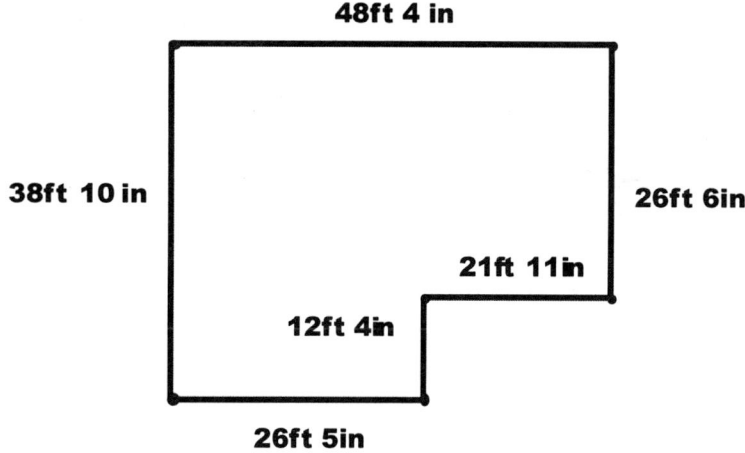

Project: National Debt
Scenario:

(c) 2006 JupiterImages Corp.

The United States National Debt is approaching $8.2 Trillion ($8,200,000,000,000)

How much money is $8.2 Trillion?

I. Use Dimensional Analysis to calculate how tall a stack of one dollar bills would be in miles to equal $8.2 Trillion. Show all your work listing the names of each dimension used.

II. How may round trips to the moon and back is the miles calculated in Problem I?

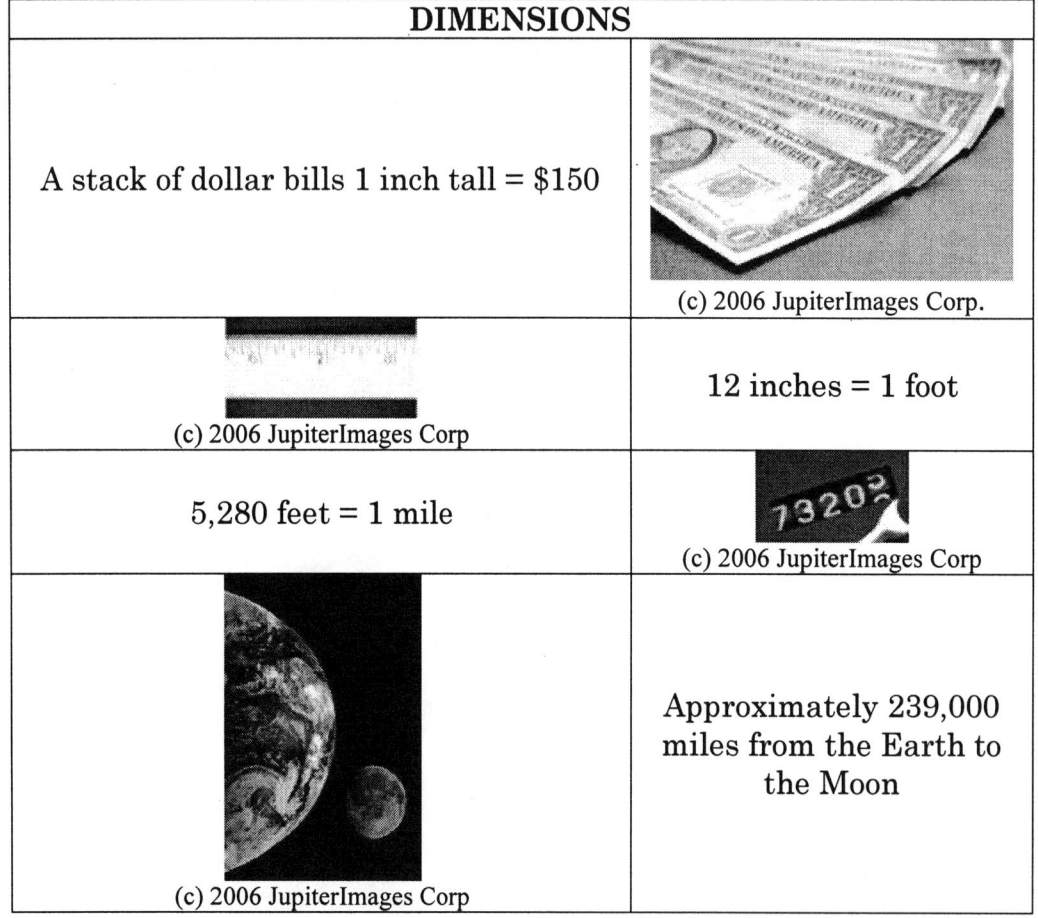

DIMENSIONS	
A stack of dollar bills 1 inch tall = $150	(c) 2006 JupiterImages Corp.
(c) 2006 JupiterImages Corp	12 inches = 1 foot
5,280 feet = 1 mile	(c) 2006 JupiterImages Corp
(c) 2006 JupiterImages Corp	Approximately 239,000 miles from the Earth to the Moon

Project: Sweat Equity

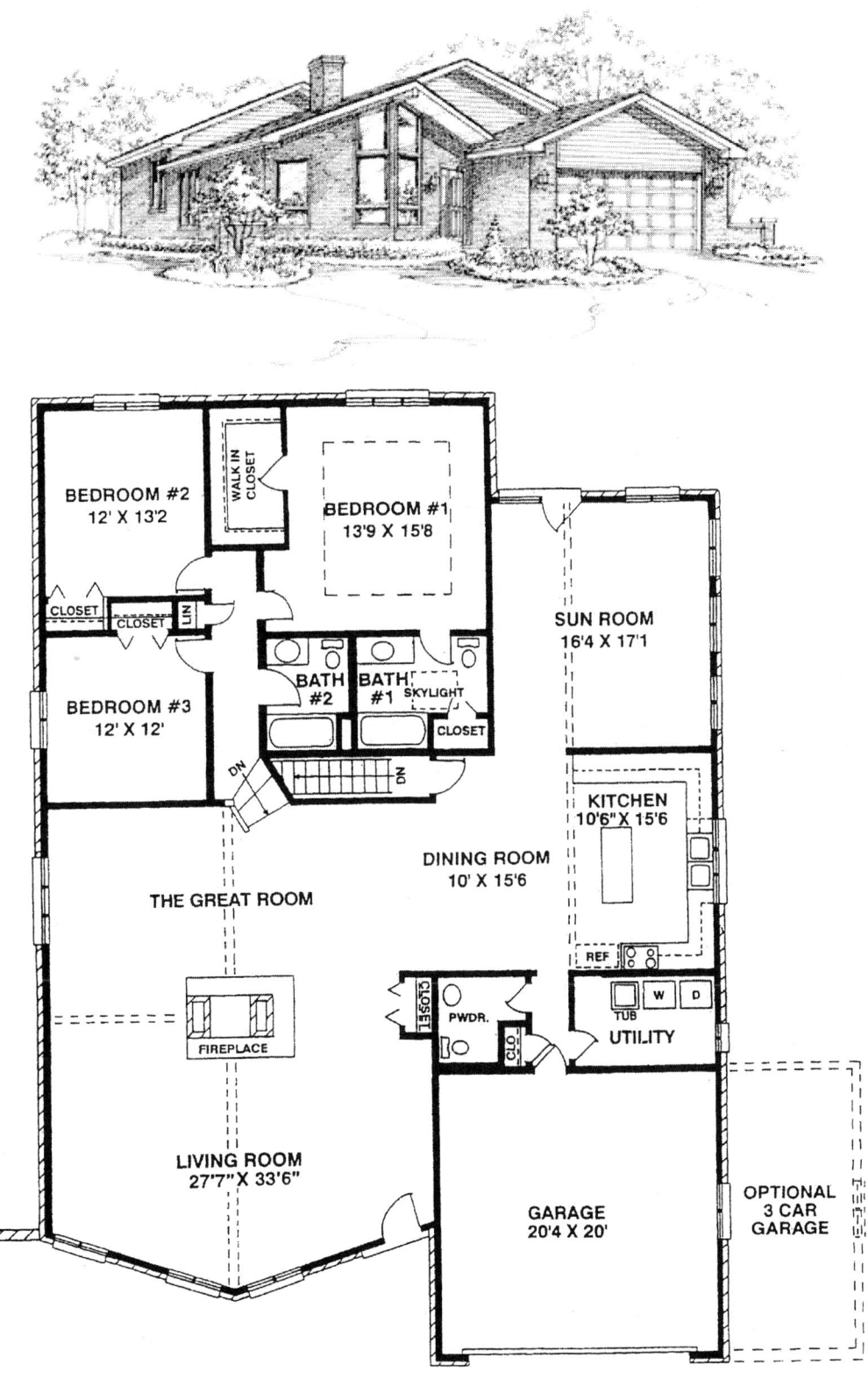

Objective: The cost of purchasing the material for home projects will be determined.

Use the floor plan on the previous page for this project. Show all formulas used and the work to support your results. Keep your work neat and organized. Work that needs deciphering will not be graded.

Note: A fraction of a tile, gallon of paint, etc. cannot be purchased. Another tile, gallon of paint, etc. must be purchased to cover this fraction. Thus, the answer to perimeter, area, and volume problems will be bumped up to the next whole number. If a shape has to be partitioned, the intermediate perimeters and areas will be rounded to tenths.

1. Determine the cost of installing a wood floor in the dining room. A carton of wood flooring costs $86.11 and covers 20.67 sq ft.

2. Use the diagram below to calculate the cost of installing carpeting in the Great Room and Living Room. (They are considered one room.) Do not worry about the see-through two-sided fireplace in the middle of the room. Carpeting costs $25.79 per square yard (padding included).

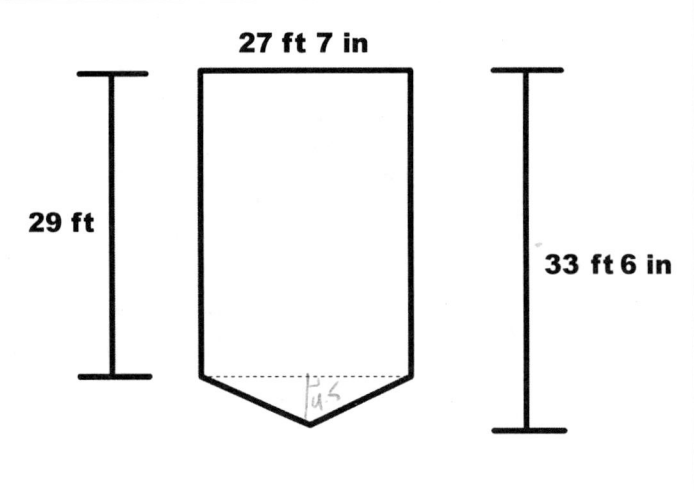

3. Determine the cost of installing a ceramic floor in the sun room. The tile is 9 inches by 9 inches and costs $34.88 per carton of 24 tiles.

4. Determine the cost of painting the walls and ceiling in Bedroom #1. A gallon of paint covers 320 sq ft and cost $24.99. (Note: The walls are 8 feet tall.)

341

5. Determine the cost of purchasing an air purifier for Bedroom #2. Choose the air purifier that is closest to, but greater than the volume of the room. (Note: The walls are 8 feet tall.)

Air Purifier Model Number	Volume (cu ft)	Cost
AP100	432	$49.99
AP200	864	$109.99
AP300	1040	$129.99
AP400	1344	$149.99
AP500	1560	$169.99
AP600	3520	$199.99
AP700	3840	$269.99

6. Antique oak ceiling moulding costs $38.85 per 12 foot section. What would it cost in material to put this moulding in Bedroom #1?

7. What is the cost of installing wood paneling in Bedroom #3? The panels are 8 ft tall and 4 ft wide and cost $24.87 per panel.

8. Find the square footage of the lot to be seeded. The house has a living area of 2,675 square feet. The 3-car garage measures 32 feet wide by 20 feet 4 inches long. The driveway is 32 feet by 50 feet. The rectangular inground swimming pool is 100 feet by 40 feet. The garden is 40 feet by 30 feet. The tennis court is 48 feet by 90 feet. The paving brick patio is 12 feet by 16 feet. The circular pond has a diameter of 12 feet.

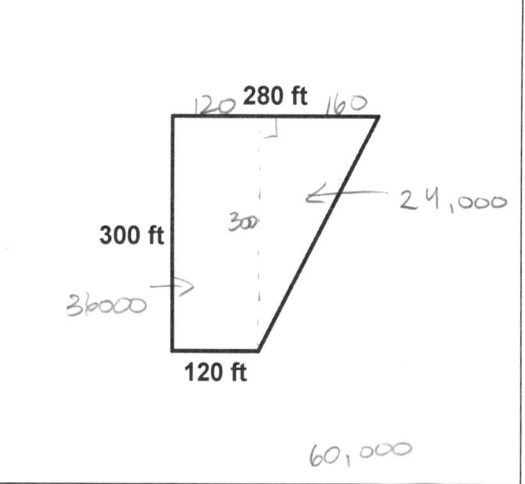

9. Concrete costs $97.79 a "*yard*". Find the cost of pouring the concrete floor for a 3 car garage. The dimensions are 32 feet wide by 20 feet 4 inches long and 6 inches thick.

10. Concrete costs $97.79 a "*yard*". Find the cost of pouring the concrete driveway. The dimensions are 32 feet wide by 50 feet long and 4 inches thick.

342

11. Use the diagram on the next page to find the cost of materials (concrete) to pour the foundation for this home. Just the walls, the floor is not included. Concrete costs $97.79 a "*yard*". The concrete foundation has a height of 7 ft 9 inches and a thickness of 10 inches

12. What is the cost of material to shingle the roof? The roof sections are:
2 sections measuring 10 feet 2 inches by 12 feet 10 inches
2 sections measuring 33 feet 6 inches by 16 feet.
2 sections measuring 25 feet by 26 feet 11 inches.
2 sections measuring 40 feet 7 inches by 11 feet 5 inches
A bundle of shingles cost $12.79 and covers $33\frac{1}{3}$ square feet.

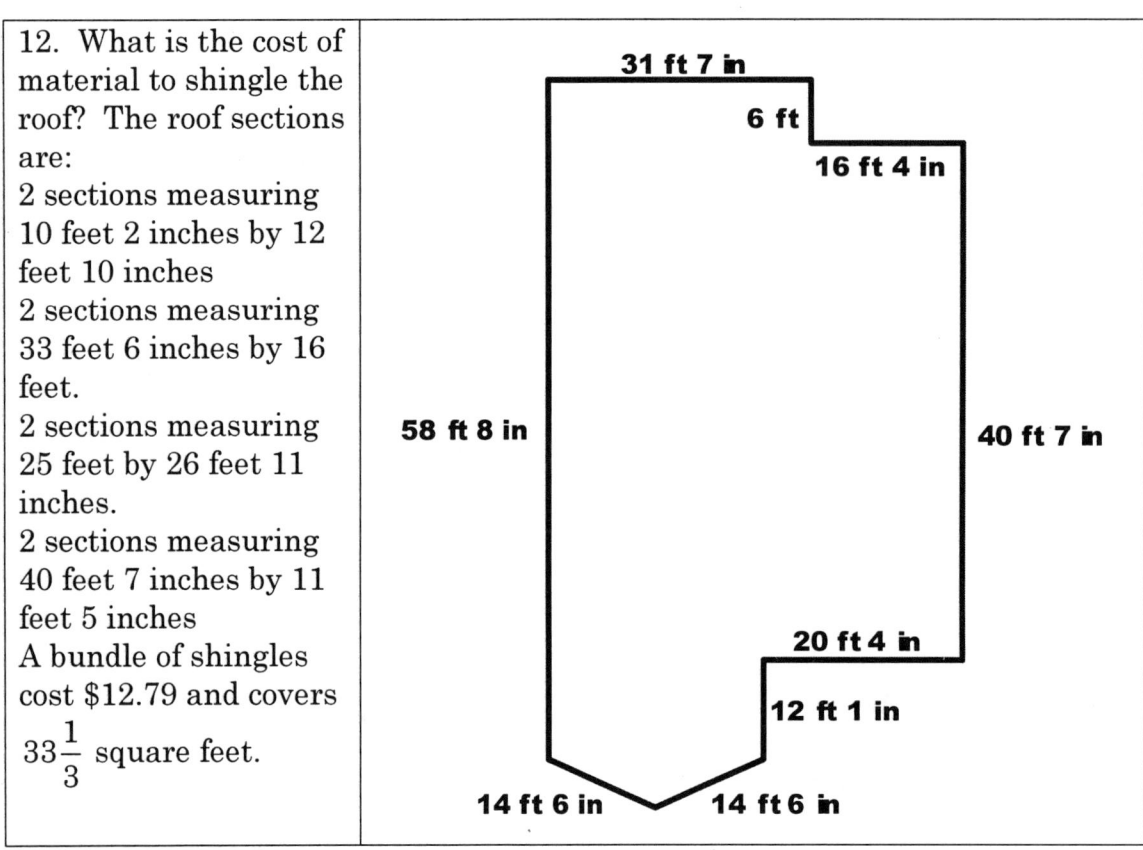

13. What is the cost of wallpapering the bedroom #1. A double bolt of wallpaper covers 56.37 square feet and costs $16.96.

14. Retaining wall bricks measuring 8" in length are used to make the walls of a pond. The pond is circular with a diameter of 12 feet. The wall will be 5 layers tall. Each brick cost $0.98. What is the cost of bricks for the pond?

15. A 12 foot by 16 foot patio is to be constructed off the sun room. Paving bricks measuring 4 inches by 8 inches will be used for the patio. What is the cost of the bricks for the patio if each brick costs $0.79.

CHAPTER 4 REVIEW

Section 4.1: Perimeter and Circumference

I. Find the perimeter (circumference) for the following shapes. Round your answers to tenths.

1.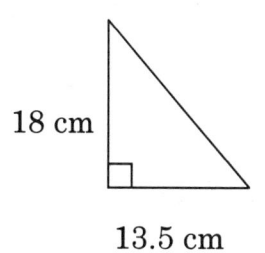

2.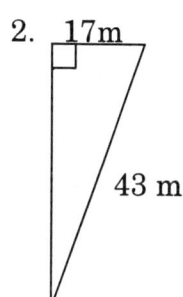

3.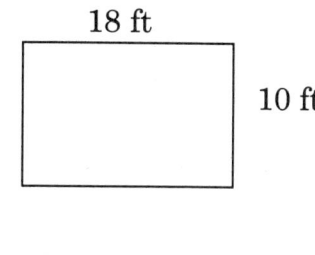

4.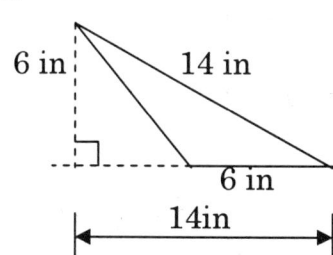

5.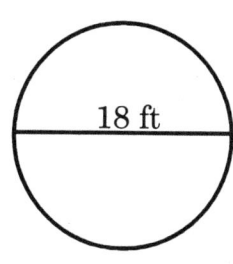

6.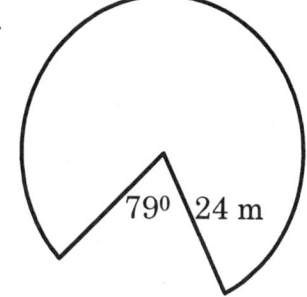

II. Use Dimensional Analysis to convert the perimeters from Problem I to the specified dimensions. Round your answer to hundredths.
   7. Convert the perimeter from Exercise 1 to meters.
   8. Convert the perimeter from Exercise 2 to centimeters.
   9. Convert the perimeter from Exercise 3 to yards.
   10. Convert the perimeter from Exercise 4 to feet.
   11. Convert the perimeter from Exercise 5 to inches.
   12. Convert the perimeter from Exercise 6 to centimeters.

Section 4.2: Area

I. Find the area for the following shapes. Bump up your answer to the next. whole number.

13.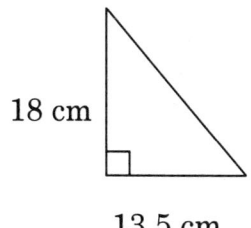

14.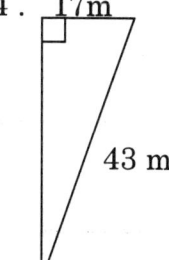

15.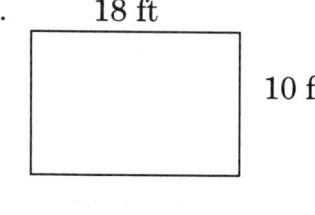

344

16.   17.   18.

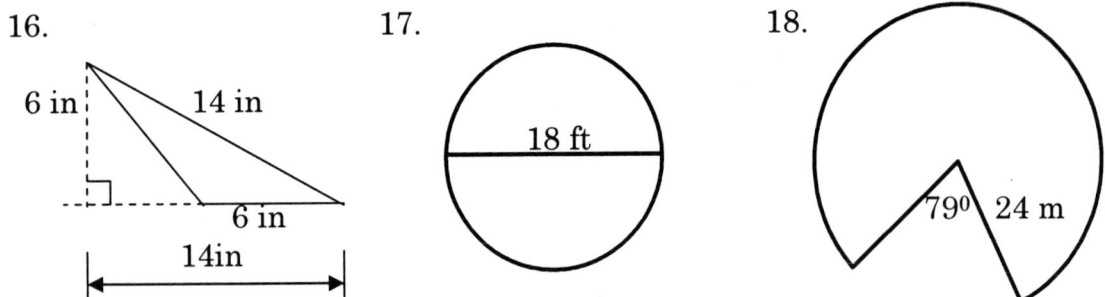

II. Use Dimensional Analysis to convert the areas problem I to the specified dimensions. Bump answers up to the next whole number.
19. Convert the area from Exercise 13 to square inches.
20. Convert the area from Exercise 14 to square centimeters.
21. Convert the area from Exercise 15 to square yards.
22. Convert the area from Exercise 16 to square centimeters.
23. Convert the area from Exercise 17 to square inches.
24. Convert the area from Exercise 18 to square centimeters.

Section 4.3: Irregular Shapes
I. Find the perimeter and area for the following shapes. Bump answers up to the next whole number. Round intermediate perimeters and areas of the partitions to tenths

25.   26.

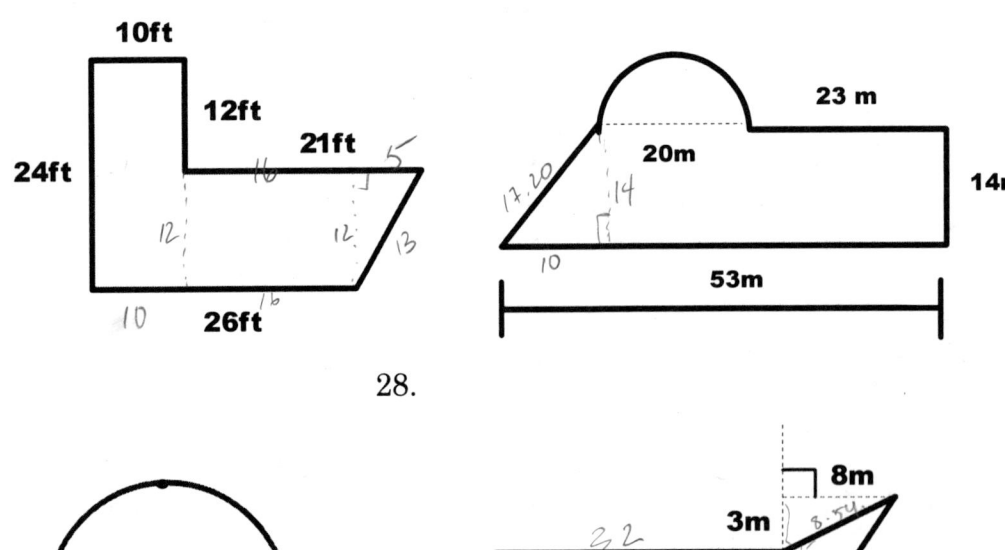

27.   28.

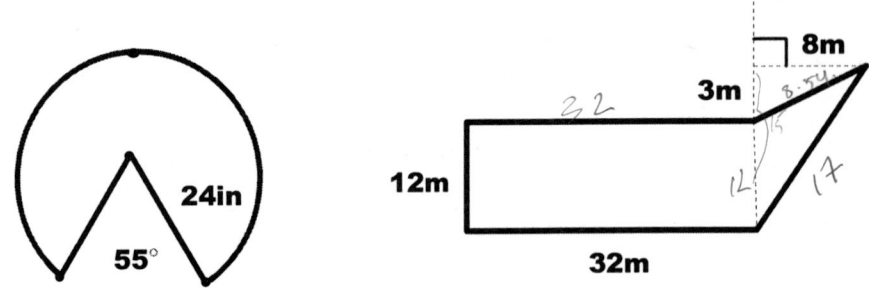

345

II. Use Dimensional Analysis to perform the following conversions. Bump answers up to the next whole number.
29. Convert the perimeter from Exercise 25 to yards.
30. Convert the perimeter from Exercise 26 to centimeters.
31. Convert the perimeter from Exercise 27 to feet.
32. Convert the perimeter from Exercise 28 to centimeters.
33. Convert the area from Exercise 25 to square yards.
34. Convert the area from Exercise 26 to square centimeters.
35. Convert the area from Exercise 27 to square feet.
36. Convert the area from Exercise 28 to square centimeters.

Section 4.4: Volume
Find the volume for the following shapes. (See Review Section 4.3 for areas.) Bump answers up to the next whole number.

37.   38.

39.   40.

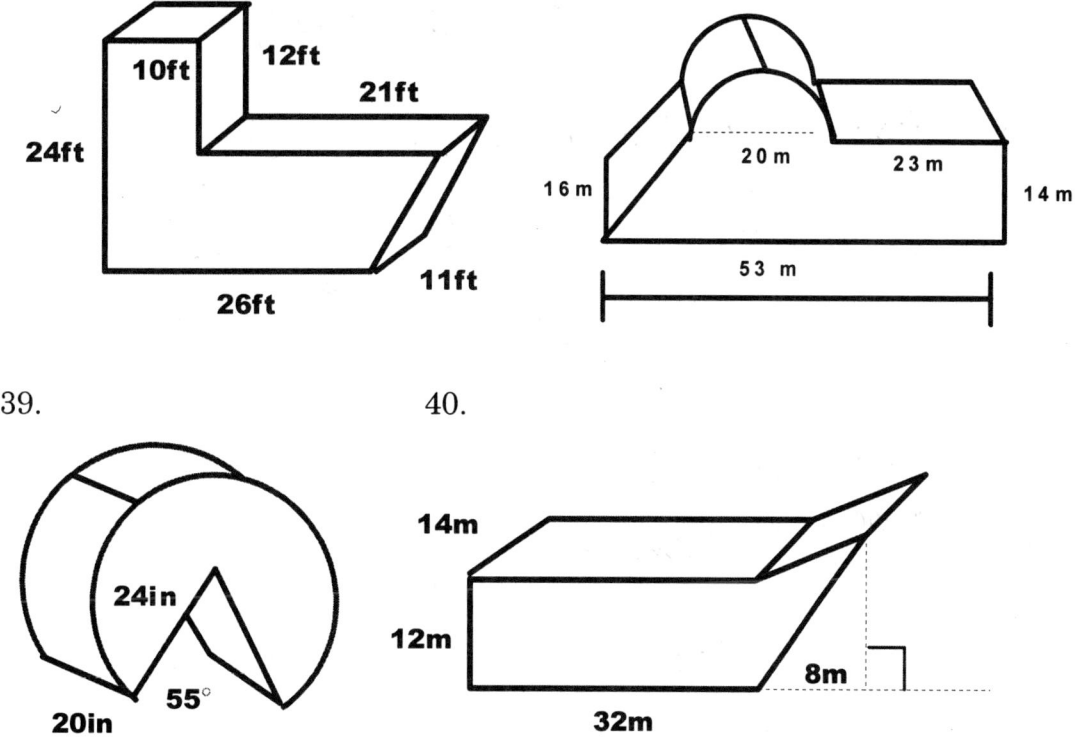

II. Use Dimensional Analysis to convert the volumes from Problem I to the specified dimension. Bump answers up to the next whole number.
41. Convert the volume of Exercise 37 to cubic yards.
42. Convert the volume of Exercise 38 to cubic centimeters.
43. Convert the volume of Exercise 39 to cubic feet.
44. Convert the volume of Exercise 40 to cubic centimeters.

346

III. Automotive conversion.
   45. Convert a 289 cubic inch engine to liters. Round to tenths.
   46. Convert a 2.8 liter engine to cubic inches. Bump answer up to the next whole number.

Section 4.5: Do it Yourself

47. The breakdown for the living room is given. Determine the cost of installing a wood floor in the family room. A carton of wood flooring costs $86.11 and covers 20.67 sq ft.

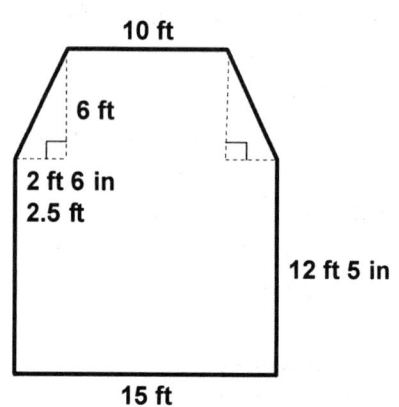

48. Determine the cost of installing carpeting in the Master Bedroom. The carpeting costs $21.98 per sq yd.

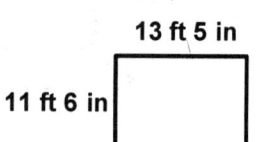

49. Determine the cost of installing a ceramic floor in the library. (It's going to be the game room; air hockey table, pinball machine, etc.) The tile is 8 in by 8 in and costs $26.88 per carton of 24 tiles.

50. Use the dimensions of the library in problems 49 to determine the cost of painting the four walls and ceiling. A gallon of paint covers 320 sq ft and costs $19.99.

51. Determine the cost of purchasing an air purifier for the Nursery (your infant child has some breathing problems). Choose the air purifier that is closest to, but greater than the volume of the room. (Note: The walls are 8 feet tall.)

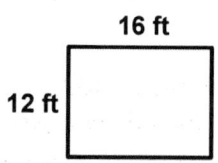

Air Purifier Model Number	Volume (cu ft)	Cost
AP100	432	$49.99
AP200	864	$109.99
AP300	1040	$129.99
AP400	1344	$149.99
AP500	1560	$169.99
AP600	3520	$199.99
AP700	3840	$269.99

52. Antique oak ceiling molding costs $3.22 a foot. Use the dimensions of the Master Bedroom in problem 48 to determine the cost in material to put antique oak molding in this room?

53. A yard of concrete cost $92.50. What will it cost in material to install a 15 ft 7 in by 18 ft 5 in by 4 in rectangular concrete patio off the dinette?

54. Find the square footage of the lot to be seeded. The house measures 2,035 square feet. The driveway is 50 feet by 30 feet. The circular above ground swimming pool is 20 feet across (diameter). The garden is 40 feet by 30 feet. The basketball area is 40 feet by 40 feet. The patio is 15 ft 7 in by 18 ft 5 in.

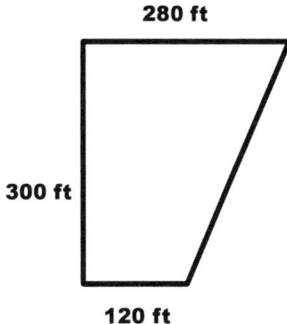

CHAPTER 5: STATISTICS

5.1 DESCRIPTIVE STATISTICS
Objective: Organize raw data into tables and graph data.

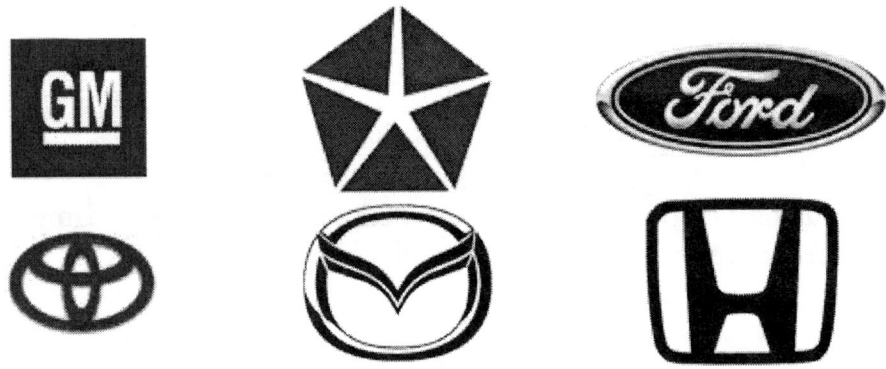

Statistics and Quality
Every industry has standards or specifications that their products must satisfy. Any industry that ignores the quality of its product will lose customers to a more quality conscious competitor. This was the harsh lesson the American auto industry learned in the 1970's.

(c) 2006 JupiterImages Corp

The 1973 oil embargo caused a gas shortage in America. Gas prices soared, gas lines were blocks long and some gas stations ran out of gas. The American car owner demanded improved gas mileage. A significant proportion of the American public, dissatisfied with the performance of the American automobile purchased Japanese automobiles. The Japanese produced an auto that provided exceptional gas mileage. This lead to the death of the "muscle cars," no one wanted to buy a car that got 8 miles or less per gallon.

Japanese car owners soon realized that their new vehicles required fewer repairs than their former American cars plus the dealerships had superior customer service. Not only did the Japanese cars get better gas mileage, they cost less to maintain. This was the beginning of the American fascination with the Japanese automobile. This perception that the American automobile is inferior to an imported automobile still exists. Some American car owners have never purchased an American made car. Today the American auto industry is very quality conscious. The quality of the American automobile equals, if not surpasses, its Japanese counterpart. Descriptive Statistics is the means used by the auto industry in pursuit of quality products.

DESCRIPTIVE STATISTICS

There are two types of data: **Qualitative (quality)** and **Quantitative (quantity)**. Qualitative data is used to describe attributes and form categories. An attribute or category is a quality that can be **counted** and **not measured**. The number of males or females in a room, how many Democrats or Republicans are in a voting district, etc. are examples of attributes or characteristics. The statistician generally assigns a counting number (no fractions) to an attribute. Fractions are inappropriate (with the exception of money.) There cannot be 8.25 males in a room or 16,423.48 Republicans in a voting district. However, there can be 8 females in a room and 16,423 Republicans in the voting district.

Quantitative data is measured. Heights, weights, distances, etc. are quantities that are measured and contain fractions. There can be a height of $5\frac{1}{2}$ feet a weight of 3.25 pounds and a distance of 26.4 miles. In this section qualitative data will be studied.

Bar Graphs and Pie Charts

Bar graphs and pie charts are two types of graphs used to display qualitative data. Before constructing bar graphs and pie charts, study the following properties for graphs of qualitative data.

Properties of Qualitative Graphs	

Applies to all graphs
 1. The graph has a definitive title describing the data being graphed.
 2. All data is accounted:
    a. The sum of the frequencies accounts for all data values.
    b. The sum of the relative frequencies adds to 1.00 (Percents sum to 100).
       Note: Due to rounding, the sum may be close to 1.00 (Percents close
           to 100.) For each category, there could be an error of $\pm 1\%$.
           100% should be contained within this interval.	

Specific to Bar Graphs	**Specific to Pie Charts**
3. Each axis has a general descriptive label. 4. Each axis has a specific label. On one axis, the bars are identified with a defining label. On the other axis, the bar's lengths are quantified with a number. An appropriate scale is chosen to display the bar's lengths.	3. Each slice is identified with a defining label. 4. The sum of the degrees in a pie chart adds to 360 degrees, but degree measures are **NOT** labeled on the pie chart. Either frequency or relative frequency label the slices. **Note:** Due to rounding, the sum may be close to 360 degrees. For each category, there could be an error of ± 1 degree. 360 degrees should be contained within this interval.

350

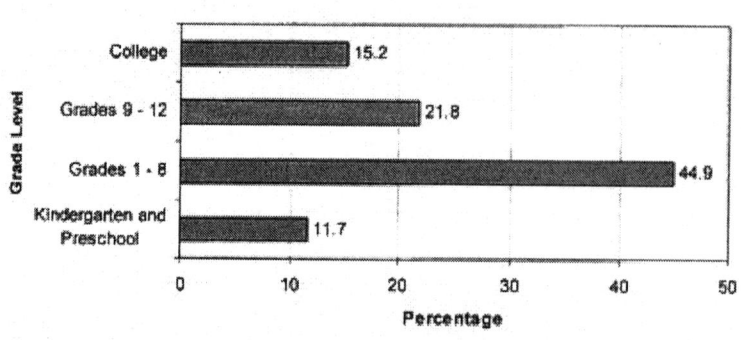

Example 1: Does the graph describing the distribution of students attending different grades satisfy the rules for a graph?

Solution:
  Step 1: Understand and picture the problem.
    Question: Does the pictured graph satisfy the properties a graph?

  Step 2: Develop a plan.
    Use the properties of a graph to determine if the properties are satisfied.

  Step 3: Execute the plan.
    1. "Where Our Students Are" is the title which defines the graph as placing students in different grades.
    2. The sum of the percentages is 93.6%. Using the 4 categories, this would lead to 93.6% ± 4%. This sum ranges from 89.6% to 97.6%. 100% is not contained in the interval, the sum is inappropriate.
    3. The graph does **not** have descriptive labels on the horizontal (x) and vertical (y) axes.
    4. The graph does have specific labels identifying the bars (pencils) and their lengths are quantified. The **scale is absent** from the horizontal axis. The graph is **not** properly constructed.

  Step 4: Check your work.
    Have the properties of a graph been satisfied?

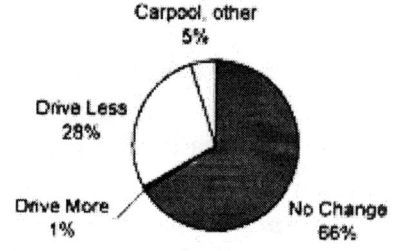

Example 2: Does the graph describing the driving habits of Americans in response to increasing gasoline prices satisfy the rules for a graph?

Solution:
Step 1: Understand and picture the problem.
Question: Does the pictured graph satisfy the properties a graph?

Step 2: Develop a plan.
Use the properties of a graph as a checklist to determine if the properties are satisfied.

Step 3: Execute the plan.
1. "Altering Driving Habits" defines the graph as drivers reactions to increased gasoline prices.
2. The numeric values are labeled as percents. The sum of the percentages is 100%. This sum is appropriate.
3. Each slice has a definitive label.
4. The chart's slices complete 360 degrees. (It's a complete circle.)
   The graph is properly constructed.

Step 4: Check your work.
Have the properties of a graph been satisfied?

Constructing Graphs for Qualitative Data

Consider the breakdown of a $1.00 donation made to the American Heart Association in the following table

RECIPIENT	CENTS per DOLLLAR
Administrative	$0.086
Fund Raising	$0.136
Community Services	$0.117
Research	$0.314
Education	$0.244
Training	$0.103
TOTALS	$1.00

Since money is counted and not measured, it can be considered an attribute.

Another way to represent the distribution of donations is with relative frequencies (decimals or percents). To calculate relative frequency, divide the frequency by the total count of data. The relative frequencies of the American Heart Association data is displayed in the next table. All graphs of this data are based on this table.

RECIPIENT	CENTS per DOLLAR	RELATIVE FREQUENCY
Administrative	$0.086	$\frac{\$0.086}{\$1.00} = 0.086 \approx 9\%$
Fund Raising	$0.136	$\frac{\$0.136}{\$1.00} = 0.136 \approx 14\%$
Community Services	$0.117	$\frac{\$0.117}{\$1.00} = 0.117 \approx 12\%$
Research	$0.314	$\frac{\$0.314}{\$1.00} = 0.314 \approx 31\%$
Education	$0.244	$\frac{\$0.244}{\$1.00} = 0.244 \approx 24\%$
Training	$0.103	$\frac{\$0.103}{\$1.00} = 0.103 \approx 10\%$
TOTALS	$1.00	$\frac{\$1.00}{\$1.00} = 1.00 = 100\%$

Bar Graphs

Rules for Constructing a Bar Graph

1. The bar graph has a definitive title describing the data being graphed.
2. Each axis has a general descriptive label.
3. Each axis also has a specific label. On one axis, the bars are identified with a defining label. On the other axis, their lengths are quantified with a number. An appropriate scale is chosen to display the bar's lengths.
4. All data is accounted.
   a. The sum of the frequencies accounts for all data values.
   b. The sum of the relative frequencies adds to 1.00 (Percents sum to 100).
   Note: Due to rounding, the sum may be close to 1.00 (Percents close to 100). For each category, there could be an error of $\pm$ 1%. 100% should be contained within this interval.

Types of bar graphs

There are two decisions to make when constructing a bar graph. The first decision concerns the placement of the bars. The bars can be displayed vertical (up and down) or horizontal (left to right). The second decision concerns the quantifying (labeling) of the bars. Either frequency or relative frequency (count, fraction, decimal, percent) may be used to quantify (label) the lengths of the bars. The following graphs display the data representing the distribution of a $1.00 donation to the American Heart Association.

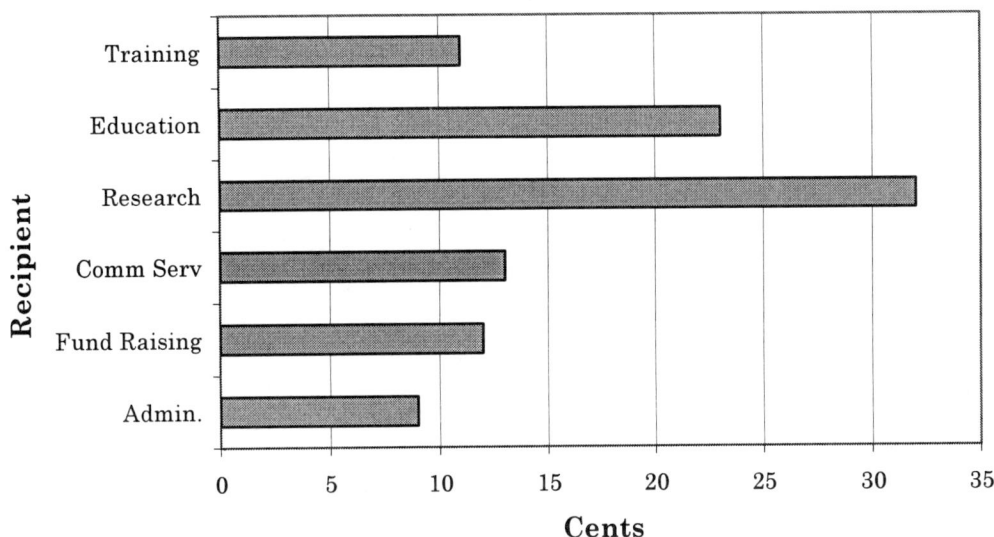

Comment: A horizontal frequency graph of the data. The title is definitive. Each axis has a general and specific label. The specific labels identifies and quantifies each bar. The scale is 5 cents. The sum of the bar's frequencies is $1.00.

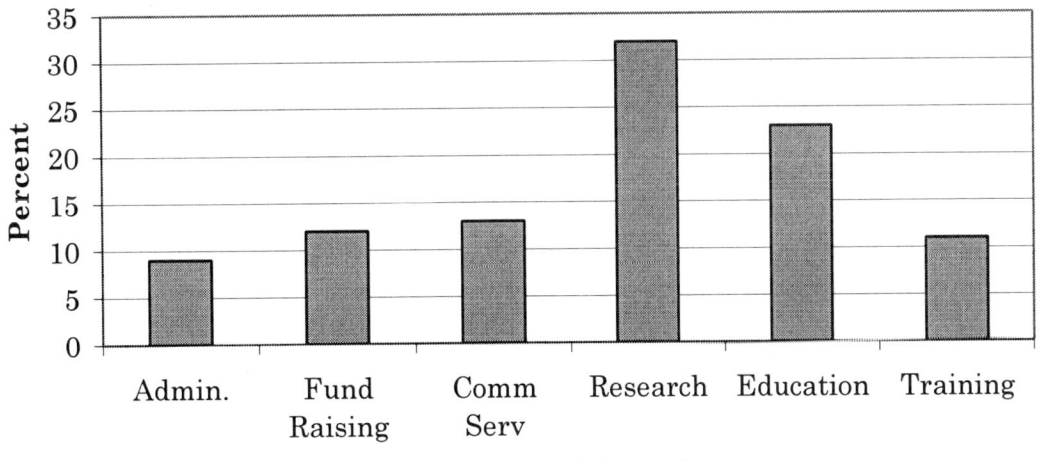

Comment: A vertical relative frequency graph of the data. The title is definitive. The specific labels identifies and quantifies each bar. The scale is 5%. The sum of the bar's percents is 100%.

Pie Chart

Another representation of qualitative data is the pie chart. The pie chart is a circle with sections cut, starting at the center. Since a complete circle has 360°, each "*slice*" of the pie represents a fraction, relative frequency, of the total attribute. The degree size of each slice is calculated by multiplying each relative frequency by 360°. This measurement gives the central angle of each slice. Using the American Heart Association data, the column labeled degrees states the degree size of each slice.

"Let's Break for Lunch. I'll order pizza."
CORNERED. Distributed by Universal Press Syndicate. Reprinted with permission. All rights reserved.

RECIPIENT	CENTS per DOLLARS	RELATIVE FREQUENCY	Degrees (Whole Number)
Administrative	$0.09	0.09	$0.09(360°) \approx 32°$
Fund Raising	$0.14	0.14	$0.14(360°) \approx 50°$
Community Services	$0.12	0.12	$0.12(360°) \approx 43°$
Research	$0.31	0.31	$0.31(360°) \approx 112°$
Education	$0.24	0.24	$0.24(360°) \approx 86°$
Training	$0.10	0.10	$0.10(360°) \approx 36°$
TOTALS	$1.00	1.00	359° *

- Note: Due to rounding, the sum of the degrees does not equal 360°. However, 359 degrees $\pm$ 6 degrees does contain 360 degrees. The sum is appropriate.

Rules for Constructing a Pie Chart
1. The pie chart has a definitive title describing the data being graphed.
2. Each slice is labeled with a descriptive label.
3. Each slice is labeled with either frequency or relative frequency (percent). **DO NOT** use degrees as a label.
4. The sum of the frequencies accounts for all data values.
5. The sum of the relative frequencies adds to 1.00 (Percents sum to 100).
   Note: Due to rounding, the sum may be close to 1.00 (Percents close to 100). For each category, there could be an error of $\pm$ 1%. 100% should be contained within this interval.
6. The sum of the degrees in a pie chart adds to 360 degrees.
   Note: Due to rounding, the sum may be close to 360 degrees. For each category, there could be an error of $\pm$ 1 degree. 360 degrees should be contained within this interval.

American Heart Association
$1.00 Donation Distribution

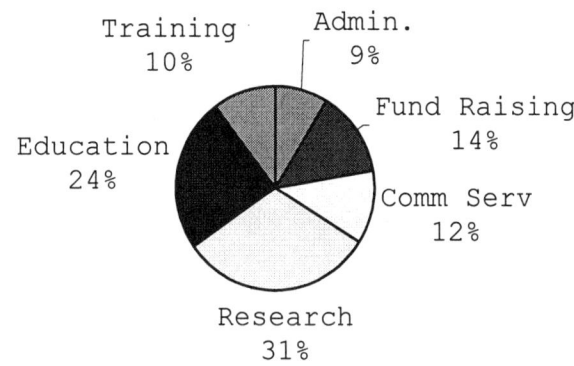

Comment: This is a pie chart of the data. The title is definitive. Each slice is labeled with a descriptive label. Each slice is labeled with its relative frequency. The percents sum to 100%.

Artistic License

In place of the traditional circle, an artist can add their creative touch to the pie chart. Here is an artist's rendering of the distribution of contributions.

Note: The percentages were removed to prevent conveying out-of-date percentages on the distribution of funds.

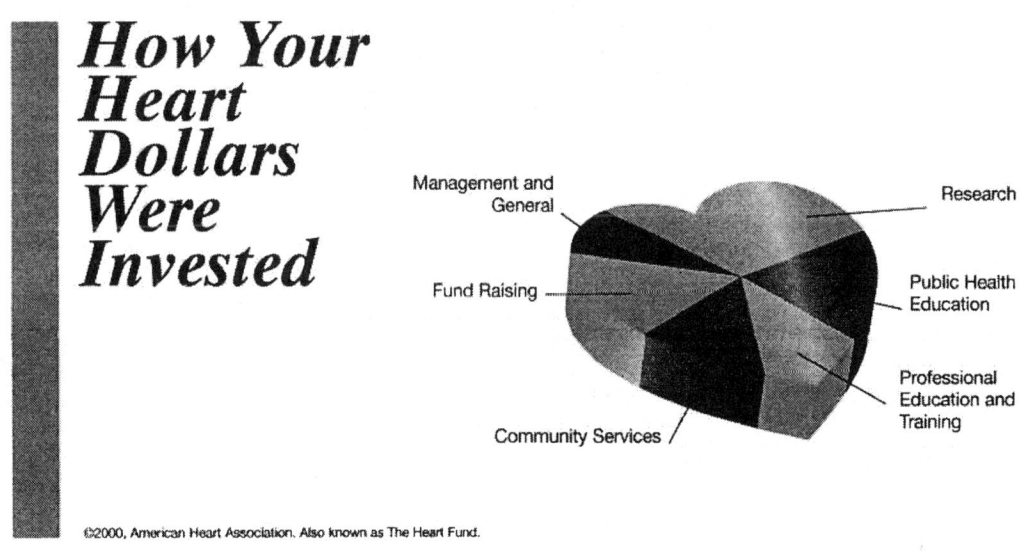

Example 3: Construct a vertical frequency bar graph displaying the different blood types. Out of 400 donors, 152 are O+, 136 are A+, 36 are B+, 12 are AB+, 28 are O-, 24 are A-, 8 are B-, and 4 are AB-. This blood drive mirrors the distribution of blood type in humans

Solution:

Step 1: Understand and picture the problem.
 1. Question: Construct a vertical frequency graph representing the various blood types.
 2. List the data: 152 are O+, 136 are A+, 36 are B+, 12 are AB+, 28 are O-, 24 are A-, 8 are B-, and 4 are AB-.

Step 2: Develop a plan.
 1. Write a descriptive title for the graph.
 2. Make a table with two columns; "Blood Types" and "Frequency." These column headings will be used to label the axes of the graph.
 3. A vertical frequency bar graph means the bars go up and down. This means the "Blood Types" and labels defining the bars will be on the horizontal (x) axis. The "Frequency" and a scale quantifying the length of the bars will be on the vertical (y) axis.

Step 3: Execute the plan.
 1. Title the graph, "Donor Blood Types."
 2. Make the table:

Blood Types	Frequency
O+	152
A+	136
B+	36
AB+	12
O-	28
A-	24
B-	8
AB-	4
Total	400

 3. Choose a scale of 20 for the length of the bars.

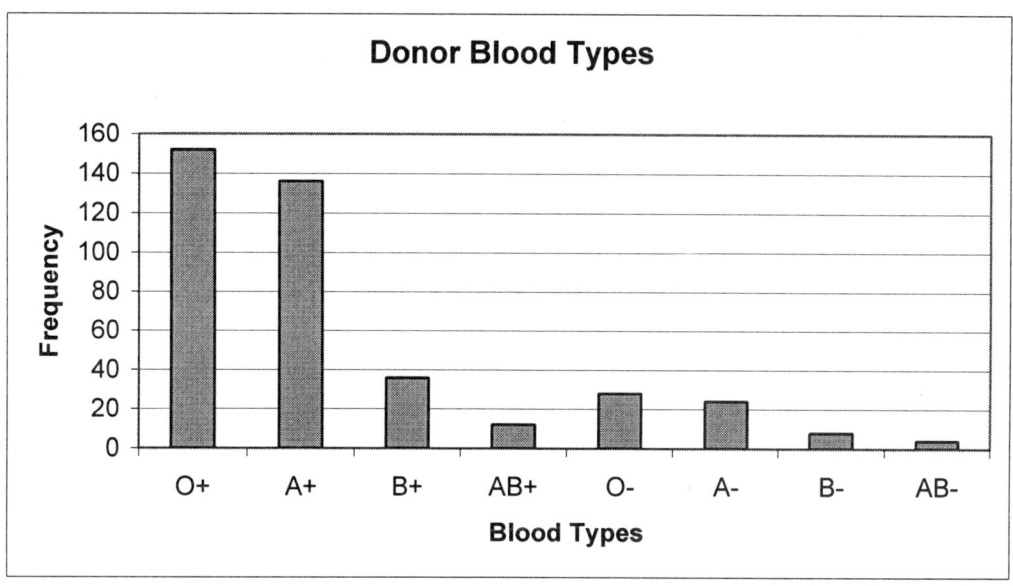

Step 4: Check your work.
 1. Does the graph satisfy the properties of a bar graph?
 2. Is the graph a vertical frequency bar graph?
 3. Is the graph readable?

Example 4: Construct a horizontal relative frequency bar graph displaying the different blood types. Out of 400 donors, 152 are O+, 136 are A+, 36 are B+, 12 are AB+, 28 are O-, 24 are A-, 8 are B-, and 4 are AB-.

Solution:
 Step 1: Understand and picture the problem.
  1. Question: Construct a horizontal relative frequency graph representing the various blood types.
  2. List the data: 152 are O+, 136 are A+, 36 are B+, 12 are AB+, 28 are O-, 24 are A-, 8 are B-, and 4 are AB-.

 Step 2: Develop a plan.
  1. Write a title for the graph.
  2. Make a table with three columns; "Blood Types," "Frequency" and "Relative Frequency. Two of these columns will be used as labels on the axes of the graph.
  3. A horizontal frequency bar graph means the bars go left to right. This means the "Blood Types" and labels defining the bars will be on the vertical (y) axis. The "Relative Frequency" and a scale quantifying the length of the bars will be on the horizontal (x) axis.

 Step 3: Execute the plan.
  1. Title the graph, "Donor Blood Types."
  2. Make the table

Blood Types	Frequency	Relative Frequency
O+	152	$152/400 = 0.38$
A+	136	$136/400 = 0.34$
B+	36	$36/400 = 0.09$
AB+	12	$12/400 = 0.03$
O−	28	$28/400 = 0.07$
A−	24	$24/400 = 0.06$
B−	8	$8/400 = 0.02$
AB−	4	$4/400 = 0.01$
Total	400	$400/400 = 1.00$

3. Choose a scale of 0.1 for the horizontal axis.

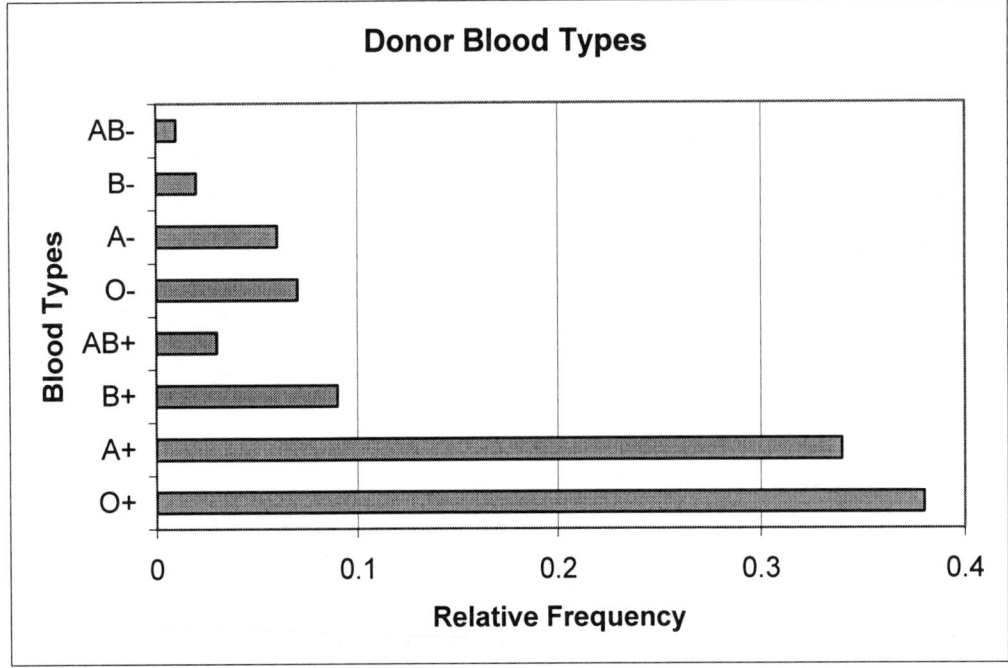

Step 4: Check your work.
1. Does the graph satisfy the properties of a bar graph?
2. Is the graph a horizontal relative frequency bar graph?
3. Is the graph readable?

Example 5: Construct a relative frequency pie chart displaying the different blood types. Out of 400 donors, 152 are O+, 136 are A+, 36 are B+, 12 are AB+, 28 are O-, 24 are A-, 8 are B-, and 4 are AB-.

Solution:

Step 1: Understand and picture the problem.

A pie chart uses a circle to represent data. Each slice of the circle represents the proportional distribution of blood types in the blood drive.

Step 2: Develop a plan.
1. Write a title for the pie chart.
2. Use the table from Example 4 and add one new column "Degrees."
3. Problem: Construct the pie chart.

Step 3: Execute the plan.
1. Title the pie chart "Donor Blood Types."

2. Add the information to the table:

Blood Types	Frequency	Relative Frequency	Degrees
O+	152	$152/400 = 0.38$	$0.38(360) = 136.8$
A+	136	$136/400 = 0.34$	$0.34(360) = 122.4$
B+	36	$36/400 = 0.09$	$0.09(360) = 32.4$
AB+	12	$12/400 = 0.03$	$0.03(360) = 10.8$
O-	28	$28/400 = 0.07$	$0.07(360) = 25.2$
A-	24	$24/400 = 0.06$	$0.06(360) = 21.6$
B-	8	$8/400 = 0.02$	$0.02(360) = 7.2$
AB-	4	$4/400 = 0.01$	$0.01(360) = 3.6$
Total	400	$400/400 = 1.00$	360

3. Using the tabled values, construct the pie chart. From the center of the circle, measure angles corresponding to these degrees. For instance, O+ measures 136.8 degrees. Draw the slice inside a circle with a central angle of about 137 degrees. **Do not label the slice with degrees**. Use percent to quantify the slices of this graph. Label each slice with an identifying label. Continue this process with the remaining slices.

Donor Blood Types

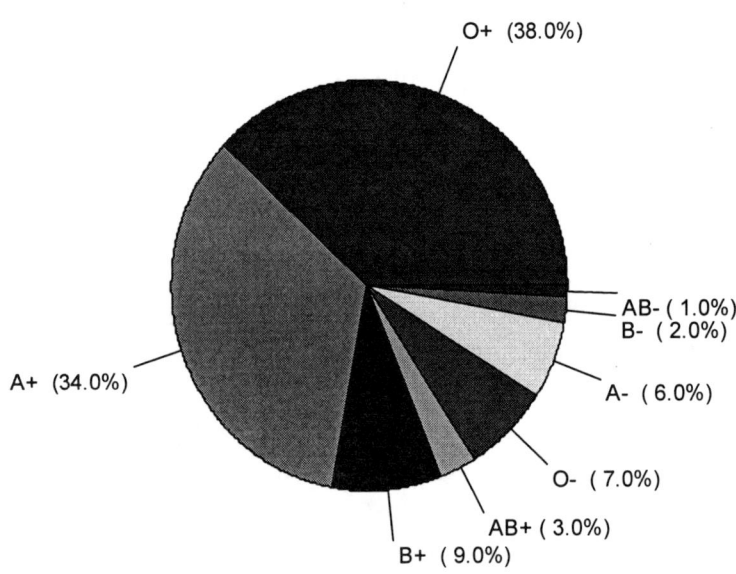

Step 4: Check your work.
1. Does the graph satisfy the properties of a pie chart?
2. Is the graph readable?

Example 6: Construct a frequency pie chart displaying the different blood types. Out of 400 donors, 152 are O+, 136 are A+, 36 are B+, 12 are AB+, 28 are O−, 24 are A−, 8 are B−, and 4 are AB−.
Solution:
Step 1: Understand and picture the problem.
A pie chart uses a circle to represent data. Each slice of the circle represents the proportional distribution of frequencies of blood types in the blood drive.

Step 2: Develop a plan.
1. Write a title for the pie chart.
2. Use the table from Example 5.
3. Construct the pie chart.

Step 3: Execute the plan.
1. Title the pie chart "Donor Blood Types."

2. Use the table information:

Blood Types	Frequency	Relative Frequency	Degrees
O+	152	152/400 = 0.38	0.38(360) = 136.8
A+	136	136/400 = 0.34	0.34(360) = 122.4
B+	36	36/400 = 0.09	0.09(360) = 32.4
AB+	12	12/400 = 0.03	0.03(360) = 10.8
O-	28	28/400 = 0.07	0.07(360) = 25.2
A-	24	24/400 = 0.06	0.06(360) = 21.6
B-	8	8/400 = 0.02	0.02(360) = 7.2
AB-	4	4/400 = 0.01	0.01(360) = 3.6
Total	**400**	400/400 = 1.00	**360**

3. Similar to the previous example, use the tabled values to construct the pie chart. From the center of the circle, measure angles corresponding to these degrees. Identify the slices. Since this is a frequency pie chart, each slice is labeled with the frequency.

Donor Blood Types

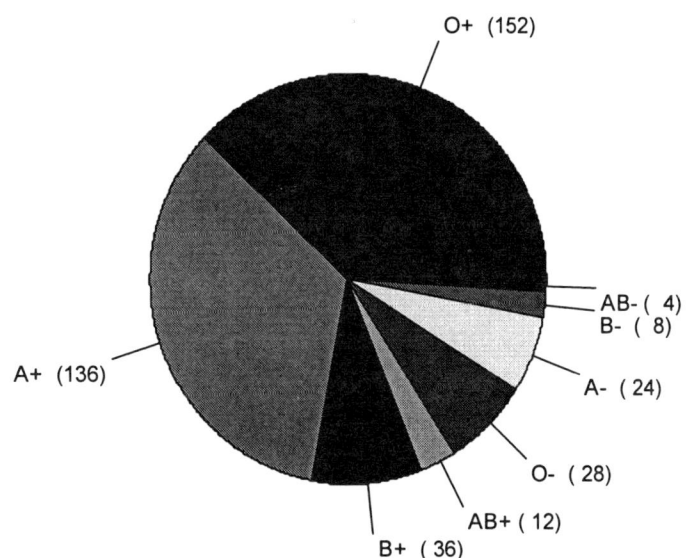

Step 4: Check your work.
1. Does the graph satisfy the properties of a pie chart?
2. Is the graph readable?

Reality check: As easy as it seems to make bar and pie charts, it is just as easy to make an error. Consider the following pie chart.

United Way Distributions

The United Way pie chart may be misinterpreted. It appears that for every dollar donated to the United Way, $0.29 goes to "Nurturing Children & Youth," $0.20 is spent on programs "Developing Self-Sufficiency," $0.16 is spent on "Fostering Health and Wellness," $0.22 is spent on "Strengthening Families and Individuals" and $0.13 is spent on "Building Strong Communities."

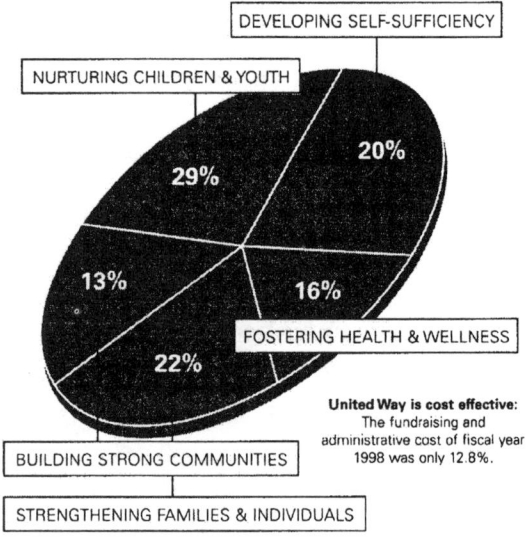

The sum of the percentages within the pie chart is 100% and accounts for every cent of a dollar. However, the caption outside of the pie chart states that the "Administrative Cost" is 12.8%. This makes the sum of the United Way expenses 112.8%, not 100%. In reality, $0.128 of every dollar donated goes toward administrative costs. The remaining $0.872 ($1.00 − $0.128 = $0.872) is distributed to the other programs.

(c) 2006 JupiterImages Corp.

Optional Student Project

"Graphs for Qualitative Data" can be found on page 430.

?   **Cognitive Problems**   **?**

1. Qualitative data is countable (discrete), not measurable (continuous). Explain the difference between countable and measurable data.
2. What is the difference between a horizontal frequency bar graph and a vertical frequency bar graph of the same data?

3. What is the difference between a frequency bar graph and a relative frequency bar graph of the same data?

(c) 2006 JupiterImages Corp.

Exercise 5.1

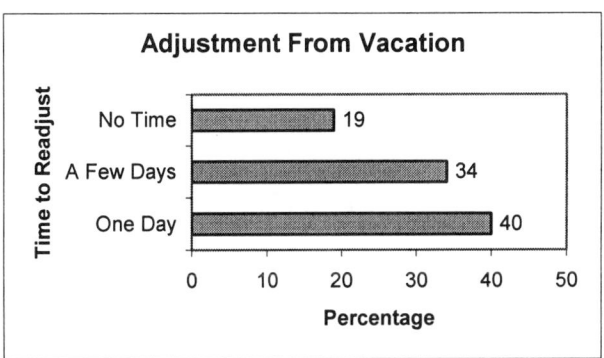

1. Does the graph "Adjustment from Vacation" satisfy the properties of a graph?

USA TODAY

2. Does the graph "How Often We Summer BBQ" satisfy the properties of a graph?

USA TODAY

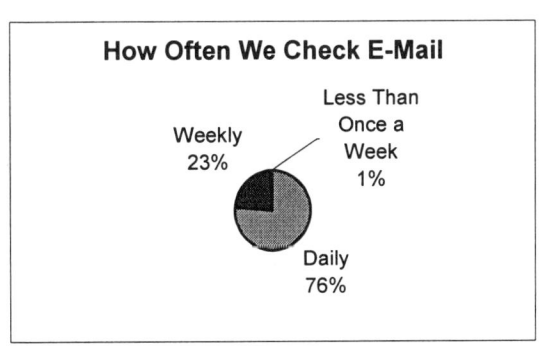

3. Does the graph "How Often We Check e-mail" satisfy the properties of a graph?

USA TODAY

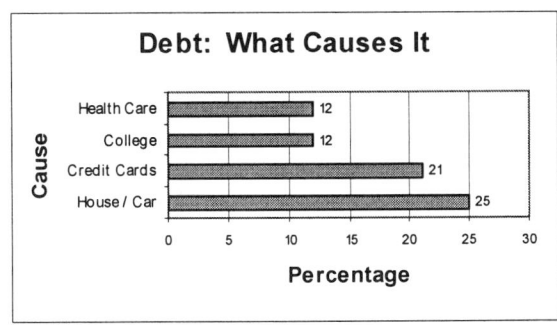

4. Does the graph "Debt: What Causes It?" satisfy the properties of a graph?

USA TODAY

5. A survey of 1,400 job applicants to produced the following results. Construct a vertical frequency bar graph for this data.

Job Candidate Mistake	Frequency
Interview	448
Resume	294
Cover Letter	126
Reference Checks	126
Interview Follow-up	98
Screening Call	84
Other	28
Don't Know	196

6. Use the data in Problem 5 to construct a horizontal relative frequency bar graph.

7. The following is the corrected table for the data presented in the "Reality Check". Construct a relative frequency pie chart for the United Way data.

Distribution	Cents / Dollar
Administrative Costs	$0.128
Nurturing Children	$0.253
Develop Self-Sufficiency	$0.174
Health & Wellness	$0.140
Strengthen Families	$0.192
Strong Communities	$0.113

8. Use the data in Problem 7 to construct a horizontal frequency bar graph.

9. A survey of American families on child care produced the following results. Construct a horizontal relative frequency bar graph.

Who's Watching the Kids?	Frequency
Center based (child care center)	3,480
Family child care homes	1,560
Babysitter	600
Care of relative	3,120
Parental care	3,240

10. Use the table in Problem 9 to construct a frequency pie chart.

11. A survey of 336, workers aged 20 – 29, yielded the following responses concerning job satisfaction. Construct a vertical relative frequency bar graph.

Response	Frequency
Happy with Career	112
Enjoy job, but its not my career choice	64
Job is OK, but it is not my career choice	64
Don't like my job, but it is my career path	20
My job just pays my expenses	76

12. Use the data in Problem 11 to construct a relative frequency pie chart.

13. The table shows how a $5 contribution to the American Cancer Society is distributed. Use the table to construct a vertical relative frequency bar graph.

Distribution	Amount
Research	$1.35
Patient Services	$1.05
Public Education	$0.90
Fund Raising	$0.90
Professional Education	$0.45
Management	$0.35

14. Use the data in Problem 13 to construct a relative frequency pie chart.

15. A survey about favorite pie flavors produced the following results. Construct a (You guessed it.) frequency pie chart for this data.

Flavor	Frequency
Cherry	200
Lemon meringue	220
Sweet Potato	170
Pumpkin	170
Apple	500
Chocolate	280
Other	460

16. Use the data in Problem 15 to construct a vertical frequency bar graph.

5.2 DESCRIPTIVE STATISTICS WITH QUANTITATIVE DATA
Objective: Organize and graph quantitative data

Unlike **"Qualitative"** data that **counts** an attribute (category). **"Quantitative"** data **measures** some aspect of an element. "Qualitative" data is *discrete,* the data is represented using countable units such as whole numbers. "Quantitative" data is *continuous*, the data is represented using measurable units such as fractions (decimals).

As an example of quantitative data, consider the Sport Utility Vehicle data set. It consists of the prices of 40 four-wheel drive (4WD) Sport Utility Vehicles (SUV). These vehicles range from the Geo Tracker to the Range Rover.

$14,655	$14,799	$63,500	$15,605	$31,985	$32,250	$26,268	$17,990
$19,300	$32,950	$33,595	$33,790	$22,708	$23,240	$23,920	$27,815
$23,405	$29,099	$29,249	$30,585	$30,645	$16,395	$16,798	$34,590
$35,550	$36,300	$27,910	$28,680	$28,950	$38,175	$41,188	$25,999
$26,185	$20,000	$25,176	$42,660	$54,950	$56,000	$21,995	$22,195

The first step is to make a table organizing the data. The construction of the table has several stages. There are three rules for constructing a table for quantitative data.

Rules for Constructing a Table from Quantitative Data
1. The class width, the size of each interval, should be the same for each class.
2. The first and last class cannot be empty.
a. The minimum data value must be contained in the first class.
b. The last class must contain the maximum data value.

When making a table for quantitative data, there are **three arbitrary decisions** that have to be made.

Arbitrary Choices
1. **Arbitrary Choice #1**: Choose the number of classes. Classes are the number of intervals, divisions, or groups of the data. Usually the number of classes ranges from 5 to 20.
2. Determine the class width. $$\text{Class Width} > \frac{\text{Maximum Data Value} - \text{Minimum Data Value}}{\text{Number of Classes}}$$ **Arbitrary Choice #2:** Choose a value for the class width that is easy to work with and makes the data easy to read.
3. **Arbitrary Choice #3:** Choose the starting point for the first class.

When tallying the data into classes, the data is then counted into a class according to the following rule.

Data Placement into a Class

Lower Limit $\leq$ *data value* $<$ Upper Limit

The *data value* is **AT LEAST** the Lower Limit and **LESS THAN** the Upper Limit.

Note: The class limits appear to overlap, but they do not. The next class starts, where the previous class ends. Because the equality condition is only on the lower class limit, the data entry is only counted in one class.

An example of a table using overlapping classes is taken from the Federal Income Tax Booklet. The first line of taxes applies to incomes of *AT LEAST $23,000 BUT LESS THAN $23,050*. The second line of taxes applies to incomes of *AT LEAST $23,050 BUT LESS THAN $23,100*, etc.

IRS

An example of a table using seemingly overlapping classes is taken from the Federal Income Tax Booklet. The first line of taxes applies to incomes of **AT LEAST $23,050 BUT LESS THAN $23,050**. The second line of taxes applies to incomes of **AT LEAST $23,050 BUT LESS THAN $23,100**, etc. A single individual with a taxable income of $23,050 would have a tax obligation of $3,161 (taken from the second row, not the first row). The table is continuous, the US government will collect taxes from everyone. They do not provide a gap where a taxpayer is exempt from paying taxes.

If Form 1040A, line 27, is—		And you are—			
At least	But less than	Single	Married filing jointly	Married filing separately	Head of a household
		Your tax is—			
23,000					
23,000	23,050	3,154	2,854	3,154	2,954
23,050	23,100	3,161	2,861	3,161	2,961
23,100	23,150	3,169	2,869	3,169	2,969
23,150	23,200	3,176	2,876	3,176	2,976
23,200	23,250	3,184	2,884	3,184	2,984
23,250	23,300	3,191	2,891	3,191	2,991
23,300	23,350	3,199	2,899	3,199	2,999
23,350	23,400	3,206	2,906	3,209	3,006

Apply these steps to the SUV data set:

1. The SUV prices are sorted from smallest to largest.

   $14,655 $14,799 $15,605 $16,395 $16,798 $17,990 $19,300 $20,000
   $21,995 $22,195 $22,708 $23,240 $23,405 $23,920 $25,176 $25,999
   $26,185 $26,268 $27,815 $27,910 $28,680 $28,950 $29,099 $29,249
   $30,585 $30,645 $31,985 $32,250 $32,950 $33,595 $33,790 $34,590
   $35,550 $36,300 $38,175 $41,188 $42,660 $54,950 $56,000 $63,500

2. **Arbitrarily Choice #1**: Choose 6 classes.

3. The maximum data value is $63,500 and the minimum data value is $14,655.

   $$\text{Class Width} > \frac{\$63,500 - \$14,655}{6} = \frac{\$48,845}{6} \approx \$8,140.83333$$

   Arbitrary Choice #2: Choose $10,000 as the class width.

4. **Arbitrary Choice #3**: Choose $10,000 as the starting point. Add the class width to your starting point. This determines both the upper limit of the current class and the lower limit of the next class. The first class includes $10,000 and is less than $20,000. The second class includes $20,000 and is less than $30,000 etc. This produces the table:

Class	Dollars ($1,000's)
1st	$10 – $20
2nd	$20 – $30
3rd	$30 – $40
4th	$40 – $50
5th	$50 – $60
6th	$60 – $70

   Note: The classes satisfy the tabled rules for quantitative data:
   a. The six classes have the same class width, $10,000.
   b. The minimum value of $14,655 falls in the first class.
   c. The maximum value of $63,500 falls in the last class.

5. To complete the table, tally the data and calculate their relative frequencies.

Class	Dollars ($1,000's)	Frequency	Relative Frequency.
1st	$10 – $20	7	$\frac{7}{40} = 0.175$
2nd	$20 – $30	17	$\frac{17}{40} = 0.425$
3rd	$30 – $40	11	$\frac{11}{40} = 0.275$

4th	$40 – $50	2	$\frac{2}{40} = 0.050$
5th	$50 – $60	2	$\frac{2}{40} = 0.050$
6th	$60 – $70	1	$\frac{1}{40} = 0.025$
Totals		40	$\frac{40}{40} = 1.000$

Note₁: The data item $20,000 is tallied (counted) in the 2nd class, 20 - 30 (thousand), not the 1st class. The 1st class, 10 - 20 (thousand) contains values that are AT LEAST 10,000 but LESS THAN 20,000. The first class does not include 20,000. The equality is only on the lower class limit. The upper limits are not counted in the class.

Note₂: Theoretically, the relative frequencies should equal 1.00. However, due to rounding, the sum may deviate slightly from 1.00.

Various tables could be constructed using this data and six classes. The following table is another legitimate arrangement of the SUV data. It has a class width of $9,000 and also satisfies the three rules for a table of quantitative data.

Class	Dollars ($1,000's)	Frequency
1st	$11 – $20	7
2nd	$20 – $29	15
3rd	$29 – $38	12
4th	$38 – $47	3
5th	$47 – $56	1
6th	$56 – $65	2
Totals		40

Note: The classes satisfy the rules for quantitative data:
    a. The six classes have the same class width, $9,000.
    b. The minimum value of $14,655 falls in the first class.
    c. The maximum value of $63,500 falls in the last class.

Graphing Quantitative Data
Two ways to graph quantitative data are histograms (bar graphs) and polygons (line graphs). Both graphs have similar properties.

370

Histograms

Properties of Histograms (Bar Graphs for Quantitative Data)

1. The histogram has a definitive title describing the data being graphed.
2. Each axis has a descriptive label.
3. Each axis has a scale. On one axis, the bars are quantified using the classes from the data table. The bars are adjacent and are continuous. On the other axis, the bar's lengths are quantified with an appropriate scale chosen to display the bar's lengths.
4. The sum of the:
   a. Frequencies accounts for all data values.
   b. Relative frequencies adds to 1.00 (Percents sum to 100).
   Note: Due to rounding, the sum may be close to 1.00 (100%). For each class, there could be an error of ± 1%. 100% should be contained within this interval.

The following graphs demonstrate the frequency and relative frequency histograms.

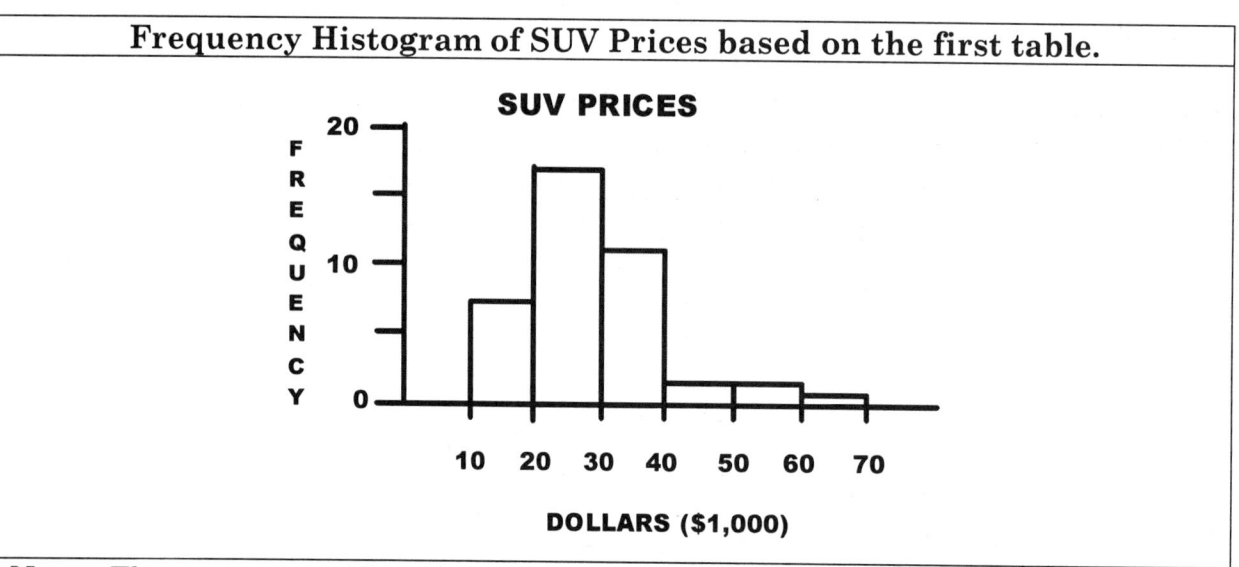

Frequency Histogram of SUV Prices based on the first table.

Note: The properties of histograms are satisfied.
1. The graph has a definitive title, "SUV Prices."
2. Each axis has a descriptive title "Dollars ($1,000's)" and "Frequency".
3. Each axis has a quantifying scale. One scale uses the class boundaries to define the bars. The bars are adjacent. The other scale quantifies the length of the bars.
4. The sum of frequencies accounts for all the data values.

Relative Frequency Histogram of SUV Prices based on the first table.

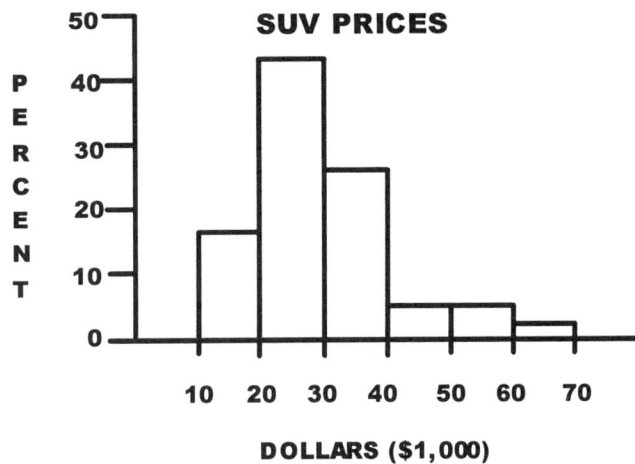

Note: The properties of histograms are satisfied.
1. The graph has a definitive title, "SUV Prices."
2. Each axis has a descriptive title "Dollars ($1,000's)" and "Percents".
3. Each axis has a quantifying scale. One scale uses the classes to define the bars. The other scale quantifies the length of the bars.
4. The sum of the percentages is 100%.

Polygons

Another type of graph for quantitative data is the polygon. In order to construct the polygon, line graph, an additional column must be added to the table. This column contains the **class mark**, which is **the midpoint of each class**. Calculate the midpoint of each class by adding the corresponding lower limit and upper limit. Then divide this sum by two. Each class mark is representative of a class.

Class	Dollars ($1,000's)	*Class Mark (Midpoints)*	Frequency	Rel. Freq.
1st	$10 – $20	*$15,000*	7	$7/40 = 0.175$
2nd	$20 – $30	*$25,000*	17	$17/40 = 0.425$
3rd	$30 – $40	*$35,000*	11	$11/40 = 0.275$
4th	$40 – $50	*$45,000*	2	$2/40 = 0.050$
5th	$50 – $60	*$55,000*	2	$2/40 = 0.050$
6th	$60 – $70	*$65,000*	1	$1/40 = 0.025$
Totals			40	$40/40 = 1.000$

371

Instead of plotting the class limits on the horizontal (x) axis, the class marks are plotted. Two additional class marks, one to the left and one to the right of the data are also recorded on the graph. These values anchor the polygon (line graph) to the horizontal axis. Frequency or relative frequency is then placed on the vertical axis.

Properties of Polygons

1. The polygon has a definitive title describing the data being graphed.
2. Each axis has a descriptive label.
3. Each axis has a scale. On one axis, the bars are quantified using the midpoints of the classes from the data table. On the other axis, the bar's lengths are quantified with an appropriate scale chosen to display the bar's lengths.
4. The sum of the:
   a. Frequencies accounts for all data values.
   b. Relative frequencies adds to 1.00 (Percents sum to 100).
   Note: Due to rounding, the sum may be close to 1.00 (100%.) For each class, there could be an error of $\pm$ 1%. 100% should be contained within this interval.
5. The line graph is anchored to the x-axis.
   Note: When anchoring the polygon to the x-axis on the left. Determine if the data can be negative. If it cannot be negative, then anchor the polygon to zero.

The following graphs demonstrate the frequency and relative frequency polygons.

Frequency Polygon of SUV Prices based on the table.

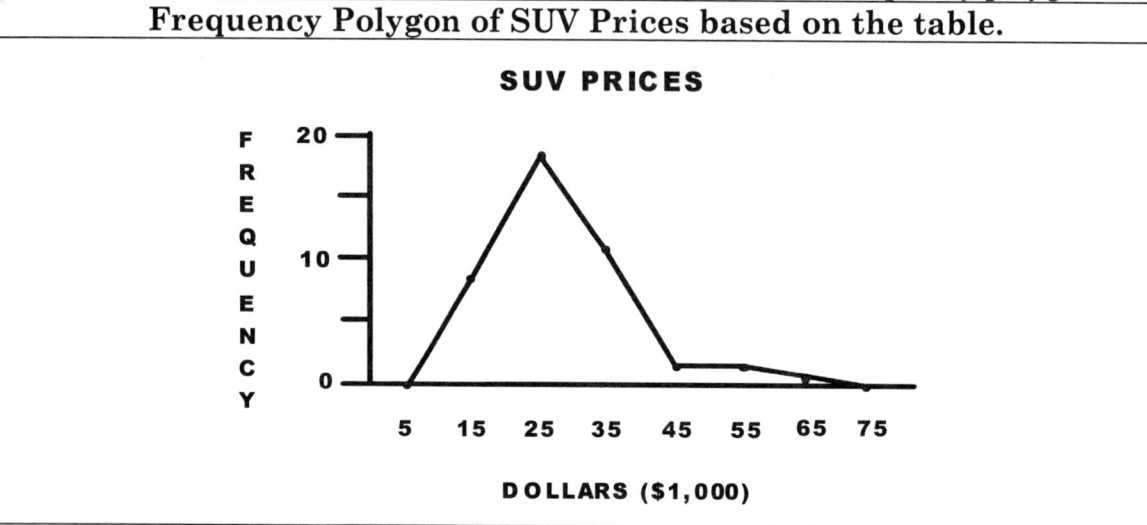

Note: The properties of polygons are satisfied.
1. The graph has a definitive title: "SUV Prices."
2. Each axis has a descriptive title: "Dollars ($1,000's)" and "Frequency".
3. Each axis has a quantifying scale. One scale uses the class marks (midpoints) to define the points in thousands of dollars. The other scale quantifies the heights of the points.
4. The sum of the frequencies accounts for all the data values.
5. The points are anchored to the x-axis.

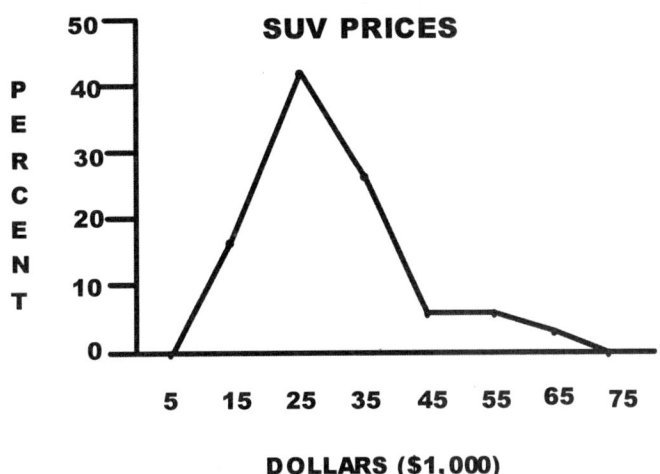

Relative Frequency Polygon of SUV Prices based on the table

Note: The properties of the graphs are satisfied.
1. The graph has a definitive title: "SUV Prices."
2. Each axis has a descriptive title: "Dollars ($1,000's)" and "Percent".
3. Each axis has a quantifying scale. One scale uses the class marks (midpoints) to define the points in thousands of dollars. The other scale quantifies the heights of the points.
4. The sum of the percentages is 100%.
5. The points are anchored to the x-axis.

Note 1: All the values for both type of graphs (histogram and polygons) originated in the table.

Note 2: Graphs should possess a "stand alone" aspect. That is the graph should have a descriptive title and each axis should be labeled. Understanding what the graph represents should not require reading the article containing the graph.

Example 1: The following data set represents the selling price of 37 new single family homes. Use five classes to construct a frequency histogram for this data.

$123,900	$130,900	$133,900	$138,900	$139,900	$146,900	$156,900
$156,900	$158,900	$159,400	$160,900	$163,900	$167,900	$167,900
$176,900	$182,900	$184,900	$186,900	$199,900	$199,900	$200,900
$204,900	$219,900	$219,900	$254,900	$256,300	$292,000	$311,750
$369,900	$385,500	$410,300	$430,500	$431,700	$436,500	$487,500
$496,500	$556,800					

Solution:
Step 1: Understand and picture the problem.
1. Question: Construct a frequency histogram using 5 classes.
2. The 37 data entries represent the prices of single family homes.

374

Step 2: Establish a plan.
  1. The data is ordered, use the formula to calculate and then choose a class width.
  2. Select a convenient starting point which is less than or equal to the minimum value. To make the data easier to work with, scale the data. Divide each price by one hundred thousand. Include "hundred thousand" in the general label.
  3. Tally the frequency for each class. (Check that the minimum value is contained in the first class and the maximum value is contained in the last class. If these conditions are not satisfied, choose another class width and/or starting point.)
  4. Place the class limits on the horizontal (x) axis and the frequencies on the vertical (y) axis. Label the axes and title the graph.

Step 3: Execute the plan.
  1. Class width $> \dfrac{\$556,800 - \$123,900}{5} = \dfrac{\$432,900}{5} = \$86,580$.
  2. Use class width = $90,000. A scale of $100,000 becomes $\dfrac{\$90,000}{\$100,000} = 0.9$
  3. Choose 1.2 (hundred thousand) as the starting point of the table.

Price ($100,000)	Frequency
1.2 – 2.1	22
2.1 – 3.0	5
3.0 – 3.9	3
3.9 – 4.8	4
4.8 – 5.7	3

4.

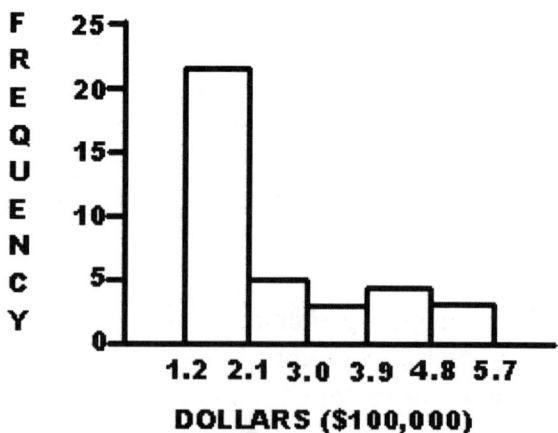

375

Step 4: Check your work.
1. Check the numbers and your calculations.
2. Have the rules for classing data and constructing a frequency histogram been satisfied?
3. Was the question answered?

Example 2: Use the data from Example 1 to construct a relative frequency polygon.
Solution:
Step 1: Understand and picture the problem.
Question: Use the data table from Example 1 to construct a relative frequency polygon.

Step 2: Establish a plan.
1. Add "class mark" to the table and find the class mark for each class.
2. Add "relative frequency" to the table and calculate the relative frequency for each class.
3. Construct the graph by placing the class marks on the horizontal (x) axis and the relative frequency on the vertical (y) axis.
4. Add one class mark to the left and the another class mark to the right of the data on the graph.
5. Label the axes and title the graph.

Step 3: Execute the plan.

Price ($100,000)	Class Mark	Frequency	Relative Frequency
1.2 – 2.1	1.65	22	$22/37 \approx 0.59$
2.1 – 3.0	2.55	5	$5/37 \approx 0.14$
3.0 – 3.9	3.45	3	$3/37 \approx 0.08$
3.9 – 4.8	4.35	4	$4/37 \approx 0.11$
4.8 – 5.7	5.25	3	$3/37 \approx 0.08$
Totals		37	$37/37 = 1.00$

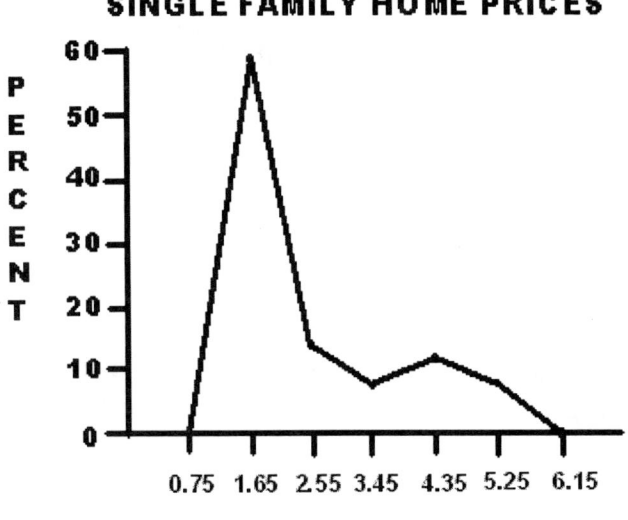

Step 4: Check your work.
1. Check the numbers and your calculations.
2. Have the rules for classing data and drawing a polygon been satisfied?
3. Was the question answered?

Time Series Polygon

Comment: Not all polygons are anchored to the x axis. There is a special polygon, "Time Series" that is not anchored to the x axis. Anchoring the polygon to the x axis states that your graph represents the entire data set. In the SUV polygon, all 40 numbers were graphed. So the polygon was anchored to the x axis.

The "Times Series" polygon on the right is NOT anchored to the x axis. This means that only part of the data set is displayed. There are data values (US population) before the year 1900 and data values (US population) after the year 2000. So this polygon is NOT anchored to the x axis. It displays part of a larger data set.

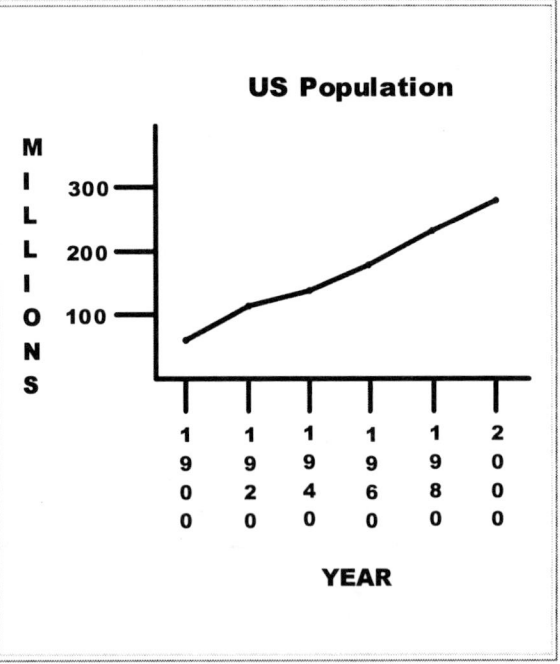

?   Cognitive Problems   **?**

Use the following table headings for questions 1 through 4.

Income ($1,000)	Class Mark	Frequency	Relative Frequency

1. For a frequency histogram, which columns are plotted on the horizontal axis and the vertical axis.
2. For a relative frequency polygon, which columns are plotted on the horizontal axis and the vertical axis.
3. For a relative frequency histogram, which columns are plotted on the horizontal axis and the vertical axis.
4. For a frequency polygon, which columns are plotted on the horizontal axis and the vertical axis.

Exercise 5.2

(c) 2006 JupiterImages Corp.

1. The price of a sport's coupe (2-door, turbocharged engine, rear spoiler, etc.) varies. Make a table with seven classes and use it to construct a frequency histogram
    $12,255 $13,567 $15,549 $16,545 $17,261 $17,266 $17,800 $18,573
    $18,875 $19,200 $19,496 $19,999 $20,829 $21,684 $22,795 $22,871
    $22,910 $25,350 $26,795 $26,975 $27,170 $27,570 $28,990 $29,238
    $32,100 $34,670 $37,345 $37,905 $39,520 $41,300 $41,409 $42,463
    $48,600 $55,050 $55,800 $56,000 $63,050 $65,900 $72,350 $82,550

2. Use the data from Problem 1 and construct a relative frequency polygon.

3. The price of a convertible varies. Make a table with six classes and use it to construct a relative frequency histogram.
    $19,735 $19,990 $20,080 $21,245 $22,015 $22,255 $23,300 $23,995
    $24,140 $25,760 $25,915 $26,710 $28,365 $31,500 $31,870 $32,000
    $32,155 $37,120 $37,470 $39,450 $41,000 $41,430 $42,995 $43,270
    $43,395 $43,500 $44,640 $44,995 $45,320 $45,500 $46,500 $48,100
    $49,930 $53,140 $55,600 $71,200 $74,970 $78,390 $80,400 $83,820

4. Use the data from Problem 3 and construct a frequency polygon.

5. Home prices vary in different neighborhoods of Chicago. Make a table with seven classes and use it to construct a frequency polygon.
$ 55,000 $ 56,000 $ 64,900 $ 77,900 $ 80,000 $ 87,500 $ 90,000
$ 95,000 $ 98,000 $109,700 $110,000 $118,000 $120,000 $128,000
$133,900 $135,000 $140,000 $142,000 $147,000 $150,000 $153,500
$159,900 $160,000 $171,000 $182,000 $187,900 $188,000 $195,000
$196,000 $220,000 $220,900 $230,000 $230,000 $245,075 $285,000
$294,000 $297,000 $349,925 $367,000

6. Use the data from Problem 5 and construct a frequency histogram.

7. Townhouse prices vary in different suburbs of Chicago. Make a table with eight classes and use it to construct a frequency polygon.
$ 74,900 $ 79,900 $ 80,000 $ 84,000 $ 89,500 $ 91,000
$ 92,500 $ 94,000 $ 95,000 $ 97,000 $ 99,500 $101,000
$101,500 $103,990 $108,000 $110,000 $113,700 $114,475
$115,000 $117,500 $118,000 $124,700 $125,250 $127,500
$135,000 $139,000 $142,000 $145,000 $146,000 $149,000
$149,900 $159,000 $161,500 $162,000 $178,500 $205,000
$224,000 $235,000 $263,000 $268,650

8. Use the data from Problem 7 and construct a frequency histogram.

9. The price for a new 1100 cc to 1200 cc motorcycle varies. Use the following data to make a table with 7 classes. Use the table to make a relative frequency polygon.
$14,500 $15,600 $16,700 $ 8,990 $15,000 $16,990 $18,200
$19,200 $16,190 $11,999 $11,999 $ 7,399 $10,449 $12,299
$ 8,995 $ 7,895 $ 8,595 $ 7,399 $ 7,999 $ 8,199 $ 9,999
$ 9,699 $10,599 $ 8,399 $ 7,899 $10,899 $ 6,999 $14,399
$11,770

10. Use the data from Problem 9 and construct a relative frequency histogram.

5.3 CENTRAL TENDENCY
Objective: Find measures of central tendency (average) data.

Average is an ambiguous term
The *"average"* is an ambiguous term and is often misinterpreted. The *"average"* is a measurement of central tendency for a set of data. Most people believe that the *"average"* of a data set is located near the center of the data. This is not always the case. Two values for the *"average"* will be investigated. These values are the **mean** and the **median**. In some situations, the values of the mean and median are extremely different. Since the average can be either the mean or median the reader may misinterpret *"average"*. Read the following article about the *average salaries of major league baseball players* that appeared during the first baseball strike.

Sanderson Sad Fans Dragged Into Dispute
By Dan Bickley Staff Writer

OAKLAND, Calif. You're sick and tired of these rich ballplayers threatening to walk out on the game aren't you? You can't fathom their argument, not when the average ballplayer makes $1.2 million a year. Well that's a misconception. The average salary in the major leagues is $395,000.
Reprinted with permission, The Chicago Sun-Times, ©1994.

Notice the two values of $1.2 million and $395,000. They are both reported as average baseball salaries. Is the center of the baseball salaries $1.2 million or $395,000? The average value of $1.2 million is the mean. The average value of $395,000 is the median. Which of these values is the most appropriate average for this data? In this situation, the median ($395,000) should be used as the appropriate average for this data. An explanation of this statement follows.

Mean vs. Median
Use the following data to calculate the mean and median. Suppose a small hardware store has twelve employees, including the owner of the store. Their annual salaries are:
$10,000 $12,000 $12,270 $12,500 $13,000 $13,600 $14,000
$15,000 $18,000 $18,750 $19,000 $98,000*   *(store owner's salary)

When asked to find an average, most people calculate the mean. To calculate the mean, add all the data values and then divide by the number of entries.

The MEAN (Average)

$$\text{MEAN} = \frac{\sum x}{n}$$

Where: $\sum$ means add
x is each data (entry) value
n is the number of data values

The mean is $\frac{\sum x}{n}$ which means add the twelve salaries and divide by 12.

$\sum x$ = $10,000 + $12,000 + $12,270 + $12,500 + $13,000 + $13,600 + $14,000 + $15,000 + $18,000 + $18,750 + $19,000 + $98,000 = $256,120

Mean = $\frac{\sum x}{n} = \frac{\$256,120}{12} \approx \$21,343.33$.

The average (mean) salary for an employee of this hardware store is $21,343.33.

This average seems higher than expected, as 11 out of 12 employees make less than the mean salary. This contradiction of an average is due to a property of the mean. ***The mean is very sensitive to extreme scores.*** A high data value, in this case $98,000, inflates the value of the mean. Thus, the mean average salary is larger than expected. Similarly, a low data value can make the mean average salary less than expected.

In situations with extreme scores, data values that are extremely larger or smaller than the majority of the data, the median should be used. The median is the middle value, half the data is above and half the data is below it. Use these steps to calculate the median:

The Median (Average)
1. Order the **n** data values, smallest to largest.
2. Find the value at the $\frac{n+1}{2}$ position.

   a. If $\frac{n+1}{2}$ is a whole number, use the data value found at that location.

   b. If $\frac{n+1}{2}$ is a not a whole number, locate the two data values corresponding to the whole numbers immediately preceding and following $\frac{n+1}{2}$. Find the midpoint (average) of these two data values.

The median for the twelve (n = 12) hardware employees is:
1. The sorted data is:
    $10,000 $12,000 $12,270 $12,500 $13,000 $13,600 $14,000
    $15,000 $18,000 $18,750 $19,000 $98,000*   *(store owner's salary)

2. The $\frac{n+1}{2}$ value is the salary found at the $\frac{12+1}{2} = \frac{13}{2} = 6.5$ location. 6.5 is not a whole number. The whole numbers immediately preceding and following 6.5 are 6 and 7 respectively. The data values at these positions are $13,600 and $14,000. The midpoint (average) of these two data values is:
$$\frac{\$13,600 + \$14,000}{2} = \frac{\$27,600}{2} = \$13,800.$$
Thus, half of the employees make less than $13,800 and half of the employee's make more than $13,800. The median is the middle (center) of the data.

Comment on the baseball salaries.
The case of the baseball players is a real situation where the average can be misinterpreted. The mean (average) of $1.2 million is inflated due to the salaries of the franchise players, the superstars of the league. Their multi-million dollar contracts are disproportionably large in comparison to the majority of baseball players. Thus the mean (average) is inflated. A reader may believe that $1.2 million is the middle of the baseball salaries. However, the median (average) is the middle salary. $395,000 is the center of the players' salaries and should be used as the average for this data set.

Example 1: Find the mean for the following data: 11.7 11.9 12.0 12.1 12.2.
Solution:
Step 1: Understand and picture the problem.
    Question: Calculate the mean using the 5 data items.

Step 2: Establish a plan.
    1. Find $\sum x$.
    2. Count the data values, n.
    3. Use the formula: Mean = $\frac{\sum x}{n}$.

Step 3: Execute the plan.
    1. $\sum x = 11.7 + 11.9 + 12.0 + 12.1 + 12.2 = 59.9$.
    2. n is 5.
    3. Substituting into the formula: mean = $\frac{59.9}{5} = 11.98$.

Step 4: Check your work.
1. Check the numbers and your calculations.
2. Was the question answered?
3. Is the answer reasonable?

Example 2: Find the median for the following data: 11.7 11.9 12.0 12.1 12.2.
Solution:
Step 1: Understand and picture the problem.
Question: Calculate the median, the middle term of the data set.

Step 2: Determine a plan.
1. The data is already ordered and let n be the number of data values.
2. Find the middle term using $\frac{n+1}{2}$ to locate its position.
3. Find the value of the term (data item) at that position.

Step 3: Execute the plan.
1. There are 5 data items, n = 5.
2. The position of the median is $\frac{n+1}{2} = \frac{5+1}{2} = \frac{6}{2} = 3$.
3. Counting from left to right, the third value, the median, is 12.0.

Step 4: Check your work.
1. Check the numbers and your calculations.
2. Was the question answered?
3. Is the answer reasonable?

Small vs. Large Data Sets

For small data sets, there is no difficulty calculating the mean or median. For large data sets, the calculator can be used to calculate the mean. Consider the 40 SUV prices from the previous section.

$14,655	$14,799	$15,605	$16,395	$16,798	$17,990	$19,300	$20,000
$21,995	$22,195	$22,708	$23,240	$23,405	$23,920	$25,176	$25,999
$26,185	$26,268	$27,815	$27,910	$28,680	$28,950	$29,099	$29,249
$30,585	$30,645	$31,985	$32,250	$32,950	$33,595	$33,790	$34,590
$35,550	$36,300	$38,175	$41,188	$42,660	$54,950	$56,000	$63,500

(c) 2006 JupiterImages Corp.

Read the instruction manual for your calculator to learn how to enter data and use the built-in STAT functions.

Example 3: Use your calculator to calculate the mean price for the SUV data.
Solution:
Step 1: Understand and picture the problem.
    Question: Calculate the mean using the SUV data.

Step 2: Determine a plan.
    Use the calculator to calculate the mean.

Step 3: Execute the plan.
    The mean, $\bar{x}$, is approximately $29,426.23.

Step 4: Check your work.
    1. Check the numbers and your calculations.
    2. Was the question answered?
    3. Is the answer reasonable?

Example 4: Find the median for the SUV data.
Solution:
Step 1: Understand and picture the problem.
    Question: Calculate the median, the middle term of the data set.

Step 2: Determine a plan.
    1. The data is already ordered and n is the number of data values.
    2. Find the middle term using $\dfrac{n+1}{2}$ to locate its position.
    3. Find the value of the term (data item) at that position.

Step 3: Execute the plan.
    1. There are 40 data items, n = 40.
    2. The position of the median is $\dfrac{n+1}{2} = \dfrac{40+1}{2} = \dfrac{41}{2} = 20.5$.
    Locate the 20th and 21st data values.
    3. The median value is halfway between the 20th and 21st values.

$14,655	$14,799	$15,605	$16,395	$16,798	$17,990	$19,300
$20,000	$21,995	$22,195	$22,708	$23,240	$23,405	$23,920
$25,176	$25,999	$26,185	$26,268	$27,815	**$27,910**	**$28,680**
$28,950	$29,099	$29,249	$30,585	$30,645	$31,985	$32,250
$32,950	$33,595	$33,790	$34,590	$35,550	$36,300	$38,175
$41,188	$42,660	$54,950	$56,000	$63,500		

The 20th value is $27,910 and the 21st value is $28,680.
The median is the midpoint (average) of these two data values.

383

384

$$\text{The median is } \frac{\$27,910 + \$28,680}{2} = \frac{\$56,590}{2} = \$28,295.$$

Step 4: Check your work.
   1. Check the numbers and your calculations.
   2. Was the question answered?
   3. Is the answer reasonable?

Example 5: Use your calculator to calculate the mean price for the single-family home data:
$123,900 $130,900 $133,900 $138,900 $139,900 $146,900 $156,900
$156,900 $158,900 $159,400 $160,900 $163,900 $167,900 $167,900
$176,900 $182,900 $184,900 $186,900 $199,900 $199,900 $200,900
$204,900 $219,900 $219,900 $254,900 $256,300 $292,000 $311,750
$369,900 $385,500 $410,300 $430,500 $431,700 $436,500 $487,500
$496,500 $556,800
Solution:
Step 1: Understand and picture the problem.
   Question: Calculate the mean using the single family home data.

Step 2: Determine a plan.
   Use the calculator to calculate the mean.

Step 3: Execute the plan.
   The mean, $\bar{x}$, is approximately $248,763.51.

Step 4: Check your work.
   1. Check the numbers and your calculations.
   2. Was the question answered?
   3. Is the answer reasonable?

Example 6: Find the median for the home price data.
Solution:
Step 1: Understand and picture the problem.
   Question: Calculate the median, the middle term of the data set.

Step 2: Determine a plan.
   1. The data is already ordered and n equals the number of data values.
   2. Find the middle term using $\frac{n+1}{2}$ to locate its position.
   3. Find the value of the term (data item) at that position.

Step 3: Execute the plan.
1. There are 37 data items, n = 37.
2. The position of the median is $\frac{n+1}{2} = \frac{37+1}{2} = \frac{38}{2} = 19$.
3. The median is value at the 19th position.

$123,900 $130,900 $133,900 $138,900 $139,900 $146,900
$156,900 $156,900 $158,900 $159,400 $160,900 $163,900
$167,900 $167,900 $176,900 $182,900 $184,900 $186,900
$199,900 $199,900 $200,900 $204,900 $219,900 $219,900
$254,900 $256,300 $292,000 $311,750 $369,900 $385,500
$410,300 $430,500 $431,700 $436,500 $487,500 $496,500
$556,800.   The median is $199,900.

Step 4: Check your work.
1. Check the numbers and your calculations.
2. Was the question answered?
3. Is the answer reasonable?

Reality Check #1: Corporate Average Fuel Economy (CAFÉ)

(c) 2006 JupiterImages Corp.

The 1973 oil embargo caused a gas shortage in America. Gas prices soared, gas lines were blocks long and some gas stations ran out of gas. The American car owner demanded better gas mileage. In 1975, Congress passed the Corporate Average Fuel Economy (CAFE) Act. CAFE required all cars average 27.5 miles per gallon (combined city-highway mileage) and all trucks average 20.7 miles per gallon (combined city-highway mileage). If the manufacturer did not meet these standards, it would be fined $55 per vehicle per gallon below the standard. That means that if a car manufacturer made 2,000,000 cars and their average was 25 mpg, their fine would be $55 times 2,000,000 cars times 2.5 mpg (below the standard number of mpg) totaling **$275,000,000! However, NO domestic automaker has ever paid a fine.** How do the automakers meet these standards? The key word in the CAFE act is "average". The automakers know Statistics and use the mean for their average. Auto manufacturers offset their gas guzzling models with more fuel-efficient models to meet the mileage standard and avoid fines. That is why the more fuel-efficient 4 cylinder engine is offered with the larger car models.

The same applies to the truck standard. The PT Cruiser is classified as a truck. A truck is defined as a vehicle where all seats, except the drivers seat, are removable or fold down to form a cargo area. The exceptional gas mileage of the PT Cruiser helps Daimler-Chrysler meet the CAFE truck standard. Chicago Tribune, October 19, 2003

? **Cognitive Problems** ?

1. An updated average salary of baseball players is $2.5 Million. Do you think this average is the mean or median?
2. What misunderstandings can arise from an article or news report that uses the ambiguous term "average" in the reporting of a story?
3. Under what conditions is the mean a poor choice for an average?
4. In finding the median, explain the difference between the position of the median and the actual value of the median.

Exercise 5.3

(c) 2006 JupiterImages Corp.

1. The price of a sport's coupe (2-door, turbocharged engine, rear spoiler, etc.) varies. Use your calculator to find the mean for this data set.
   $12,255 $13,567 $15,549 $16,545 $17,261 $17,266 $17,800 $18,573
   $18,875 $19,200 $19,496 $19,999 $20,829 $21,684 $22,795 $22,871
   $22,910 $25,350 $26,795 $26,975 $27,170 $27,570 $28,990 $29,238
   $32,100 $34,670 $37,345 $37,905 $39,520 $41,300 $41,409 $42,463
   $48,600 $55,050 $55,800 $56,000 $63,050 $65,900 $72,350 $82,550

2. Use the data from Problem 1 and find the median.

3. The price of a convertible varies. Find the median for this data set
   $19,735 $19,990 $20,080 $21,245 $22,015 $22,255 $23,300 $23,995
   $24,140 $25,760 $25,915 $26,710 $28,365 $31,500 $31,870 $32,000
   $32,155 $37,120 $37,470 $39,450 $41,000 $41,430 $42,995 $43,270
   $43,395 $43,500 $44,640 $44,995 $45,320 $45,500 $46,500 $48,100
   $49,930 $53,140 $55,600 $71,200 $74,970 $78,390 $80,400 $83,820

4. Use the data from Problem 3 and your calculator to find the mean.

5. Home prices vary in different neighborhoods of Chicago. Use your calculator to find the mean for this data.
   $ 55,000 $ 56,000 $ 64,900 $ 77,900 $ 80,000 $ 87,500 $ 90,000
   $ 95,000 $ 98,000 $109,700 $110,000 $118,000 $120,000 $128,000
   $133,900 $135,000 $140,000 $142,000 $147,000 $150,000 $153,500
   $159,900 $160,000 $171,000 $182,000 $187,900 $188,000 $195,000
   $196,000 $220,000 $220,900 $230,000 $230,000 $245,075 $285,000
   $294,000 $297,000 $349,925 $367,000

6. Use the data from Problem 5 and find the median.

7. Townhouse prices vary in different suburbs of Chicago. Use your calculator to find the mean for this data.

$ 74,900	$ 79,900	$ 80,000	$ 84,000	$ 89,500	$ 91,000
$ 92,500	$ 94,000	$ 95,000	$ 97,000	$ 99,500	$101,000
$101,500	$103,990	$108,000	$110,000	$113,700	$114,475
$115,000	$117,500	$118,000	$124,700	$125,250	$127,500
$135,000	$139,000	$142,000	$145,000	$146,000	$149,000
$149,900	$159,000	$161,500	$162,000	$178,500	$205,000
$224,000	$235,000	$263,000	$268,650		

8. Use the data from Problem 7 and find the median.

9. The price for a new 1100 cc to 1200 cc motorcycle varies. Find the median for for this data.

$14,500	$15,600	$16,700	$ 8,990	$15,000	$16,990	$18,200
$19,200	$16,190	$11,999	$11,999	$ 7,399	$10,449	$12,299
$ 8,995	$ 7,895	$ 8,595	$ 7,399	$ 7,999	$ 8,199	$ 9,999
$ 9,699	$10,599	$ 8,399	$ 7,899	$10,899	$ 6,999	$14,399
$11,770						

10. Use your calculator to find the mean for the data in Problem 9.

5.4 IQR Boxplot and Standard Deviation
Objective: Boxplots, outliers and standard deviation will be studied.

Boxplots

Boxplots are used to pinpoint suspicious data. Suspicious data is data that is unusually small or large when compared to the other values in the data set. These unusual data values are called *outliers* and need to be investigated before further statistical analysis is performed on the data set.

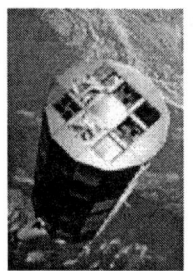

Read the following report that verifies the importance of investigating the origins of outliers.

(c) 2006 JupiterImages Corp.

Ozone and Outliers

The 'Ozone hole' above Antarctica provides the setting for one of the most infamous outliers in recent history. It is a great story demonstrating why **NOT** to delete outliers from a data set merely because they are outliers.

In 1985 three researchers (Farman, Gardinar and Shanklin) were puzzled by the results of the data gathered by the British Antarctic Survey. It showed that the ozone levels for Antarctica had dropped 10% below normal January levels. Why didn't the Nimbus 7 satellite, which recorded ozone levels, record these low ozone concentrations? Upon examination of the data, they realized that the satellite was in fact recording these low ozone concentration levels and had been doing so for years. Since these recorded ozone concentrations were so low they were being treated as outliers by a computer program and discarded! The Nimbus 7 satellite had in fact been gathering evidence of low ozone levels since 1976. **The damage to our atmosphere caused by chlorofluorocarbons went undetected and untreated for up to nine years because outliers were erroneously discarded and never examined.**

Moral: Don't just toss out outliers. They may be the most valuable members of a dataset!

CONSTRUCTING AN IQR BOXPLOT

Boxplots are also called box-and-whiskers plots. They display the main features of a data set. The boxplot described in this section is the Inner-Quartile Range (IQR) boxplot. The box represents the middle half of the data from the first quartile (the 25% position in the data) to the third quartile (the 75% position in the data). The left edge of the box (hinge) is essentially the first quartile. The right edge of the box (hinge) is essentially the third quartile. The median (the center or middle of the data) is displayed as a plus symbol (+) within the box.

The whiskers are used to display any data that is unusually low or high from the other values in the data set. These unusual values are called outliers. Outliers are represented by an asterisk (*). The length of the whiskers is determined by the following calculations:
1. IQR = Third Quartile − First Quartile
2. Left whisker = First Quartile − 1.5 *times* (IQR)
3. Right whisker = Third Quartile + 1.5 *times* (IQR)

Consider the Sport Utility Data (SUV) data set used in the quantitative data section. The data consists of the prices of 40 four-wheel drive (4WD) vehicles.

$14,655	$14,799	$15,605	$16,395	$16,798	$17,990	$19,300	$20,000
$21,995	$22,195	$22,708	$23,240	$23,405	$23,920	$25,176	$25,999
$26,185	$26,268	$27,815	$27,910	$28,680	$28,950	$29,099	$29,249
$30,585	$30,645	$31,985	$32,250	$32,950	$33,595	$33,790	$34,590
$35,550	$36,300	$38,175	$41,188	$42,660	$54,950	$56,000	$63,500

Applying the quartile formulas to the SUV Price table; the first, second (median) and third quartiles are:

From the previous section, the median is the value at the $\frac{n+1}{2}$ location, where n is the number of values in the data set. In this case, n = 40. Locate the $\frac{40+1}{2} = \frac{41}{2} = 20.5$ value. This is the midpoint of the 20th and 21st values. These values are highlighted in the data set below.

$14,655	$14,799	$15,605	$16,395	$16,798	$17,990	$19,300	$20,000
$21,995	$22,195	$22,708	$23,240	$23,405	$23,920	$25,176	$25,999
$26,185	$26,268	$27,815	**$27,910**	**$28,680**	$28,950	$29,099	$29,249
$30,585	$30,645	$31,985	$32,250	$32,950	$33,595	$33,790	$34,590
$35,550	$36,300	$38,175	$41,188	$42,660	$54,950	$56,000	$63,500

The median is $\frac{\$27,910 + \$28,680}{2} = \frac{\$56,590}{2} = \$28,295$.

The median of $28,295 has half the data, 20 values below it, and the other half of the data, the other 20 values above it. The median is shown in the data set.

$14,655 $14,799 $15,605 $16,395 $16,798 $17,990 $19,300 $20,000
$21,995 $22,195 $22,708 $23,240 $23,405 $23,920 $25,176 $25,999
$26,185 $26,268 $27,815 $27,910

Median = $28,295

$28,680 $28,950 $29,099 $29,249 $30,585 $30,645 $31,985 $32,250
$32,950 $33,595 $33,790 $34,590 $35,550 $36,300 $38,175 $41,188
$42,660 $54,950 $56,000 $63,500

First Quartile

The median is not used in the calculation of the first quartile. The first quartile utilizes the first half of the data set, the first 20 values. The first quartile is the median of the first 20 values, values in positions 1 through 20 of the data set. Locate the $\frac{20+1}{2} = \frac{21}{2} = 10.5$ value. This is the midpoint of the 10th and 11th values. These values are highlighted in the data set below.

$14,655 $14,799 $15,605 $16,395 $16,798 $17,990 $19,300 $20,000
$21,995   **$22,195**   **$22,708** $23,240 $23,405 $23,920 $25,176 $25,999
$26,185 $26,268 $27,815 $27,910

The first quartile is $\frac{\$22,195 + \$22,708}{2} = \frac{\$44,903}{2} = \$22,451.50$.

The first quartile of $22,451.50 has half of the data, 10 values below it, and the other half of the data, the other 10 values above it. The first quartile is shown in the data set of the first 20 values below:

$14,655 $14,799 $15,605 $16,395 $16,798 $17,990 $19,300 $20,000
$21,995 $22,195

First Quartile = $22,451.50

$22,708 $23,240 $23,405 $23,920 $25,176 $25,999 $26,185 $26,268
$27,815 $27,910

Third Quartile

The median is not used in the calculation of the third quartile. The third quartile utilizes the second half of the data set, the second 20 values. The third quartile is the median of the second 20 values, values in positions 21 through 40, of the data set. Locate the $\frac{20+1}{2} = \frac{21}{2} = 10.5$ value. This is the midpoint of the 10th and 11th values for the second set of 20 values. These values are highlighted in the data set.

$28,680 $28,950 $29,099 $29,249 $30,585 $30,645 $31,985 $32,250
$32,950   **$33,595 $33,790** $34,590 $35,550 $36,300 $38,175 $41,188
$42,660 $54,950 $56,000 $63,500

The third quartile is $\dfrac{\$33,595 + \$33,790}{2} = \dfrac{\$67,385}{2} = \$33,692.50$.

The third quartile of $33,692.50 has half of the data, 10 values below it and the other 10 values above it. The third quartile is shown in the data set of the second 20 values below:

$28,680 $28,950 $29,099 $29,249 $30,585 $30,645 $31,985 $32,250
$32,950 $33,595

Third Quartile = $33,692.50

$33,790 $34,590 $35,550 $36,300 $38,175 $41,188 $42,660 $54,950
$56,000 $63,500

Calculating the whiskers:
1. IQR = Third Quartile − First Quartile = $33,692.50 − $22,451.50 = $11,241.00
2. 1.5 *times* (IQR) = 1.5($11,241.00) = $16,861.50
3. Left Whisker = First Quartile − 1.5 *times* (IQR) =
    $22,451.50 − $16,861.50 = $5,590.00
4. Right Whisker = Third Quartile + 1.5 *times* (IQR) =
    $33,692.50 + $16,861.50 = $50,554.00

The boxplot is:

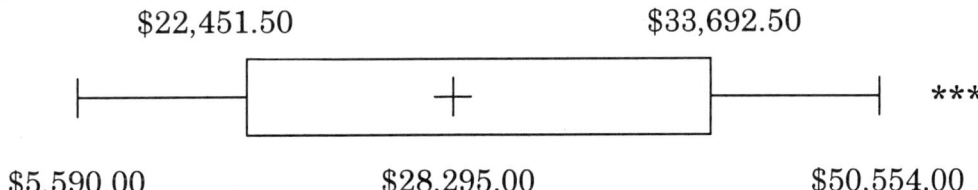

$5,590.00 $28,295.00 $50,554.00

Note: The three asterisks represent the three prices ($54,950 $56,000 and $63,500) of SUVs that are unusually high compared to the other prices in the data set. These prices should be investigated before further analysis of the data is performed.

FYI: Why 1.5 IQR's?
Why it is customary to multiply 1.5 * IQR to find outliers? In particular, why is it 1.5 and not 2?" From Paul Velleman, who interviewed John Tukey. The "official" answer given by John Tukey (creator of the IQR boxplot) is: "Because 1 is too small and 2 is too large."

Example 1: The following data set represents the selling price of 37 new single family homes. Construct an IQR boxplot to determine if any of the home prices are suspicious (outliers)?

$123,900	$130,900	$133,900	$138,900	$139,900	$146,900	$156,900
$156,900	$158,900	$159,400	$160,900	$163,900	$167,900	$167,900
$176,900	$182,900	$184,900	$186,900	$199,900	$199,900	$200,900
$204,900	$219,900	$219,900	$254,900	$256,300	$292,000	$311,750
$369,900	$385,500	$410,300	$430,500	$431,700	$436,500	$487,500
$496,500	$556,800

Solution:
Step 1: Understand and picture the problem.
  1. Question: Are any of the 37 home prices suspicious? Are any of the home prices outliers?
  2. Construct an IQR boxplot to locate outliers in the data.

Step 2: Establish a plan.
  1. If needed, sort the data from smallest to largest.
  2. Calculate the five values associated with an IQR boxplot:
     a. Median   b. First Quartile   c. Third Quartile
     d. Left Whisker   e. Right Whisker
  3. Draw the IQR boxplot.
  4. Compare the home prices to the left and right whiskers to determine if any of the data is suspicious.

Step 3: Execute the plan.
  1. The data is sorted.
  2a. Calculate the median.
      The data set has 37 values, n = 37. The median is the value at the $\frac{n+1}{2}$ location. The median is the $\frac{37+1}{2} = \frac{38}{2} = 19^{th}$ value.

      $123,900 $130,900 $133,900 $138,900 $139,900 $146,900 $156,900
      $156,900 $158,900 $159,400 $160,900 $163,900 $167,900 $167,900
      $176,900 $182,900 $184,900 $186,900 **$199,900** $199,900 $200,900
      $204,900 $219,900 $219,900 $254,900 $256,300 $292,000 $311,750
      $369,900 $385,500 $410,300 $430,500 $431,700 $436,500 $487,500
      $496,500 $556,800

      The median is $199,900.

  2b. Calculate the first quartile.
      The first quartile is the median of the first 18 values, values in positions 1 through 18 of the data set.

Locate the $\frac{18+1}{2} = \frac{19}{2} = 9.5$ value. This is the midpoint of the 9th and 10th values for the first 18 values. These values are highlighted in the data set.

$123,900 $130,900 $133,900 $138,900 $139,900 $146,900
$156,900 $156,900 **$158,900 $159,400** $160,900 $163,900
$167,900 $167,900 $176,900 $182,900 $184,900 $186,900

The first quartile is the midpoint of the 9th and 10th values,

The first quartile is $\frac{\$158,900 + \$159,400}{2} = \frac{\$318,300}{2} = \$159,150$.

2c. Calculate the third quartile.

The third quartile is the median of the second 18 values, values 20 through 37 of the data set. Locate the $\frac{18+1}{2} = \frac{19}{2} = 9.5$ value. The third quartile is the midpoint of the 9th and 10th values for the second 18 values. These values are highlighted in the data set.

$199,900 $200,900 $204,900 $219,900 $219,900 $254,900
$256,300 $292,000 **$311,750 $369,900** $385,500 $410,300
$430,500 $431,700 $436,500 $487,500 $496,500 $556,800

The third quartile is $\frac{\$311,750 + \$369,900}{2} = \frac{\$681,650}{2} = \$340,825$.

2d. Calculate the left whisker.
   i. IQR = Third Quartile – First Quartile
      IQR = $340,825 – $159,150 = $181,675
   ii. 1.5 *times* (IQR) = 1.5($181,675) = $272,512.50
   iii. Left Whisker = First Quartile – 1.5 *times* (IQR)
      Left Whisker = $159,150 – $272,512.50 = –$113,362.50.
      The price of a home cannot be negative. The left whisker is $0.

2e. Calculate the right whisker.
   Right Whisker = Third Quartile + 1.5 *times* (IQR)
   Right Whisker = $340,825 + $272,512.50 = $613,337.50.

3. The IQR boxplot is:

   $159,150 $340,825

   |———[+   ]———|

   $0 $199,900 $613,337.50

4. None of the home prices are less than $0 or greater than $613,337.50. None of the home prices are suspicious or outliers.

Step 4: Check your work.
1. Check the numbers and your calculations.
2. Is the answer reasonable?
3. Has the question been answered?

Measure of Dispersion, Standard Deviation

(c) 2006 JupiterImages Corp.

Dispersion is the scattering or spread of the data values. These values can be clustered close together or can be widely varied in value. The measure of dispersion communicates this relationship with a number. The consistency of the data is another way of interpreting dispersion. As an example of the significance of dispersion, consider the contents of six 12 oz. bottles taken from two different bottling machines.

Machine 1 (ounces):	10.0, 11.0, 12.0, 12.0, 13.0, 14.0
Machine 2 (ounces):	11.8, 11.9, 12.0, 12.0, 12.1, 12.2

If measures of central tendency were the only measures used to determine the quality of production, both machines would be equal. Calculate their means:

Mean of Machine 1: $\frac{10.0+11.0+12.0+12.0+13.0+14.0}{6} = \frac{72.0}{6} = 12.0$ ounces

Mean of Machine 2: $\frac{11.8+11.9+12.0+12.0+12.1+12.2}{6} = \frac{72.0}{6} = 12.0$ ounces

Calculate their medians. Find the value at the $\frac{n+1}{2} = \frac{6+1}{2} = \frac{7}{2} = 3.5$ position.

Find the midpoint of the 3rd and 4th data values.

Median of Machine 1: $\frac{12.0+12.0}{2} = \frac{24.0}{2} = 12.0$ ounces

Median of Machine 2: $\frac{12.0+12.0}{2} = \frac{24.0}{2} = 12.0$ ounces

Both machines have the same mean and median! Both machines fill the bottles with an average of 12 ounces. Both measures of average, the mean and median equal 12 ounces. However, the consistency of production is obviously not the same for the two machines. The bottles from Machine 2 are filled with less error (closer to the 12 oz. goal) than the bottles from Machine 1. This consistency, or inconsistency, is expressed by a measure of dispersion. The **standard deviation** is the measure of dispersion that will now be addressed.

Standard Deviation

Standard deviation is a dispersion measurement representing the average distance EACH data value varies from the mean. The formula for sample standard deviation is:

Standard Deviation

$$s = \sqrt{\frac{\Sigma(x-\bar{x})^2}{n-1}}$$

where: Σ means add
x is a data value
$\bar{x}$ is the sample mean.
n is the number of values.

The following tables demonstrate the calculations required in finding the standard deviation, s, for the two bottling machines. The data from Machine 1 produces the table.

Data x	Mean $\bar{x}$	$x - \bar{x}$	$(x - \bar{x})^2$
10.0	12.0	−2.0	4.0
11.0	12.0	−1.0	1.0
12.0	12.0	0.0	0.0
12.0	12.0	0.0	0.0
13.0	12.0	1.0	1.0
14.0	12.0	2.0	4.0
Totals			10.0

1. From the table, $\Sigma(x-\bar{x})^2 = 10$ and n = 6.

2. $s = \sqrt{\dfrac{\Sigma(x-\bar{x})^2}{n-1}} = \sqrt{\dfrac{10}{6-1}} = \sqrt{2} \approx 1.41421356$

The standard deviation for Machine 1 is 1.41421356.

The data from Machine 2 produces the table.

Data x	Mean $\bar{x}$	$x - \bar{x}$	$(x - \bar{x})^2$
11.8	12.0	−0.2	0.04
11.9	12.0	−0.1	0.01
12.0	12.0	0.0	0.00

12.0	12.0	0.0	0.00
12.1	12.0	0.1	0.01
12.2	12.0	0.2	0.04
Totals			0.10

1. From the table, $\Sigma(x-\bar{x})^2 = 0.10$ and n = 6.

2. $s = \sqrt{\dfrac{\Sigma(x-\bar{x})^2}{n-1}} = \sqrt{\dfrac{0.10}{6-1}} = \sqrt{0.02} \approx 0.141421356$

3. The standard deviation for Machine 2 is 0.141421356.

Since both machines fill bottles with an average of 12 ounces, a dispersion measurement, standard deviation will determine the better machine. In an ideal situation, all the values in a data set would be equal. This would produce a standard deviation of zero (0). It is unlikely for every item produced to be identical. The goal of manufacturing is to get the standard deviation as close to zero as possible. **Since 0.141421356 is less than 1.41421356, Machine 2 is the better bottling machine.**

In a large data set, the calculation of the standard deviation is very tedious. To simplify this process, the calculator is used to find the standard deviation.

(c) 2006 JupiterImages Corp.

Read the instruction manual for your calculator to learn how to enter data and use the built-in STAT functions. For standard deviation use s_x.

Example 2: Use your calculator to find the standard deviation for the prices of 37 single family homes.

$123,900	$130,900	$133,900	$138,900	$139,900	$146,900	$156,900
$156,900	$158,900	$159,400	$160,900	$163,900	$167,900	$167,900
$176,900	$182,900	$184,900	$186,900	$199,900	$199,900	$200,900
$204,900	$219,900	$219,900	$254,900	$256,300	$292,000	$311,750
$369,900	$385,500	$410,300	$430,500	$431,700	$436,500	$487,500
$496,500	$556,800

Solution:
Step 1: Understand and picture the problem.
    Question: Calculate the standard deviation for the home price data.

Step 2: Determine a plan.
  Use your calculator to calculate the standard deviation.

Step 3: Execute the plan.
  The standard deviation, s, is approximately $123,194.39.

Step 4: Check your work.
  1. Check the numbers and your calculations.
  2. Is the answer reasonable?
  3. Was the question answered?

Optional Student Project
"Descriptive Statistics for Quantitative Data" can be found on page 431.

? Cognitive Problems ?

1. Explain the significance of an outlier.
2. Explain the consequences of automatically dropping outliers from the data set.
3. Explain why standard deviation alone does not give a complete description of a data set.

Exercise 5.4

1. The price of sport coupes (2-door, turbocharged engine, rear spoiler, etc.) vary. Use your calculator to find the standard deviation for this data set.

   $12,255 $13,567 $15,549 $16,545 $17,261 $17,266 $17,800 $18,573
   $18,875 $19,200 $19,496 $19,999 $20,829 $21,684 $22,795 $22,871
   $22,910 $25,350 $26,795 $26,975 $27,170 $27,570 $28,990 $29,238
   $32,100 $34,670 $37,345 $37,905 $39,520 $41,300 $41,409 $42,463
   $48,600 $55,050 $55,800 $56,000 $63,050 $65,900 $72,350 $82,550

2. Construct an IQR boxplot to determine any suspicious values (outliers) in the data set from Problem 1.

3. The price of convertibles also vary. Construct an IQR boxplot to determine if any of the convertible prices are suspicious (outliers).
   $19,735 $19,990 $20,080 $21,245 $22,015 $22,255 $23,300 $23,995
   $24,140 $25,760 $25,915 $26,710 $28,365 $31,500 $31,870 $32,000
   $32,155 $37,120 $37,470 $39,450 $41,000 $41,430 $42,995 $43,270
   $43,395 $43,500 $44,640 $44,995 $45,320 $45,500 $46,500 $48,100
   $49,930 $53,140 $55,600 $71,200 $74,970 $78,390 $80,400 $83,820

4. Use the data from Problem 3 and your calculator to find the standard deviation.

5. Home prices vary in different neighborhoods of Chicago. Use your calculator to find the standard deviation for this data.
   $ 55,000 $ 56,000 $ 64,900 $ 77,900 $ 80,000 $ 87,500 $ 90,000
   $ 95,000 $ 98,000 $109,700 $110,000 $118,000 $120,000 $128,000
   $133,900 $135,000 $140,000 $142,000 $147,000 $150,000 $153,500
   $159,900 $160,000 $171,000 $182,000 $187,900 $188,000 $195,000
   $196,000 $220,000 $220,900 $230,000 $230,000 $230,000 $245,075
   $285,000 $294,000 $297,000 $349,925 $367,000

6. Construct an IQR boxplot to determine any suspicious values (outliers) in the data set from Problem 5.

7. Townhouse prices vary in different suburbs of Chicago. Construct an IQR boxplot to determine if any of the townhouse prices are suspicious (outliers).

   $ 74,900 $ 79,900 $ 80,000 $ 84,000 $ 89,500 $ 91,000
   $ 92,500 $ 94,000 $ 95,000 $ 97,000 $ 99,500 $101,000
   $101,500 $103,990 $108,000 $110,000 $113,700 $114,475
   $115,000 $117,500 $118,000 $124,700 $125,250 $127,500
   $135,000 $139,000 $142,000 $145,000 $146,000 $149,000
   $149,900 $159,000 $161,500 $162,000 $178,500 $205,000
   $224,000 $235,000 $263,000 $268,650

8. Use the data from Problem 7 and your calculator to find the standard deviation.

9. The price for a new 1100 cc to 1200 cc motorcycle varies. Construct an IQR boxplot to determine if any of the motorcycle prices are suspicious (outliers).
   $14,500 $15,600 $16,700 $ 8,990 $15,000 $16,990 $18,200 $19,200
   $16,190 $11,999 $11,999 $ 7,399 $10,449 $12,299 $ 8,995 $ 7,895
   $ 8,595 $ 7,399 $ 7,999 $ 8,199 $ 9,999 $ 9,699 $10,599 $ 8,399
   $ 7,899 $10,899 $ 6,999 $14,399 $11,770

10. Use the data from Problem 9 and your calculator to find the standard deviation.

5.5 STANDARD NORMAL CURVE
Objective: The Standard Normal Curve with Mean and Standard Deviation equal to 0 and 1, respectively, will be investigated.

The mean and standard deviation are used to define the standard normal curve, a bell-shaped pattern.

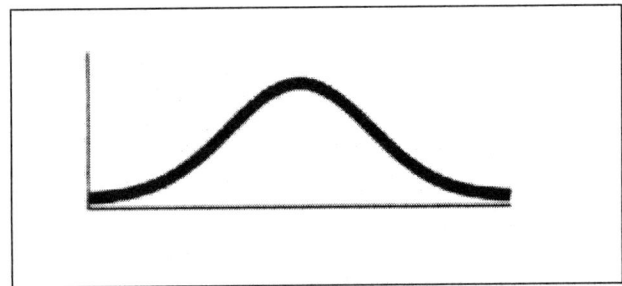

The standard normal curve is the basis for inferential statistics. Inferential statistics uses data taken from samples to make predictions about the population. Consider the unemployment figures that are released every month. It is impossible to interview everyone in the country (population) every month. However, by interviewing a representative part of the population (sample), inferential statistics can be used to predict the percentage of unemployment present in the population. The standard normal curve is fundamental to this calculation.

Also many situations in nature and manufacturing follow a bell-shaped pattern when graphed:
- A. The weights of newborn babies
- B. The scores on IQ tests
- C. The heights of females
- D. The weights of males
- E. The shoe sizes of 12 year old children
- F. The weight of 14.5 ounce bags of potato chips
- G. The liquid contents of 12 ounce cans of cola
- H. The life expectancies of AA alkaline batteries

Because bell-shaped curves occur in many common situations, it has been exhaustively studied by statisticians. The result of these studies is summarized in the following table.

> **Properties of the Standard Normal Curve:**
>
> A. The Standard Normal Curve is symmetric. 50% of the area of the curve is to the left side of its center and 50% of the area of the curve is to the right of its center.
> B. The **mean**, represented by the symbol "mu" (μ), divides the Standard Normal Curve into two equal portions. The mean, μ, is the center mark on the horizontal axis.
> C. The Standard Normal Curve has a mean of zero and a **standard deviation** of one. The standard deviation, represented by the symbol **sigma (σ)**, represents the unit of measure for scaling the horizontal axis of the Standard Normal Curve. Positive values are located to the right of the mean and negative values are located to the left of the mean. These values are called **z-scores or standard scores**.
> D. The total area under the Standard Normal Curve equals 1.00. The area under the Standard Normal Curve is a graphical representation of a probability distribution.
> **Note:** Probability will be studied in a later chapter. The purpose now is to become familiar with the Standard Normal Curve.

These properties of the Standard Normal Curve are displayed in the following figure:

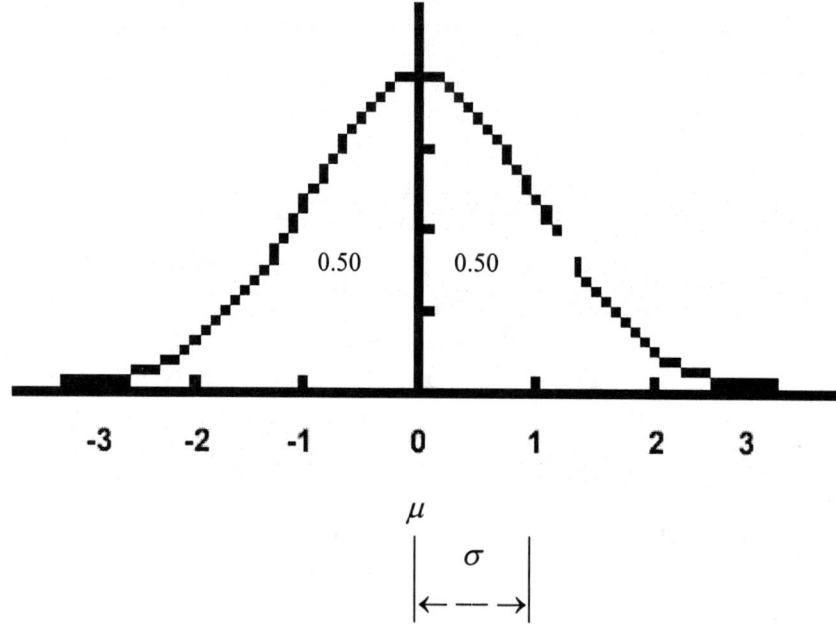

The table on the **inside back cover** is the result of an exhaustive study by statisticians of the Standard Normal Curve. The values in the table represent the area under the curve starting at the mean, 0, and extending to the z score. This table is called the "Standard Normal (z) Distribution Table."

The interpretation of the Standard Normal (z) Distribution Table, is clarified by the Empirical Rule. This rule describes the area underneath the curve for various z-scores.

The Empirical Rule

A. About 68% of the area under the curve falls within 1 standard deviation of the mean.
B. About 95% of the area under the curve falls within 2 standard deviations of the mean.
C. About 99.7% of the area under the curve falls within 3 standard deviations of the mean.

To read the Standard Normal (z) Distribution Table find the value at z = 1.00. This value represents the area under the bell-shaped curve starting at the Mean (Zero) and extending to 1.00 (one standard deviation from the mean) as shown in the figure below.

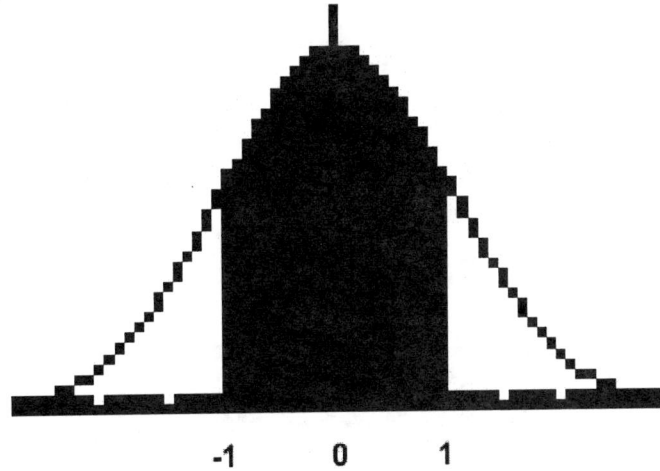

The first column of the table lists the z score to one decimal place, look for 1.0. The value to the right of 1.0 in the column .00 is 0.3413. Since the Standard Normal Curve is symmetrical, 0.3413 is also the value from the Mean (Zero) to z = –1.00 (one standard deviation to the left of the mean). Thus, 0.3413 + 0.3413 = 0.6826 (about 68%) of the area under the curve falls within one standard deviation of the mean.

The following bell-shaped curve displays plus or minus (±) one standard deviation from mean. These standard units are referred to as z-scores.

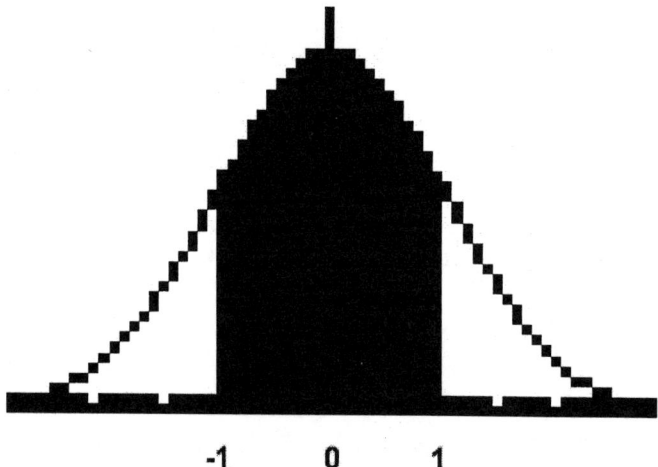

Similarly, check the table to verify that the value at z = 2.00 yields 0.4772. This value represents the area under the curve from zero to two. Due to the symmetry of the Standard Normal Curve, the value for z = −2.00 is the same.

Thus, 0.4772 + 0.4772 = 0.9544 (about 95%) of the area under the curve falls within 2 standard deviations of the mean. The following bell-shaped curve displays plus or minus (±) two standard deviations of the mean.

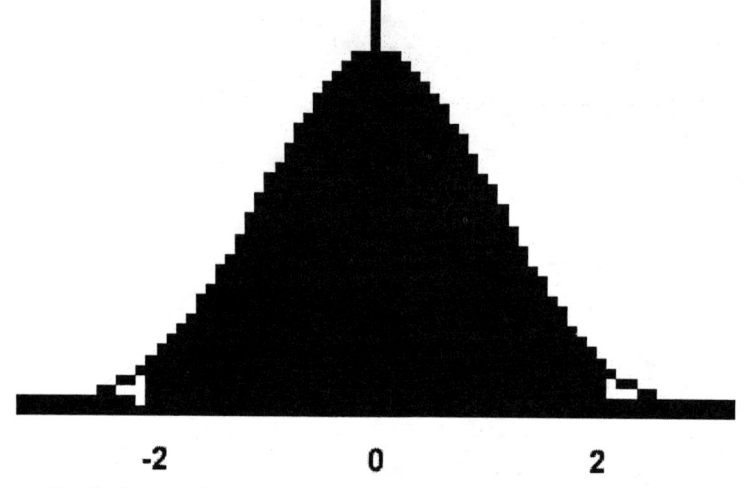

Use the table to find the value at z = 3.00. This point is three standard deviations to the right of the mean. The Standard Normal Table yields 0.4987. This value represents the area under the curve from zero to three. Due to the symmetry of the Standard Normal Curve, the value for z = −3.00 is the same. Thus, 0.4987 + 0.4987 = 0.9974 (about 99.7%) of the area under the curve falls within 3 standard deviations of the mean. The following bell-shaped curve displays plus or minus (±) three standard deviations of the mean.

403

Working With the Normal Curve

When working with the Normal Probability Distribution, there are only five types of problems that you will encounter. The five scenarios have the following diagrams.

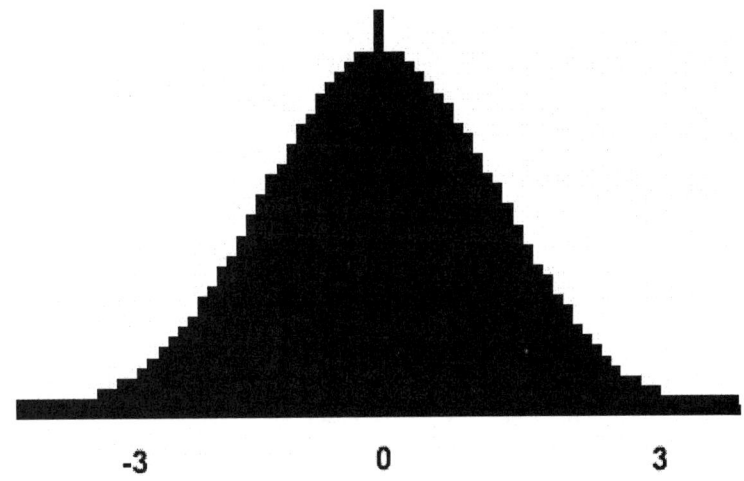

Five possible scenarios for Standard Normal Curve problems.

Scenario 1: The area under the curve extends from the mean to a standard score. See Example 1.

Scenario 2: The area under the curve extends from one side of a standard score involving only the tail of the Normal curve. See Example 2.

Scenario 3: The area under the curve includes one half of the Normal curve and extends to a standard score on the other side of the mean. See Example 3.

Scenario 4: The area under the curve involves two standard scores on the same side of the mean. See Example 4.

Scenario 5: The area under the curve involves two standard scores on opposite sides of the mean. See Example 5.

Scenario 1: The area under the curve extends from the mean to a standard score.

Example 1: Find the area under the curve for $0 < z < 1.48$.

Solution:

Step 1: Understand and Picture the Problem.
1. Question: What is the area under the curve for z between 0 and 1.48?
2. Draw a bell-shaped curve above the horizontal axis and label its center zero.
3. Label the point 1.48 on the horizontal axis and shade in the region under the curve from zero to 1.48.

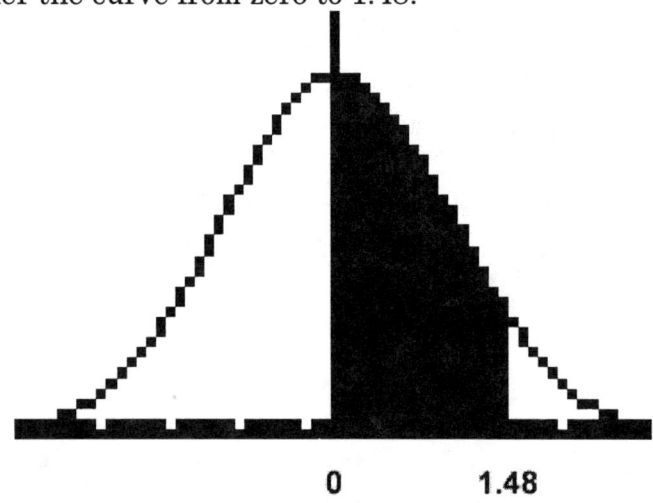

Step 2: Develop a plan.
1. Locate the z-score of 1.48 in the Standard Normal Distribution Table.
2. The corresponding value represents the area starting at zero and ending at 1.48.

Step 3: Execute the plan.
Use the Standard Normal Table to find the probability at $z = 1.48$. Look down the z column to 1.4. Then move to the right to the column 0.08 which yields 0.4306. The area under the curve between 0 and 1.48 is 0.4306.

Step 4: Check your work.
1. Check the numbers and your calculations.
2. Was the question answered?
3. Is your answer reasonable?

Scenario 2: The area under the curve extends from one side of a standard score involving only the tail of the Normal curve.

Example 2: Find the area under the curve for z > 1.48.

Solution:

Step 1: Understand and Picture the Problem.
1. Question: What is the area under the curve for z greater than 1.48?
2. Draw a bell-shaped curve above a horizontal axis and label its center zero.
3. Label the point 1.48 on the horizontal axis and shade in the region under the curve to the right of 1.48.

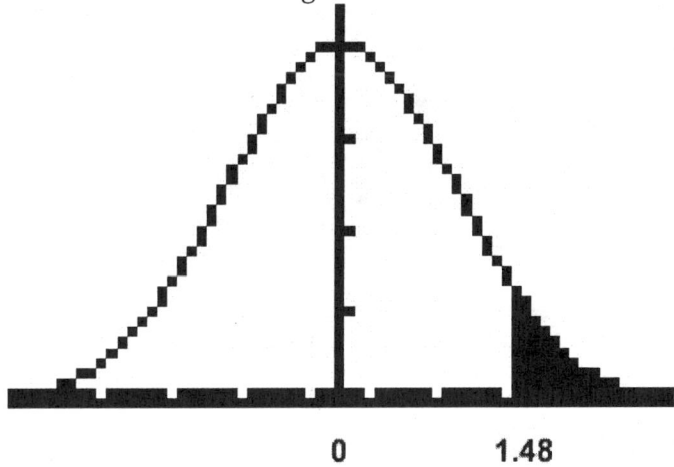

Step 2: Develop a plan.
1. Locate the z-score of 1.48 in the Standard Normal Dist. Table.
2. The corresponding value represents the area starting at zero and ending at 1.48. Since this value does not equate to the shaded region, an adjustment is required.
3. Since the entire area under the curve to the right of the mean is 0.5000 and the problem's region is in the tail of the upper half of the Standard Normal Curve, subtract the table value from 0.5000 to find the area of the region beyond z = 1.48.

Step 3: Execute the plan.
1. From the Standard Normal Table, z = 1.48 yields 0.4306.
2. 0.4306 represents the area from zero to 1.48.
3. 0.5000 − 0.4306 = 0.0694. Area under the curve z > 1.48 is 0.0694.

Step 4: Check your work.
1. Check the numbers and your calculations.
2. Was the question answered?
3. Is your answer reasonable?

Scenario 3: The area under the curve includes one half of the Normal curve and extends to a standard score on the other side of the mean.
Example 3: Find the area under the curve for z < 1.48.
Solution:
Step 1: Understand and Picture the Problem.
1. Question: What is the area under the curve for z less than 1.48?
2. Draw a bell-shaped curve above a horizontal axis and label its center zero.
3. Label the point 1.48 on the horizontal axis and shade the entire region under the curve to the left of 1.48.

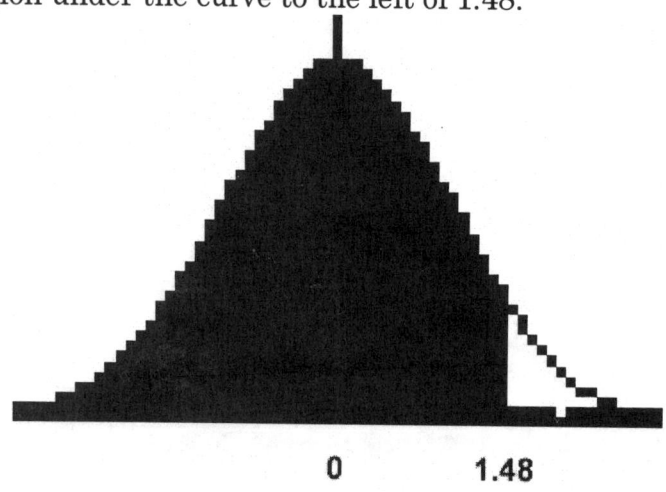

Step 2: Develop a plan.
1. Locate the z-score 1.48 in the Standard Normal Distribution Table.
2. The corresponding value represents the area starting at zero and ending at 1.48. Since this value does not equate to all of the shaded region, an adjustment is required.
3. The shaded region includes the entire lower half of the Standard Normal Curve, which has an area of 0.5000 plus the region from the mean to 1.48. Adding 0.5000 to the corresponding table value for z = 1.48 will define the shaded region.

Step 3: Execute the plan.
1. From the Standard Normal Table, z = 1.48 yields 0.4306.
2. 0.4306 represents the area from zero to 1.48.
3. 0.5000 + 0.4306 = 0.9306
   Thus, the area under the curve for z < 1.48 is 0.9306.

Step 4: Check your work.
1. Check the numbers and your calculations.
2. Was the question answered?
3. Is your answer reasonable?

Scenario 4: The area under the curve involves two standard scores on the same side of the mean.

Example 4: Find the area under the curve for 0.53 < z < 1.48.

Solution:

Step 1: Understand and Picture the Problem.
1. Question: What is the area under the curve for z between 0.53 and 1.48?
2. Draw a bell-shaped curve above a horizontal axis and label its center zero.
3. Label the points 0.53 and 1.48 on the horizontal axis and shade in the region under the curve between 0.53 and 1.48.

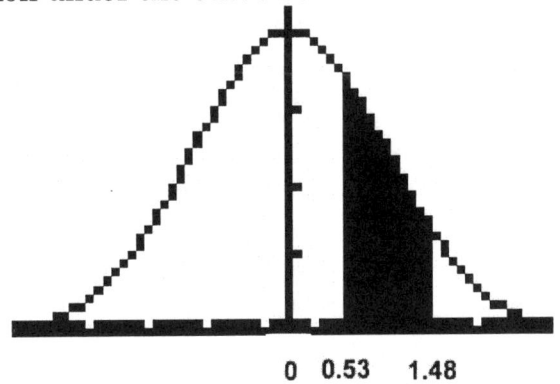

Step 2: Develop a plan.
1. Locate the z-scores 0.53 and 1.48 in the Standard Normal Table.
2. The corresponding values represents the area starting at zero for both z-scores and ending at 0.53 and 1.48, respectively. Since these values do not equate to the shaded region, an adjustment needs to be made.
3. Since the problem's region is between these two values, on the same side of the mean, subtract the smaller area from the larger area to define the shaded region.

Step 3: Execute the plan.
1. From the Standard Normal Table, z = 0.53 yields 0.2019 and z = 1.48 yields 0.4306.
2. Both areas start at the mean and extend to the right.
3. 0.4306 − 0.2019 = 0.2287.
   Thus the area under the curve for 0.53 < z < 1.48 is 0.2287.

Step 4: Check your work.
1. Check the numbers and your calculations.
2. Was the question answered?
3. Is your answer reasonable?

Scenario 5: The area under the curve involves two standard scores on opposite sides of the mean.

Example 5: Find the area under the curve for –0.53 < z < 1.48.

Solution:

Step 1: Understand and Picture the Problem.
1. Question: What is the area under the curve for z between –0.53 and 1.48?
2. Draw a bell-shaped curve above a horizontal axis and label its center zero.
3. Label the points –.53 and 1.48 on the horizontal axis and shade in the region under the curve between –0.53 and 1.48.

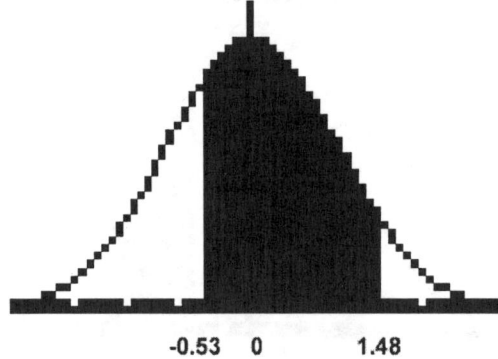

Step 2: Develop a plan.
1. Locate the z-scores 0.53 and 1.48 in the Standard Normal Table.
2. The corresponding values represents the area starting at zero for both z-scores and ending at –0.53 and 1.48, respectively. Since these values do not equate to the shaded region, an adjustment needs to be made.
3. Since the problem's region consists of both of these two values, on opposite sides of the mean, add these areas to define the shaded region.

Step 3: Execute the plan
1. From the Standard Normal Table, z = 0.53 yields 0.2019 and z = 1.48 yields 0.4306.
2. 0.2019 represents the area to the left of the mean. 0.4306 represents the area to the right of the mean.
3. 0.2019 + 0.4306 = 0.6325.
   Thus the area under the curve for –0.53 < z < 1.48 is 0.6325.

Step 4: Check your work.
1. Check the numbers and your calculations.
2. Was the question answered?
3. Is your answer reasonable?

❓ Cognitive Problems ❓

1. Why does the table for the Standard Normal Curve contain only positive z-scores?
2. How do you determine if the z-score is positive or negative?

Exercise 5.5
Find the area under the standard normal curve for the following standard scores.

1. $0 < z < 1.09$ 2. $-2.97 < z < 0$ 3. $0 < z < 0.25$ 4. $-0.94 < z < 0$ 5. $z > 1.45$

6. $z < -1.51$ 7. $z < -0.44$ 8. $z > 1.58$ 9. $z < 1.58$ 10. $z > -1.58$

11. $z > -0.25$ 12. $z < 2.02$ 13. $1.00 < z < 2.00$ 14. $-1.45 < z < -0.98$

15. $-2.09 < z < -0.61$ 16. $0.35 < z < 2.84$ 17. $-2.05 < z < 2.05$

18. $-1.44 < z < 2.09$ 19. $-1.76 < z < 0.54$ 20. $-2.75 < z < 1.97$

5.6 STANDARD SCORE (Z) BASED ON AN AREA
Objective: Find the standard score (z-score) for a given area under the Standard Normal Curve.

Another type of problem involving the Standard Normal Curve is the inverse operation of the problems just discussed. Given an area under the curve, find the corresponding z-score.

Percentile is another way of expressing area under the Standard Normal Curve. The 70th percentile, P_{70}, represents that z-score that has 70% of the area under the Standard Normal Curve from the left to the right. (Remember, 100% is the complete area under the Standard Normal Curve.) This also implies that the remaining 30% of the data is to the right of that same z-score. For example, find the value of z that corresponds to the 70th percentile, P_{70}. The following figure displays this information.

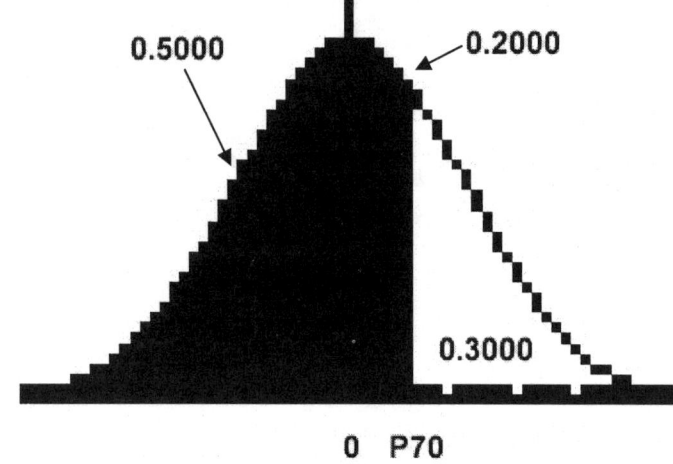

The mean (z = 0) marks the middle, the 50th Percentile, P_{50}. Half of the data is below or to the left of this mark. Since 70% is greater than the mean, 20% more is needed to equal 70%. From the Standard Normal Distribution Table, locate the two numbers closest to 0.2000 in the body. Choose consecutive rows and columns that come closest to containing 0.2000.

$$0.1985 \quad \mathbf{0.2000} \quad 0.2019$$
$$\downarrow \quad\quad\quad\quad\quad\quad \downarrow$$
$$z = 0.52 \quad\quad\quad z = 0.53$$

Since 0.1985 is closer to 0.2000 than 0.2019, then P_{70} has z = 0.52.

Example 1: Find the value of z associated with the 10th Percentile, P_{10}.
Solution:
  Step 1: Understand and Picture the Problem.

1. Problem: Find the z-score corresponding to P_{10}.
2. Since an area under the curve is given, a "z-score" has to be found.
3. Draw a bell-shaped curve above a horizontal axis and label its center with a zero.
4. P_{10} is the z-score where 0.1000 of the area under the curve is located in the left tail.
5. Locate and label the lower 10% (.1000) on the curve.

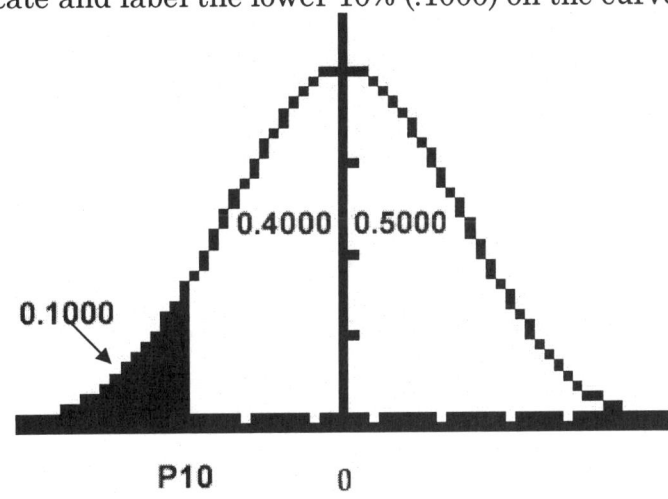

Step 2: Develop a plan.
1. Since the Standard Normal Distribution Table reads from the mean and goes to a point z, subtract 0.1000 from 0.5000 to find the value that has to be located in the Standard Normal Table.
2. Identify the row and column that comes closest (above or below) to this area value.

Step 3: Execute the plan.
1. $0.5000 - 0.1000 = 0.4000$
2. From the Standard Normal Distribution Table, choose consecutive rows and columns that come closest to containing 0.4000.

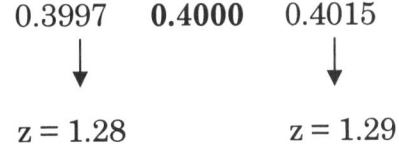

The value 0.3997 is closer to 0.4000 than 0.4015. The z score of 1.28 is used. Observe P_{10} is to the left of the mean. For this reason the z score is negative. Thus, P_{10} has $z = -1.28$

Step 4: Check your work.
1. Check the numbers and your calculations.
2. Was the question answered?

3. Is your answer reasonable?

Example 2: Find the value of z associated with the 95th Percentile, P_{95}.
Solution:
Step 1: Understand and Picture the Problem.
1. Problem: Find the z-score corresponding to P_{95}.
2. Since an area under the curve is given, a "z-score" has to be found.
3. Draw a bell-shaped curve above a horizontal axis and label its center with a zero.
4. P_{95} is the z-score where 0.9500 of the area under the curve is to the left of it.
5. Locate and label the 95% (.9500) on the curve.

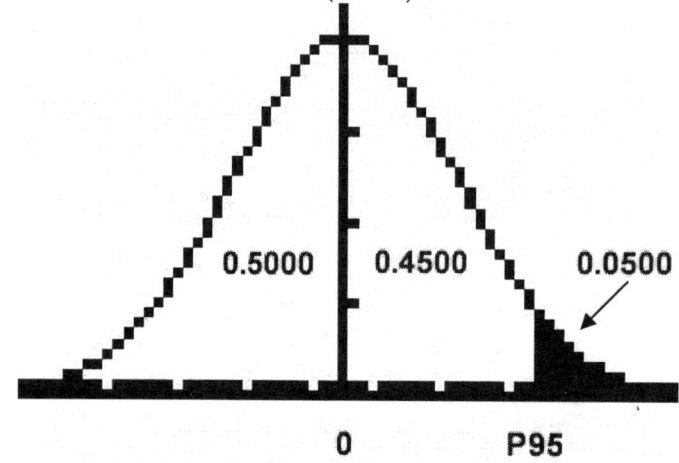

Step 2: Develop a plan.
1. Since the Standard Normal Distribution Table reads from the mean and goes to a point z, subtract 0.5000 from 0.9500 to find the value that has to be located in the Standard Normal Table.
2. Identify the row and column that comes closest (above or below) to containing this value.

Step 3: Execute the plan.
1. $0.9500 - 0.5000 = 0.4500$
2. From the Standard Normal Distribution Table, identify the consecutive rows and columns that come closest to containing 0.4500.

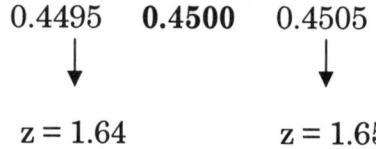

$$z = 1.64 \qquad z = 1.65$$
Notice that, 0.4500 is the midpoint of 0.4495 and 0.4505. Neither

point z = 1.64 or z = 1.65 is closer. When the value being located is the midpoint of the interval, the midpoint of the z-scores is taken as the answer. Add the two z-scores and divide by two, $\frac{1.64 + 1.65}{2} = 1.645$. Thus, P_{95} has z = 1.645.

Step 4: Check your work.
1. Check the numbers and your calculations.
2. Was the question answered?
3. Is your answer reasonable?

Example 3: Find the z-scores associated with the middle 95 percent
Solution:
Step 1: Understand and Picture the Problem.
1. Problem: Find the z-scores corresponding to an area under the curve where 95% of the area is centered about the mean.
2. Since an area under the curve is given in this problem, a "z-score" has to be found.
3. Draw a bell-shaped curve above a horizontal axis and label its center with a zero.
4. 0.9500 of the area under the curve is in the middle of the Normal curve. Since the Normal curve is symmetric, then 0.4750 is to the left and right of the center. There is also a total of 0.0500 in the tails. This means 0.0250 is in each tail
5. Locate and label the middle 95% on the curve.

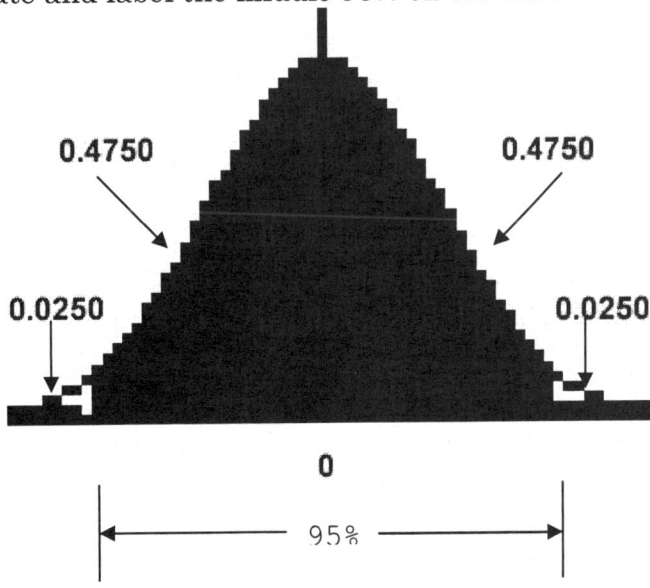

Step 2: Develop a plan.
1. Since the Standard Normal Distribution Table reads from the

mean and goes to a point z, the value 0.4750 that has to be located in the Standard Normal Table.
2. Identify the consecutive rows and columns that come closest to containing 0.4750.

Step 3: Execute the plan.
1. From the Standard Normal Distribution Table, identify the row and column that comes closest (above or below) to containing 0.4750.
2. In rare instances, the value is found exactly in the table. For 0.4750 the z-score is 1.96.
3. There are two z-scores, one to the left of the mean and the other on the right of the mean. The z-score on the left is − 1.96. The z-score on the right is 1.96.
4. The z-scores associated with the middle 95% is $z = \pm 1.96$.

Step 4: Check your work.
1. Check the numbers and your calculations.
2. Was the question answered?
3. Is your answer reasonable?

Cognitive Problems

1. In your own words explain what is meant by the 95th percentile. Give one example using the 95th percentile.
2. Is there a z-score associated with the 100th percentile. Explain your answer.

Exercise 5.6

(c) 2006 JupiterImages Corp.

1. Find the z-score corresponding to P_{30}.
2. Find the z-score associated with the 80th percentile.
3. Find the z-score corresponding to P_{75}.
4. Find the z-score associated with the 25th percentile.
5. Find the z-scores that mark the lower and upper 15% of the data.
6. Find the z-scores that mark the lower and upper 20% of the data.
7. Find the z-scores corresponding to the middle 70% of the data.
8. Find the z-scores corresponding to the middle 80% of the data.
9. Find the z-scores corresponding to the middle 90% of the data.
10. Find the z-scores corresponding to the middle 98% of the data.

5.7 CONVERTING TO STANDARD SCORES
Objective: Convert real data into standard scores.

The Standard Normal Curve is the model for all bell-shaped probability distributions.

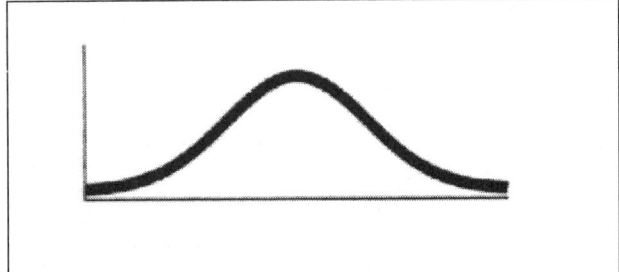

The Standard Normal Curve studied in section 5.5 had a mean of zero and a standard deviation of one. Z-scores were used to find the area under the Normal curve.

However, there are many distributions that are bell-shaped and do not have a mean of zero and a standard deviation of one.

      A. The weights of newborn babies
      B. The scores on IQ tests
      C. The heights of females
      D. The weights of males
      E. The shoe sizes of 12 year old children
      F. The weights of 14.5 ounce bags of potato chips
      G. The liquid contents of 12 ounce cans of cola
      H. The life expectancies of AA alkaline batteries

These problems follow the same five scenarios presented in Section 5.5. In these problems the area under the curve represents the probability of an event taking place.

Five possible scenarios for Normal Probability Distribution problems.

Scenario 1: The probability region (area) extends from the mean to a standard score. See following problem.

Scenario 2: The probability region (area) extends from one side of a standard score involving only the tail of the Normal curve. See Example 1.

Scenario 3: The area under the curve includes one half of the Normal curve and extends to a standard score on the other side of the mean. See Example 2.

Scenario 4: The probability region (area) involves two standard scores on the same side of the mean. See Example 3.

Scenario 5: The probability region (area) involves two standard scores on opposite sides of the mean. See Example 4.

416

Scenario 1: Suppose the time required to assemble a computer is normally distributed with a mean of 78 minutes and a standard deviation of 7 minutes. What is the probability it takes at from 78 to 85 minutes to assemble the computer?

In order to solve this problem, draw a bell-shaped curve with two horizontal axes. The first scale represents the "data" and the second scale represents the "z-scores". An adjustment (standardization) needs to be made to use the Standard Normal Curve. This standardization or conversion assigns z-scores, standard scores, to the data. Once this standardization has been performed, the Standard Normal Distribution Table can be used. Data is converted to standard scores by the following conversion formula:

STANDARD SCORE
$$z = \frac{x - \mu}{\sigma}$$
Where: x is the data value.
     μ is the mean of the data set.
     σ is the standard deviation of the data set.
     z is the standard score, z-score.

Working with two scales is not unusual. Consider a thermometer, Fahrenheit is often on one side of the mercury and Celsius is on the other side. AM/FM radios are another example where we use two scales. Even the speedometer of your car is calibrated in miles per hour (mph) and kilometers per hour (kph). Once the bell-shaped curve is drawn, label its center with 78 on the "data" scale (x) and 0 on the "z-score" scale. Then locate 85 on the "data" scale and shade in the region between 78 to 85 to account for the probability.

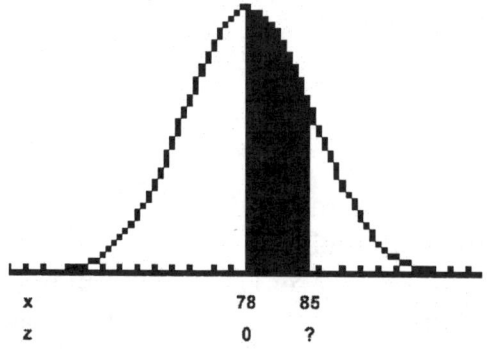

The drawing shows that the time 78 minutes corresponds to standard score of 0 having a probability of 0.0000 Next convert the data point, 85 to a standard score, z. Use the conversion formula:

$$z = \frac{x - \mu}{\sigma} = \frac{85 - 78}{7} = \frac{7}{7} = 1.00$$

Look up z = 1.00 in the Standard Normal Distribution Table. The area is 0.3413. There is a 34.13% chance the computer's assembly time takes from 78 to 85 minutes.

Example 1 (Scenario 2): The time required to assemble a computer is normally distributed with a mean of 78 minutes and a standard deviation of 7 minutes. What is the probability it takes at least 85 minutes to assemble the computer?

Solution:

Step 1: Understand and Picture the Problem.
1. Question: What is the probability the assembly time is at least 85 minutes? P(assembly time greater than or equal to 85 minutes)?
2. Since the assembly time is normally distributed, the Standard Normal Curve can be used as a model.
3. Standard deviation is 7 and the mean is 78.
4. To represent the data, draw a bell-shaped curve with "data" and "z-score" scales.
5. Label its center with 78 on the "data" scale and 0 on the "z-score" scale.
6. Locate and label 85 on the curve, according to the "data" scale.
7. To represent scores of "at least 85", shade in the region under the curve to the right of 85.

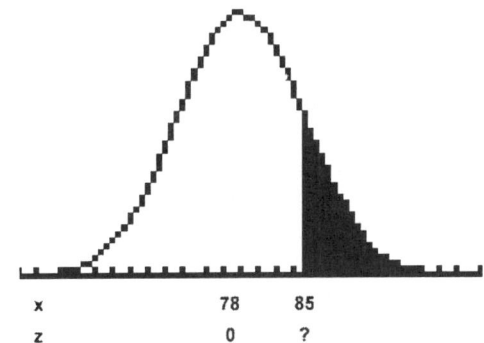

Step 2: Develop a Plan.
1. Convert 85 to a standard score, z-score.
2. Find this standard score (z) in the Standard Normal Distribution Table.
3. Since this value represents the area from 78 to 85, subtract this value from 0.5000.

Step 3: Execute the Plan.
1. $z = \dfrac{x - \mu}{\sigma} = \dfrac{85 - 78}{7} = \dfrac{7}{7} = 1.00$.
2. A standard score of 1.00 corresponds to an area of 0.3413.
3. 0.5000 − 0.3413 = 0.1587.
   Thus, P(Assembly time is at least 85 minutes) = 0.1587.

Step 4: Check your work.
1. Check the numbers and your calculations.
2. Was the question answered?
3. Is your answer reasonable?

Example 2 (Scenario 3): the time required to assemble a computer is normally distributed with a mean of 78 minutes and a standard deviation of 7 minutes. What is the probability it takes at most 85 minutes to assemble the computer?
Solution:
Step 1: Understand and Picture the Problem.
1. Question: What is the probability the assembly time is at most 85 minutes? P(Assembly time is less than or equal to 85 minutes)?
2. Since the assembly time is normally distributed, the Standard Normal Curve can be used as a model.
3. Standard deviation is 7 and the mean is 78.
4. To represent the data, draw a bell-shaped curve with "data" and "z-score" scales.
5. Label its center with 78 on the "data" scale and 0 on the "z-score" scale.
6. Locate and label 85 on the curve, according to the "data" scale.
7. To represent scores of "at most 85", shade in the region under the curve to the left of 85.

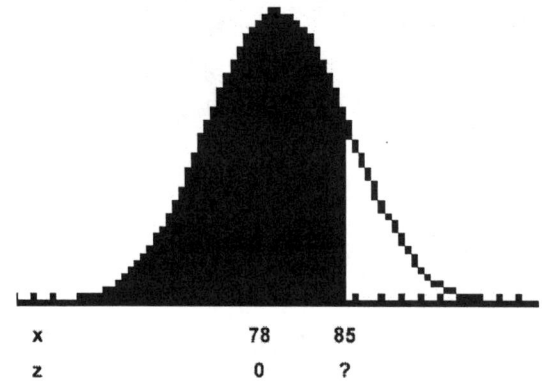

Step 2: Develop a Plan.
1. Convert 85 to a standard score, z-score.
2. Find this standard score (z) in the Standard Normal Distribution Table.
3. Since this value represents the area from 78 to 85, add this value to 0.5000.

Step 3: Execute the Plan.
1. $z = \dfrac{x - \mu}{\sigma} = \dfrac{85 - 78}{7} = \dfrac{7}{7} = 1.00.$

2. A standard score of 1.00 corresponds to an area of 0.3413.
3. 0.5000 + 0.3413 = 0.8413.
   Thus, P(Assembly time is at most 85 minutes) = 0.8413.

Step 4: Check your work.
1. Check the numbers and your calculations.
2. Was the question answered?
3. Is your answer reasonable?

Example3 (Scenario 4): The time required to assemble a computer is normally distributed with a mean of 78 minutes and a standard deviation of 7 minutes. What is the probability it takes them 80 to 85 minutes to assemble a computer?
Solution:
Step 1: Understand and Picture the Problem.
1. Question: What is the probability that the assembly process will take between 80 to 85 minutes to complete? P(Assembly time is from 80 to 85 minutes)?
2. Since the assembly time is normally distributed, the Standard Normal Curve can be used as a model.
3. Standard deviation is 7 minutes and the mean is 78 minutes.
4. Draw a bell-shaped curve with "data" and "z-score" scales.
5. Label its center with 78 on the "data" scale and 0 on the "z-score" scale.
6. Locate and label the points 78 and 85 on the curve using the "data" scale.
7. To represent the times from 80 to 85, shade in the region under the curve between 80 and 85.

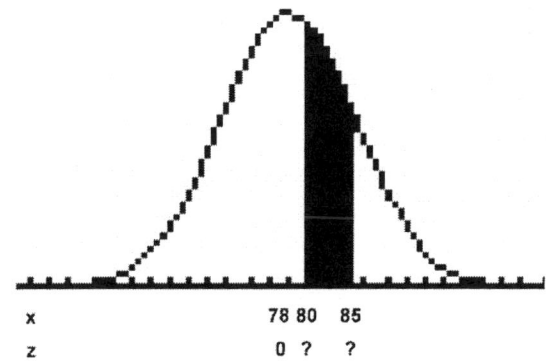

Step 2: Develop a Plan.
1. Convert 80 and 85 to standard scores, z-scores.
2. Find these standard scores in the Standard Normal Distribution Table.
3. The values represent areas on the same side of the mean, subtract them.

420

Step 3: Execute the Plan.
1. $z = \dfrac{x - \mu}{\sigma} = \dfrac{80 - 78}{7} = \dfrac{2}{7} \approx 0.29,$ $z = \dfrac{x - \mu}{\sigma} = \dfrac{85 - 78}{7} = \dfrac{7}{7} = 1.00.$
2. A standard score of 0.29 corresponds to an area of 0.1141. A standard score of 1.00 corresponds to an area of 0.3413.
3. $0.3413 - 0.1141 = 0.2272$.
   Thus, P(Assembly time is from 80 to 85 minutes) = 0.2272.

Step 4: Check your work.
   1. Check the numbers and your calculations.
   2. Was the question answered?
   3. Is your answer reasonable?

Example 4 (Scenario 5): The time required to assemble a computer is normally distributed with a mean of 78 minutes and a standard deviation of 7 minutes. What is the probability it takes from 70 to 85 minutes to assemble the computer?
Solution:
Step 1: Understand and Picture the Problem.
   1. Question: What is the probability that the assembly process will take between 70 to 85 minutes to complete? P(Assembly time is from 70 to 85 minutes)?
   2. Since the assembly time is normally distributed, the Standard Normal Curve can be used as a model.
   3. Standard deviation is 7 minutes and the mean is 78 minutes.
   4. Draw a bell-shaped curve with "data" and "z-score" scales.
   5. Label its center with 78 on the "data" scale and 0 on the "z-score" scale.
   6. Locate and label the points 70 and 85 on the curve using the "data" scale.
   7. To represent the times from 70 to 85, shade in the region under the curve between 70 and 85.

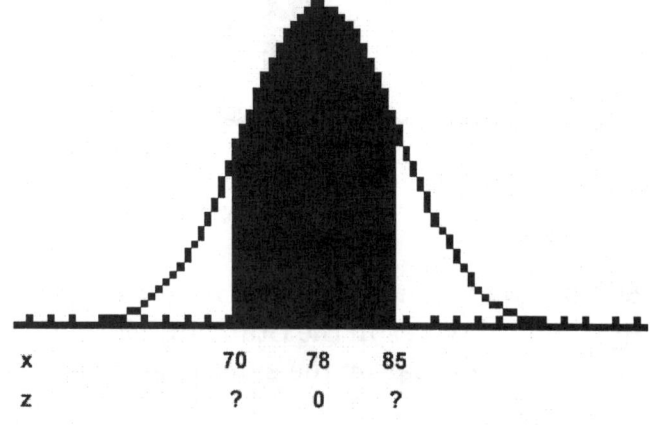

Step 2: Develop a Plan.
   1. Convert 70 and 85 to standard scores, z-scores.

2. Find these standard scores in the Standard Normal Distribution Table.
3. Since these values represent areas on both sides of the mean, add them.

Step 3: Execute the Plan.
1. $z = \dfrac{x - \mu}{\sigma} = \dfrac{70 - 78}{7} = \dfrac{-8}{7} \approx -1.14,$ $z = \dfrac{x - \mu}{\sigma} = \dfrac{85 - 78}{7} = \dfrac{7}{7} = 1.00.$
2. A standard score of –1.14 corresponds to an area of 0.3729. A standard score of 1.00 corresponds to an area of 0.3413.
3. 0.3729 + 0.3413 = 0.7142.
   P(Assembly time is from 70 to 85 minutes) = 0.7142.

Step 4: Check your work.
1. Check the numbers and your calculations.
2. Was the question answered?
3. Is your answer reasonable?

(c) 2006 JupiterImages Corp.

Historical Perspective
Abraham de Moivre (1667 – 1754) France

The *normal curve* was developed mathematically in 1733 by Abraham DeMoivre as an approximation to the binomial distribution. Gauss used the normal curve to analyze astronomical data in 1809. The normal curve is often called the Gaussian distribution. The term bell-shaped curve is also used in everyday usage to describe the normal curve.

Legend has it that DeMoivre predicted the day of his death. He found that he was sleeping 15 minutes longer each night and summing the minutes he calculated that he would die on the day that he slept more than 24 hours. **He was right!**

❓ Cognitive Problems ❓

1. Describe the attributes of the bell-shaped curve.
2. A data set is normally distributed with a mean of 240 and standard deviation of 36. Restate the "Empirical Rule" using this data set.
3. Why does the mean of a normally distributed data set labeled as the center mark on the "data" scale correspond to zero on the "z-score" scale?
4. Give an example of a data set that can be modeled by a bell shaped curve.

Exercise 5.7

Use the following information for problem 1 through 5. The contents of cartons of orange juice are normally distributed with a mean of 64 ounces and a standard deviation of 2.1 ounces.

1. If a carton is randomly selected, what is the probability that its contents will be at least 66 ounces?
2. If a carton is randomly selected, what is the probability that its contents will be at most 66 ounces?
3. If a carton is randomly selected, what is the probability that its contents will be between 64 and 66 ounces?
4. If a carton is randomly selected, what is the probability that its contents will be between 62 and 64 ounces?
5. If a carton is randomly selected, what is the probability that its contents will be between 66 and 68 ounces?

Use the following information for problems 6 through 10.
A automobile tire manufacturer produces an all-weather radial passenger tire whose distances for tire life are normally distributed with a mean of 59,200 miles and a standard deviation of 1,400 miles.

6. What is the probability that the tire will last between 59,200 and 60,000 miles?
7. What is the probability that the tire will last less than 60,000 miles?
8. What is the probability that the tire will last more than 60,000 miles?
9. What is the probability that the tire will last between 58,000 and 60,000 miles?
10. What is the probability that the tire will last between 60,000 and 62,000 miles?

Use the following information for problems 11 through 15.
The weights of a pure chocolate candy bar is normally distributed with a mean of 1.65 ounces and a standard deviation of 0.2 ounce. If a candy bar is randomly selected, what is the probability that its weight will be:

11. Greater than 1.8 ounces?
12. Less than 1.8 ounces?
13. Between 1.65 and 1.8 ounces?
14. Between 1.5 and 1.8 ounces?
15. Between 1.8 and 2.0 ounces?

5.8 DATA (X) BASED ON A PROBABILITY
Objective: Find the value (x) corresponding to a given probability.

Another type of problem involving the Standard Normal Curve is the inverse operation of the problems in the previous section. Given a probability, find the corresponding data value. For example, a test has a mean of 500 and a standard deviation of 60, find the data value, x that corresponds to the 70th percentile, P_{70}. In this situation the following conversion formula is used.

FINDING SCORES FROM PROBABILITIES

$$x = \mu + z(\sigma)$$

Where: x is the data value.
  μ is the mean of the data set.
  σ is the standard deviation of the data set.
  z is the standard score, z-score.

Since the mean (μ) is 500 and the standard deviation (σ) is 60, only the z score for the 70th percentile needs to be found. (Remember, 100% is the complete area under the curve.) The 70th percentile on the Standard Normal Curve describes the picture from left to right. 70% of the data is to the left of the z-score. This also implies that the remaining 30% of the data is to the right of that same z-score. The following figure displays this information.

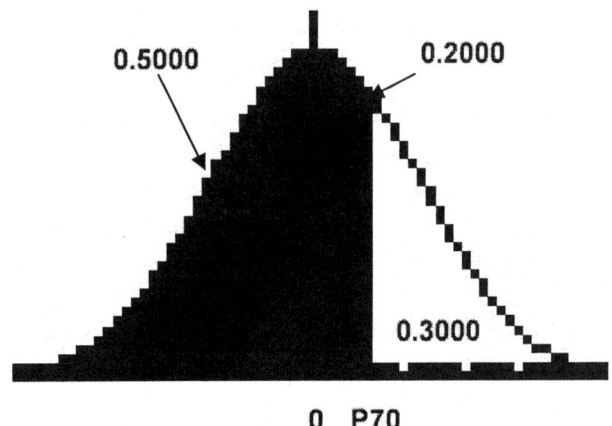

The mean (z = 0) marks the middle, the 50th Percentile, P_{50}. Half of the data is below this mark. Since 70% is greater than the mean, 20% more is needed to equal 70%, P_{70}. From the Standard Normal Distribution Table, identify the consecutive rows and columns that come closest to containing 0.2000.

424

$$0.1985 \quad \mathbf{0.2000} \quad 0.2019$$
$$\downarrow \qquad\qquad\qquad \downarrow$$
$$z = 0.52 \qquad\qquad z = 0.53$$

Since 0.1985 is closer to 0.2000 than 0.2019, then use $z = 0.52$ for P_{70}.

Substituting these values into the formula for x yields
$$x = \mu + z(\sigma)$$
$$x = 500 + (0.52)(60) = 531.2$$
Thus the value 531.2 marks the 70th percentile score for this exam.

Example 1: The amount of alkaline phosphatase, an indicator for liver disease, in the blood has a mean of 72.5 and a standard deviation of 35. Find the data value (x) associated with the 10th Percentile, P_{10}.

Solution:
 Step 1: Understand and Picture the Problem.
  1. Problem: Find the data value (x) corresponding to P_{10}.
  2. Since the mean, standard deviation and probability are given in this problem, a "z-score" has to be found.
  3. Draw a bell-shaped curve above a horizontal axis and label its center with a zero.
  4. P_{10} is the z-score where 0.1000 of the area under the curve is in the left tail.
  5. Locate and label the lower 10% (.1000) on the curve.

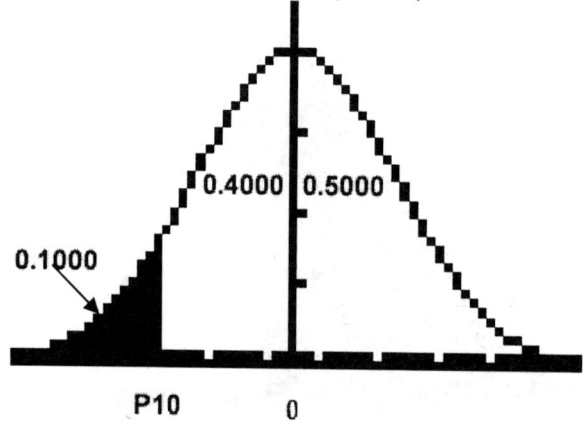

 Step 2: Develop a plan.
  1. Since the Standard Normal Distribution Table reads from the mean and goes to a point z, subtract 0.1000 from 0.5000 to find the value that has to be located in the Standard Normal Table.
  2. Identify the consecutive rows and columns that come closest to this value.
  3. Substitute the corresponding values into the formula: $x = \mu + z(\sigma)$.

Step 3: Execute the plan.
1. 0.5000 − 0.1000 = 0.4000
2. From the Standard Normal Distribution Table, identify the consecutive rows and columns that come closest to containing 0.4000.

   0.3997   **0.4000**   0.4015
   ↓   ↓
   z = 1.28   z = 1.29

   The value 0.3997 is closer to 0.4000 than 0.4015. The z score of 1.28 is used. Observe P_{10} is to the left of the mean. For this reason the z score is negative. Thus, $P_{10} = -1.28$
3. Substituting these corresponding values into the formula for x yields
$$x = \mu + z(\sigma)$$
$$x = 72.5 + (-1.28)(350) = 27.7$$
Thus the value 27.7 marks the 10th percentile for this indicator.

Step 4: Check your work.
1. Check the numbers and your calculations.
2. Was the question answered?
3. Is your answer reasonable?

Example 2: An operation has a mean of 3 hours and a stand. dev. of 30 minutes $\left(\frac{1}{2} hr\right)$. Find the value (x) associated with the 95th percentile, P_{95}.

Solution:
Step 1: Understand and Picture the Problem.
1. Problem: Find the z-score corresponding to P_{95}.
2. Since the mean, standard deviation and probability are given in this problem, a "z-score" has to be found.
3. Draw a bell-shaped curve above a horizontal axis and label its center with a zero.
4. P_{95} is the z-score where 0.9500 of the area under the curve is below or to the left of it.
5. Locate and label the 95% (.9500) on the curve.

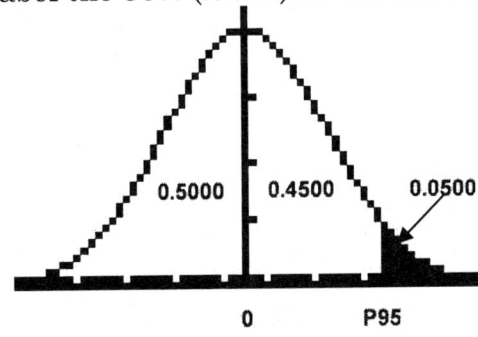

426

Step 2: Develop a plan.
1. Since the Standard Normal Distribution Table reads from the mean and goes to a point z, subtract 0.5000 from 0.9500 to find the value that has to be located in the Standard Normal Table.
2. Identify the consecutive rows and columns that come closest to containing this value.
3. Substitute these values into the formula: $x = \mu + z(\sigma)$.

Step 3: Execute the plan.
1. $0.9500 - 0.5000 = 0.4500$.
2. From the Standard Normal Distribution Table, identify the consecutive rows and columns that come closest to containing 0.4500.

$$0.4495 \quad \mathbf{0.4500} \quad 0.4505$$
$$z = 1.64 \qquad\qquad z = 1.65$$

Notice that, 0.4500 is the midpoint of 0.4495 and 0.4505. Neither point z = 1.64 or z = 1.65 is closer. When the value being located is the midpoint of the interval, the midpoint of the z-scores is taken as the answer. Add the two z-scores and divide by two
$$\frac{1.64 + 1.65}{2} = 1.645. \text{ Thus, } P_{95} = 1.645.$$

3. Substituting the corresponding values into the formula to calculate x yields
$$x = \mu + z(\sigma)$$
$$x = 3 + (1.645)(0.5) = 3.8225$$
Thus the value 3.8225 hours about 3 hours 49 minutes marks the 95th percentile for this operation.

Step 4: Check your work.
1. Check the numbers and your calculations.
2. Was the question answered?
3. Is your answer reasonable?

Example 3: The amount of glucose, an indicator for diabetes, in the blood has a mean of 174 and a standard deviation of 15. Find the z-scores associated with the middle 95 percent.
Solution:
Step 1: Understand and Picture the Problem.
1. Problem: Find the z-score corresponding to the middle 95%
2. Since the mean, standard deviation and probability are given in this problem, a "z-score" has to be found.

3. Draw a bell-shaped curve above a horizontal axis and label its center with a zero.
4. 0.9500 of the area under the curve is in the middle of the Normal curve. Since the Normal curve is symmetric, then 0.4750 is to the left and right of the center. There is also 0.0500 in the tails. This means 0.0250 is in each tail
5. Locate and label the middle 95% on the curve.

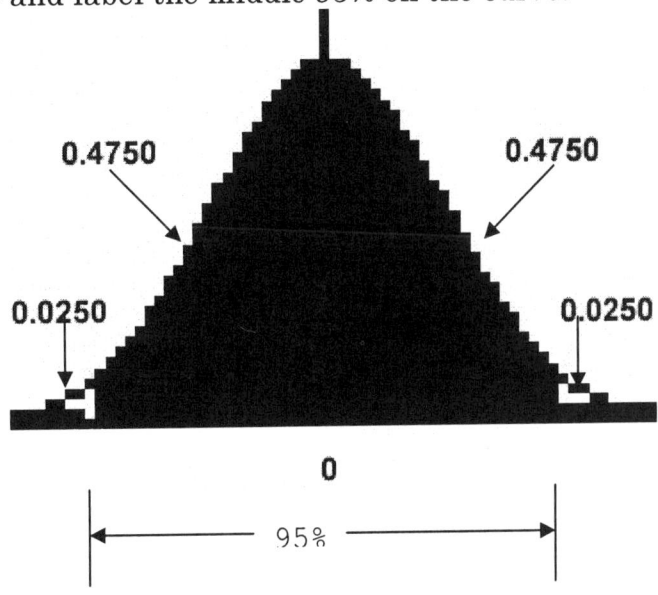

Step 2: Develop a plan.
1. Since the Standard Normal Distribution Table reads from the mean and goes to a point z, the value 0.4750 that has to be located in the Standard Normal Table.
2. Identify the consecutive rows and columns that come closest to containing this value.
3. Substitute then corresponding values into the formula: $x = \mu + z(\sigma)$.

Step 3: Execute the plan.
1. From the Standard Normal Distribution Table, identify the consecutive rows and columns that come closest to containing 0.4750.
2. In rare instances, the value is found exactly in the table. For 0.4750 the z-score is 1.96.
3. There are two z-scores, one to the left of the mean and the other on the right of the mean. The z-score on the left is – 1.96. The z-score on the right is 1.96.
4. Substituting the corresponding values into the formula for x yields
$$x = \mu + z(\sigma)$$
$$x = 174 + (-1.96)(15) = 144.6$$

$$x = 174 + (1.96)(15) = 203.4$$

Thus the middle 95% falls between the values 144.6 and 203.4 for this indicator.

Step 4: Check your work.
1. Check the numbers and your calculations.
2. Was the question answered?
3. Is your answer reasonable?

Optional Student Project

"Normal Probability Distribution" can be found on page 432.

Chapter Review

Chapter review problems can be found on page 433.

Cognitive Problems

1. A data set is normally distributed with a mean of 240 and standard deviation of 36. Restate the "Empirical Rule" using this data set.
2. Why does the mean of a normally distributed data set labeled as the center mark in the "data" scale correspond to zero on the "z-score" scale?

Exercise 5.8

Use the following information for problems 1 and 2. The weight of a pure chocolate candy bar is normally distributed with a mean of 1.65 ounces and a standard deviation of 0.2 ounce. If a candy bar is randomly selected,
1. Find the value corresponding to P_{15}. Any candy bar falling in the bottom 15% will be melted down and reprocessed.
2. Find the value corresponding to P_{90}. Any candy bar falling in the upper 10% will be melted down and reprocessed.

3. The scores on a certification exam are normally distributed with a mean of 900 and a standard deviation of 66. Find the value associated with the 40th percentile. The certification board has determined that this will be the lowest passing score on the exam. What is the minimum passing score on this certification exam?

4. The scores on the police department's lieutenant exam are normally distributed with a mean of 600 and a standard deviation of 102. Find the value corresponding to the 80th percentile. The police advisory board advances candidates who score above this mark to the next level of testing in their pursuit of earning a lieutenant's commission in the police department. What is the minimal score needed to qualify for further testing?

Use the following information for problems 5 and 6. The contents of cartons of orange juice are normally distributed with a mean of 64 ounces and a standard deviation of 2.1 ounces.

5. Find the value that corresponds to the lower 15% of this distribution. Cartons that fall in this region will be emptied and properly refilled.

6. Find the value that corresponds to the upper 15% of this distribution. Cartons that fall in this region will be emptied and properly refilled.

7. A automobile tire manufacturer produces an all-weather radial passenger tire whose distances for tire life are normally distributed with a mean of 59,200 miles and a standard deviation of 1,400 miles. Find the value associated with the lower 14% of this distribution. Any tire that falls within this region will be replaced by the company.

Use the following information for problems 8 and 9. The amount of white blood cells, serves as an indicator of disease in the blood, has a mean of 7.3 and a standard deviation of 2.3.

8. Find the data value (x) associated with the 12th Percentile.

9. Find the data value (x) associated with the 94 Percentile.

10. An operation has a mean of 7 hours and a standard deviation of 45 minutes $\left(\dfrac{3}{4}\text{hr}\right)$. Find the data value (x) associated with the 91st percentile.

PROJECT: GRAPHS FOR QUALITATIVE DATA

Objective: Construct graphs for categorical data.

US GOVERNMENT'S EXPENDITURES 2001
Income for 2001 was $2,000 Billion
Expenditures were $1,900 Billion

PROGRAM	$ BILLIONS
Social Security	684
National Defense	342
Interest on the National Debt	190
Communities	190
Social Programs	342
Law Enforcement	38
Pay Down the National Debt	114

US GOVERNMENT'S EXPENDITURES 2002
Income for 2002 was $1,900 Billion
Expenditures were $2,100 Billion

PROGRAM	$ BILLIONS
Social Security	798
National Defense	420
Interest on the National Debt	168
Communities	210
Social Programs	441
Law Enforcement	63

I. Make a separate table organizing the US Expenditures for 2001 and 2002.
II. Use the tables to construct the following graphs.
   A. Horizontal frequency bar graph for 2001.
   B. Vertical frequency bar graph for 2002.
   C. Relative frequency pie chart for 2001.
   D. Relative frequency pie chart for 2002.
III. Compare expenditures for 2001 and 2002. What differences do you find? Write at least one paragraph describing what you feel are the factors responsible for these differences.

PROJECT: DESCRIPTIVE STATISTICS FOR QUANTITATIVE DATA

Objective: Use Descriptive Statistics to organize raw data into a table, make graphs, find measures of central tendency, dispersion, and outliers.

A sample of truck prices produced the following results.

$16,900	$19,020	$27,025	$17,601	$13,149	$33,617	$16,850	$42,185
$22,689	$12,260	$31,500	$19,035	$35,660	$14,199	$31,165	$19,165
$35,660	$15,955	$23,405	$13,365	$17,910	$37,515	$16,399	$19,065
$35,370	$40,240	$16,983	$19,525	$14,115	$37,625	$22,175	$29,935
$14,996	$19,320	$22,595	$33,205	$28,855	$16,270	$27,525	$30,415

Perform the following operations:
I. Sort the data in ascending order.

II. Make a table with seven classes.

III. Use the table in Part II to construct a:
   A. Relative frequency polygon.　　B. Frequency histogram.

IV. Use your calculator to calculate the following descriptive statistics.
   A. Mean　　B. Standard Deviation

V. Find the:
   A. Median　　B. First Quartile　　C. Third Quartile

VI. Use the values from Part V to
   A. Construct an IQR boxplot.
   B. Determine which data values, if any, are outliers.

PROJECT: NORMAL PROBABILITY DISTRIBUTION

Objective: Use the Normal Probability Distribution to make decisions about a population.

An apparel company makes blue jeans and leather pants.

I.

Female Parameters

The average (μ) height of a female adult is 65.5 inches with a standard deviation (σ) of 3 inches.

A. What percent of female adults are taller than 6 feet (72 inches)?
B. What percent of female adults are taller than 5 feet (60 inches)?
C. What percent of female adult heights are between 60 inches and 72 inches?
D. Because of the high cost of leather, the company has decided they cannot profitably make leather pants in all sizes. Find the heights corresponding to the following percentages. These are the heights of the shortest and tallest females who can purchase leather pants from this company.
　1. The bottom 8%　　　　　2. The upper 6%

II.

Male Parameters

The average (μ) height of a male adult is 68.5 inches with a standard deviation (σ) of 3.25 inches.

A. What percent of male adults are shorter than 6 feet (72 inches)?
B. What percent of male adults are shorter than 5 feet (60 inches)?
C. What percent of male adult heights are between 60 inches and 72 inches?
D. Because of the high cost of leather, the company has decided they cannot profitably make leather pants in all sizes. Find the heights corresponding to the following percentages. These are the heights of the shortest and tallest males who can purchase leather pants from this company.
　1. The bottom 9%　　　　　2. The upper 7%

CHAPTER 5 REVIEW

Section 5.1

1. Does the graph "How Closely Do You Read an Employment Contract" satisfy the properties of a graph?

   USA TODAY

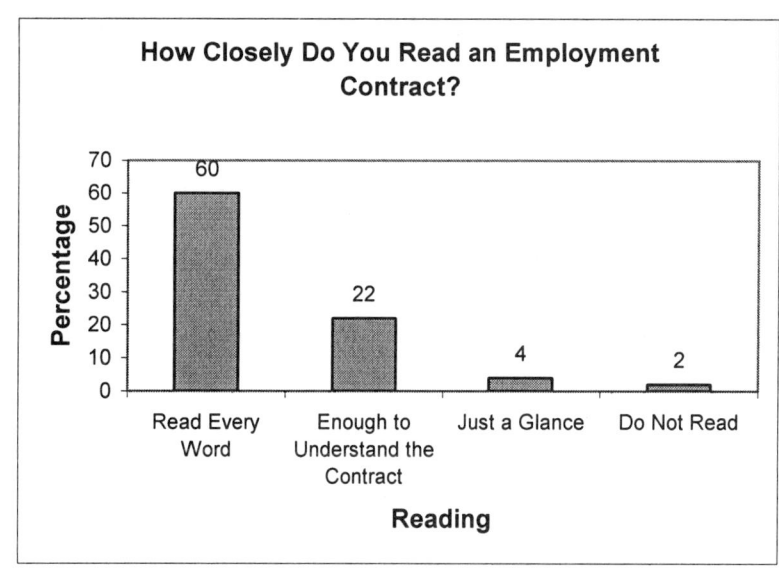

2. Does the graph "What Stresses Us Out?" satisfy the properties of a graph?

   USA TODAY

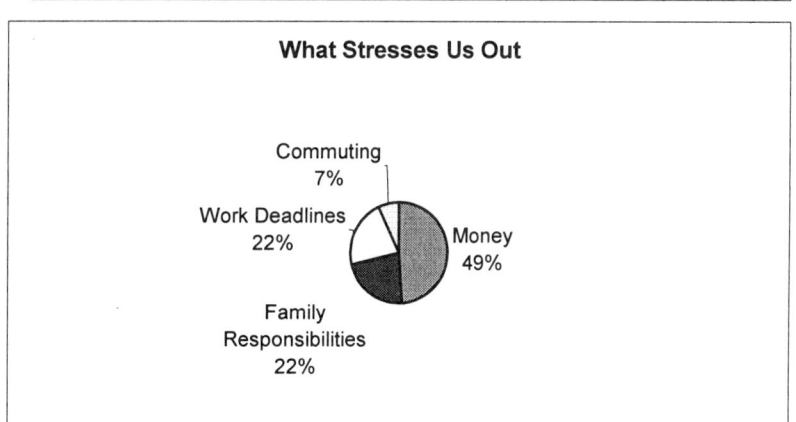

3. Make a relative frequency pie chart for the 2,000 vehicle owners surveyed.

Personal Vehicle of Choice	Frequency
Car / Station Wagon	1,140
Pickup Truck	360
Sport Utility Vehicle (SUV)	240
Van	180
Other	80

4. Using the data in problem 3, make a frequency pie chart.

5. Using the data in problem 3, make a vertical frequency bar graph.

6. Using the data in problem 3, make a horizontal relative frequency bar graph.

Section 5.2
7. A survey of watercraft (jet-ski) produced the following results
   $9,999 $8,999 $5,699 $6,999 $8,499 $8,999 $9,599 $7,999
   $8,599 $6,999 $6,499 $9,199 $7,999 $5,999 $10,199 $6,399
   $10,399 $11,999 $7,399 $7,799 $6,499 $9,499 $9,299 $7,899
   Sort the data in ascending order and make a table with seven classes.

8. Use the table in problem 7 to construct a relative frequency polygon.

9. Use the table in problem 7 to construct a frequency histogram

Section 5.3
10. Use the information from problem 7 to calculate the
    A. Mean (Use your calculator's built in function.)
    B. Median

Section 5.4
11. Use the information from problem 7 to calculate the
    A. Standard Deviation (Use your calculator's built in function.)
    B. First Quartile
    C. Third Quartile

12 Use the values from Problems 10 and 11 to construct an IQR boxplot

13. Are any of the data values in problem 7 outliers?

Section 5.5
14. Find $P(0 < z < 1.49)$
15. $P(-2.22 < z < 0)$
16. Find $P(z > 1.49)$
17. Find $P(z < 1.49)$
18. Find $P(1.49 < z < 2.22)$
19. Find $P(-2.22 < z < 1.49)$

Section 5.6
20. Find the z-score associated with the 79th percentile.
21. Find the z-score associated with the 24th percentile.
22. Find the z-scores that marks the lower 22% and the upper 11% of the data.
23. Find the z-scores corresponding to the middle 86%

Section 5.7
Use the following information for problems 24 through 27. A automobile manufacturer produces a battery whose useable life is normally distributed with a mean of 60 months and a standard deviation of 4 months.
24. What is the probability that the battery will last longer than 63 months?
25. What is the probability that the battery will last longer than 54 months?
26. What is the probability that the battery will last between 54 and 63 months?
27. What is the probability that the battery will last between 63 and 68 months?

Exercise 5.8
28. The scores on a certification exam are normally distributed with a mean of 1200 and a standard deviation of 200. Find the value associated with the 45th percentile. The certification board has determined that this will be the lowest passing score on the exam. What is the minimum passing score on this certification exam?

29. The scores on the police department's lieutenant exam are normally distributed with a mean of 600 and a standard deviation of 102. Find the value corresponding to the 85th percentile. The police advisory board advances candidates who score above this mark to the next level of testing in their pursuit of earning a lieutenant's commission in the police department. What is the minimal score needed to qualify for further testing?

Use the following information for problems 30 and 31.
The contents of a carton of milk is normally distributed with a mean of 128 ounces and a standard deviation of 2.4 ounces. The gallon carton is weighed and the amount of milk in the container is computed.
30. Find the value that corresponds to the lower 18% of this distribution. Cartons that fall in this region will be emptied and properly refilled.

31. Find the value that corresponds to the upper 12% of this distribution. Cartons that fall in this region will be emptied and properly refilled.

CHAPTER 6: PROBABILITY

6.1 PROBABILITY
Objective: Determine the probability of a simple event.

Probabilities are frequently used in newscasts, weather forecasting, advertising and gambling. Radio, television and newspapers regularly use probabilities to report the likelihood of events taking place.

- In a court trial, the DNA expert testifies that the probability the blood found at the crime scene came from the defendant, is 0.999995834. This means that it is very unlikely, 0.000004166, that someone else committed the crime.
- A weather forecast predicts a 90% chance for rain. This means that you should take an umbrella, there is only a 10% chance that it will not rain.
- The accuracy of a home pregnancy test is advertised as 99.5%. This means that it is very unlikely, 0.5%, that the test results are incorrect.

(c) 2006 JupiterImages Corp.

Probability values are written as fractions, decimals or percents. These numbers represent the likelihood of the named event happening and are always between 0 and 1, inclusive.

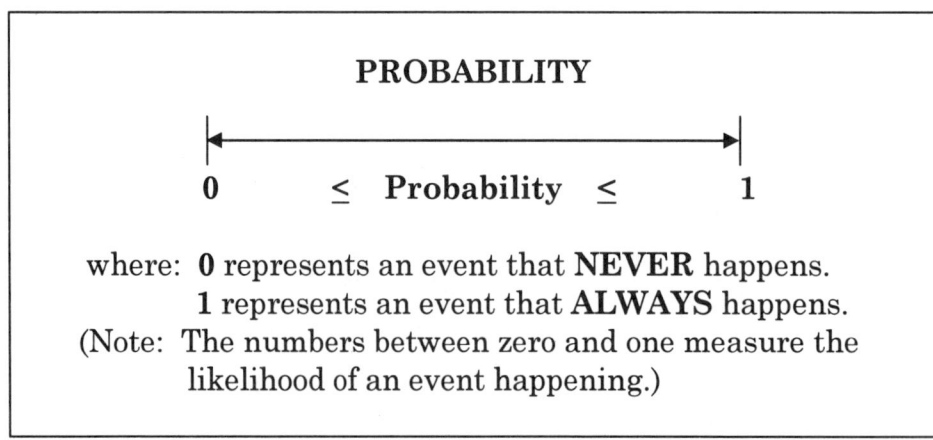

PROBABILITY

$0 \leq \text{Probability} \leq 1$

where: **0** represents an event that **NEVER** happens.
1 represents an event that **ALWAYS** happens.
(Note: The numbers between zero and one measure the likelihood of an event happening.)

(c) 2006
JupiterImages Corp.

Hold your pen above the desk. When released, what is the probability of the pen remaining suspended in the air? Due to the law of gravity, this will **NEVER** happen in our environment, thus the probability is **ZERO**. And what is the probability of the pen falling to the desk or floor? Again, due to the law of gravity, this will **ALWAYS** happen, thus the probability is **ONE**. These events may be written using probability notation, P(event). P(event) means the probability of the event described.

Thus, P(Released pen remains suspended in the air) = 0.
P(Released pen falls to the desk or floor) = 1.

Realistically, most situations do not have probabilities of zero or one. They have probabilities between zero and one, exclusive. A weather forecast predicts a 20% chance of rain for the evening. This means that there is a good chance that it won't rain. The 20% is closer to zero (never happening) than it is to one (always happening).

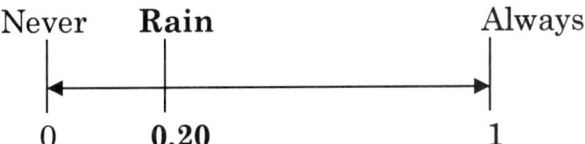

Similarly, a 90% success rate for an operation is not a guarantee that the operation will succeed. This means the operation has a good success rate. The 90% is closer to one (always happening) than it is to zero (never happening).

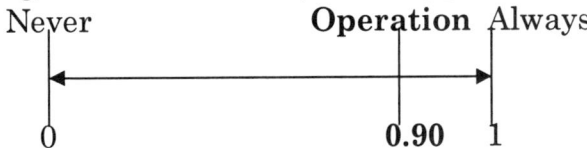

Consider going to your car. What is the probability of it starting? The probability of this event is dependent upon several factors. If your car is tuned and the weather is warm, then the probability should be near one (*). However, if your car is out-of-tune and the temperature is below freezing, then the probability should be near zero (#).

where: # represents an out-of-tuned car, with an old battery, in an Alaskan Winter.
* represents a tuned car, with a new battery on a sunny day in Florida.

Probability Vocabulary
1. **Experiment:** A planned activity performed to identify results.
2. **Trial:** Each repetition of the experiment.
3. **Event:** A single result, occurrence, or outcome from a single trial of the experiment.
4. **Sample Space:** A list of all possible events from an experiment.

Consider the tossing of a single die, one cube from a set of dice.
  1. Experiment: Toss a single die to identify the number of dots
     shown face up.
  2. Trial: A toss of a single die.
  3. Event: 3 dots are shown face up.
  4. Sample Space: 1, 2, 3, 4, 5, or 6.

(c) 2006 JupiterImages Corp

The Probability of an Event, A, Written as P(A) is:

$$P(A) = \frac{\text{Count of the occurence of Event A}}{\text{Total count of all possible events in the Sample Space}}.$$

Returning to the die toss, the P(3) = $\frac{1}{6} \approx 0.167$.

(Note: It is customary to round probabilities to the thousandths.)

The event "3" appears only once in the sample space of the die toss. The die has only one side with three dots. The sample space has 6 events. The die consists of six sides, each side with a different number of dots from 1 to 6. Consider an experiment where a pair of dice are tossed.

Example 1: Toss a pair of dice and find P(Sum of the dots is 5).
Solution:
  Step 1: Understand and picture the problem.
    1. Question: What is the probability of getting a sum of 5 dots on a pair of
       dice ? The experiment consists of responding to the question.
    2. Two dice are tossed at the same time (trial). The tossing of a pair of dice
       has several sums. These sums occur from different pairs of dots. The
       possible sums are: 2, 3, 4, 5, 6, 7, 8, 9, 10, 11, 12 (sample space).
    3. The requested event is the sum of 5 dots.

  Step 2: Establish a plan.
    1. Determine the sample space. List all possible sums of dots resulting
       from tossing a pair of dice.

2. Count the number of ways the sum of 5 dots occurs.
3. Count all possible ways the sum of dots occur.
4. Use the definition of probability to find

   P(Sum of dots is 5) = $\dfrac{\text{The number of ways the "sum of 5 dots" occurs}}{\text{The number of ways all possible sums occur}}$

Step 3: Execute the plan.
1. Sample space is:

DIE 1	DIE 2					
	1	2	3	4	5	6
1	2	3	4	**5**	6	7
2	3	4	**5**	6	7	8
3	4	**5**	6	7	8	9
4	**5**	6	7	8	9	10
5	6	7	8	9	10	11
6	7	8	9	10	11	12

   All possible sums 2, 3, 4, 5, 6, 7, 8, 9, 10, 11, 12, occur in 36 ways.
2. The event, sum is 5, happens 4 different ways:
   a. 1 on the first die and a 4 on the second die.
   b. 2 on the first die and a 3 on the second die.
   c. 3 on the first die and a 2 on the second die.
   d. 4 on the first die and a 1 on the second die.
3. The P(sum of dots is 5) = $\dfrac{4}{36} = \dfrac{1}{9} \approx 0.111$.

Step 4: Check your work.
1. Check the numbers and your calculations.
2. Was the question answered?
3. Is the answer reasonable? Remember, probability is a fraction or decimal between zero and one, inclusive. It may also be written as a percent.

Example 2: A small package (1.74 oz.) of candy contains 7 red, 4 yellow, 7 orange, 2 green and 3 brown candies. Find the P(Selecting a green candy).
Solution:
Step 1: Understand and picture the problem.
1. Question: What is the probability of selecting a green candy? The experiment consists of responding to the question.
2. The selection of a single candy (trial) has several different outcomes: Red, Yellow, Orange, Green, Brown (sample space).
3. The requested event is choosing a Green candy.

Step 2: Establish a plan.
1. Determine the total number of candies in the package, sample space.
2. Find the number of green candies, the requested event.
3. Use the definition of probability to find the P(Green).
P(Selecting a green candy) = $\frac{\text{Number of green candies}}{\text{Total number of candies}}$.

Step 3: Execute the plan.
1. Sample space is: 7 red, 4 yellow, 7 orange, 2 green and 3 brown candies. There is a total of 23 candies.
2. There are 2 green candies.
3. The P(Green) = $\frac{2}{23} \approx 0.087$.

Step 4: Check your work.
1. Check the numbers and your calculations.
2. Was the question answered?
3. Is the answer reasonable? Remember, probability is a fraction or decimal between zero and one, inclusive. It may also be written as a percent.

Example 3: A multiple-choice question has 4 responses. If the person guesses the answer, find P(An incorrect response).
Solution:
Step 1: Understand and picture the problem.
1. Question: What is the probability of guessing incorrectly? The experiment consists of responding to the question.
2. Random selecting (guessing) a response (event) results in 3 incorrect responses and 1 correct response (sample space).
3. The requested event is choosing an incorrect response.

Step 2: Establish a plan.
  1. Determine the number of possible responses, the sample space.
  2. Find the number of incorrect responses, the requested event.
  3. Use the definition of probability to find the P(An incorrect response).
     P(An incorrect response) = <u>Number of incorrect responses</u> .
                                  Total number of responses

Step 3: Execute the plan.
  1. There are 4 possible responses (3 incorrect and 1 correct) to the question.
  2. There are 3 ways to incorrectly respond to the question.
  3. The P(An incorrect response) = $\frac{3}{4}$ = 0.750.

Step 4: Check your work.
  1. Check the numbers and your calculations.
  2. Was the question answered?
  3. Is the answer reasonable? Remember, probability is a fraction or decimal between zero and one, inclusive. It may also be written as a percent.

Example 4: A committee is comprised of 8 men and 4 women. Find the P(Female) if one person is selected from the committee.
Solution:
Step 1: Understand and picture the problem.
  1. Question: What is the probability of selecting a female? The experiment consists of responding to the question.
  2. The random selection of a person (trial) is made from a group comprised of 8 males and 4 females (sample space).
  3. The requested event is selecting a female.

Step 2: Establish a plan.
  1. Determine the number of people on the committee.
  2. Find the number of females on the committee.
  3. Use the definition of probability to find:
     P(Female) = <u>Number of females on the committee</u> .
                    Total number of committee members

Step 3: Execute the plan.
  1. The 8 men and 4 women totals 12 committee members.
  2. There are 4 women on the committee.
  3. The P(Female) = $\frac{4}{12}$ = $\frac{1}{3}$ ≈ 0.333.

Step 4: Check your work.
1. Check the numbers and your calculations.
2. Was the question answered?
3. Is the answer reasonable? Remember, probability is a fraction or decimal between zero and one, inclusive. It may also be written as a percent.

Example 5: The following table describes the clients of an auto insurance agency.

Age	Frequency
Under 20	600
$20 \leq x < 30$	640
$30 \leq x < 40$	440
$40 \leq x < 50$	260
$50 \leq x < 60$	200
$60 \leq x < 70$	55
$70 \leq x < 80$	4
$80 \leq x < 90$	1

Find the P(A driver is under 20 years of age) if a client is chosen from this company.
Solution:
Step 1: Understand and picture the problem.
1. Question: What is the probability of selecting a person under 20? The experiment consists of responding to the question.
2. The random selection of a person is categorized by six age groups (trial): Under 20, $20 \leq x < 30$, $30 \leq x < 40$, $40 \leq x < 50$, $50 \leq x < 60$, $60 \leq x < 70$, $70 \leq x < 80$, $80 \leq x < 90$ (sample space).
3. The requested event is choosing a person under 20 year of age.

Step 2: Establish a plan.
1. Determine the number of people that are insured.
2. Find the number of people under 20 years of age.
3. Use the definition of probability to find the P(Driver under 20).
   P(Driver under 20) = Number of drivers under 20 years of age .
                                             Total number of drivers

Step 3: Execute the plan.
1. Adding the frequency column produces 2,200 drivers.
2. There are 600 drivers under the age of 20.
3. The P(Driver under 20) = $\dfrac{600}{2,200} \approx 0.273$.

Step 4: Check your work.
1. Check the numbers and your calculations.
2. Was the question answered?
3. Is the answer reasonable?

Relative Frequency and Probability

Example 5 demonstrates probability using a concept we already know, relative frequency. Relative frequency is probability. Add a relative frequency column to the table in Example 5.

Age	Frequency	Relative Frequency
Under 20	600	$\frac{600}{2,200} \approx 0.273$
$20 \leq x < 30$	640	$\frac{640}{2,200} \approx 0.291$
$30 \leq x < 40$	440	$\frac{440}{2,200} = 0.200$
$40 \leq x < 50$	260	$\frac{260}{2,200} \approx 0.118$
$50 \leq x < 60$	200	$\frac{200}{2,200} \approx 0.091$
$60 \leq x < 70$	55	$\frac{55}{2,200} = 0.025$
$70 \leq x < 80$	4	$\frac{4}{2,200} \approx 0.002$
$80 \leq x < 90$	1	$\frac{1}{2,200} \approx 0.000$ *
Totals	2,200	1.000

* **Note:** The 0.000 is not equivalent to 0. The 0 means never. 0.000 means that due to rounding to three decimal places, the result is 0.000. However, there is a value beyond the third decimal place.

The relative frequency column can be used to find probabilities such as:
A. P(A driver who is at least 40 years old) ≈ 0.236.
B. P(A driver who is at less than 30 years old) ≈ 0.564.
C. P(A driver is less than 80 years old) ≈ 1.000.
D. P(Driver is at least 80 years old but less than 90 years old) ≈ 0.000.
Note: The 0.000 and the 1.000 are not equivalent to 0 and 1. The 0 means never and 1 means always. 0.000 means that due to rounding to three decimal places, the result is 0.000. However, there is a value beyond the third decimal place. Similarly, 1.000 means that due to rounding, the result is 1.000.

All the problems in the previous chapter (qualitative and quantitative) that had a table with a relative frequency column can be reworded into probability problems. Probabilities are relative frequencies.

Example 6: Consider the 40 SUV prices given in section 5.2.
The SUV prices were sorted from smallest to largest.

$14,655 $14,799 $15,605 $16,395 $16,798 $17,990 $19,300 $20,000
$21,995 $22,195 $22,708 $23,240 $23,405 $23,920 $25,176 $25,999
$26,185 $26,268 $27,815 $27,910 $28,680 $28,950 $29,099 $29,249
$30,585 $30,645 $31,985 $32,250 $32,950 $33,595 $33,790 $34,590
$35,550 $36,300 $38,175 $41,188 $42,660 $54,950 $56,000 $63,500

Find the P(an SUV costs at least $20,000 but less than $30,000).

Solution:
Step 1: Understand and picture the problem.
1. Question: P(an SUV costs at least $20,000 but less than $30,000)? The experiment consists of responding to the question.
2. Probability is a fraction (relative frequency). The numerator is the frequency for the event and the sample space is the sum of the frequencies.
3. The sample space is the six classes. The trial is selecting on of these classes. The event is the specific class " an SUV costs at least $20,000 but less than $30,000".

Step 2: Establish a plan.
1. Use the table for the SUV prices.
2. Find the relative frequency for an SUV costing at least $20,000 but less than $30,000.

Step 3: Execute the plan.

Dollars ($1,000's)	Frequency	Relative Frequency.
$10 – $20	7	$\frac{7}{40} = 0.175$
$20 – $30	17	$\frac{17}{40} = 0.425$
$30 – $40	11	$\frac{11}{40} = 0.275$
$40 – $50	2	$\frac{2}{40} = 0.050$
$50 – $60	2	$\frac{2}{40} = 0.050$
$60 – $70	1	$\frac{2}{40} = 0.025$
	40	$\frac{40}{40} = 1.000$

P(an SUV costs at least $20,000 but less than $30,000) = 0.425 = 42.5%.

Step 4: Check your work.
1. Check the numbers and your calculations.
2. Was the question answered?
3. Is the answer reasonable? Remember, probability is a fraction or decimal between zero and one, inclusive. It may also be written as a percent.

Example 7: Consider the breakdown of a $1.00 donation made to the American Heart Association in the following table

RECIPIENT	CENTS per DOLLLAR
Administrative	$0.086
Fund Raising	$0.136
Community Services	$0.117
Research	$0.314
Education	$0.244
Training	$0.103
TOTALS	$1.00

Find P(Money will be given to heart research).
Solution:
Step 1: Understand and picture the problem.
1. Question: P(Money will be given to heart research)?)? The experiment consists of responding to the question.
2. Probability is a fraction (relative frequency). The numerator is the frequency for the event and the sample space is the sum of the frequencies.
3. The sample space is the six recipients of the American Heart Association money. The trial is selecting on of these recipients. The event is the specific recipient " heart research".

Step 2: Establish a plan.
1. Use the table for the American Heart Association and add a relative frequency column.
2. Find the relative frequency for heart research.

Step 3: Execute the plan.

RECIPIENT	CENTS per DOLLAR	RELATIVE FREQUENCY
Administrative	$0.086	$0.086/$1.00 = 0.086 ≈ 8.6%
Fund Raising	$0.136	$0.136/$1.00 = 0.136 ≈ 13.6%

445

Community Services	$0.117	$0.117/$1.00 = 0.117 ≈ 11.7%
Research	$0.314	$0.314/$1.00 = 0.314 ≈ 31.4%
Education	$0.244	$0.244/$1.00 = 0.244 ≈ 24.4%
Training	$0.103	$0.103/$1.00 = 0.103 ≈ 10.3%
TOTALS	$1.00	$1.00/$1.00 = 1.00 = 100%

P(Money given to heart research) = 31.4%.

Step 4: Check your work.
1. Check the numbers and your calculations.
2. Was the question answered?
3. Is the answer reasonable? Remember, probability is a fraction or decimal between zero and one, inclusive. It may also be written as a percent.

COMPLEMENTARY EVENTS
Some examples of events and their complements.

Event	Complement of the Event
1. Winning	1. Losing
2. On	2. Off
3. True	3. False
4. Correct	4. Incorrect
5. Female	5. Male

The complement of an event written as $\overline{E}$ is the negation of that event, E. The probability of the complement of an event is the probability that event **NOT** happening. From Example 3, the P(E), P(an incorrect response on a 4 item multiple-choice question) = 0.75. The complement of this event is the probability of **NOT** getting an incorrect response, which is the same as getting a correct response. Since only 1 of the 4 responses is correct,

$P(\overline{E}) = P(\textbf{NOT}$ getting an incorrect response$) = \dfrac{1}{4} = 0.25$.

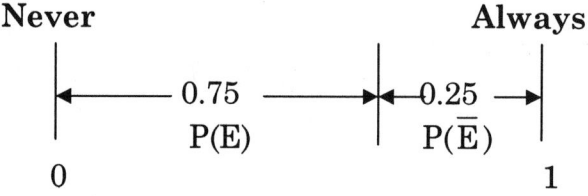

Either an event occurs or it doesn't. This demonstrates a fundamental property of an event and its complement. The sum of their probabilities must equal 1. Notice that P(Incorrect) = 0.75 and P(Not incorrect) = 0.25.

$$P(E) + P(\overline{E}) = 0.75 + 0.25 = 1.00.$$

Thus, the relationship between an event E and its complement $\overline{E}$ can be expressed as:

Complementary Events
$P(E) + P(\overline{E}) = 1$
$1 - P(E) = P(\overline{E})$
$1 - P(\overline{E}) = P(E)$

Example 8:
A small package (1.74 oz.) of candy contains 7 red, 4 yellow, 7 orange, 2 green and 3 brown candies. Find the P(NOT Green) on the selection of a single candy.
Solution:
  Step 1: Understand and picture the problem.
    1. Question: What is the probability of **NOT** selecting a green candy? The experiment consists of responding to the question.
    2. The selection of a single candy (trial) has several different outcomes: Red, Yellow, Orange, Green, Brown (sample space).
    3. The requested event is **NOT** choosing a Green candy.

  Step 2: Establish a plan.
    1. Determine the total number of candies in the package.
    2. Find the number of green candies.
    3. Use the definition of probability to find the
       P(Green) = Number of times that Green occurs .
                  Total number of candies
    4. Since P(Green) and P(NOT Green) are complements.
       We may use the definition of complements to find the P(NOT Green).
       P(NOT Green) = 1 − P(Green).

  Step 3: Execute the plan.
    1. The Sample space is: 7 red, 4 yellow, 7 orange, 2 green and 3 brown candies. There are a total of 23 candies.
    2. There are 2 Green candies.
    3. The P(Green) = $\frac{2}{23} \approx 0.087$.
    4. P(NOT Green) $\approx 1 - 0.087 \approx 0.913$.

  Step 4: Check your work.
    1. Check the numbers and your calculations.

2. Is the answer reasonable?
3. Is the answer reasonable? Remember, probability is a fraction or decimal between zero and one, inclusive. It may also be written as a percent.

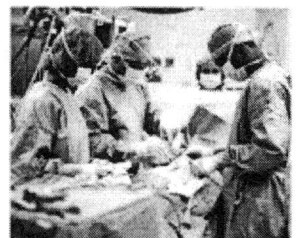
(c) 2006JupiterImages Corp.

Example 9: Suppose a heart transplant operation is 80% successful. What is the probability of the heart transplant operation NOT being successful.

Solution:

Step 1: Understand and picture the problem.
1. Question: What is the probability of the operation **NOT** being successful? The experiment consists of responding to the question.
2. The selection of a heart transplant operation (trial) has two outcomes: successful and not successful (sample space).
3. The requested event is **NOT** successful.

Step 2: Establish a plan.
Since P(Operation being successful) and P(Operation NOT being successful) are complements. We may use the definition of complements to find the P(Operation NOT being successful).
P(Operation NOT being successful) = 1 – P(Operation being successful)

Step 3: Execute the plan.
1. P(Operation being successful) = 0.80.
2. P(Operation NOT being successful) = 1 – 0.80 = 0.20.

Step 4: Check your work.
1. Check the numbers and your calculations.
2. Was the question answered?
3. Is the answer reasonable?

 (c) 2006 JupiterImages Corp.	**Historical Perspective** **Blaise Pascal** (1623 – 1662) France **Pierre de Fermat** (1601- 1665) France
In response to the gambling questions of French courtier, Cevalier de Méré, Blaise Pascal and Pierre de Fermat collaborated and founded the **theory of probability**. Their correspondence produced **"expected value."** Expected value will be discussed later in this chapter.	

❓ Cognitive Problems ❓

1. Explain the difference between an event with a probability of zero and an event with a probability near zero.
2. Give an example of an event that has a probability of zero.
3. Give an example of an event that has a probability near zero.
4. Give an example of an event that has a probability of one.
5. Give an example of an event that has a probability near one.
6. What can be said of an event that has a probability of 1.25 or 125%?
7. Give an example of complementary events.

(c) 2006 JupiterImages Corp.

Exercise 6.1
Write the answers to three decimal places.

1. Use the sample space from Example 1 for the tossing of a pair of dice.

DIE 1 \ DIE 2	1	2	3	4	5	6
1	2	3	4	5	6	7
2	3	4	5	6	7	8
3	4	5	6	7	8	9
4	5	6	7	8	9	10
5	6	7	8	9	10	11
6	7	8	9	10	11	12

Find: A. P(sum of the dots is 7)   B. P(sum of the dots is 11)
       C. P(sum of the dots is NOT 7)   D. P(sum of the dots is NOT 11)
       E. P(sum of the dots is 15)   F. P(sum of the dots is NOT 15)

2. A package of candy contains 12 Orange, 4 Red, 5 Yellow, 6 Green, and 4 Brown candies. If one candy is selected, find:
   A. P(Green)   B. P(Orange)   C. P(Not Yellow)   D. P(Not Purple)

3. In a clinical study, 116 patients out of 1,630 patients experienced drowsiness after taking the medication. If one patient is selected, find the probability that the patient experiences drowsiness as a side effect of this medication.

4. In a clinical study, 109 patients out of 1,630 patients experienced headaches after taking the medication. If one patient is selected, find the probability that the patient did not suffer a headache after taking this medication.

5. In a clinical study, 32 patients reported dizziness as a side effect of the medication and 1,598 patients did not experience dizziness. If one patient is selected, find the probability that the patient experiences dizziness.

6. In a same clinical study, 85 patients reported dry mouth as a side effect of the medication and 1,545 patients did not experience dry mouth. If one patient is selected, find the probability that the patient does not experience dry mouth.

7. The National Safety Council reported the following accidental deaths.

Motor Vehicles	Falls	Poison Liquid	Poison Gas	Drowning	Fire	Choking	Firearms
42,000	13,500	6,500	700	4,800	4,000	2,900	1,600

If one death is taken from this report, find:
A. P(Drowning)
B. P(Not Firearms)
C. P(Poison due to liquids)
D. P(Not Motor Vehicles)

8. The National Safety Council reported the following deaths due to motor vehicles.

Cause of Motor Vehicle Death	Frequency
Collision with another motor vehicle	17,900
Collision with a fixed object	11,900
Pedestrian accidents	6,200
Noncollision accidents	4,500
Collision with a pedalcycle	800
Collision with a train	600
Other (hitting a deer, etc.)	100

If one death is taken from this report, find:
A. P(Collision with a train).
B. P(Not a pedestrian accident).
C. P(Collision with another motor vehicle).
D. P(Not a collision with a fixed object).

9. The FBI reported the following crimes.

Crime	Frequency
Violent Crime	1,932,270
Property crime	12,505,900
Murders	23,760
Rape	109,060
Robbery	672,480
Burglary	2,979,900
Larceny theft	7,915,200

If one crime is taken from this report, find:

A. P(Murder) B. P(Not robbery) C. P(Property crime) D. P(Not Larceny theft)

10. The Bureau of Labor Statistics of the US Department of Labor reported that the distribution of wages paid to hourly workers was

Gender and Age	Wage < $10/hour	Wage at least $10/hour
Male, 16 years and older	17,512,000	14,187,000
Female, 16 years and older	23,126,000	8,491,000

If one person is taken from this report, find:

A. P(Male and earns less than $10.00 per hour)
B. P(Female and earns at least $10.00 per hour)
C. P(Not a Male and earns at least $10.00 per hour)

11. A survey of the number of children living at home with their parents in a small town produced the following table.   If one home is selected, find:

Number of Children Living at Home With Their Parents

x	Frequency
0	523
1	434
2	658
3	235
4	117
5	21
6	8
7	0
8	2

A. P(3 children live with their parents).
B. P(Not 3 children live with their parents).
C. P(No children live with their parents).
D. P(Children live with their parents).
E. P(10 children live with their parents)
F. P(Not 12 children live with their parents).

12. A survey of the number of people living at a residence in a small town produced the following table. If one home is selected, find:

Number of People Living at the Same Residence

x	Frequency
1	110
2	328
3	400
4	289
5	149
6	74
7	12
8	14
9	1
10	1

A. P(5 people live in a residence).
B. P(Not 7 people live in a residence).
C. P(9 people live in a residence).
D. P(Not 3 people live in a residence).
E. P(12 people live in a residence).
F. P(Not 12 people live in a residence).

13. The movie industry interviewed people as to how often they went to the movies, within the last year. The results are displayed in the following table:

Number of Times a Person Went to the Movies

x	Frequency
0	125
1	446
2	365
3	442
4	276
5	228
6	325
7	409
8	289
9	126
10	56
11	87
12	26

If one person is selected from this survey, find:
A. P(person has attended 6 movies in the last year).
B. P(person has attended 8 movies in the last year).
C. P(person has attended not 5 movies in the last year).
D. P(person has attended no movies in the last year).

14. The telephone company is interested in how many different phone numbers a household has. Since the invention of computers, the fax machine, beepers, cellular phones, etc., people have more than one phone number.

Number of Different Phone Numbers Used by a Household

x	Frequency
0	24
1	275
2	654
3	543
4	449
5	165
6	90

If one household is selected from this survey, find:
A. P(household has 4 different phone numbers).
B. P(household does not have 3 different phone numbers).
C. P(household has 6 different phone numbers).
D. P(household has no phone numbers).
E. P(household does not have 10 phone numbers).

15. The auto industry is interested in how many motor vehicles a household has.

Number of Motor Vehicles in a Household

x	Frequency
0	135
1	878
2	2,412
3	1,716
4	623
5	201

If one household is selected from this survey, find:
A. P(household has 2 motor vehicles).
B. P(household does not have 3 motor vehicles).
C. P(household has 6 motor vehicles).
D. P(household does not have 7 motor vehicles).
E. P(household has no motor vehicles).

16. A raffle sells 3,800 tickets ($100 each). The prize table is:
   1st Prize: $290,000   2nd Prize: $5,000   3rd Prize: $2,500
   4th Prize: $1,500   5th Prize: $1,000
   If one ticket is chosen, find:
   A. P(Winning $290,000)   B. P(Not winning $290,000)   C. P(Winning $1,000)
   D. P(Not winning $2,500)   E. P(Losing)

17. A raffle sells 1,200 tickets ($100 each). The prize table is:
    1st Prize: $25,000   2nd Prize: $10,000
    3rd Prize: $5,000   4th Prize: $2,000
    5th through 13th Prizes: $1,000   14th through 20th Prizes: $500
    21st through 60th Prizes: $150
    If one ticket is chosen, find:
    A. P(Winning $25,000)   B. P(Not winning $5,000)
    C. P(Winning $1,000)   D. P(Not winning $500)
    E. P(Winning $150)   F. P(Losing)
    Note: This raffle will be analyzed more closely in another section of this chapter.

18. One year, the federal government's received $700 billion from Social Security, $1,000 billion from Personal Income Taxes, $160 billion from Excise Taxes and $140 billion from Corporate Income Taxes. Consider the percentages of the federal budget contributed by these sources, find:
    A. P(Contribution to the budget did not come from Social Security).
    B. P(Contribution to the budget came from Personal Income Taxes).
    C. P(Contribution to the budget came from Law Enforcement).
    D. P(Contribution to the budget did not come from the National Debt).

19. A raffle sells 3,300 tickets ($100 each). The prize table is:
    1st Prize: $135,000   2nd Prize: $10,000
    3rd Prize: $2,000   4th and 5th Prizes: $1,000
    6th through 10th Prizes: $500   11th through 20th Prizes: $200
    If one ticket is chosen find:
    A. P(Winning $135,000)   B. P(Not winning $200)   C. P(Winning $125)
    D. P(Losing)   E. P(Winning $500)

20. One year the federal government's spent $814 billion on Social Security, $484 billion on National Defense, $462 billion on Social Programs, $220 billion on Human and Community Development, $154 billion on Interest paid on the National Debt, and $66 billion on Law Enforcement. Consider the percentages of the federal budget spent on these programs, find:
    A. P(Money funded Social Security).
    B. P(Money did not fund paying the interest on the nation debt).
    C. P(Money funded Law Enforcement).
    D. P(Money did not fund Corporate Income Taxes).
    E. P(Money funded National Defense).

6.2 ODDS
Objective: Odds and probability will be compared.

Odds ≠ Probability
Both odds and probability use ratios (a comparison of two values) to describe the chance of an event occurring, but they are not interchangeable. Probability is defined as a fraction comparing the number of times an event occurs to the size of the sample space. Probability is always a value between zero and one, inclusive.

There are two types of odds: **"odds against"** and **"odds in favor"** of an event. The first number mentioned in an odds problem is the number of chances associated with the words describing it. For instance, in an "odds against" statement, the first number represents the number of times the event will not occur, while the second number represents the complement, the number of times the event will occur. In an "odds in favor" statement, the first number represents the number of times an event will occur, while the second number represents the complement, the number of times an event will not occur.

Types of Odds

Odds Against an Event:
Number of ways an event does not occur : Number of ways it can occur.

Odds in Favor of an Event:
Number of ways an event can occur : Number of ways it does not occur.
Note: The colon is read as "to"

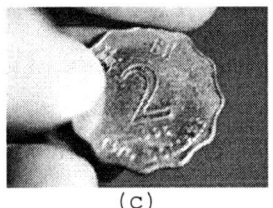
(c) 2006JupiterImages Corp

Consider a coin toss. The sample space is: Heads, Tails. If the coin is fair, then the $P(\text{Heads}) = \frac{1}{2}$ and the $P(\text{Tails}) = \frac{1}{2}$.

Odds are a comparison of the number of times an event occurs in the sample space to the number of times an event does not occur in the sample space. These values are usually separated by a colon (:) or hyphen (–) and read as "to". The odds in favor of tails on the coin toss is:

Number of tails in the sample space **"to"** Number of not tails in the sample space
1 : 1.

Thus the odds in favor of tails on a coin toss is 1 : 1.

Note: Odds of 1 : 1 describe a fair competition. Each event has a 50% chance of occurring and a 50% chance of not occurring. Odds where the values are different favor one event occurring over the other event.

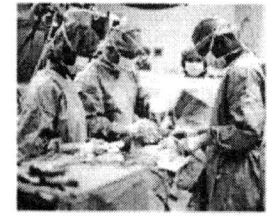

Returning to probability. In the previous section
P(Transplant operation being successful) = 0.80 = 4 / 5.
Its complement is:
P(Transplant operation NOT being successful) =
1 − P(Transplant operation being successful) =
1 − 0.80 = 0.20 = 1 / 5.

Using odds to describe the transplant operation produces 4 successful operations and 1 unsuccessful operation for every 5 operations. The odds in favor of a successful heart transplant operation is 4 : 1 (4 to 1). The odds against a successful heart transplant operation is 1 : 4 (1 to 4).

Another example of odds is taken from a chainsaw manual. The owner's manual recommends a mixture of gasoline to engine oil in a ratio of 40 : 1. This means that 40 ounces of gasoline are mixed with 1 ounce of engine oil. A 3.2 ounce bottle of engine oil is marketed to be mixed with a gallon of gasoline.

Note: A gallon of gasoline is 128 ounces, verify that 128 : 3.2 is equivalent to 40 : 1.

$$\frac{\text{Gasoline}}{\text{Engine Oil}} = \frac{128}{3.2} = \frac{40}{1} \qquad \frac{\text{Gasoline}}{\text{Engine Oil}} = \frac{128}{3.2} \times \frac{40}{1} \qquad 128 \times 1 = 3.2 \times 40$$

Since 128 = 128, the proportions are equal.

Converting the odds 40 : 1 to probabilities:

$$P(\text{gasoline}) = \frac{40}{41} \approx 0.976 \text{ and } P(\text{Oil}) = \frac{1}{41} \approx 0.024.$$

Consider the field of horses that ran in a Kentucky Derby.

Number	Horse	Jockey	Odds
1	Crypto Star	Pat Day	10 – 1
2	Phantom On Tour	Jerry Baley	12 – 1
3	Concerto	Carlos H. Marquez	8 – 1
4	Captain Bodgit	Alex Solis	5 – 2
5	Silver Charm	Gary Stevens	5 – 1
6	Celtic Warrior	Francisco Torres	50 – 1
7	Pulpit	Shane Sellers	2 – 1
8	Hello	Mike Smith	12 – 1
9	Jack Flash	Craig Perret	30 – 1
10	Shammy Davis	Willie Martinez	30 – 1
11	Deeds Not Words	Corey Nakatani	50 – 1
12	Crimson Classic	Robbe Albarado	50 – 1
13	Free House	David Flores	8 – 1

The odds against horse #2, "Phantom on Tour", winning the Kentucky Derby are listed 12 : 1 (read twelve to one). This means that out of a total of 13 chances, there are 12 chances that "Phantom on Tour" will lose the Kentucky Derby (Not come in first) and 1 chance that "Phantom on Tour" will win (come in first) the race.

The complement of the "odds against" an event is the "odds in favor" of an event. The odds in favor of "Phantom On Tour" winning are 1 : 12. This means that out of a total of 13 chances, there is 1 chance that "Phantom On Tour" will win the Kentucky Derby (come in first) and 12 chances that "Phantom On Tour" will lose the race (NOT come in first).

Similarly, the **odds against** horse #11, "Deeds Not Words", winning the Kentucky Derby are 50 : 1. There are 50 chances that "Deeds Not Words" will lose the Kentucky Derby and 1 chance that "Deeds Not Words" will win the race. Using the other type of odds, the **odds in favor** of "Deeds Not Words", winning the Kentucky Derby were 1 : 50. There is 1 chance that "Deeds Not Words" will win the Kentucky Derby and 50 chances that "Deeds Not Words" will lose the race. In both representations of the race, "Deeds Not Words" has 1 chance of winning and 50 chances of losing the Kentucky Derby. In order to express this event as a probability, the total number of chances, 51, is written in the denominator. Thus,

$$P(\text{"Deeds Not Words" winning the Kentucky Derby}) = \frac{1}{51}$$

$$P(\text{"Deeds Not Words" losing the Kentucky Derby}) = \frac{50}{51}$$

Both odds and probability can be used to describe the chance of an event taking place.

Example 1: The odds against the horse "Captain Bodgit", winning the Kentucky Derby are 5 : 2. What is the probability that "Captain Bodgit" will lose the race?
Solution:
  Step 1: Understand and picture the problem.
    1. Question: What is probability "Captain Bodgit" will lose the Derby?
    2. The 5 : 2 odds given are the "odds against" Captain Bodgit winning the race. That means there were 5 chances that Captain Bodgit will lose the race and 2 chances that Captain Bodgit will win the race.

  Step 2: Establish a plan.
    1. Convert the odds to a probability (fraction).
    2. Add the two numbers in the odds to determine the denominator, the relative size of the sample space.
    3. Choose the appropriate value for the numerator, the chances of losing.
    4. Calculate P("Captain Bodgit" losing the Kentucky Derby).

  Step 3: Execute the plan.
    1. The odds against winning are 5 : 2.
    2. In probability, the denominator is the sum of the odds, 5 + 2 = 7.
    3. The numerator is 5, the chances of "Captain Bodgit" losing.
    4. P(Captain Bodgit losing the Kentucky Derby) = $\frac{5}{7} \approx 0.7142857 \approx 71.4\%$.

    The probability "Captain Bodgit" will lose the Derby is about 71.4%.

  Step 4: Check your work.
    1. Check the numbers and your calculations.
    2. Was the question answered?
    3. Is the answer reasonable? Remember, probability is a fraction or decimal between zero and one, inclusive. It may also be written as a percent.
    Note: "Captain Bodgit" came in second in a photo finish. "Silver Charm", with odds 5 : 1, won the Kentucky Derby. Probability only gives an idea of what might occur. There are no guarantees.

Example 2: In a pre-election survey, the odds in favor of a candidate winning the election were 3 : 2. What is the probability of the candidate losing the election?
Solution:
  Step 1: Understand and picture the problem.
    1. Question: What is probability of the candidate losing the election?
    2. The odds given are the odds in favor of the candidate winning, 3 : 2. This means that the candidate has 3 chances of winning the election and 2 chances of losing the election.

Step 2: Establish a plan.
    1. Convert the odds to a probability, fraction.
    2. Add the two numbers in the odds to determine the denominator, the relative size of the sample space.
    3. Choose the appropriate value for the numerator, the chances of losing the election.
    4. Calculate P(The candidate losing the election).

Step 3: Execute the plan.
    1. Odds in favor of winning are 3 : 2.
    2. The denominator is the sum of the odds, 3 + 2 = 5.
    3. The numerator is 2, the chances of the candidate losing the election.
    4. P(The candidate losing the election) = $\frac{2}{5}$ = 0.40.

    There is a 40% chance the candidate will lose the election.

Step 4: Check your work.
    1. Check the numbers and your calculations.
    2. Was the question answered?
    3. Is the answer reasonable? Remember, probability is a fraction or decimal between zero and one, inclusive. It may also be written as a percent.

Example 3: The probability of an electrical part functioning properly is 0.997. What are the odds in favor of the electrical part functioning properly?
Solution:
Step 1: Understand and picture the problem.
    1. Question: What are the odds in favor of the electrical part functioning?
    2. The probability given is the probability the electrical part functions properly, $0.997 = \frac{997}{1,000}$.
    3. The numerator, 997, represents the chances the electrical part functions properly and the denominator, 1,000 is the relative size of the sample space.

Step 2: Establish a plan.
    1. Convert the probability to odds.
    2. Subtract the numerator from the denominator to determine the complement, the number of chances that the electrical part does not function (fail).
    3. Write the numbers in the correct order.

460

Step 3: Execute the plan.
1. The probability the electrical part functions is $\frac{997}{1{,}000}$.
2. 1,000 (sample space) − 997 (functions) = 3 (fails).
3. The odds in favor of the electrical part functioning properly are 997 : 3.

Step 4: Check your work.
1. Check the numbers and your calculations.
2. Was the question answered?
3. Is the answer reasonable?

Example 4: The probability of winning the "Pick 4 Lottery Game" (picking 4 numbers in the exact order) is 0.0001. What are the odds against winning the "Pick 4" game?

Solution:
Step 1: Understand and picture the problem.
1. Question: What are the odds against winning the "Pick 4" game?
2. The given probability is the probability of winning the Pick 4 game, $0.0001 = \frac{1}{10{,}000}$.
3. The numerator, 1, represents the chance of winning the "Pick 4" game in a sample space size of 10,000 (the denominator).

Step 2: Establish a plan.
1. Convert the probability to odds.
2. Subtract the numerator from the denominator to determine the complement, the number of chances of losing the Pick 4 game.
3. Write the numbers in the correct order.

Step 3: Execute the plan.
1. The probability of winning is $\frac{1}{10{,}000}$.
2. 10,000 (sample space) − 1 (chance to win) = 9,999 (chances of losing).
3. The odds against winning the "Pick 4" game are 9,999 : 1.

Step 4: Check your work.
1. Check the numbers and your calculations.
2. Was the question answered?
3. Is the answer reasonable?

Simplifying Odds

Suppose 312 people bet on "Phantom on Tour" to win the Kentucky Derby and 3,692 people bet on other horses to win the Kentucky Derby. What are the odds against "Phantom on Tour" winning the Kentucky Derby? From the bets made, the odds against "Phantom on Tour" winning the Kentucky Derby are 3,692 to 312. These are not the values posted at the racetrack. The posted odds are simplified. Use the following strategy to simplify odds.

Simplify Odds
1. Divide the larger value of the odds by the smaller value of the odds.
2. Round the quotient to the tenths place, if the **tenths part of the quotient** ranges from four tenths (0.4) to six tenths (0.6) inclusive, rewrite the decimal portion as five tenths (0.5). Otherwise, round to the nearest whole number.
3. When rounding to the nearest whole number, the odds are either **whole number : 1** or **1 : whole number**. When rounding with five tenths (0.5), double the number. The odds are **doubled number : 2** or **2 : doubled number**.

Simplifying the odds for "Phantom on Tour":
1. Divide the larger value of the odds by the smaller value of the odds.

$$\frac{3,692}{312} \approx 11.83333333$$

2. Round the quotient to the tenths place, if the **tenths part of the quotient** ranges from four tenths (0.4) to six tenths (0.6) inclusive, rewrite the decimal portion as five tenths (0.5). Otherwise, round to the nearest whole number.

$$11.8 \approx 12$$

3. When rounding to the nearest whole number, the odds are either **whole number : 1** or **1 : whole number**. When rounding with five tenths (0.5), double the number. The odds are **doubled number : 2** or **2 : doubled number**.

The Odds against "Phantom on Tour" winning the Kentucky Derby is 12 : 1. The 12 is written first because it represents the number of bets made on other horses. This represents the larger number 3,692.

Example 5: Suppose 1,123 people bet on "Captain Bodgit" to win the Kentucky Derby and 2,879 people bet on other horses to win the Kentucky Derby. What are the odds against "Captain Bodgit" winning the Kentucky Derby?
Solution:
Step 1: Understand and picture the problem.
1. Question: What are the odds against "Captain Bodgit" winning the Kentucky Derby?

2. 2,879 people bet on other horses to win the Kentucky Derby.
   1,123 people bet on "Captain Bodgit" to win the Kentucky Derby.
3. The odds against "Captain Bodgit" winning the Kentucky Derby is 2,879 : 1,123.

Step 2: Establish a plan.
1. Divide the larger value of the odds by the smaller value of the odds.
2. Round the quotient to the tenths place, if the **tenths part of the quotient** ranges from four tenths (0.4) to six tenths (0.6) inclusive, rewrite the decimal portion as five tenths (0.5). Otherwise, round to the nearest whole number.
3. When rounding to the nearest whole number, the odds are either **whole number : 1 or 1 : whole number.** When rounding to five tenths (0.5), double the number. The odds are **doubled number : 2 or 2 : doubled number**.

Step 3: Execute the plan.
1. Divide the larger value of the odds by the smaller value of the odds.
   $$\frac{2,879}{1,123} \approx 2.563668744.$$
2. Round the quotient to the tenths place, if the **tenths part of the quotient** ranges from four tenths (0.4) to six tenths (0.6) inclusive, round the decimal part to five tenths (0.5). Otherwise, round to the nearest whole number.
   $2.563668744 \approx 2.6$ 2.6 is rewritten as 2.5.
3. When rounding to the nearest whole number, the odds are either **whole number : 1 or 1 : whole number.** When rounding to five tenths (0.5), double the number. The odds are **doubled number : 2 or 2 : doubled number.** Since 2.5 is the rounded value, double 2.5. The odds against "Captain Bodgit" winning the Kentucky Derby is $(2 \times 2.5) : 2$ which is 5 : 2.

Step 4: Check your work.
1. Check the numbers and your calculations.
2. Was the question answered?
3. Is the answer reasonable?

?   **Cognitive Problems**   **?**

1. Explain the similarity between odds and probability.
2. Explain how the odds against an event being 1,000,000 : 1 are the same as the odds in favor of an event being 1 : 1,000,000.

3. If the odds in favor of an event are 1:2, are they the same as the odds in favor of an event being 2:4? Explain. (Hint: What are their probabilities of occurring?)

Exercise 6.2

1. The odds against "Silver Charm" winning the Kentucky Derby were 5 : 1. Find
   A. Odds in favor of "Silver Charm" winning the Kentucky Derby.
   B. P(Silver Charm losing the Kentucky Derby).
   C. P(Silver Charm winning the Kentucky Derby).
   Note: Silver Charm did win.

2. A public opinion poll shows that if the election was held today, the candidate's probability of being elected is 37%. Find the:
   A. P(candidate losing). B. Odds in favor of the candidate winning.
   C. Odds in favor of the candidate losing.
   D. Odds against the candidate winning.
   E. Odds against the candidate losing.

3. A medication claims a 0.7% chance of having a headache as a side effect. Find
   A. P(not experiencing a headache). B. Odds in favor of having a headache.
   C. Odds in favor of not having a headache.
   D. Odds against having a headache.
   E. Odds against not having a headache.

4. A committee is comprised of 8 females and 3 males. A chairperson is to be chosen from this committee. Find
   A. Odds in favor of a female being chosen chairperson.
   B. P(Male not being chosen as the chairperson).
   C. P(Male being chosen as the chairperson).

5. In a 2.6 oz. Package of candy there are 12 Orange, 4 Red, 5 Yellow, 6 Green, and 4 Brown candies. If one candy is selected, find the:
   A. P(Green). B. Odds against selecting a green candy.
   C. P(Orange). D. Odds in favor of selecting an orange candy.

6. A public opinion poll surveyed 250 voters. 145 of the surveyed voters responded that they would vote in favor of a tax increase for their fire protection district. Find the:
   A. P(voting against the tax increase) B. Odds in favor of the tax increase.
   C. Odds against the tax increase.

7. In the same opinion poll used in problem 6, 155 out of 250 surveyed voters responded that they would vote against a tax increase for their library district. Find the:
   A. P(voting in favor of the tax increase)
   B. Odds in favor of the tax increase.
   C. Odds against the tax increase.

8. The National Safety Council reported the following accidental deaths.

Motor Vehicles	Falls	Poison Liquid	Poison Gas	Drowning	Fire	Choking	Firearms
42,000	13,500	6,500	700	4,800	4,000	2,900	1,600

   If one death is taken from this report, find the:
   A. P(Drowning).
   B. Odds against drowning.
   C. P(Poison due to liquids).
   D. Odds in favor of a lethal fall.

9. The National Safety Council reported the following deaths due to motor vehicles.

Conditions of the Fatality	Frequency
Collision with another motor vehicle	17,900
Collision with a fixed object	11,900
Pedestrian accidents	6,200
Noncollision accidents	4,500
Collision with a pedalcycle	800
Collision with a train	600
Other (hitting a deer, etc.)	100

   If one death is taken from this report, find the:
   A. P(Collision with a train).
   B. Odds against a pedestrian accident.
   C. P(Collision with another motor vehicle).
   D. Odds in favor of a collision with a fixed object.

10. The FBI reported the following crimes.

Crime	Frequency
Violent Crime	1,932,270
Property Crime	12,505,900
Murders	23,760
Rape	109,060
Robbery	672,480
Burglary	2,979,900
Larceny Theft	7,915,200

   If one crime is taken from this report, find the:
   A. P(Murder).
   B. Odds against Murder.
   C. P(Property crime).
   D. Odds in favor of a violent crime.

6.3 PROBABILITY DISTRIBUTIONS
Objective: Construct probability distributions and calculate their expected value, average.

Many businesses, medical professions, and government agencies store information in computerized databases as records. Each record contains data pertaining to their operation. These records are then used for making predictions or identifying trends. For example, at one time the average American family had 2.3 children, down from 2.7 children a 10 years earlier. This average value reflects a drop in the number of children in the American family. Even though the idea of having 2.3 children is bizarre, the average value conveys information that most families have between two and three children. Since the average dropped from 2.7 children to 2.3 children, the trend in American families is to have 2 children rather than 3 children. These averages were calculated from a probability distribution.

(c) 2006JupiterImages Corp

Previously the vocabulary terms for probability (experiment, trial, event and sample space) were defined.

PROBABILITY VOCABULARY
1. **Experiment:** A planned activity performed to identify results.
2. **Trial:** Each repetition of the experiment.
3. **Event:** A result, occurrence, or outcome from a single trial of the experiment.
4. **Sample Space:** A list of all possible events from an experiment.

A probability distribution summarizes an experiment by listing the sample space and their corresponding probabilities. Random variables, numbers assigned to each event, identify all events of the experiment. The occurrence of each event is counted and their corresponding probabilities are calculated. This information is listed in a table format called the probability distribution table.

PROBABILITY DISTRIBUTION VOCABULARY
1. **Random Variable:** A number assigned to an event that represents an outcome of the experiment.
2. **Probability Distribution**: Lists all values of the random variable from a specified point of view. These values represent all possible singular events that comprise an experiment and their corresponding probabilities.

Revisit the sample space for tossing a pair of dice.

DIE 1	DIE 2					
		1	2	3	4	5	6
1	2	3	4	5	6	7
2	3	4	5	6	7	8
3	4	5	6	7	8	9
4	5	6	7	8	9	10
5	6	7	8	9	10	11
6	7	8	9	10	11	12

There are 36 possible outcomes comprising the sample space.

In this experiment, the point of view is the sum of the dots on a pair of dice. The random variable has values from 2 to 12, inclusive. The number of different ways that each event occurs is counted and a probability distribution table is made. The outcomes for the sum of dots is listed in the column labeled x (random variable). The count for each outcome is displayed in the frequency column and the probability of each event (outcome) is listed in the column labeled P(x).

Probability Distribution Table for Tossing a Pair of Dice

x	Frequency	P(x)
2	1	$1/36$
3	2	$2/36 = 1/18$
4	3	$3/36 = 1/12$
5	4	$4/36 = 1/9$
6	5	$5/36$
7	6	$6/36 = 1/6$
8	5	$5/36$
9	4	$4/36 = 1/9$
10	3	$3/36 = 1/12$

11	2	$2/36 = 1/18$
12	1	$1/36$
Totals	36	$36/36 = 1$

"x" represents the random variable (sum of dots on the dice), frequency denotes the count of each event, and P(x) indicates the probability of that specified event. For example, P(4) denotes the probability of a sum of 4, (1 and 3, 2 and 2, 3 and 1). Since a 4 can occur three different times out of the possible 36 events, $P(4) = 3/36 = 1/12$.

CHARACTERICS OF A PROBABILITY DISTRIBUTION

1. For each event, $0 \leq P(x) \leq 1$
2. The sum of the probabilities column equals one, $\Sigma P(x) = 1$.
   Note: Σ is the Greek letter sigma and indicates the sum of the probabilities.

The characteristics of a probability distribution are consistent with the definition of probability:

1. For each event, $0 \leq P(x) \leq 1$.

   The probability of any individual event falls between 0 (impossible) and 1 (certain), inclusive.

2. The sum of the probabilities column equals one, $\Sigma P(x) = 1$.

   The second characteristic of a probability distribution signifies that the complete sample space for an experiment is listed. Since a sample space lists all possible events for an experiment and each event of the probability distribution table has a corresponding probability, the sum must equal 1 (100%). If the sum of the probabilities does not equal 1, then the sample space is incomplete or a probability is miscalculated.

   Note: When the probabilities are rounded, the sum will be close to 1.00.

Construct a Probability Distribution Table

Probability distribution tables are easily constructed from databases. An experiment may consist of scanning a set of records from a database. The scan focuses on counting the occurrence of an event. Each record scan is considered a trial of the experiment. A number is allocated to represent each unique event. The frequency of each event is counted and a table is made listing the tally of each random variable. These tallies are then converted to probabilities.

Consider the experiment of scanning 500 traffic court records. For example, each traffic record is one trial. The number of prior speeding tickets listed on that record is the random variable. The court records indicated that the number of prior speeding tickets ranged from 0 to 5. Counting the number of times these random variables occurred produced the following probability distribution table:

Prior Speeding Tickets

x	Frequency	P(x)
0	120	120/500 = 0.24
1	200	200/500 = 0.40
2	70	70/500 = 0.14
3	60	60/500 = 0.12
4	40	40/500 = 0.08
5	10	10/500 = 0.02
Totals	500	500/500 = 1.00

The probability distribution table produces a visual summary of the 500 court records. From this table, a variety of probabilities can be found.

1. P(Exactly 2 prior speeding tickets) = P(2) = 0.14.
   Note: Taken directly from the table.

2. P(More than 2 prior speeding tickets) =
   P(3) or P(4) or P(5) = P(3) + P(4) + P(5) = 0.12 + 0.08 + 0.02 = 0.22.
   Note: Find the sum of the specified probabilities.

3. P(Less than 2 prior speeding tickets) =
   P(0) or P(1) = P(0) + P(1) = 0.24 + 0.40 = 0.64.
   Note: Find the sum of the specified probabilities.

4. P(At most 2 prior speeding tickets) =
   P(0) or P(1) or P(2) = P(0) + P(1) + P(2) = 0.24 + 0.40 + 0.14 = 0.78.
   Note: Find the sum of the specified probabilities.

5. P(At least 2 prior speeding tickets) =
   P(2) or P(3) or P(4) or P(5) = P(2) + P(3) + P(4) + P(5) =
   0.14 + 0.12 + 0.08 + 0.02 = 0.36.
   Note₁: Find the sum of the specified probabilities.
   Note₂: An alternate approach uses the complement.
   $\quad$ 1 − (P(0) + P(1)) = 1 − (0.24 + 0.40) = 1 − 0.64 = 0.36.

Example 1: A new allergy medicine is administered to 200 groups of people. Each group is comprised of 6 people (totaling 1200 people). The experiment records the number of people, in each group of six, who experienced relief from their allergy symptoms. Let x represent the number of people relieved of their allergy symptoms. The value of the random variable, x, ranges from zero to six. A group may have no one experience relief from their allergy symptoms (x = 0) to everyone experience relief (x = 6). The following probability distribution table displays the results of this experiment.

Allergy Relief

x	Frequency	P(x)
0	1	0.005
1	8	0.040
2	20	0.100
3	64	0.320
4	82	0.410
5	15	0.075
6	10	0.050
Totals	200	1.000

Find the probability that more than half of the group will demonstrate relief of their allergy symptoms.

Solution:

Step 1: Understand and Picture the problem.
1. Question: Find the P(more than half of the group is allergy free).
2. More than half of a group with six people means 4, 5, or 6 people show relief of their allergy symptoms.
3. The probability distribution table contains all random events and their corresponding probabilities.

Step 2: Develop a plan.
1. Find the P(4), P(5) and P(6) from the table.
2. Find their sum.

Step 3: Execute the plan.
1. P(4) = 0.410, P(5) = 0.075, and P(6) = 0.050.
2. P(more than half demonstrate relief) =
   P(4) + P(5) + P(6) = 0.410 + 0.075 + 0.050 = 0.535.

Step 4: Check your work.
1. Check the numbers and your calculations.
2. Was the question answered?
3. Is the answer reasonable?

Example 2: Use the table from Example 1, find the probability that at least half of the group members demonstrate relief from their allergy symptoms.
Solution:
  Step 1: Understand and Picture the problem.
      1. Question: Find the P(At least half of the group is allergy free).
      2. At least half of a group with six people means 3, 4, 5, or 6 people show relief of their allergy symptoms.
      3. The probability distribution table contains all random events and their corresponding probabilities.

  Step 2: Develop a plan.
      1. Find P(3), P(4), P(5), and P(6) from the table.
      2. Add these probabilities.

  Step 3: Execute the plan.
      1. P(3) = 0.320, P(4) = 0.410, P(5) = 0.075, and P(6) = 0.050.
      2. P(at least half demonstrate relief) =
         P(3) + P(4) + P(5) + P(6) = 0.320 + 0.410 + 0.075 + 0.050 = 0.855.

  Step 4: Check your work.
      1. Check the numbers and your calculations.
      2. Was the question answered?
      3. Is the answer reasonable?

Example 3: A study of the weekly car sales from last year resulted in the following information. Only once did the dealership sell no cars in a week. Three times they sold only 1 car, five times they sold 2 cars, eight times they sold 3 cars, they never sold 4 cars, eleven times they sold 5 cars, twenty times they sold 6 cars, they never sold 7 cars, three times they sold 8 cars and one time they sold 9 cars. Find the probability that at most 4 cars were sold in a week.
Solution:
  Step 1: Understand and picture the problem.
      1. Question: What is the probability that at most 4 (4 cars or less) will be sold in a week?
      2. The point of perspective is the number of cars sold in a week.
      3. The range of the random variable is 0 to 9 car sales per week.
         One time no cars (**0**) were sold in a week.
         Three times **1** car was sold in a week.
         Five times **2** cars were sold in a week.
         Eight times **3** cars were sold in a week.
         Zero times **4** cars were sold in a week.
         Eleven times **5** cars were sold in a week.
         Twenty times **6** cars were sold in a week.

Zero times **7** cars were sold in a week.
Three times **8** cars were sold in a week.
One time **9** cars were sold in a week.

Step 2: Develop a plan.
1. Make a probability distribution table to organize this data.
2. Find P(0), P(1), P(2), P(3), and P(4).
3. Add these probabilities.

Step 3: Execute the plan.
1. The probability distribution table is:

Weekly Car Sales

x	Frequency	P(x)
0	1	$1/52$
1	3	$3/52$
2	5	$5/52$
3	8	$8/52 = 2/13$
4	0	$0/52 = 0$
5	11	$11/52$
6	20	$20/52 = 5/13$
7	0	$0/52 = 0$
8	3	$3/52$
9	1	$1/52$
Totals	52	$52/52 = 1$

2. P(0) = $1/52$, P(1) = $3/52$, P(2) = $5/52$, P(3) = $8/52$, and P(4) = $0/52$.
3. P(At most 4 cars were sold) = P(0) + P(1) + P(2) + P(3) + P(4) = $1/52 + 3/52 + 5/52 + 8/52 + 0/52 = 17/52 \approx 0.327$.

Step 4: Check your work.
1. Check the numbers and your calculations.
2. Was the question answered?
3. Is the answer reasonable?

Example 4: Use the probability distribution from Example 3 to find the probability that no less than 5 cars were sold weekly.
Solution:
  Step 1: Understand and picture the problem.
    1. Question: What is the probability that no less than 5 cars are sold in a week?
    2. The perspective is the number of cars sold in a week.
    3. The range of the random variable is 0 to 9 cars sold.
    4. No less than 5 means 5 or more cars are sold in a week. That is 5, 6, 7, 8 or 9 cars are sold weekly.

  Step 2: Develop a plan.
    1. Find P(5), P(6), P(7), P(8) and P(9) from the table.
    2. Add these probabilities.

  Step 3: Execute the plan.
    1. P(5) = $11/52$, P(6) = $20/52$, P(7) = $0/52$, P(8) = $3/52$, and P(9) = $1/52$.
    2. P(No less than 5 cars sold) = P(5) + P(6) + P(7) + P(8) + P(9) = $11/52 + 20/52 + 0/52 + 3/52 + 1/52 = 35/52 \approx 0.673$.

  Step 4: Check your work.
    1. Check the numbers and your calculations.
    2. Was the question answered?
    3. Is the answer reasonable?

EXPECTED VALUE

An emergency room needs to be prepared for an expected number of patients with heart problems per day. A restaurant wants to serve all customers and needs to know how many hamburgers are expected to be sold per day. As a safety issue, a city manager wants to predict how many fires are expected in the next year and if there is a need for another fire station. Many situations require an expected value. The expected value is the average (mean) of the probability distribution. It is denoted with an "E" and is defined by the formula:

EXPECTED VALUE

$$E = \sum(x \cdot P(x))$$

where: $\sum$ is the Greek letter sigma, meaning add.
   x is the random variable.
   P(x) is the random variable's corresponding probability.

Consider the 500 traffic court records from the problem presented earlier in this section. The probability distribution is:

Prior Speeding Tickets

x	Frequency	P(x)
0	120	$120/500 = 0.24$
1	200	$200/500 = 0.40$
2	70	$70/500 = 0.14$
3	60	$60/500 = 0.12$
4	40	$40/500 = 0.08$
5	10	$10/500 = 0.02$
Totals	500	$500/500 = 1.00$

To find the average number of speeding tickets for the people in traffic court, add a new column, "x • P(x)". For each random variable, x, find the product of "x" and its corresponding probability, "P(x)." Lastly, find the sum of this column.

Prior Speeding Tickets

x	Frequency	P(x)	x • P(x)
0	120	$120/500 = 0.24$	**0.00**
1	200	$200/500 = 0.40$	**0.40**
2	70	$70/500 = 0.14$	**0.28**
3	60	$60/500 = 0.12$	**0.36**
4	40	$40/500 = 0.08$	**0.32**
5	10	$10/500 = 0.02$	**0.10**
Totals	500	$500/500 = 1.00$	**1.46**

The average number of speeding tickets for the people in this traffic court is approximately 1.46 tickets. On the average people in this court have a record of one or two prior speeding tickets.

Example 5: Find the average number of people per group from Example 1 that were relieved of their allergy symptoms.

Solution:
Step 1: Understand and Picture the Problem.
Use the table from Example 1. Find the expected value. (This value is calculated from the probability distribution table and is also known as the average.)

Step 2: Develop a Plan.
1. Add the new column, "x • P(x)" to the probability distribution table from Example 1.
2. Calculate the product "x • P(x)" by multiplying the random variable, "x," by its corresponding probability, "P(x)."
3. Find the sum of this column.

Step 3: Execute the Plan.
1-2.

Allergy Relief

x	Frequency	P(x)	x • P(x)
0	1	0.005	**0.000**
1	8	0.040	**0.040**
2	20	0.100	**0.200**
3	64	0.320	**0.960**
4	82	0.410	**1.640**
5	15	0.075	**0.375**
6	10	0.050	**0.300**
Totals	200	1.000	**3.515**

3. Thus, in the test groups of six people, the average number of patients relieved of their allergy symptoms is approximately 3.515 patients. Hence the average number per group is 3 to 4 people were relieved of their allergy symptoms.

Step 4: Check Your Work.
1. Check the numbers and your calculations.
2. Was the question answered?
3. Does 3 or 4 people out of six experiencing allergy relief seem reasonable?

Example 6: Find the average number of weekly car sales from Example 3.
Solution:
Step 1: Understand and Picture the Problem.
Use the table from Example 3. Find the expected value. (This value is calculated from the probability distribution table and is also known as the average.)

Step 2: Develop a Plan.
1. Add the new column, "x • P(x)", to the probability distribution table from Example 3.
2. Calculate the product "x • P(x)" by multiplying the random variable, "x", by its corresponding probability, "P(x)".
3. Find the sum of this column.

Step 3: Execute the Plan.
1-2.

Weekly Car Sales

x	Frequency	P(x)	x • P(x)
0	1	$1/52$	$0/52$
1	3	$3/52$	$3/52$
2	5	$5/52$	$10/52$
3	8	$8/52$	$24/52$
4	0	$0/52$	$0/52$
5	11	$11/52$	$55/52$
6	20	$20/52$	$120/52$
7	0	$0/52$	$0/52$
8	3	$3/52$	$24/52$
9	1	$1/52$	$9/52$
Totals	52	$52/52 = 1$	$245/52$

3. Thus, the average weekly car sales is $245/52$, which is approximately 4.7115385, cars per week. This predicts that 4 or 5 cars were sold per week.

Step 4: Check Your Work.
1. Check the numbers and your calculations.
2. Was the question answered?
3. Does 4 or 5 car sales per week seem reasonable?

Example 7: Find the expected value (average) for the following raffle:

"HIGHROLLERS RAFFLE"

LAS VEGAS NIGHTS Friday, October 16th
Saturday, October 17th

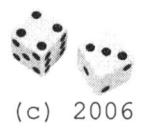
(c) 2006 JupiterImages Corp

7:00 P.M. until 1:00 A.M.

"One in every 20 tickets sold wins"!

(c) 2006 JupiterImages Corp

PRIZES:

1st Prize:..........$25,000   4th Prize:..........."Big Screen TV"
2nd Prize: $10,000   5th – 13th Prize:................$1,000
3rd Prize:.........$ 5,000   14th – 20th Prize:..............$.. 500
                21st – 60th Prize:...............$150

Donation $100   ONLY 1,200 TICKETS AVAILABLE

Solution:

Step 1: Understand and Picture the Problem.
  Make a probability distribution table to calculate the expected value. This value is also known as the average.

Step 2: Develop a Plan.
  1a. Make a table labeling one column "Event" (How much money can be won).
   b. Label another column "x", this is the random variable, it is the winnings MINUS the cost of the ticket.
   c. Label the next column "Frequency", this is the number of winners for each prize.
   d. Label the next column "P(x)". This is the probability for each event.
   e. Label the last "x * P(x)". This column will be used to calculate the expected value (Average).
  2. Calculate the product "x * P(x)" by multiplying the random variable, "x", by its corresponding probability, "P(x)".
  3. Find the sum of this column.

Step 3: Execute the Plan.
  1-2.

High Rollers Raffle

Event	x	Frequency	P(x)	x * P(x)
Win $25,000	$24,900	1	1 / 1,200	$24,900 / 1,200
Win $10,000	$9,900	1	1 / 1,200	$9,900 / 1,200
Win $5,000	$4,900	1	1 / 1,200	$4,900 / 1,200
Win Big Screen TV (worth $2,000)	$1,900	1	1 / 1,200	$1,900 / 1,200
Win $1,000	$900	9	9 / 1,200	$8,100 / 1,200
Win $500	$400	7	7 / 1,200	$2,800 / 1,200
Win $150	$50	40	40 / 1,200	$2,000 / 1,200
Lose, (Win $0)	−$100	1,140	1,140 / 1,200	−$114,000 / 1,200
	Totals	**1,200**	**1,200/1,200**	**−$59,500 / 1,200**

3. The average money loss by a ticket buyer is −$59,500 / 1,200 ≈ −$49.58. This means that the charity made about $49.58 per ticket sold. Since there were 1,200 ticket sold, the charity made a profit of $59,500.

   Note: $\frac{\$59,500}{1,200}(1,200) = \$59,500$.

Step 4: Check Your Work.
1. Check your numbers and calculations.
2. Was the question answered?
3. At $100 a ticket, selling 1,200 tickets produces:
   a. Revenue of 100(1,200) = $120,000.
   b. The Costs of the prizes are:
      $25,000+$10,000+$5,000+$2,000+$9,000+$3,500+$6,000 = $60,500.
   c. Profit = Revenue − Cost = $120,000 − $60,500 = $59,500.

   Note: This matches the profit calculated using the expected value.

(c) 2006 JupiterImages Corp.

Optional Student Project:
The project "Raffle" can be found on page 528.

? Cognitive Problems ?

1. Does the following information represent a probability distribution? Explain your response. If it is not a legitimate probability distribution, what needs to be corrected to make it a probability distribution?

Number of Defective Cans in a Case of Cola

x	Frequency	P(x)
0	150	0.60
1	40	0.12
2	60	0.24

2. Explain the difference between "at most 2 defective cans" and "at least 2 defective cans" taken from a 12 pack of cans. State the random variables that satisfy each condition.
3. Suppose the random variables of a probability distribution range from 0 to 36. Suppose you need to find the P(x is not less than 8). Explain how you would go about finding this probability?

(c) 2006 JupiterImages Corp.

Exercise 6.3

1. A survey of the number of children living at home with their parents in a small town produced the following table:

Number of Children Living at Home With Their Parents

x	Frequency
0	523
1	434
2	658
3	235
4	117
5	21
6	8
7	0
8	2

Complete the probability distribution table and answer the following questions. If one home is selected, find:
A. P(At least 3 children live with their parents). 3,4,5,6,7,8
B. P(At most 3 children live with their parents). 0,1,2,3
C. P(More than 3 children live with their parents). 4,5,6,7,8
D. P(No more than 3 children live with their parents). 0,1,2,3
E. What is the average number of children living at home with their parents?

2. A survey of the number of people living at a residence in a small town produced the following table:

Number of People Living at the Same Residence

x	Frequency
1	110
2	328
3	400
4	289
5	149
6	74
7	12
8	14
9	1
10	1

Complete the probability distribution table and answer the following questions. If one home is selected, find:

A. P(At least 5 people live in a residence).
B. P(At most 7 people live in a residence).
C. P(Less than 3 people live in a residence).
D. P(No less than 6 people live in a residence).
E. What is the average number of people in a residence?

3. The movie industry interviewed 3200 people as to how often they went to the movies, within the last year. The results are displayed in the following table:

Number of Times a Person Went to the Movies

x	Frequency
0	125
1	446
2	365
3	442
4	276
5	228
6	325
7	409
8	289
9	126
10	56
11	87
12	26

Complete the probability distribution table and answer the following questions. If one person is selected from this survey, find:

A. P(person has attended at least 6 movies in the last year).
B. P(person has attended at most 8 movies in the last year).
C. P(person has attended more than 5 movies in the last year).

D. P(person has attended no more than 7 movies in the last year).
E. Find the average number of movies seen last year by the people surveyed.

4. The telephone company is interested in how many different phone numbers a household has. Since the invention of computers, the fax machine, beepers, cellular phones, etc., people have more than one phone number. They surveyed 2,200 households with the following results:

Number of Different Phone Numbers Used by a Household

x	Frequency
0	24
1	275
2	654
3	543
4	449
5	165
6	90

Complete the probability distribution table and answer the following questions. If one household is selected from this survey, find:
A. P(household has at least 4 different phone numbers).
B. P(household has at most 3 different phone numbers).
C. P(household has more than 4 different phone numbers).
D. P(household has no more than 2 different phone numbers).
E. Find the average number of phone numbers for the households surveyed.

5. The automobile industry is interested in how many motor vehicles a household has. They surveyed 5,965 households with the following results:

Number of Motor Vehicles in a Household

x	Frequency
0	135
1	878
2	2,412
3	1,716
4	623
5	201

Complete the probability distribution table and answer the following questions. If one household is selected from this survey, find:
A. P(household has at least 2 motor vehicles).
B. P(household has at most 3 motor vehicles).
C. P(household has more than 2 motor vehicles).
D. P(household has no more than 2 motor vehicles).
E. Find the average number of motor vehicles for the households surveyed.

6.4 Decision Trees, Decision–Making Under Uncertainty
Objective: Use a decision tree in an application of "Expected Value" for a probability distribution.

Making decisions is an integral part of life. Sometimes these decisions are immediate and routine. Do I order the burger or salad? What am I going to wear today? What are my weekend plans? Can I stay in bed a little longer? Other times the decisions are long term and complex. What type of car, domestic or foreign, should I purchase? Should I rent an apartment or purchase a home? Should I take the position in town or accept the position in another state? Do I get married? Do we have children? These decisions depend on one's personal feelings, financial situation and relationships with other people. These decisions do not easily allow themselves to be objectively analyzed and solved. The subjective, emotional, part of the decision has to be factored into the solution.

However, in many business decisions, it is advisable to objectively plan a course of action. Should a business produce their product by the conventional method or invest in new technology to assist in the production? Should a popular restaurant keep their dining area the same size or expand their dining room? Should an industry stay in the same area or relocate their factory to another state or even another country? These business decisions involve choices that should be objectively investigated.

In making a business decision, a manager must select a course of action that optimizes some goal. Common goals are maximizing profits, and minimizing costs. Making a choice (decision) is choosing a course of "action". The decision results in a set of uncontrollable factors (fate) called "states of nature".

ELEMENTS COMMON TO DECISION-MAKING:

1. **Actions (choices)**
   The decision maker must choose one action from a set of actions.
2. **States of Nature (consequences)**
   Results of the actions are beyond the control of the decision maker and are estimated with a payoff table.
3. **Payoff Table (Organizes actions and states of nature)**
   A table showing the benefits of each action. These tables may be in the form of profits, revenues, costs, time, etc.
4. **Measure of Uncertainty (probabilities)**
   Probabilities are the recorded occurrences of each "state of nature".

Consider a small company that makes ceiling fans, they must choose from the following choices:
A) Continue to make their product by the conventional method.
B) Invest in the purchase of new machinery.

A consulting firm is hired to assist the company with their decision. The consulting firm studies the company, their product, their competition and the public's response (demand) for their product. As a result of their investigation, they construct two tables, a payoff table and probability table.

The payoff table displays each "action" (choice), and each possible "state of nature" in terms of projected profits. The payoff table predicts what the company's profits will be for each decision (conventional method vs. new machinery) based on the demand for their product. Demand can decrease, remain the same or increase. Using the conventional method the company is predicted to make $10,000,000 if demand decreases, $24,000,000 if demand remains the same, and $42,000,000 if demand increases. Investing in new machinery, the company is predicted to lose $30,000,000 if demand decreases, make $45,000,000 if demand remains the same, and make $70,000,000 if demand increases. This is summarized in the payoff table.

PAYOFF TABLE OF PROFITS ($ MILLIONS)

ACTION	STATE OF NATURE (DEMAND)		
	Decreases	Constant	Increases
Conventional Method	10	24	42
Invest in New Machinery	−30	45	70

The probability table displays each "state of nature" and its corresponding probability. The probability table displays the consulting firm's analysis of the market for the company's product. It predicts whether the company's market share will increase, remain the same or decrease. The consulting firm predicts there is a 14% chance the company's market share will increase, 54% chance it will remain the same and a 32% chance the market share will decrease. This is summarized in the probability table.

DEMAND	PROBABILITY
Increases	0.14
Constant	0.54
Decreases	0.32

A decision tree summarizes this information in an orderly fashion. The schematic representation displays the decision-making process. The square (□) indicates the start and branches to the possible decisions. At the end of each action choice (decision) is a circle (O). Branches extending from the circle are uncontrollable factors or "states of nature". The decision tree for the data is:

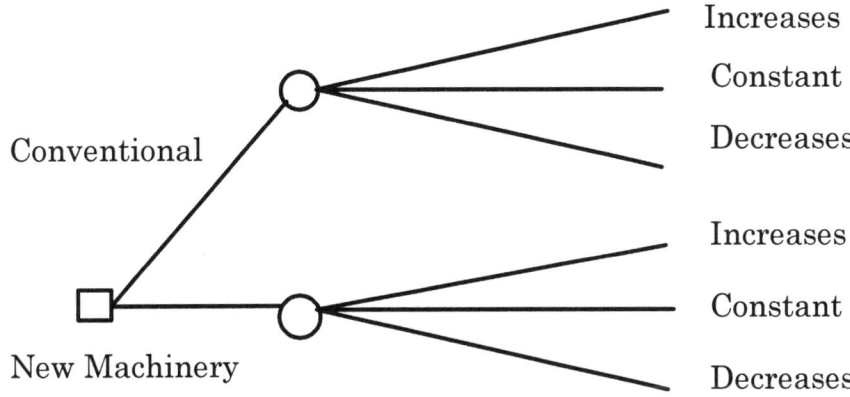

The decision, the type of investments that the company makes in their assembly process, is initiated by the square (□). Each branch extending from the square represents the possible "actions" that the company can decide to follow. At the right of these branches is a circle (O). The circle represents the "states of nature". This is a set of uncontrollable factors that the company cannot control or manipulate. In this case, the public's demand for their product is the "state of nature". Public demand is not controllable.

Using both the payoff table and the probabilities for each "state of nature", an expected value or mean, is calculated for each decision. The expected value is calculated according to the formula

---

Expected Value (Mean)

$$E(x) = \Sigma(x \bullet P(x))$$

where: Σ means add the following values.
  x is the payoff for each "action" corresponding to its "state of nature".
  P(x) is the corresponding probability for each "state of nature".

The decision path with the appropriate expected value (largest value if the objective is a maximum and the smallest value if the objective is a minimum.) is recommended by the consulting firm as the action the company should pursue. Following is the complete decision tree:

484

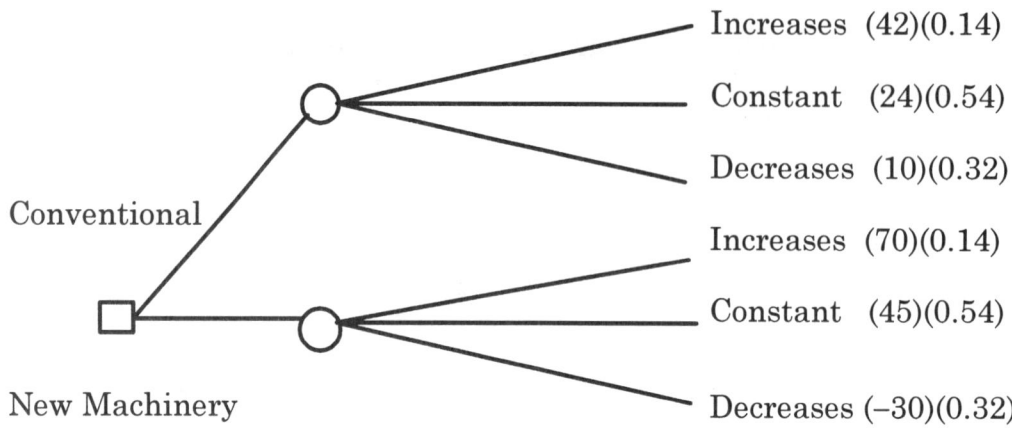

The expected value for each choice is:

Conventional Method	New Machinery
(42)(0.14) = 5.88	(70)(0.14) = 9.80
(24)(0.54) = 12.96	(45)(0.54) = 24.30
(10)(0.32) = + 3.20	(−30)(0.32) = + −9.60
22.04	24.50

In order to maximize profits, the consulting firm recommends purchasing new machinery. Their expected profit is $24.5 million ($24,500,000). This decision may earn the company an expected profit of $24.5 million, whereas remaining the conventional method has an expected profit of $22.04 million.

Example 1: Suppose the probabilities change and the demand decreases is 0.5, remains the same is 0.3 and increases is 0.2. Using these new probabilities, along with the actions and the payoff table from the previous problem, make a decision tree. Recommend the course of action that the company should follow to maximize their profits.

Solution:
Step 1: Understand and Picture the Problem.
  1. Question: What decision should the company make to maximize profits?
  2. The two choices from the previous example are:
     a. continue with the conventional method of assembly.
     b. invest in new machinery.
  3. The payoff table is:

PAYOFF TABLE OF PROFITS ($ MILLIONS)

ACTION	STATE OF NATURE (DEMAND)		
	Decreases	Constant	Increases
Conventional Method	10	24	42
Invest in New Machinery	−30	45	70

4. The probabilities for the "states of nature" for the public's demand are:

DEMAND	PROBABILITY
Increases	0.20
Constant	0.30
Decreases	0.50

Step 2: Develop a Plan.
1. Draw a decision tree displaying each choice (action) as a branch originating from the square.
2. At the end of each choice, draw a circle and display each state of nature as a branch originating from the circle.
3. At the end of each state of nature branch, list the payoff with its corresponding probability.
4. Use the formula for expected value (mean) to calculate the expected value for each decision.

$$E(x) = \Sigma x \bullet P(x)$$

where: Σ means add the following values.
x is the payoff for each "action" corresponding to its "state of nature".
P(x) is the corresponding probability.

5. Choose the largest expected value, it represents the maximum expected profit, as the appropriate recommendation.

Step 3: Execute the plan.
1 – 4, the decision tree is:

Following is the complete decision tree for the small company:

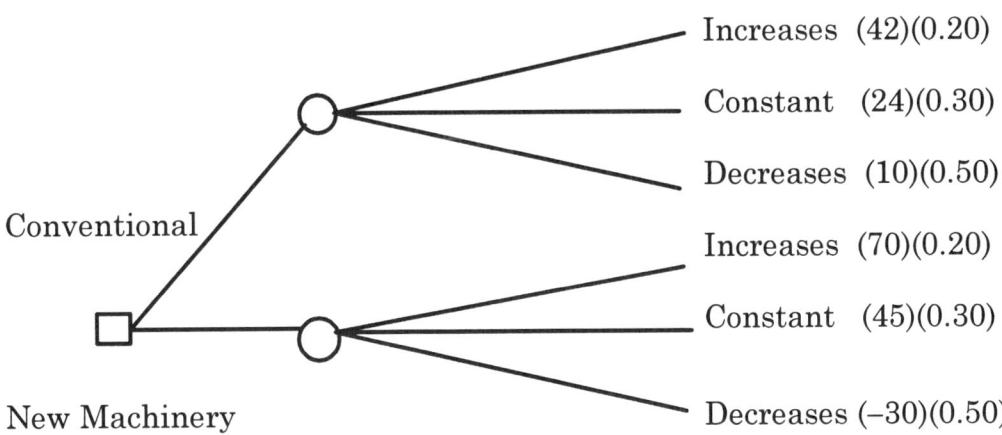

The expected value for each choice is:

486

Conventional Method	New Machinery
(42)(0.20) = 8.4	(70)(0.20) = 14.0
(24)(0.30) = 7.2	(45)(0.30) = 13.5
(10)(0.50) = + 5.0	(−30)(0.50) = + −15.0
20.6	12.5

5. Thus, the consulting firm recommends remaining with the conventional method. This decision may earn an expected profit of $20.6 million ($20,600,000). While investing in new machinery has an expected profit of $12.5 million ($12,500,000).

Step 4: Check your work.
1. Check the numbers and your calculations.
2. Was the question answered?
3. Is the answer reasonable?

COSTS

Another way to maximize profits is to minimize company costs. In situations where a business cannot increase the price of their product (Revenue) due to intense competition, a company can increase their profit by reducing the cost of producing their product. The following example demonstrates maximizing profits by minimizing costs.

Example 2: An auto parts supplier wants to minimize the cost of making their car batteries. It has a 3 year contract with a major automobile corporation to provide car batteries at a fixed price. If it increases the price of its batteries, the contract is broken and the automobile manufacturer is free to purchase its batteries from another supplier. A consulting firm is hired and provides the battery maker with the following actions:
  A. Make no changes, keep the same procedures.
  B. Stay in their current location, but upgrade their machinery.
  C. Relocate and build a new facility with state of the art equipment.

The consultant also provides the following payoff and probability tables based on the cost of making batteries, the demand for their batteries, and the competing car battery manufacturers.

PAYOFF TABLE OF COSTS ($ MILLIONS)

ACTION	STATE OF NATURE (DEMAND)		
	Low	Medium	High
Same Procedures	7,800	8,000	8,400
Upgrade Machinery	7,900	7,800	7,600
New Facility	8,100	7,700	7,200

487

The probability table for the "states of nature" is:

DEMAND	PROBABILITY
High	0.31
Medium	0.44
Low	0.25

Should the owners make no changes, upgrade their machinery or build a new facility in order to minimize their costs?

Solution

Step 1: Understand and Picture the Problem.
1. Question: What decision should the company make to minimize costs?
2. The three choices are:
   a. continue with the same method of assembly.
   b. upgrade the machinery.
   c. relocate and build a state of the are facility.
3. The payoff table is:

PAYOFF TABLE OF COSTS ($ MILLIONS)

	STATE OF NATURE (DEMAND)		
ACTION	Low	Medium	High
Same Procedures	7,800	8,000	8,400
Upgrade Machinery	7,900	7,800	7,600
New Facility	8,100	7,700	7,200

4. The probability table for the "state of nature" is:

DEMAND	PROBABILITY
High	0.31
Medium	0.44
Low	0.25

Step 2: Develop a Plan.
1. Draw a decision tree displaying each choice (action) as a branch originating from the square.
2. At the end of each choice, draw a circle and display each state of nature as a branch originating from the circle.
3. At the end of each state of nature branch, list the payoff with its corresponding probability.
4. Use the formula for expected value (mean) to calculate the expected value for each decision.

$$E(x) = \Sigma x \cdot P(x)$$

where: Σ means add the following values.

x is the payoff for each "action" corresponding to its "state of nature".

P(x) is the corresponding probability.

5. Choose the decision with the smallest expected value as the recommendation.

Step 3: Execute the plan, the decision tree is:

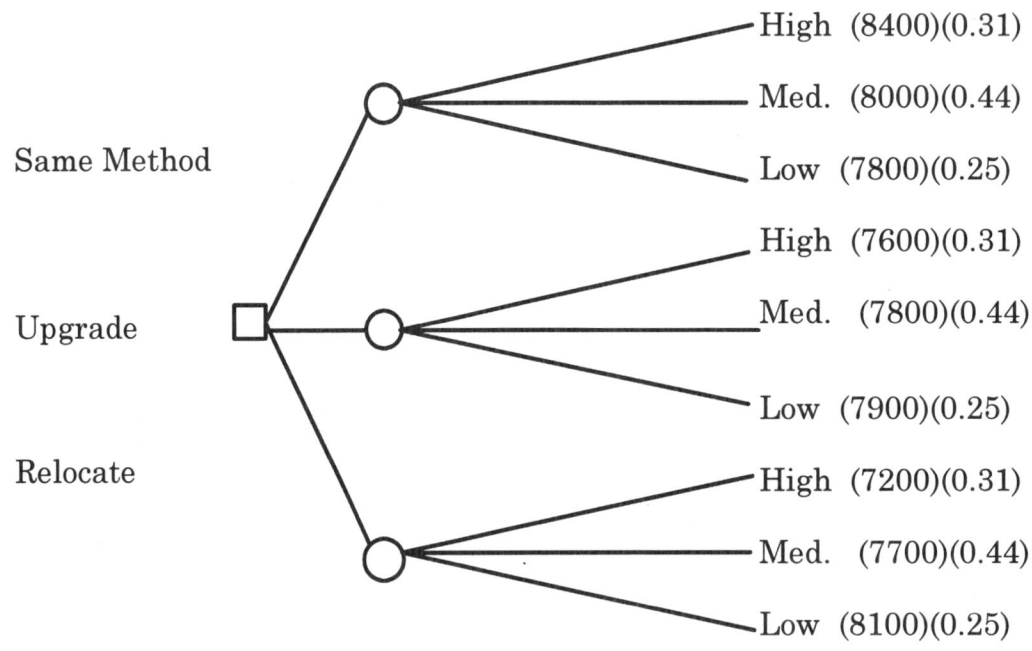

The expected value for each choice is:

Same Method	Upgrade	Relocate
(8400)(0.31) = 2604	(7600)(0.31) = 2356	(7200)(0.31) = 2232
(8000)(0.44) = 3520	(7800)(0.44) = 3432	(7700)(0.44) = 3388
(7800)(0.25) = + 1950	(7900)(0.25) = + 1975	(8100)(0.25) = + 2025
8074	7763	7645

Thus, the consulting firm recommends relocating. This decision may earn an expected cost of $7,645 million ($7,645,000,000).

Step 4: Check your work.
1. Check the numbers and your calculations.
2. Was the question answered?
3. Is the answer reasonable?

(c) 2006 JupiterImages Corp.

Optional Student Project:
The project "Decision Tree" can be found on page 530.

? Cognitive Problems ?

1. Explain the effects the payoff table has in the making of a decision?
2. Why are probabilities assigned to each "state of nature" in the construction of a decision tree?
3. Is the choice which optimizes the goal, maximizes the profit or minimizes the cost, guaranteed to happen?
4. Explain why the "States of Nature" are considered uncontrollable.

Exercise 6.4

(c) 2006 JupiterImages Corp.

1. Consider a small company that makes glass beakers for laboratories, they must choose from the following choices:
    A) Continue to make their product by the conventional method.
    B) Invest in the purchase of new machinery.

    A consulting firm is hired to assist the company with their decision. The payoff table is:

    PAYOFF TABLE OF PROFITS ($ MILLIONS)

ACTION	STATE OF NATURE (DEMAND)		
	Decreases	Constant	Increases
Conventional Method	5	18	40
Invest in New Machinery	−30	50	80

    The probability table is:

DEMAND	PROBABILITY
Increases	0.35
Constant	0.45
Decreases	0.20

    What decision path should the consulting firm recommend to maximum profits?

2. Using the information from Problem 1, change the probability table and answer the same question.

DEMAND	PROBABILITY
Increases	0.12
Constant	0.38
Decreases	0.50

3. Consider a small company that makes seat belts for automobiles. They must choose from the following choices:
   A) Continue to make their product by the conventional method.
   B) Invest modestly in the purchase of new machinery.
   C) Invest substantially in the purchase of new machinery.

   A consulting firm is hired to assist the company with their decision. The payoff table is:

   PAYOFF TABLE OF PROFITS ($ MILLIONS)

ACTION	STATE OF NATURE (DEMAND)		
	Decreases	Constant	Increases
Conventional Method	10	24	42
Modest Investment	-30	45	70
Substantial Investment	-95	60	80

   The probability table is:

DEMAND	PROBABILITY
Increases	0.14
Constant	0.54
Decreases	0.32

   What decision path should the consulting firm recommend to maximum profits?

4. Using the information from Problem 3, change the probability table and answer the same question.

DEMAND	PROBABILITY
Increases	0.4
Constant	0.4
Decreases	0.2

   Answer the same question.

5. A small family-operated restaurant is trying to decide if they should make no changes or expand their current location with a small or large addition. A consulting firm analyzed their business and the payoff table is:

   PAYOFF TABLE OF PROFITS ($ THOUSANDS)

ACTION	STATE OF NATURE (DEMAND)		
	Low	Medium	High
No Addition	30	40	50
Small Addition	24	55	60
Large Addition	-40	70	80

   The probability table for the "states of nature" is:

DEMAND	PROBABILITY
High	0.20
Medium	0.58
Low	0.22

What size of expansion, if any, should the family pursue to maximize their profits?

6. Using the information from Problem 5, change the probability table and answer the same question.

DEMAND	PROBABILITY
High	0.20
Medium	0.28
Low	0.52

7. A plastics company wants to minimize the cost of making their CD cases. A consulting firm is hired and provides the company with the following actions:
   A. Make no changes, keep the same procedures.
   B. Upgrade their machinery.

The payoff table is:

PAYOFF TABLE OF COSTS ($ MILLIONS)

ACTION	STATE OF NATURE (DEMAND)		
	Low	Medium	High
Same Procedures	22	35	50
Upgrade Machinery	4	38	66

The probability table for the "states of nature" is:

DEMAND	PROBABILITY
High	0.02
Medium	0.79
Low	0.19

Should the owners make no changes or upgrade their machinery to minimize their costs?

8. Using the information from Problem 7, change the probability table and answer the same question.

DEMAND	PROBABILITY
High	0.32
Medium	0.49
Low	0.19

9. A chemical company wants to minimize the cost of making gasoline. A consulting firm is hired and provides the company with the following actions:
   A. Make no changes, keep the same procedures.
   B. Upgrade their facility.
   C. Relocate operations to another facility.

The payoff table is:

PAYOFF TABLE OF COSTS ($ MILLIONS)

ACTION	STATE OF NATURE (DEMAND)		
	Low	Medium	High
Same Procedures	120	240	400
Upgrade	88	210	440
Relocate	60	300	600

The probability table for the "states of nature" is:

DEMAND	PROBABILITY
High	0.31
Medium	0.48
Low	0.21

Should the owners make no changes, upgrade, or relocate to minimize their costs?

10. Using the information from Problem 9, change the probability table and answer the same question.

DEMAND	PROBABILITY
High	0.22
Medium	0.60
Low	0.18

6.5 VENN DIAGRAMS
Objective: Construct Venn Diagrams to help understand multiple events and calculate their probabilities.

Mutually Exclusive

Consider the following year end survey of 100 students taking this mathematics course.

   48 students earned an "A" for this mathematics course.
   6 students earned an "F" for this mathematics course.
   No one earned an "A" and an "F" for this mathematics course.

The above information can be written as:
   A = students earning an "A" in this mathematics course = 48 students.
   F = students earning an "F" in this mathematics course = 6 students.

This situation is readily represented by a Venn Diagram. A Venn Diagram uses a rectangle to portray the universe (the sample space), the 100 students being surveyed. Within the rectangle are circles for each event, an "A" in this mathematics course and an "F" in this mathematics course. Since there are no students earning both an A and an F for this course, these circles do not overlap.

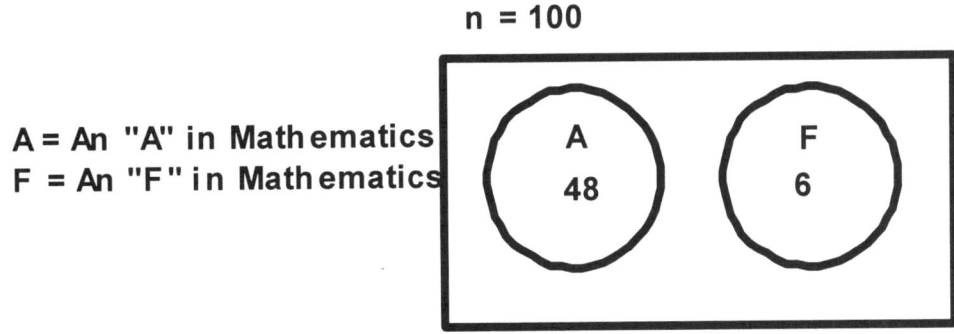

When the circles do not overlap, there is no intersection. The events are called *MUTUALLY EXCLUSIVE*. The two events do not exist at the same time.

Of the 100 students surveyed, only 54 students (48 earning an "A" and 6 earning an "F") have been placed in the circles. There are 46 students (100 − 48 − 6 = 46) missing. Place 46 outside of the "A" and "F" circles, but inside the rectangle. These 46 students earned grades different from an "A" and an "F". The complete Venn diagram is

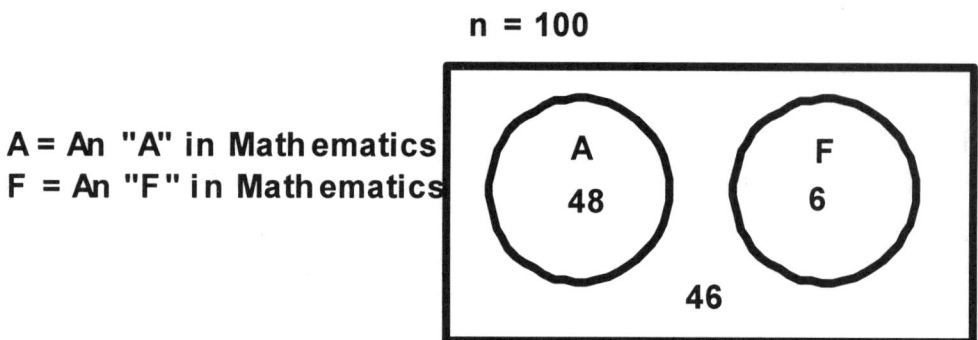

For this survey, the following probabilities can be found from the above Venn Diagram.

P(A student earning both an "A" and an "F" in this math course.) = $\frac{0}{100}$ = 0%

P(A student earning an "A" in this math course.) = $\frac{48}{100}$ = 48%

P(A student earning neither an "A" nor an "F" in this math course.) = $\frac{46}{100}$ = 46%

P(A student earning an "F" in this math course.) = $\frac{6}{100}$ = 6%

Not Mutually Exclusive
Consider the following survey of 100 students.
    60 were taking a Mathematics course.
    50 were taking a Writing course.
    20 were taking both Mathematics and Writing courses.

There appears to be a contradiction in this information. 100 students were surveyed, yet it appears that 130 students (60 Math + 50 Writing + 20 Math and Writing) were surveyed. This apparent contradiction is explainable. Some students were counted more than once. These students satisfied more than one event. This overlapping of events is called the intersection, "∩". The above information can be written as:

  M = students taking a Mathematics course = 60 students.
  W = students taking a Writing course = 50 students.
  M ∩ W = students taking both Mathematics and Writing courses = 20 students.

This situation is readily represented by a Venn Diagram. A Venn diagram uses a rectangle to portray the universe (the sample space), the 100 students being surveyed. Within the rectangle are circles for each event, Mathematics and Writing courses. Since there are students taking both courses, these circles must overlap.

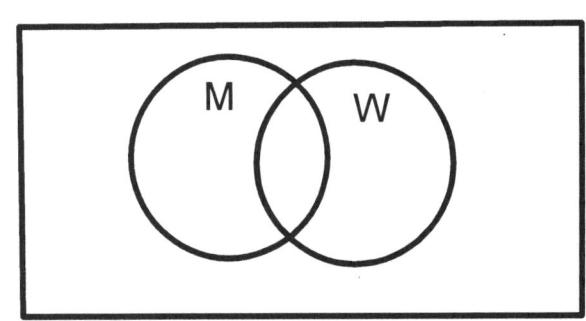

M = Mathematics
W = Writing

When the events overlap, the intersection is not zero, is called **NOT MUTUALLY EXCLUSIVE**. The events can exist simultaneously.

When entering data on the Venn diagram, list the universe, total number of students outside the rectangle. Then start with the intersection, the overlap. Place twenty students taking both Mathematics and Writing at the intersection (overlap) of the two circles. These twenty students are included in the sixty students taking a Mathematics course and the fifty students taking a Writing course. The difference indicates only forty students take Math only (60 Math – 20 Math and Writing = 40 Math only). Place forty in the remaining portion of Mathematics circle. Similarly, thirty students take Writing only (50 Writing – 20 Math and Writing = 30 Writing only). Place thirty in the remaining portion of the Writing circle.

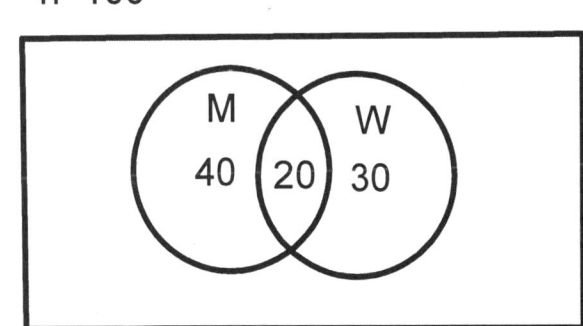

M = Mathematics
W = Writing

Of the 100 students surveyed, only 90 students (20 Mathematics and Writing + 40 Mathematics only + 30 Writing only) have been placed in Mathematics and/or Writing. Thus, ten students are missing. Place ten outside of the Mathematics and Writing circles, but inside the rectangle. These ten students are taking courses other than Mathematics and Writing. The complete Venn diagram is

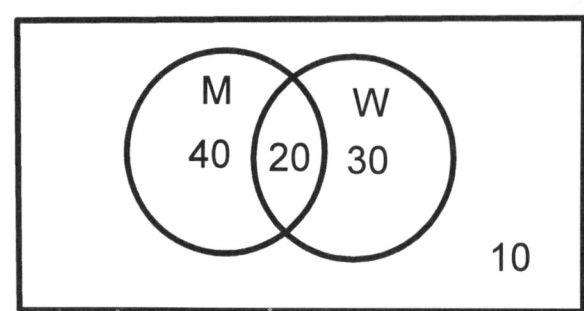

M = Mathematics
W = Writing

For this survey, the following probabilities can be found.

P(A student taking both Math and Writing) = $\frac{20}{100}$ = 20%

P(A student taking only Mathematics) = $\frac{40}{100}$ = 40%

P(A student taking neither Mathematics nor Writing) = $\frac{10}{100}$ = 10%

P(A student taking only Writing) = $\frac{30}{100}$ = 30%

Example 1: 300 people were asked about various political issues. The results were:
  120 were in favor of a balanced budget.
  110 were in favor of reducing farming subsidies.
   92 were in favor of reducing the funding of Medicare.
   65 were in favor of a balanced budget and reducing farming subsidies.
   50 were in favor of a balanced budget and reducing funding of Medicare.
   48 were in favor of reducing farming subsidies and the funding of Medicare.
   30 were in favor of all three issues.

Find the probability that an interviewed person will favor:
A. only the balanced budget.   B. only reducing farm subsidies.
C. only reducing Medicare.   D. none of these three issues.

Solution:
  Step 1: Understand and picture the problem.
     1. Find the probabilities for questions A through D.
     2. Three issues are presented: balanced budget, farm subsidies and Medicare.
     3. There are 30 people in favor of all three issues, the three issues overlap.
     4. Find the number of people who favor a particular issue or hold no opinion about these issues.

Step 2: Establish a plan.
1. Draw a Venn diagram to organize this information.
2. Draw a rectangle and place the sample space value 300 people, outside the rectangle.
3. Inside the rectangle, draw three circles that overlap. Label them Budget (B), Farming (F), and Medicare (M).
4. Start with the intersection of all three circles, label it 30 for the people favoring all three issues.
5. Calculate the values in the remaining portions of the circles by subtracting the corresponding intersection values from the value representing the issue.
6. Calculate the value outside of the circles by adding the values within each section. Then subtract this sum from the size of the sample space (universe).
7. Use the Venn Diagram to answer the questions.

Step 3: Execute the plan.
1 – 4.

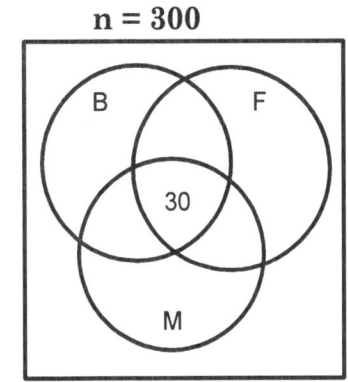

B = Balanced Budget
F = Reduce Farm Subsidies
M = Reduce Medicare

Note: The label 30 identifies that 30 people favored all three issues.

5. Calculate the remaining overlapping regions. Focus only on the circles labeled.
   a. Reduced farm subsidies, F, and Medicare, M, = 48 – 30 = 18.
   b. Balanced budget, B, and reduced Medicare, M, = 50 – 30 = 20.
   c. Balanced budget, B, and reduced farm subsidies, F, = 65 – 30 = 35.

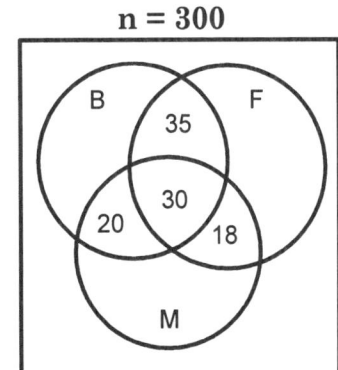

B = Balanced Budget
F = Reduce Farm Subsidies
M = Reduce Medicare

Calculate the remaining portions of the circles. Focus only on the circles labeled.
   d. Reducing Medicare, M, only = 92 − (20 + 30 + 18) = 24.
   e. Reducing farm subsidies, F, only = 110 − (35 + 30 + 18) = 27.
   f. Balanced budget, B, only = 120 − (20 + 30 + 35) = 35.

B = Balanced Budget
F = Reduce Farm Subsidies
M = Reduce Medicare

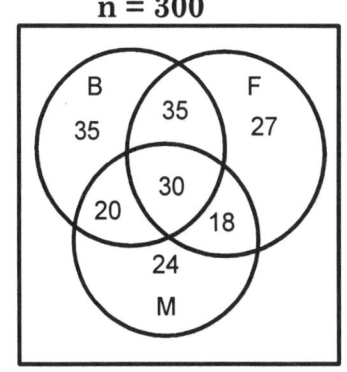

6. Subtract all values in the circles from 300.
   Other = 300 − (35 + 20 + 30 + 35 + 24 + 18 + 27) = 111

B = Balanced Budget
F = Reduce Farm Subsidies
M = Reduce Medicare

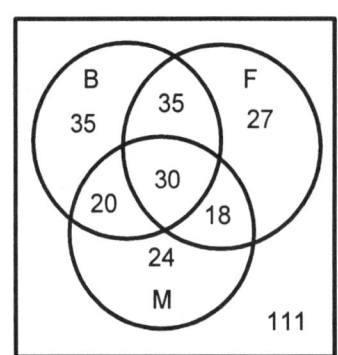

View the final Venn Diagram to read these results.
7A. 35 people favor only the balancing of the budget.

   P(person favors only the balanced budget) = $\frac{35}{300} \approx 11.7\%$

B. 27 people favor only the reduction of farm subsidies.

   P(person favors only reducing farm subsidies) = $\frac{27}{300} = 9\%$

C. 24 people favor only the reduction of Medicare.

   P(person favors only reducing Medicare) = $\frac{24}{300} = 8\%$

D. 111 people have no opinion pertaining to these issues.

   P(person has no opinion on these three issues) = $\frac{111}{300} = 37\%$

Step 4: Check your work.
- 1. Check the numbers and your calculations.
- 2. Has everyone been accounted for?
- 3. Were the questions answered?
- 4. Are the answers reasonable?

Example 2: 300 people were asked about various political issues. The results were:
  120 were in favor of a balanced budget.
  110 were in favor of reducing farming subsidies.
  92 were in favor of reducing the funding of Medicare.
  65 were in favor of a balanced budget and reducing farming subsidies.
  50 were in favor of a balanced budget and reducing funding of Medicare.
Find the probability that an interviewed person will favor:
A. only the balanced budget.   B. only reducing the farm subsidies.
C. only reducing Medicare.   D. all three of these issues.
E. none of these issues.

Solution:
Step 1: Understand and picture the problem.
- 1. Find the probabilities for questions A through E.
- 2. Three issues are presented: balanced budget, farm subsidies and Medicare.
- 3. No one favors all three issues. The three circles, representing the issues, do not share a common intersection region.
- 4. Since someone favors a balanced budget and reduced farm subsidies, the balanced budget circle overlaps the reduced farming subsidies circle.
- 5. Since someone favors a balanced budget and reduced Medicare, the balanced budget circle overlaps the reduced Medicare circle.
- 6. Answer the questions.

Step 2: Establish a plan.
- 1. Draw a Venn diagram to organize this information.
- 2. Draw a rectangle and list the 300 people outside the rectangle.
- 3. Draw three circles within the rectangle. Put Budget in the middle with Farming overlapping on one side and Medicare overlapping on the other side. Make sure that the Farming circle does not overlap the Medicare circle.
- 4. Start with the intersections of two circles. Enter the values representing those people who favor two of the three issues.
- 5. Calculate the values in the remaining portions of the circles by subtracting the corresponding intersection values from the value representing the issue.

500

6. Calculate the value outside of the circle. Add the values within each section. Then subtract this sum from the size of the sample space (universe).
7. Use the Venn Diagram to answer the questions.

Step 3: Execute the plan, 1 – 3.

B = Balanced Budget
F = Reduce Farm Subsidies
M = Reduce Medicare

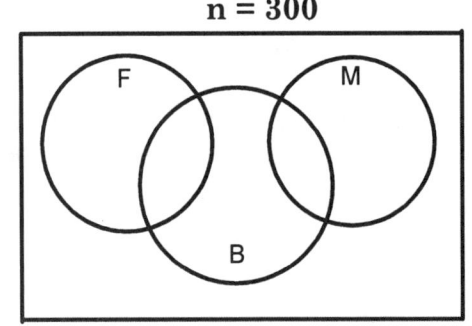

4.

B = Balanced Budget
F = Reduce Farm Subsidies
M = Reduce Medicare

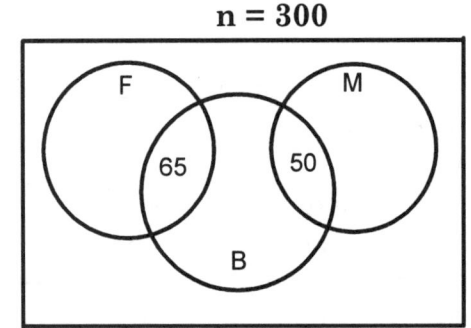

5. Focusing on the labeled circles.
   a. Reducing Medicare, M, only = 92 – 50 = 42.
   b. Reducing farm subsidies, F, only = 110 – 65 = 45.
   c. Balanced budget, B, only = 120 – (65 + 50) = 5.

B = Balanced Budget
F = Reduce Farm Subsidies
M = Reduce Medicare

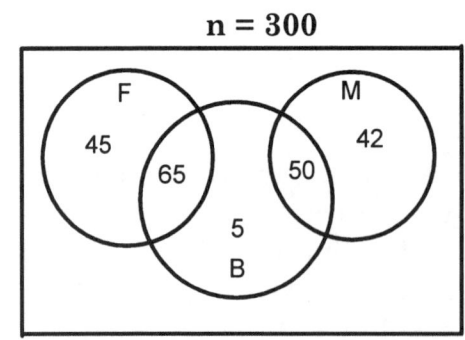

6. Subtract all values in the circles from 300.
   Other issues = 300 − (45 + 65 + 5 + 50 + 42) = 93

B = Balanced Budget
F = Reduce Farm Subsidies
M = Reduce Medicare

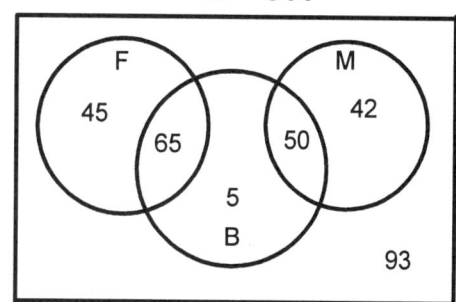

View the final Venn Diagram to read these results.

7A. 5 people favor only the balancing of the budget.

   P(person favors only the balanced budget) = $\frac{5}{300} \approx 1.7\%$

B. 45 people favor only the reduction of farm subsidies.

   P(person favors only reducing farm subsidies) = $\frac{45}{300} = 15\%$

C. 42 people favor only the reduction of Medicare.

   P(person favors only reducing Medicare) = $\frac{42}{300} = 14\%$

D. No one favors all three issues.

   P(person favors all three issues) = $\frac{0}{300} = 0\%$

E. 93 people favor none of these issues.

   P(person favors none of these issues) = $\frac{93}{300} = 31\%$

Step 4: Check your work.
   1. Check the numbers and your calculations.
   2. Has everyone been accounted for?
   3. Were the questions answered?
   4. Are the answers reasonable?

Example 3: 300 people were asked about various political issues. The results were:
120 were in favor of a balanced budget.
110 were in favor of reducing farming subsidies.
92 were in favor of reducing the funding of Medicare.
65 were in favor of a balanced budget and reducing farming subsidies.

502

Find the probability that an interviewed person will favor:
  A. only the balanced budget.
  B. only reducing farm subsidies.
  C. both the balanced budget and reducing Medicare.
  D. none of these three issues.

Solution:
  Step 1: Understand and picture the problem.
    1. Find the probabilities for questions A through D.
    2. Three issues are presented: balanced budget, farm subsidies and Medicare.
    3. No one favors all three issues. The three circles, representing the issues, do not share a common intersection region.
    4. Since someone favors a balanced budget and reduced farm subsidies, these circles overlap.
    5. Medicare is not referenced with any other issue. This circle stands alone.
    6. Find the number of people who favor a particular issue or hold no opinion about these issues.

  Step 2: Establish a plan.
    1. Draw a Venn diagram to organize this information.
    2. Draw a rectangle and list the 300 people outside the rectangle.
    3. Draw three circles within the rectangle. The Budget circle overlaps the Farming circle. The Medicare circle does not intersect the other circles.
    4. Start with the intersection of the two circles, enter the given value.
    5. Calculate the values in the remaining portions of the circles by subtracting the corresponding intersection values from the value representing the issue.
    6. Calculate the value outside of the circles by adding the values within each section. Then subtract this sum from the size of the sample space (universe).
    7. Use the Venn Diagram to answer the questions.

  Step 3: Execute the plan, 1 – 3.

B = Balanced Budget
F = Reduce Farm Subsidies
M = Reduce Medicare

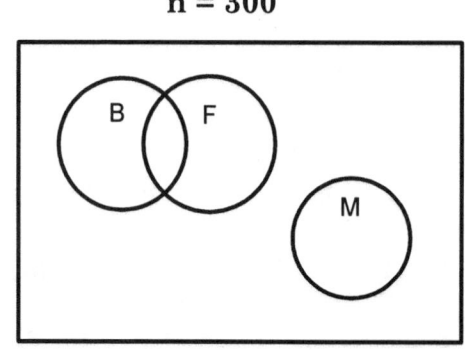

4.

B = Balanced Budget
F = Reduce Farm Subsidies
M = Reduce Medicare

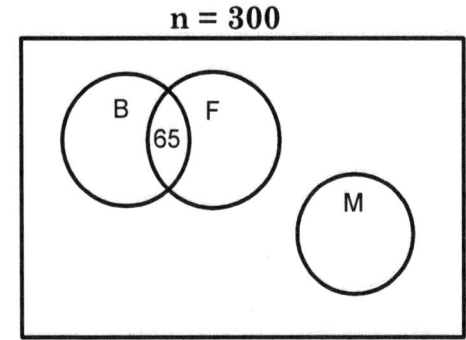

5. Focusing on the labeled circles.
   a. Reducing Medicare, M, only = 92.
   b. Reducing farm subsidies, F, only = 110 – 65 = 45.
   c. Balanced budget, B, only = 120 – 65 = 55.

B = Balanced Budget
F = Reduce Farm Subsidies
M = Reduce Medicare

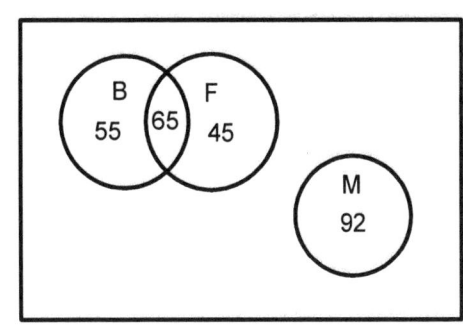

6. Other issues = 300 – (45 + 65 + 55 + 92) = 43

B = Balanced Budget
F = Reduce Farm Subsidies
M = Reduce Medicare

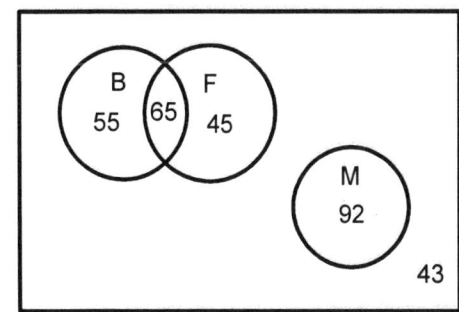

504

View the final Venn Diagram to read these results.

7A. 55 people favor only the balancing of the budget.

P(A person favoring a balanced budget) = $\frac{55}{300} \approx 18.3\%$

B. 45 people favor only the reduction of farm subsidies.

P(A person favoring reducing farm subsidies) = $\frac{45}{300} = 15\%$

C. No one favors both the balancing of the budget and the reduction of Medicare.

P(A person favoring both a balanced budget and reducing farm subsidies) = $\frac{0}{300} = 0\%$

D. 43 people do not favor any of these three issues.

P(A person does not favor any of these issues) = $\frac{43}{300} \approx 14.3\%$

Step 4: Check your work.
1. Check the numbers and your calculations.
2. Has everyone been accounted for?
3. Were the questions answered?
4. Are the answers reasonable?

MUTUALLY EXCLUSIVE

In Example 3, the Medicare circle did not overlap with either the Farming or the Balanced Budget circles. Medicare and Farming are **MUTUALLY EXCLUSIVE** events. **MUTUALLY EXCLUSIVE** events have no common (shared) elements. The Venn diagram displays them as disjoint sets, they do not overlap. Similarly, Medicare and Balanced Budget are mutually exclusive events (disjoint sets). However, Farming and Balanced Budget are not mutually exclusive, they do share common elements which is displayed as an intersection.

  Cognitive Problems

1. Draw a Venn Diagram for the following problem and find the contradiction that makes the problem impossible to solve.

In a recent survey, 250 people were asked about various political issues. The results were:

120 were in favor of a balanced budget.
110 were in favor of reducing farming subsidies.
92 were in favor of reducing the funding of Medicare.
65 were in favor of a balanced budget and reducing farming subsidies.

2. Give an example of two sets that are mutually exclusive?
3. Given an example of two sets that are not mutually exclusive?

Exercise 6.5

(c) 2006 JupiterImages Corp.

1. In a recent television survey for 8:00 PM Sunday evening, the following data was found for 200 viewers:
   120 people watched a "Science Fiction" show regularly.
   105 people watched "The Movie of the Week" regularly.
   62 people watched the "Science Fiction" show and "The Movie of the Week" regularly (Viewed one show at its regular time and watched the other show later on a VCR/DVD/TiVo.)
   Find the probability that a randomly selected viewer watches
   A. only the "Science Fiction" show on a regular basis.
   B. only "The Movie of the Week" on a regular basis.
   C. neither of these two shows.

2. In a clinical study of a prescription medication for high blood pressure, 607 people were given the medication with the following results:
   33 people complained of headaches.
   18 people experienced dizziness.
   11 people had difficulties breathing.
   8 people had both headaches and dizziness.
   6 people had both headaches and difficulty breathing.
   5 people had both dizziness and difficulty breathing.
   2 people suffered all three symptoms.
   Find the probability that a randomly selected participant experienced:
   A. only headaches.
   B. only dizziness.
   C. only difficulty breathing.
   D. none of these symptoms.

3. In a clinical study of a prescription medication for high blood pressure, 607 people were given the medication with the following results:
   33 people complained of headaches.
   18 people experienced dizziness.
   11 people had difficulties breathing.
   8 people had both headaches and dizziness.
   6 people had both headaches and difficulty breathing.

Find the probability that a randomly selected participant experienced:
A. only headaches.
B. only dizziness.
C. only difficulty breathing.
D. none of these symptoms.

4. A survey of 120 franchised fast food restaurants produced the following results:
   25 have indoor play areas for small children.
   50 have indoor public telephones.
   50 were open for breakfast.
   32 have a salad bar.
   20 have a indoor play area and indoor phone.
   14 have a salad bar and indoor phone.
   12 have a salad bar and indoor play area.
   21 have an indoor phone and were open for breakfast.
   10 have a salad bar, indoor play area and indoor phone.
   Find the probability that a randomly selected restaurant has:
   A. only an indoor phone.
   B. only a salad bar.
   C. only open for breakfast.
   D. only an indoor play area for children.
   E. none of these four categories.

5. A survey of 120 franchised fast food restaurants produced the following results:
   25 have indoor play areas for small children.
   50 have indoor public telephones.
   50 were open for breakfast.
   32 have a salad bar.
   20 have a indoor play area and indoor phone.
   14 have a salad bar and indoor phone.
   10 have an indoor phone and were open for breakfast.
   Find the probability that a randomly selected restaurant has:
   A. only an indoor phone.
   B. only a salad bar.
   C. only open for breakfast.
   D. only an indoor play area for children.
   E. none of these four categories.

6.6 "OR", THE ADDITION RULE
Objective: Find the probability of multiple events connected by an "OR".

The conjunctions "OR" and "AND" are not interchangeable. In electronics, as well as in legal documents and probability, "or" and "and" have different meanings.

Electronic Perspective
From an electronics point of review, consider the following two circuit diagrams. The light is OFF in both of these diagrams. In order to turn on the light, the electrons must follow a complete path (solid line) from the power source (battery), to the light, and back to the power source.

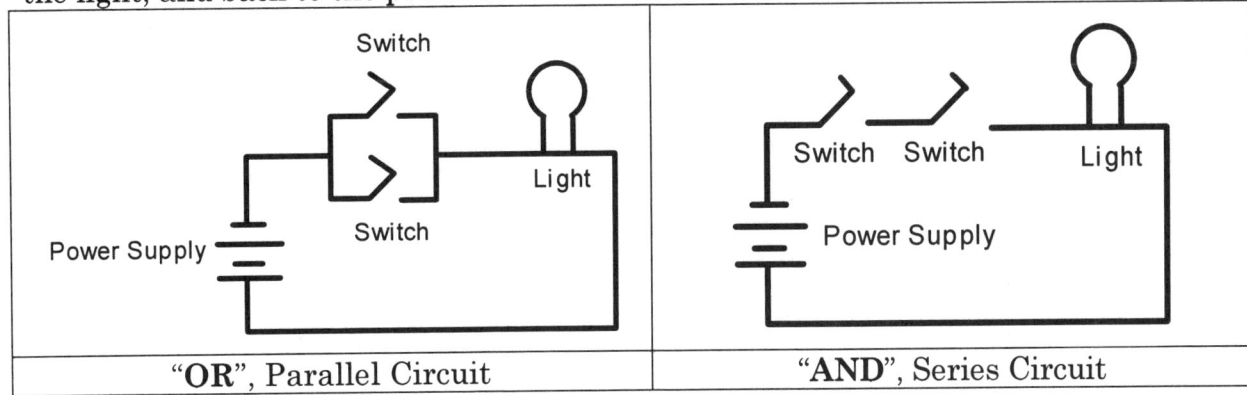

"OR", Parallel Circuit   "AND", Series Circuit

The "OR", Parallel Circuit
In the parallel **("OR")** circuit, either switch or both switches can be closed (pushed down) to complete the path (filling in the gap caused by the switch being open). This turns ON the light as illustrated in the next three figures.

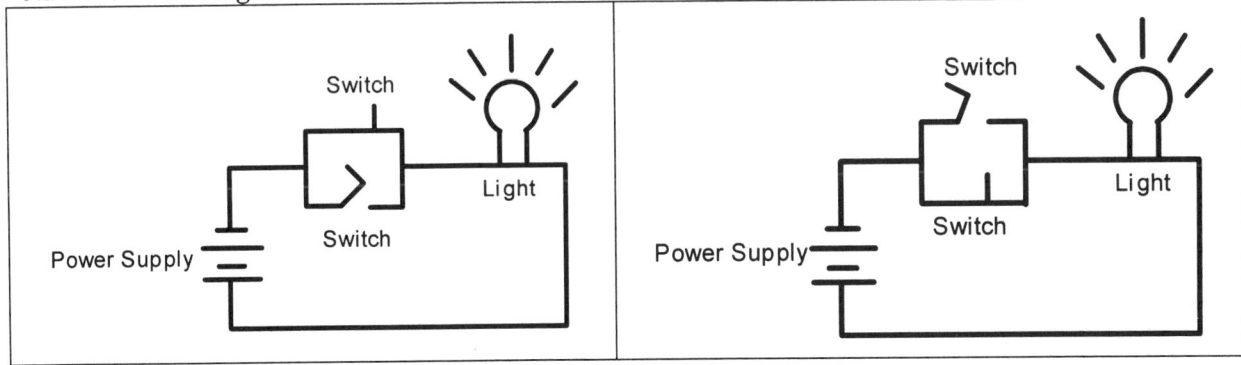

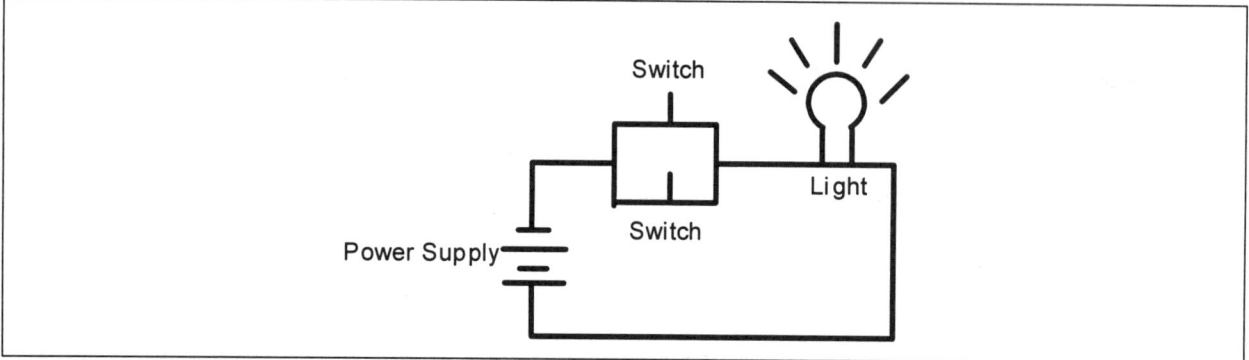

Note: In the "OR" circuit, a minimum of **ONE** switch needs to be closed to turn on the light.

The "AND", Series Circuit

In the series **("AND")** circuit **BOTH** switches need to be closed (pushed down) to turn on the light. Closing a single switch is insufficient, it does not complete the path (there is still a gap or break caused by one switch being open. This is depicted in the three figures below. Only one switch is closed in the first figure and the light is "OFF". . Only one switch is closed in the second figure and the light is "OFF". In the third figure, **BOTH** switches are closed and the light is "ON".

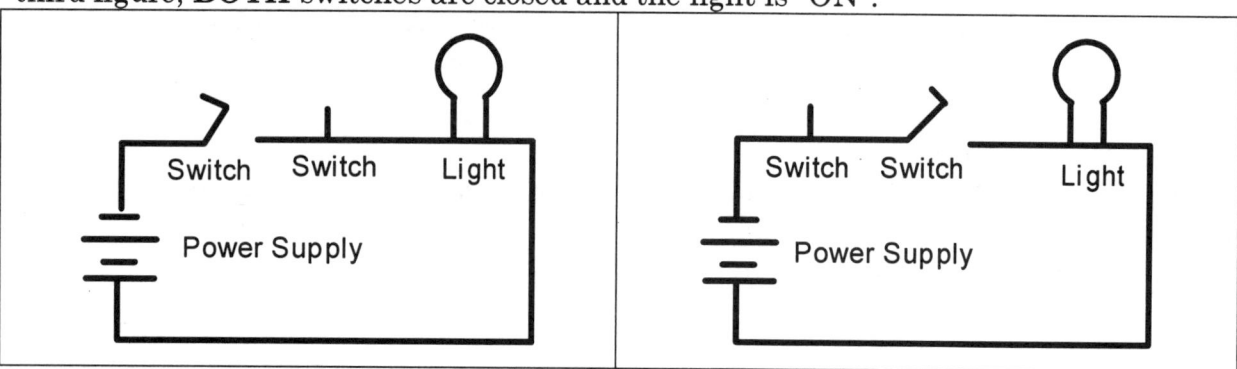

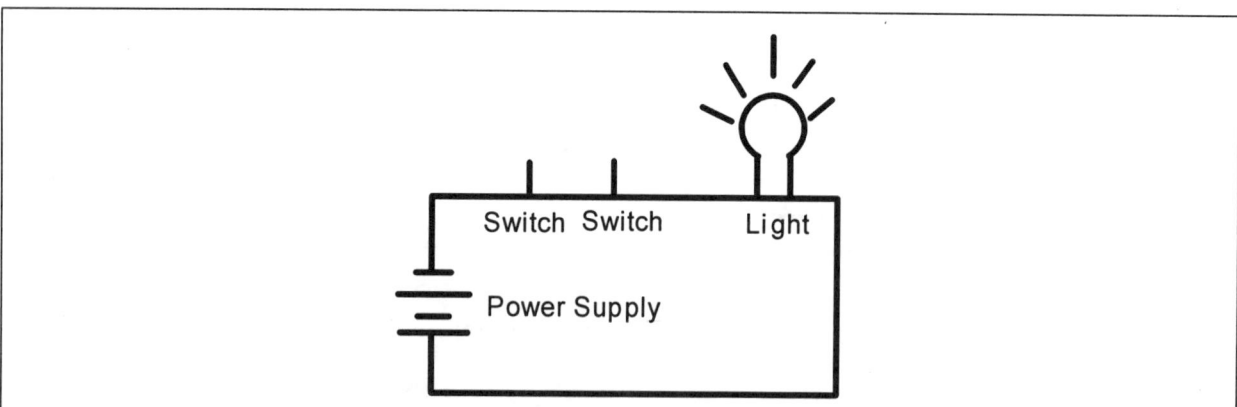

Note: In the "AND" circuit, **ALL** switches need to be closed to turn on the light.

Legal Perspective or Interpretation
The same relationship applies to legal documents. A married couple may open a joint savings account using either "Mr. **OR** Mrs." or "Mr. **AND** Mrs.". In the "Mr. **OR** Mrs." account, only 1 signature is needed to withdraw money or close the account. Either signature is sufficient authorization for the removal of money from an "**OR**" account. In the "Mr. **AND** Mrs." account, both signatures are needed to withdraw money from the account. A single signature is legally insufficient authorization for the withdrawal of money from an "**AND**" account.

Probability Perspective or Interpretation
Probability follows the same structure. P(A or B) implies the probability of either "event A" or "event B" occurring. P(A and B) implies the probability of both "event A" and "event B" occurring. Consider the two scenarios:

Scenario 1:
One hundred students were surveyed.
    34 were taking a Mathematics course.
    48 were taking a Writing course.
    No one was taking both Mathematics and Writing courses.

Since no student is enrolled in both Mathematics and Writing courses, the events are **mutually exclusive** (there is no overlap). The Venn Diagram is

M = Mathematics
W = Writing

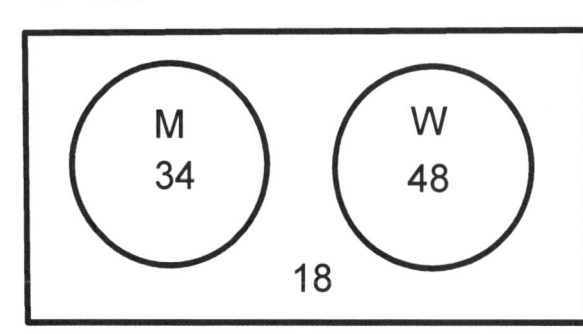

P(A student enrolled in a Mathematics AND Writing course) = 0

Find P(A student enrolled in a Mathematics OR Writing course).
P(A student enrolled in a Mathematics OR Writing course) =
The sum of the numbers of students in the Mathematics and Writing circles.
Using the Venn Diagram: P(Mathematics OR Writing) = $\frac{34+48}{100} = \frac{82}{100} = 0.82$.

Scenario 2:
One hundred students were surveyed.
   60 were taking a Mathematics course.
   50 were taking a Writing course.
   20 were taking both Mathematics and Writing courses.

Twenty students are enrolled in both Mathematics and Writing course. There is an overlap, the events are not mutually exclusive.. The Venn diagram is

M = Mathematics
W = Writing

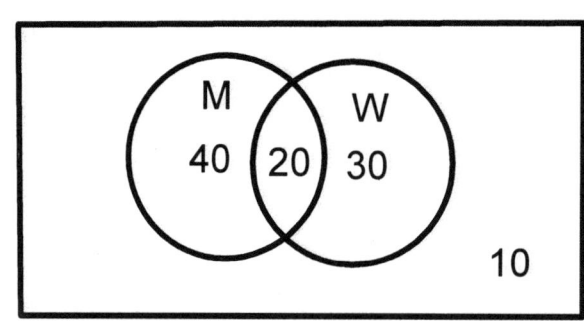

P(A student enrolled in a Mathematics AND Writing course) = $\dfrac{20}{100} = \dfrac{1}{5}$.

Find P(A student enrolled in a Mathematics OR Writing course)
P(A student enrolled in a Mathematics OR Writing course) =
The sum of the number of students in the Mathematics and Writing circles.
Using the Venn Diagram: P(Mathematics **or** Writing) = $\dfrac{40+20+30}{100} = \dfrac{90}{100} = 0.90$.

An alternative method to solving this problem is to use the formula

---
P(A OR B), The Addition Rule

If event A and event B are mutually exclusive,
then P(A **OR** B) = P(A) + P(B)

If event A and event B are not mutually exclusive, then
P(A **OR** B) = P(A) + P(B) — P(A **AND** B)

Refer back to Scenario 1.
Find P(A student enrolled in a Mathematics OR Writing course).
Since the events are mutually exclusive, the formula is P(A **OR** B) = P(A) + P(B).
P(A student enrolled in a Mathematics OR Writing course) =
P(Math) + P(Writing) = $\dfrac{34+48}{100} = \dfrac{82}{100} = 0.82$.

Refer back to Scenario 2.
Find P(A student enrolled in a Mathematics OR Writing course)
Since the events are not mutually exclusive, the formula is
P(A **OR** B) = P(A) + P(B) — P(A **AND** B).

P(Math) = $\dfrac{60}{100}$,   P(Writing) = $\dfrac{50}{100}$,   P(Math and Writing) = $\dfrac{20}{100}$.

P(Mathematics **or** Writing) = P(Math) + P(Writing) − P(Math and Writing).

P(Mathematics **or** Writing) = $\dfrac{60 + 50 - 20}{100} = \dfrac{90}{100} = 0.90$.

The subtraction of P(A **AND** B) removes the one count of the 20 students that was counted twice. These 20 students are part of the 60 students taking Mathematics and they are also part of the 50 students taking Writing. Each student must only be counted once. This double counting is corrected by subtracting the shared count.

Example 1:
300 people were asked about various political issues. The results were:
   120 were in favor of a balanced budget.
   110 were in favor of reducing farming subsidies.
   92 were in favor of reducing the funding of Medicare.
   65 were in favor of a balanced budget and reducing farming subsidies.
Use a Venn Diagram to find the P(Balanced budget **OR** reducing farming subsidies) when a single person is selected.
Solution:
 Step 1: Understand and picture the problem.
   1. Three issues are presented: balanced budget, farm subsidies and Medicare.
   2. No one favors all three issues. The three circles, representing the issues, do not share a common intersection region.
   3. Since someone favors a balanced budget and reduced farm subsidies, these circles overlap.
   4. Medicare is not referenced with any other issue. This circle stands alone.
   5. Find the sum value of people in the balanced budget and reducing farming subsidies circles.

 Step 2: Establish a plan.
   1. Draw a Venn diagram to organize this information.
   2. Draw a rectangle and list the 300 people outside the rectangle.
   3. Draw three circles within the rectangle. The Budget circle overlaps the Farming circle. The Medicare circle does not intersect the other circles.
   4. Start with the intersection of the two circles, enter the given value.

5. Calculate the values in the remaining portions of the circles by subtracting the corresponding intersection values from the value representing the issue.
6. Calculate the value outside of the circles by adding the values within each section. Then subtract this sum from the size of the sample space (universe).
7. Use the Venn Diagram to answer the question.

Step 3: Execute the plan, 1 – 3.

B = Balanced Budget
F = Reduce Farm Subsidies
M = Reduce Medicare

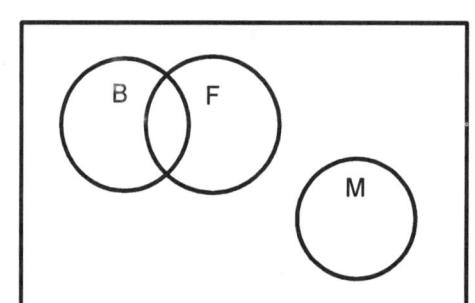

4.

B = Balanced Budget
F = Reduce Farm Subsidies
M = Reduce Medicare

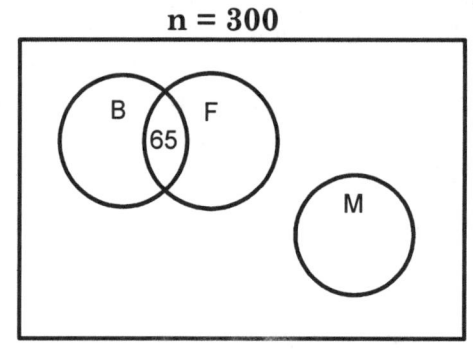

5. Reducing Medicare only = 92.
Reducing farm subsidies only = 110 – 65 = 45.
Balanced budget only = 120 – 65 = 55.

B = Balanced Budget
F = Reduce Farm Subsidies
M = Reduce Medicare

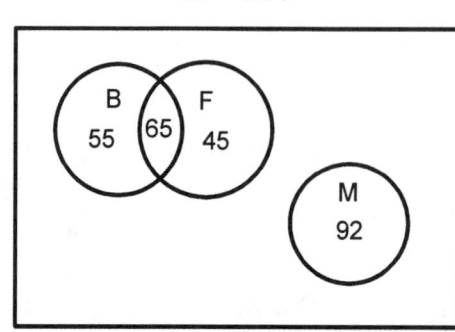

6. Other issues = 300 − (45 + 65 + 55 + 92) = 43.

B = Balanced Budget
F = Reduce Farm Subsidies
M = Reduce Medicare

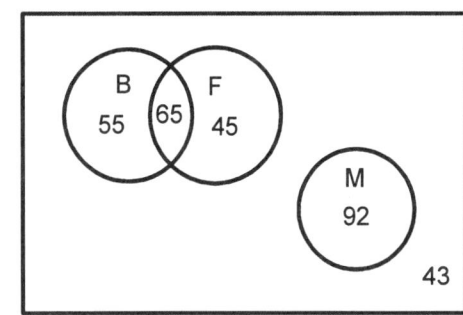

n = 300

7. Using the Venn Diagram:
P(Balanced budget **or** reducing farm subsidies) =
$\frac{55+65+45}{300} = \frac{165}{300} = 0.55$.

Step 4: Check your work.
1. Check the numbers and your calculations.
2. Was the question answered?
3. Is the answer reasonable?

Example 2:
Repeat Example 1 using the Addition Rule to find the P(Balanced budget **OR** reducing farming subsidies) when a single person is selected.
Solution:
Step 1: Understand and picture the problem.
1. Question: Find P(Balanced budget or reducing farm subsidies).
2. An "OR" problem implies that the student favors either the balanced budget or reducing farm subsidies or favors both issues.
3. Since balanced budget and reducing farm subsidies are not mutually exclusive, their overlap must be taken into account.

Step 2: Devise a plan.
The formula or the Venn diagram can be used to solve this problem.
Use the formula: P(A or B) = P(A) + P(B) − P(A and B)

Step 3: Execute the plan.
P(Balanced budget or reducing farm subsidies) =
P(Balanced budget) = $\frac{120}{300}$, P(Farm) = $\frac{110}{300}$,
P(Balanced Budget and Farm) = $\frac{65}{300}$.

P(Balanced) + P(Farm) − P(Balanced and Farm) = $\frac{120+110-65}{300} = \frac{165}{300}$.

P(Balanced) + P(Farm) − P(Balanced and Farm) = 0.55.

There is a 55% chance that a selected person favors a balanced budget or reducing farm subsidies.

Step 4: Check your work.
1. Check the numbers and your calculations.
2. Was the question answered?
3. Is the answer reasonable?

Example 3:
Use the Venn Diagram drawn in Example 1 to find the P(Balanced budget **OR** reducing Medicare) when a single person is selected.

Solution:
Step 1: Understand and picture the problem.
1. Question: Find P(Balanced budget or reducing Medicare).
2. An "OR" problem implies that the student favors either the balanced budget or reducing farm subsidies or favors both issues.

Step 2: Devise a plan.
Find the sum value of people in the balanced budget and Medicare circles.

Step 3: Execute the plan.

P(Balanced budget or Medicare) =

P(Balanced budget) = $\frac{120}{300}$, P(Reducing Medicare) = $\frac{92}{300}$,

P(Balanced budget) + P(Reducing Medicare) = $\frac{120+92}{300} = \frac{212}{300}$.

P(Balanced budget) + P(Reducing Medicare) ≈ 0.7066667.

There is a 70.7% chance that a selected person favors a balanced budget or reducing Medicare.

Step 4: Check your work.
1. Check the numbers and your calculations.
2. Was the question answered?
3. Is the answer reasonable?

Example 4:
Repeat Example 1 using the Addition Rule to find the P(Balanced budget **OR** reduce Medicare) when a single person is selected.

Solution:
Step 1: Understand and picture the problem.
1. Question: Find P(Balanced budget or reducing Medicare).
2. An "OR" problem implies that the student favors either the balanced budget or reducing farm subsidies or favors both issues.
3. Since balanced budget and reducing farm subsidies are mutually exclusive, their overlap is zero.

Step 2: Devise a plan.
The formula or the Venn diagram can be used to solve this problem.
Use the formula: P(A or B) = P(A) + P(B).

Step 3: Execute the plan.
P(Balanced budget or Medicare) =

P(Balanced budget) = $\frac{120}{300}$, P(Reducing Medicare) = $\frac{92}{300}$,

P(Balanced budget) + P(Reducing Medicare) = $\frac{120+92}{300} = \frac{212}{300}$,

P(Balanced budget) + P(Reducing Medicare) ≈ 0.7066667.
There is a 70.7% chance that a selected person favors a balanced budget or reducing Medicare.

Step 4: Check your work.
1. Check the numbers and your calculations.
2. Was the question answered?
3. Is the answer reasonable?

Example 5:
A survey of 1,000 people produced the following results:

Gender	Smoker	Non-smoker
Male	280	220
Female	210	290

If a single person is selected, find the P(Female **OR** Smoker).
Solution:
Step 1: Understand and picture the problem.
1. Question: P(Female or Smoker).
2. An "OR" problem implies that the person is a female or a smoker or a female smoker.
3. Since Female and Smoker are not mutually exclusive, their overlap must be taken into account.

Step 2: Devise a plan.
1. Find the totals of each row and column in the table.
2. Use the formula: P(A or B) = P(A) + P(B) − P(A and B)

Step 3: Execute the plan.
1.

Gender	Smoker	Non-smoker	Totals
Male	280	220	500
Female	*210*	290	*500*
Totals	*490*	510	1,000

2. $P(\text{Female}) = \dfrac{500}{1,000}$, $P(\text{Smoker}) = \dfrac{490}{1,000}$,

$P(\text{Female and Smoker}) = \dfrac{210}{1,000}$,

P(Female or Smoker) = P(Female) + P(Smoker) − P(Female and smoker)

$P(\text{Female or Smoker}) = \dfrac{500 + 490 - 210}{1,000} = \dfrac{780}{1,000} = 0.780.$

The probability of selecting a female or smoker is 78%.

Step 4: Check your work.
1. Check the numbers and your calculations.
   P(Female) + P(Smoker) − P(Female and smoker) =
2. Was the question answered?
3. Is the answer reasonable?

(c) 2006 JupiterImages Corp.

Optional Student Project

Project "Venn Diagrams and Probability" can be found on page 532.

? Cognitive Problems ?

1. Explain the difference between "and" and "or" and give one example of each.
2. What are the implications of a car warranty worded, 5 years **AND** 60,000 miles? Would your car still be under warranty after:
   a. 3 years, if it had 72,000 miles? Explain.
   b. 7 years, if it had 46,000 miles? Explain.

3. The premise of a movie requires that the key of the captain **AND** the key of the second in command be used in order to launch a nuclear missile. Would a single key be sufficient in launching the missile? Explain.

Exercise 6.6

(c) 2006 JupiterImages Corp.

For problem 1 through 4, use a Venn Diagram to find the following probabilities. For probabilities using "or", use either the Venn Diagram or the formula for "or".

1. In a recent television survey for 8:00 PM Sunday evening, the following data was found for 200 viewers:

    120 people watched a "Science Fiction" show regularly.
    105 people watched "The Movie of the Week" regularly.
    62 people watched the "Science Fiction" show and "The Movie of the Week" regularly. (They watched one show at its regular time and watched the other show later on a VCR.)

    If a person is randomly selected from this group, find:
    A. P("Science Fiction" show).
    B. P("Movie of the Week").
    C. P("Science Fiction" or "Movie of the Week").
    D. P(Neither of these two shows).

2. In a clinical study of a prescription medication for high blood pressure, 607 people were given the medication:

    33 people complained of headaches.
    18 people experienced dizziness.
    11 people had difficulties breathing.
    8 people had both headaches and dizziness.
    6 people had both headaches and difficulty breathing.
    5 people had both dizziness and difficulty breathing.
    2 people suffered all three symptoms.

    If a person is randomly selected from this group, find:
    A. P(only headaches).
    B. P(only dizziness).
    C. P(only difficult breathing).
    D. P(headaches or dizziness).
    E. P(headaches or difficulty breathing).
    F. P(dizziness or difficulty breathing).
    G. P(none of these symptoms).

3. In a clinical study of a medication for high blood pressure, 607 people were tested.
   33 people complained of headaches.
   18 people experienced dizziness.
   11 people had difficulties breathing.
   8 people had both headaches and dizziness.
   6 people had both headaches and difficulty breathing.
   If a person is randomly selected from this group, find:
   A. P(only headaches).
   B. P(only dizziness).
   C. P(only difficulty breathing).
   D. P(headaches or dizziness).
   E. P(headaches or difficulty breathing).
   F. P(dizziness or difficulty breathing).
   G. P(none of these symptoms).

4. A survey of 120 fast food restaurants produced the following results:
   25 had indoor play areas for small children.
   50 had indoor public telephones.
   50 were open for breakfast.
   32 had a salad bar.
   20 had a indoor play area and indoor phone.
   14 had a salad bar and indoor phone.
   10 had an indoor phone and were open for breakfast.
   If a restaurant is randomly selected from this group, find:
   A. P(only an indoor phone).
   B. P(only a salad bar).
   C. P(only were open for breakfast).
   D. P(only an indoor play area for children).
   E. P(indoor phone or indoor play area).
   F. P(indoor phone or open for breakfast).
   G. P(indoor phone or salad bar).
   H. P(indoor play area or open for breakfast).
   I. P(indoor play area or salad bar).
   J. P(open for breakfast or salad bar).
   K. P(none of these four categories).

5. In a clinical study of a new allergy medicine, the following reactions were noted:

Reactions to the Medication

Medication	Drowsiness	Headache	Dizziness	Nausea
New Medication	116	110	32	41
Old Medication	67	11	3	4
Placebo	71	103	20	32

If a patient is randomly selected from this study group, find:
A. P(Drowsy reaction or took a Placebo).
B. P(Took the new prescribed medication or had a headache).
C. P(Became Nauseous or took the old medication).
D. P(Had a headache or became dizzy).

6. In a clinical study of a new prescription allergy medicine, the following reactions to medications were noted:

Reaction to the Medication

Medication	Insomnia	Headache	Blurred Vision	Rash
New Medication	97	65	4	4
Old Medication	77	50	1	0
Placebo	12	64	2	0

If a patient is randomly selected from this study group, find:
A. P(Suffered from Blurred Vision or took a Placebo).
B. P(Took the newly prescribed medication or had a headache).
C. P(Experienced a rash or took the old medication).
D. P(Had a headache or suffered from insomnia).

7. The Bureau of Labor Statistics of the US Department of Labor reported that the distribution of wages paid to hourly workers was

Gender and Age	Wage < $10/hour	Wage at least $10/hour
Male, 16 years and older	17,512,000	14,187,000
Female, 16 years and older	23,126,000	8,491,000

If one person is randomly selected from this report, find:
A. P(Male or earns less than $10.00 per hour)
B. P(Female or earns at least $10.00 per hour)
C. P(Male or earns at least $10.00 per hour)
D. P(Female or earns less than $10.00 per hour)

6.7 "AND", THE MULTIPLICATION RULE
Objective: Find the probability of multiple events connected by an "AND"

There are many words in the English language that possess different definitions. The only way to comprehend how the word is being used is to study the word in its context. Consider the word **"scale"**.
1. It can be a verb meaning to climb, *They will attempt to "scale" Mt. Everest.*
2. It can be a noun referring to a:
   a. part of a fish, *The fish "scale" has to be removed before cooking.*
   b. measuring device, *The infant was placed on a "scale" to determine her weight.*
   c. graduated series:
      i. *the numbers on the horizontal axis of the graph is in thousands.*
      ii. *a musical scale.*

Without a sentence containing "scale", it is impossible to distinguish which definition of "scale" to use. In the context of a sentence, the particular meaning of "scale" can be determined.

In Statistics, "and" is such a word with multiple definitions. In the previous section, "and" meant events that happen at the **SAME** instant, like a student who took both Mathematics and Writing courses in the same semester. In this section, another definition for "and" is presented. "And" means distinct events that happen at **DIFFERENT** times. First Event A happens followed by Event B, in that order. There are two rules for "and" in this context. One rule is for independent events and another rule is for dependent events.

Independent Events

> **DEFINITION OF INDEPENDENT EVENTS**
> If the occurrence of the first event (A) has no influence on the occurrence of the second event (B), then events A and B are independent.
> **Note:** In this case it does not matter which event occurs first.

As an example of two independent events, consider the flipping of a coin and the tossing of a single die: Find P(Heads on a coin and a 3 on the die). One way to solve this problem is to list the sample space:

(c) 2006 JupiterImages Corp

Heads and 1	Heads and 2	***Heads and 3***	Heads and 4
Heads and 5	Heads and 6	Tails and 1	Tails and 2
Tails and 3	Tails and 4	Tails and 5	Tails and 6

$$P(\text{Heads on the coin and a 3 on the die}) = \frac{1}{12} \approx 0.083$$

The outcome of the coin flip (heads or tails) has no influence on the outcome of the die toss (1, 2, 3, 4, 5, or 6). If events are "independent", then the following rule applies for "and."

> **MULTIPLICATION RULE FOR INDEPENDENT EVENTS**
> If A and B are independent, then P(A and B) = P(A) × P(B)

Use the Multiplication Rule for Independent Events:
P(Heads on a coin and 3 on the die) = P(Heads on a coin) × P(3 on a die).

$$\frac{1}{12} = \left(\frac{1}{2}\right)\left(\frac{1}{6}\right)$$

$$\frac{1}{12} = \frac{1}{12}$$

The multiplication rule for independent events yields the same result as the sample space approach.

Dependent Events
If two events are not independent, then they are dependent.

> **DEFINITION OF DEPENDENT EVENTS**
> If the occurrence of the first event (A) does influence the occurrence of the second event (B), then events A and B are dependent.
> **Note:** In this case it does matter which event occurs first.

(c) 2006 JupiterImages Corp

To demonstrate the difference between independent and dependent events, consider a jar of candy. A small package (1.74 oz.) of candy contains 7 red, 4 yellow, 7 orange, 2 green and 3 brown candies.

The probabilities of the different colors are:
P(Red) = $\frac{7}{23}$, P(Yellow) = $\frac{4}{23}$, P(Orange) = $\frac{7}{23}$, P(Green) = $\frac{2}{23}$, P(Brown) = $\frac{3}{23}$.

Find P(Selecting a red candy and a second red candy).

The selection of 2 red candies will be presented in two scenarios:
1. With replacement, the independent events model.
2. Without replacement, the dependent events model.

Independent, with replacement
A candy is selected, it is then returned to the package before the second selection is made (with replacement). Since the first candy is returned, the composition of the package and the probabilities of the different colors remain the same. The first selection had no influence on the second pick. Thus, these events are independent. Use the Multiplication Rule for Independent Events.

$$P(\text{Red candy and red candy}) = P(\text{Red candy}) \times P(\text{Red candy}) = \frac{7}{23} \times \frac{7}{23} = \frac{49}{529} \approx 0.093.$$

Dependent, without replacement
A candy is selected and is discarded (set aside or eaten). The composition of the candy for the second selection is different from the composition of the candy for the first selection. The first selection influences the second selection. The first selection and the second selection are dependent events. The following rule applies for "and" when the events are dependent.

THE MULTIPLICATION RULE FOR DEPENDENT EVENTS

If A and B are dependent, then $P(A \text{ and } B) = P(A) \times P(B|A)$
where: 1. A is the first event
2. $P(B|A)$ is read "the probability of event B GIVEN that event A has taken place."

In this situation the candy is not replaced. The "without replacement" probabilities for the first pick are still:

$$P(\text{Red}) = \frac{7}{23}, \quad P(\text{Yellow}) = \frac{4}{23}, \quad P(\text{Orange}) = \frac{7}{23}, \quad P(\text{Green}) = \frac{2}{23}, \quad P(\text{Brown}) = \frac{3}{23}.$$

These probabilities change for the second pick. Since the focus is on the red candy, assume a red candy has been removed. Adjustments must be made to the sample size and the number of red candies. There is one less candy in the sample space (23 − 1 = 22) and there is one less red candy (7 − 1 = 6). Thus, the probabilities for the second pick are:

$$P(\text{Red}) = \frac{6}{22} = \frac{3}{11}, \quad P(\text{Yellow}) = \frac{4}{22} = \frac{2}{11}, \quad P(\text{Orange}) = \frac{7}{22},$$

$$P(\text{Green}) = \frac{2}{22} = \frac{1}{11}, \quad P(\text{Brown}) = \frac{3}{22}.$$

P(Red candy and red candy) =

$$P(\text{First candy red}) \times P(\text{Second candy red | first candy red}) = \frac{7}{23} \times \frac{3}{11} = \frac{21}{253} \approx 0.083$$

Example 1: A couple plans on having four children. What is the probability that all of the children will be female?
Solution:
  Step 1: Understand and picture the problem.
   1. Question: Find the probability of the couple having four females out of four children.
   2. Assuming the gender of one child has no influence on the gender of another child, the events are independent.

  Step 2: Establish a plan.
   Use the multiplication rule of independent events.
   P(Female) × P(Female) × P(Female) × P(Female)

  Step 3: Execute the plan.
   P(Female and female and female and female) =
   $\frac{1}{2} \times \frac{1}{2} \times \frac{1}{2} \times \frac{1}{2} = \left(\frac{1}{2}\right)^4 = \frac{1}{16} \approx 0.063$.
   The probability of having 4 females out of 4 children is approximately 6.3%.

  Step 4: Check your work.
   1. Check your numbers and calculations.
   2. Was the question answered?
   3. Is the answer reasonable?

Example 2: Twelve applicants, 8 males and 4 females, apply for 2 similar jobs. What is the probability that the 2 job vacancies will be filled by men?
Solution:
  Step 1: Understand and picture the problem.
   1. Question: Find the probability of 2 men being hired from 12 applicants consisting of eight men.
   2. Since the hiring of one person changes the composition of the remaining applicants, the events are dependent.

  Step 2: Establish a plan.
   Use the multiplication rule for dependent events.
   P(Male and male) =
   P(Male hired first) × P(Male hired second | male hired first).

  Step 3: Execute the plan.
   P(male and male) = $\frac{8}{12} \times \frac{7}{11} = \frac{56}{132} \approx 0.424$.
   The P(The two job vacancies are filled by men) = 42.4%.

Step 4: Check your work.
: 1. Check your numbers and calculations.
: 2. Was the question answered?
: 3. Is the answer reasonable?

Example 3: Based upon records from last year, a fire department found that 8% of its calls were false alarms. Find the probability that the next three calls will **all** be legitimate.
Solution:
Step 1: Understand and picture the problem.
: 1. Question: When 8% of the alarms are false, what is the probability that the next three alarms are not false alarms?
: 2. Since responding to one alarm has no bearing on the legitimacy of the next alarm, the events are independent.

Step 2: Establish a plan.
: 1. Use the complement rule, 1 − P(False alarm) = P(Legitimate).
: 2. Use the multiplication rule for independent events.
: P(Legit. and legit. and legit.) = P(Legit.) × P(Legit.) × P(Legit.)

Step 3: Execute the plan.
: 1. P(Legitimate alarm) = 1 − 0.08 = 0.92.
: 2. P(Legit. and legit. and legit.) = (0.92)(0.92)(0.92) = $(0.92)^3 \approx 0.779$.
: The P(The next three calls will be legitimate) = 77.9%.

Step 4: Check your work.
: 1. Check your numbers and calculations.
: 2. Was the question answered?
: 3. Is the answer reasonable?

Example 4: Four video cameras in a shipment of 50 cameras are inadvertently broken during shipping. The store receiving the shipment selects 2 different cameras for testing. If both cameras pass the test, then the shipment is accepted. What is the probability the store accepts the shipment of cameras.
Solution:
Step 1: Understand and picture the problem.
: 1. Question: In a shipment of 50 cameras, 4 are broken, find the probability of selecting two different cameras and both are good, not broken.
: 2. Since the first camera tested is set aside, (That means 2 different cameras are tested), the events are dependent.

Step 2: Establish a plan.
1. Use the complement rule, 1 − P(Broken camera) = P(Good camera).
2. Find P(Second camera is good | first camera is good)
3. Use the multiplication rule for dependent events.
P(Good and good) =
P(First camera is good) × P(Second camera is good | first camera is good)

Step 3: Execute the plan.
1. P(Good camera) = $1 - \frac{4}{50} = \frac{46}{50}$.

2. P(Second camera is good | first camera is good) = $\frac{45}{49}$.

3. P(Good and good) = $\frac{46}{50} \times \frac{45}{49} = \frac{2{,}070}{2{,}450} \approx 0.845$.

The P(The store accepts the camera shipment) = 84.5%.

Step 4: Check your work.
1. Check your numbers and calculations.
2. Was the question answered?
3. Is the answer reasonable?

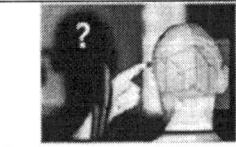
(c) 2006 JupiterImages Corp.

Chapter Review
Chapter review problems can be found on page 534.

? Cognitive Problems ?

1. What is meant by 2 events being independent?
2. What is meant by 2 events being dependent?
3. Give one example of 2 independent events.
4. Give one example of 2 dependent events.
5. Explain when to use the addition rule vs. the multiplication rule. Give examples of each situation.

Exercise 6.7

(c) 2006 JupiterImages Corp.

Write your answer in decimal form rounded to the thousandths place.
1. Use the sample space for the tossing of a pair of dice.

525

	DIE 2					
DIE 1						
	2	3	4	5	6	7
	3	4	5	6	7	8
	4	5	6	7	8	9
	5	6	7	8	9	10
	6	7	8	9	10	11
	7	8	9	10	11	12

A. P(First toss is 7 and then another 7)   B. P(First toss is 7 and then an 11).
C. P(First toss is 7 and then Not a 7).   D. P(First toss is 4 and then another 4).

2. In a 2.6 oz. box of candy, there are 12 Orange, 4 Red, 5 Yellow, 6 Green, and 4 Brown candies. If two candies are selected, **with replacement**, find the:
A. P(First Green and then another Green).   B. P(First Orange and then a Green).
C. P(First Red and then Not a Yellow).   D. P(First Not Red and second Not Red).

3. In a 2.6 oz. box of candy, there are 12 Orange, 4 Red, 5 Yellow, 6 Green, and 4 Brown candies. If two candies are selected, **without replacement**, find the:
A. P(First Green and then another Green).   B. P(First Orange and then a Green).
C. P(First Red and then Not a Yellow).   D. P(First Not Red and then Not Red).

4. In a clinical study of a prescription allergy medicine, 116 patients out of 1,630 patients experienced drowsiness after taking the medication. If two different patients are taken from this study group, find the probability they suffered these side effects.
A. P(Both patients experienced drowsiness).
B. P(Neither patient experienced drowsiness).

5. In a clinical study of the prescribed allergy medicine, 109 patients out of 1,630 patients experienced headaches after taking the medication. If two different patients are taken from this study group, find the probability they suffered these side effects.
A. P(Both patients had headaches).   B. P(Neither experienced headaches).

6. In a clinical study of the prescribed allergy medicine, 32 patients reported dizziness as a side effect of the medication and 1,598 patients did not experience dizziness. If two different patients are taken from this group, find the probability they suffered these side effects.
   A. P(Both patients suffered dizziness).   B. P(Neither patient suffered dizziness).

7. In a clinical study of the prescribed allergy medicine, 85 patients reported dry mouth as a side effect of the medication and 1,545 patients did not experience dry mouth. If two different patients are taken from this study group, find the probability they suffered these side effects.
   A. P(Both patients experienced dry mouth).
   B. P(Neither patient experienced dry mouth).

8. The National Safety Council reported the following deaths due to motor vehicles.

Cause of Motor Vehicle Death	Frequency
Collision with another motor vehicle	17,900
Collision with a fixed object	11,900
Pedestrian accidents	6,200
Noncollision accidents	4,500
Collision with a pedalcycle	800
Collision with a train	600
Other (hitting a deer, etc.)	100

   If two different deaths are taken from this report, without replacement, find the:
   A. P(Both deaths involve collisions with a train).
   B. P(Neither death is a pedestrian accident).
   C. P(Both deaths involve collisions with another motor vehicle).
   D. P(Neither death involve collisions with a fixed object).
   E. P(Both deaths are due to other circumstances).

9. The Bureau of Labor Statistics of the US Department of Labor reports that 10% of fatal occupational accidents are caused by falls. If three different reports on fatal occupational accidents are selected, find the:
   A. P(All three deaths are due to falls).
   B. P(None of the three deaths are due to falls).

10. The US Department of Education reports that 71.2% of students graduate from high school. If four people apply for a job, find the:
    A. P(All four of the people graduated from high school).
    B. P(None of the four people graduated from high school).

Project: Raffle

Objective: Calculate probabilities and odds for a fund raising raffle.

Many churches, social organizations and schools use raffles as fund raisers. Study the raffle described below. Tickets cost $100.00. There are 50,000 tickets available.

Prize List

(c) 2006 Jupiter Images Corp

Prizes	Number of Winners
$1,000,000 Dream House or $1,000,000 cash	1
Cadillac STS or Mercedes-Benz SLK 280 or $50,000	1
Pontiac Solstice or Mustang Convertible or $25,000	1
Harley Davidson Softail Deluxe or $15,000	1
His and Hers Rolex Watches or $10,000	1
Samsung Digital Camera / Camcoder or Panasonic 9" Portable DVD Player or $500	140
Polaroid Under the Counter 7" LCD TV or Supersonic DVD Home Theater System or $200	635
Samsung DVD VCR Player or Casio 2.5" LCD TV or $100	635

529

Use the raffle information to answer the following questions.

1. Make a probability distribution table for this raffle. Use the prize money as the events. Remember to subtract the price of the ticket from the events when determining the random variables.

2. If all 50,000 tickets are sold, how much money will this charity make on their raffle?   Add winnings − from $mil = profit.

3. Using the prize money, not the random variable to define an event, find:
    A. P(Winning the $1,000,000 Dream Home). $1/50,000$

    B. Odds against winning the $1,000,000 Dream Home. $49,999 : 1$

    C. Odds in favor of winning $50,000 or a Mercedes-Benz or a Cadillac. $1 : 49,999$

    D. P(Winning a prize worth at most $500).

    E. Odds against winning at a prize worth at least $200.

    F. P(Winning a prize worth more than $500). $5/$

    G. Odds in favor of winning any prize.

    H. P(Not Winning something).

    I. P(Winning a prize worth less than $25,000).

    J. Odds against winning a prize worth at least $15,000.

4. Find the expected value of this raffle.

5. How is the expected value related to the profit the charity makes on this raffle?

PROJECT: DECISION TREES

Objective: Use a decision tree to make a recommendation based on payoff and probability tables.

A computer manufacturer offers a line of notebook computers. They hire a consulting firm to evaluate their company's procedures.

(c) 2006 JupiterImages Corp

The consulting firm studies their line of laptop computers, how their product compares to the competitions, the potential sales growth of their product and the manufacturing process.

The consulting firm compiles a profit payoff table and a cost payoff table that outlines four courses of action for the company.

Choice 1: Stay at the same location and continue to make the notebook computers in the same manner.

Choice 2: Stay at the same location, but outsource some of the components. Rather than produce most of the parts, purchase more of the components from other suppliers.

Choice 3: Relocate. Build a new facility in another location, but assemble the laptop computers the same manner as choice 1.

Choice 4: Relocate and outsource. Build a new facility in another location and purchase more of the components from other suppliers.

I. The profit payoff table is

PROFIT PAYOFF TABLE ($ MILLIONS)

ACTION	STATE OF NATURE (DEMAND)		
	Decreases	Constant	Increases
Stay and Same Procedures	4,200	5,200	6,600
Stay and Outsource	5,000	5,900	7,100
Relocate and Same Procedures	3,000	5,300	7,900
Relocate and Outsource	−1,200	5,000	9,700

The consulting firm also constructs a probability table predicting the demand for notebook computers have three possible paths.

Path 1: Demand for notebook computers increases.
Path 2: Demand for notebook computers remains the same (constant).
Path 3: Demand for notebook computers decreases.

The probability table for demand is:

DEMAND	PROBABILITY
Increases	0.39
Constant	0.37
Decreases	0.24

As the consultant hired by the computer manufacturer
1. Construct a tree diagram from this information.
2. Calculate the expected value for each choice.
3. Based on the tree diagram, what course of action would you recommend to maximize the company's profits.

II. The cost payoff table is

COST PAYOFF TABLE ($ MILLIONS)

ACTION	STATE OF NATURE (DEMAND)		
	Decreases	Constant	Increases
Stay and Same Procedures	3,000	2,400	3,600
Stay and Outsource	2,200	2,000	1,800
Relocate and Same Procedures	3,600	2,600	3,000
Relocate and Outsource	3,500	2,400	2,100

Use the same probability table for demand given in Problem I.
As the consultant hired by the computer manufacturer
1. Construct a tree diagram from this information.
2. Calculate the expected value for each choice.
3. Based on the tree diagram, what course of action would you recommend to minimize the company's expenditures.

PROJECT: VENN DIAGRAMS AND PROBABILITY

Objective: Find probabilities and odds using a Venn Diagram and the Addition Rule for Probability.

A clinical study of a new, non-drowsy, prescription medication for allergy relief was tested on 400 patients. The following table details the adverse effects of this new medication and the number of patients who experienced these effects.

(c) 2006 JupiterImages Corp

Adverse Effects	Number
Headache	24
Insomnia	11
Dry Mouth	10
Drowsiness	12
Nervousness	20
Dizziness	13
Fatigue	19
Nausea	15
Sore Throat	22
Anorexia	9
Thirst	10
Fatigue and Headache	3
Insomnia and Anorexia	5
Nausea and Dizziness	4
Nervousness and Dizziness	3
Nervousness and Nausea	4
Dry Mouth and Nervousness	7
Dry Mouth and Headaches	4
Headaches and Nervousness	5
Nervousness, Nausea and Dizziness	1
Headaches, Dry Mouth and Nervousness	3

Use this table to make a Venn Diagram. Then use the Venn diagram or probability formulas, where applicable, to calculate the following probabilities and odds.

1. P(Experiencing a Headache)

2. P(Experiencing only a Headache)

3. P(Not experiencing Dry Mouth)

4. Odds in favor of experiencing Dizziness.

5. Odds against experiencing Drowsiness.

6. Odds against not experiencing Nausea.

7. P(Experiencing Nervousness and Nausea)

8. P(Experiencing Nervousness or Nausea)

9. P(Experiencing a Headache and becoming Drowsy)

10. P(Experiencing a Headache or becoming Drowsy)

11. Odds in favor of experiencing an adverse reaction.

12. Odds against experiencing an adverse reaction.

13. P(Experiencing a Headache and Fatigue and Dry Mouth)

14. P(Experiencing a Headache or Fatigue or Dry Mouth)

15. P(Experiencing Nervousness and Nausea and Dizziness)

16. P(Experiencing Nervousness or Nausea or Dizziness)

17. Odds against experiencing a Sore Throat and being Thirsty.

18. Odds in favor of experiencing Nervousness or Nausea.

19. Odds against experiencing Anorexia or Fatigue.

20. P(Experiencing Anorexia and Fatigue)

CHAPTER 6 REVIEW

Section 6.1

1. In a clinical study of a new prescription allergy medicine, 28 patients out of 215 patients experienced headaches after taking the medication. If one patient is taken from this study group, find
   A. P(patient experiences a headache).
   B. P(patient does not experience a headache).

2. After taking a new allergy medicine, 16 patients experienced nausea and 199 patients did not experience nausea. If one patient is taken from this study group, find
   A. P(patient experiences nausea).
   B. P(patient does not experience nausea).

3. The Bureau of Labor Statistics reported the following workplace deaths.

Cause of Workplace Death	Frequency
Transportation (vehicle accidents, plane crashes)	2,767
Violence (homicides, suicides)	1,318
Hit by equipment, objects	988
Harmful substances, environment	659
Falls	409
Fire, explosions	198

If one death is taken from this report, find:
A. P(Violent death)
B. P(Death was not caused by a fall)
C. P(Transportation death)
D. P(Death was not caused by a fire (explosion))

Section 6.2

4. The lineup for a previous Breeder's Cup Classic was

Number	Horse	Jockey	Odds
1	Evening Attire	John Velazquez	15 – 1
2	Pleasantly Perfect	Alex Solis	10 – 1
3	Volponi	Santos	15 – 1
4	Funny Cide	Julie Krone	8 – 1
5	Hold that Tiger	Edgar Prado	12 – 1
6	Dynever	Corey Nakatani	15 – 1
7	Perfect Drift	Gary Stevens	7 – 2
8	Medaglia d'Oro	Jerry Bailey	3 – 1
9	Congaree	Pat Valenzuela	6 – 1
10	Ten Most Wanted	Pat Day	4 – 1

Find
   A. Odds against Funny Cide winning.
   B. Odds in favor of Ten Most Wanted Winning.
   C. P(Hold That Tiger winning)
   D. P(Perfect Drift not winning)

5. A public opinion poll surveyed 400 voters. 289 of the surveyed voters responded that they would vote in favor of a tax increase for their fire protection district. Find the:
   A. P(voting in favor of the tax increase).
   B. P(voting against the tax increase)
   C. Odds in favor of the tax increase.
   D. Odds against the tax increase.

Section 6.3

6. Study the raffle described below. Tickets cost $20.00. There are only 2,000 tickets available. The prizes are

Places	Prizes
First Place	$10,000
Second Place	$1,000
Third Place	$750
Fourth Place	$500
Fifth Place	$400
Sixth Place	$300
Seventh Place	$200
Eighth through Twelfth Place	$100

Find: A. P(Winning any prize)   B. Odds against winning $100
      C. Odds in favor of losing money   D. P(winning at least $500)
      E. The expected value of this raffle.

Section 6.4

7. A small family-operated restaurant is trying to decide if they should make no changes or expand their current location with a small or large addition. A consulting firm analyzed their business and the payoff table is:

PAYOFF TABLE OF PROFITS ($ THOUSANDS)

ACTION	STATE OF NATURE (DEMAND)		
	Low	Medium	High
No Addition	32	48	55
Small Addition	17	55	70
Large Addition	−40	60	80

The probability table for the "states of nature" is:

DEMAND	PROBABILITY
High	0.25
Medium	0.48
Low	0.27

What size of expansion, if any, should the family pursue to maximize their profits?

8. Using the information from Problem 7, change the probability table and answer the same question.

DEMAND	PROBABILITY
High	0.20
Medium	0.38
Low	0.42

9. Use the results from Problem 8. However, the payoff table is a COST table, not a profit table. What size of expansion, if any, should the family pursue to maximize their profits?

Section 6.5

10. In a clinical study of a prescription medication for high cholesterol, 500 people were given the medication with the following results:
    - 12 people complained of headaches.
    - 24 people experienced dizziness.
    - 15 people had high blood pressure.
    - 18 people experience anxiety
    - 8 people had both high blood pressure and dizziness.
    - 10 people had both dizziness and anxiety.

    Find the probability that a randomly selected participant experienced:
    A. only headaches.
    B. only dizziness.
    C. only high blood pressure
    D. only anxiety
    E. none of these symptoms.

11. In a clinical study of a prescription medication for high cholesterol, 500 people were given the medication with the following results:
    - 12 people complained of headaches.
    - 24 people experienced dizziness.
    - 15 people had high blood pressure.
    - 18 people experience anxiety
    - 10 people had both dizziness and anxiety.

8 people had both high blood pressure and dizziness.
   7 people had headaches and dizziness
   5 people had high blood pressure and headaches
   3 people had high blood pressure, headaches and dizziness
Find the probability that a randomly selected participant experienced:
A. only headaches.
B. only dizziness.
C. only high blood pressure
D. only anxiety
E. none of these symptoms.

Section 6.6

12. Use the "OR" Probability Rule to solve the following problem.
    In a clinical study of a prescription medication for high cholesterol,
    500 people were given the medication with the following results:
       12 people complained of headaches.
       24 people experienced dizziness.
       15 people had high blood pressure
       18 people experience anxiety
        8 people had both high blood pressure and dizziness.
       10 people had both dizziness and anxiety.
    Find for a randomly selected participant find
    A. P(Headaches OR high blood pressure)
    B. P(High blood pressure OR dizziness)
    C. P(Headaches OR anxiety)
    D. P(Anxiety OR dizziness)

13. Use the "OR" Probability Rule to solve the following problem.
    In a clinical study of a prescription medication for high cholesterol,
    500 people were given the medication with the following results:
       12 people complained of headaches.
       24 people experienced dizziness.
       15 people had high blood pressure.
       18 people experience anxiety
       10 people had both dizziness and anxiety.
        8 people had both high blood pressure and dizziness.
        7 people had headaches and dizziness
        5 people had high blood pressure and headaches
        3 people had high blood pressure, headaches and dizziness
    Find the probability that a randomly selected participant experienced:
    A. P(Headaches OR high blood pressure)
    B. P(High blood pressure OR dizziness)

C. P(Headaches OR anxiety)
D. P(Anxiety OR dizziness)

14. In a clinical study of a new prescription high blood pressure medicine, the following reactions to medications were noted:

Reaction to the Medication

Medication	Dizziness	Headache	Swelling	Irregular Heartbeat
Prescribed Medication	18	33	16	10
Placebo	18	31	8	9

If a patient is randomly selected from this study group, find:
A. P(Suffered from Dizziness OR took a Placebo).
B. P(Took the prescribed medication OR had a headache).
C. P(Experienced irregular heartbeat OR took the medication).
D. P(Had a headache OR suffered from swelling).

Section 6.7

15. In a clinical study of the prescribed rash medicine, 32 patients reported burning sensations as a side effect of the medication and 568 patients did not experience burning sensations. If two different patients are taken from this group, find
    A. P(Both patients suffered burning sensations).
    B. P(Neither patient suffered burning sensations).

16. In a clinical study of the prescribed coughing medicine, 24 patients reported nausea as a side effect of the medication and 148 patients did not experience nausea. If two different patients are taken from this group, find
    A. P(Both patients suffered nausea).
    B. P(Neither patient suffered nausea).

17. The Bureau of Labor Statistics of the US Department of Labor reports that 20% of fatal occupational accidents are caused by violence. If three different reports on fatal occupational accidents are selected, find the:
    A. P(All three deaths are due to violence).
    B. P(None of the three deaths are violence).

18. A survey reported that 33% of people interviewed were happy with their jobs. If four different people are selected., find the:
    A. P(All four people are happy with their jobs).
    B. P(None of the four people are happy with their jobs).

Appendix A

Solution to the Chapter Reviews

CHAPTER 1 REVIEW

Section 1.1: Signed Numbers

1. $(-\$6.00)+(-\$24.00)+(-\$12.00)+(\$4.42)+(\$7.24)+(\$0.80)+(\$16.60)+(\$2.55) =$
   $(-\$42.00)+(\$31.61) = -\$10.39$

2. $48\left(\dfrac{-1}{6}\right)(72)\left(\dfrac{1}{-2}\right) = \overset{8}{\cancel{48}}\left(\dfrac{-1}{\cancel{6}}\right)\overset{36}{\cancel{(72)}}\left(\dfrac{1}{\cancel{-2}}\right) = 288$

3. $(-8)+12-6-(-7)+(-2) =$
   $(-8)+12+(-6)+(-2)+12+7 = (-8)+(-6)+(-2)+12+7 = (-16)+19 = 3$

4. $\dfrac{-7}{18}\left(\dfrac{35}{42}\right)\left(\dfrac{-6}{-21}\right)\left(\dfrac{2}{5}\right) = \left(\dfrac{\cancel{-7}}{\cancel{18}}\right)\left(\dfrac{\cancel{35}}{\cancel{42}}\right)\left(\dfrac{\cancel{-6}}{\cancel{-21}}\right)\left(\dfrac{\cancel{2}}{\cancel{5}}\right) = \dfrac{-1}{27}$

5. $(\$1{,}258.55)+(\$748.16)+(\$24.55)+(-\$27.17)+(-\$46.18)+(-\$256.79)+(-\$1{,}480.18) =$
   $(\$2021.26)+(-\$1910.31) = \$110.95$

Section 1.2: Order of Operations

6. $\sqrt{4}\,[7-3(4)]+8 = \sqrt{4}\,[7-12]+8 \quad \sqrt{4}\,[-5]+8 \quad 2[-5]+8 \quad -10+8 = -2$

7. $\sqrt{4}\,[7-3^2(4)]+8 =$
   $\sqrt{4}\,[7-9(4)]+8 \quad \sqrt{4}\,[7-36]+8 \quad \sqrt{4}\,[-29]+8 \quad 2[-29]+8$
   $-58+8 = -50$

8. $\sqrt{4}\,[7-3(4)]^2+8 =$
   $\sqrt{4}\,[7-12]^2+8 \quad \sqrt{4}\,[-5]^2+8 \quad \sqrt{4}\,[25]+8 \quad 2[25]+8 \quad 50+8=58$

9. $\sqrt{4}\,[(7-3)(4)]+8 = \quad \sqrt{4}\,[(4)(4)]+8 \quad \sqrt{4}\,[16]+8 \quad 2[16]+8 \quad 32+8 = 40$

10. $\sqrt{4}[7-3^3(4)]+8=$
$\sqrt{4}[7-27(4)]+8 \quad \sqrt{4}[7-108]+8 \quad \sqrt{4}[-101]+8 \quad 2[-101]+8$
$-202+8=-194$

Section 1.3: Ratio and Proportion

11. $\dfrac{3.5}{6}=\dfrac{21}{x}$ $3.5x = 126$ $\left(\dfrac{1}{3.5}\right)(3.5)x = \left(\dfrac{1}{3.5}\right)(126)$ $x=\dfrac{126}{3.5}=36$

12. $\dfrac{x}{3}=\dfrac{20}{24}$ $24x = 60$ $\left(\dfrac{1}{24}\right)(24)x = \left(\dfrac{1}{24}\right)(60)$ $x=\dfrac{60}{24}=2.5$

13. $\dfrac{8.25}{100}=\dfrac{x}{\$24{,}286.44}$ $100x = \$200{,}363.18$
$\left(\dfrac{1}{100}\right)(100)x = \left(\dfrac{1}{100}\right)(\$200{,}363.18) = \dfrac{(\$200{,}363.18)}{100} \approx \$2{,}003.63$

14. $1{,}250 - 60 = 1{,}190$ student DO have e-mail accounts.
$\dfrac{x}{100}=\dfrac{1{,}190}{1{,}250}$ $1{,}250x = 119{,}000$
$\left(\dfrac{1}{1{,}250}\right)(1{,}250)x = \left(\dfrac{1}{1{,}250}\right)(119{,}000) = \dfrac{(119{,}000)}{1{,}250} = 95.2$ $x = 95.2\%$

15. $\dfrac{35}{100}=\dfrac{434}{x}$ $35x = 43{,}400$
$\left(\dfrac{1}{35}\right)(35)x = \left(\dfrac{1}{35}\right)(43{,}400) = \dfrac{(43{,}400)}{35} = 1{,}240$ 1,240 people were interviewed.

16. $\dfrac{4.5}{100}=\dfrac{x}{400}$ $100x = 1{,}800$ $\left(\dfrac{1}{100}\right)(100)x = \left(\dfrac{1}{100}\right)(1{,}800) = \dfrac{(1{,}800)}{100} = 18$
18 people suffered extreme nausea.

Section 1.4: Proportion Applications

17. $\dfrac{1}{7.42}=\dfrac{324}{x}$ $1x = 2{,}404.08$
$324 US is equivalent to 2,404.08 South African rands.

A3

18. $\dfrac{1}{0.88} = \dfrac{x}{1{,}248}$ $0.88x = 1{,}248$

$\left(\dfrac{1}{0.88}\right)(0.88)x = \left(\dfrac{1}{0.88}\right)(1{,}248) = \dfrac{(1{,}248)}{0.88} \approx 1{,}418.181818$

1,248 Euros is approximately $1,418.18 US.

19. $\dfrac{1}{80} = \dfrac{x}{670}$ $80x = 670$ $\left(\dfrac{1}{0.80}\right)(0.80)x = \left(\dfrac{1}{0.80}\right)(670) = \dfrac{(670)}{0.80} = 8.375$

The two cities are 8.375 inches, $8\dfrac{3}{8}$ inches, apart on the map

20. $\dfrac{1}{6} = \dfrac{11.5}{x}$ $1x = 69$ $\dfrac{1}{6} = \dfrac{9.75}{x}$ $1x = 58.5$

The house measure 69 feet by 58ft 6 in.

21. Calories consumed are 740 + 310 = 1,050.

$\dfrac{\text{Activity}}{\text{Unit Distance or Unit Time}} = \dfrac{\text{Calories in Food}}{\text{Distance or Time}}$ $\dfrac{\text{calories}}{\text{hour}} = \dfrac{422}{1} = \dfrac{1{,}050}{x}$

$422x = 1{,}050$ $\left(\dfrac{1}{422}\right)(422x) = \left(\dfrac{1}{422}\right)(1{,}050)$ $x = \dfrac{1{,}050}{422} \approx 2.49$

0.49(60 minutes) ≈ 29 minutes. It would take 2 hours, 40 minutes of intense weightlifting to burn the calories from lunch.

22. Calories consumed are 740 + 310 = 1,050.

$\dfrac{\text{Activity}}{\text{Unit Distance or Unit Time}} = \dfrac{\text{Calories in Food}}{\text{Distance or Time}}$ $\dfrac{\text{calories}}{\text{hour}} = \dfrac{704}{1} = \dfrac{1{,}050}{x}$

$704x = 1{,}050$ $\left(\dfrac{1}{704}\right)(704x) = \left(\dfrac{1}{704}\right)(1{,}050)$ $x = \dfrac{1{,}050}{704} \approx 1.49$

0.49(60 minutes) ≈ 29 minutes. It would take 1 hours, 40 minutes of rope jumping to burn the calories from lunch.

23. (FS)(FA) = (OS)(OA),   16(108) = 24x   1,728 = 24x

$\left(\dfrac{1}{24}\right)(1{,}728) = \left(\dfrac{1}{24}\right)(24)x$ $\dfrac{(1{,}728)}{24} = x = 72$

72 ml of 24% solution and (108 – 72), 36 ml of distilled water are needed.

A4

24. $(FS)(FA) = (OS)(OA)$, $8(40) = 100x$ $320 = 100x$

$\left(\dfrac{1}{100}\right)(320) = \left(\dfrac{1}{100}\right)(100)x$ $\dfrac{(320)}{100} = x = 3.2$

3.2 ml of pure solution and $(40 - 3.2)$, 36.8 ml of distilled water are needed.

Section 1.5: Solving Linear Equations

25. $24x - 36 = 79.2$
$24x - 36 + 36 = 79.2 + 36$
$24x = 115.2$
$\left(\dfrac{1}{24}\right)(24)x = \left(\dfrac{1}{24}\right)(115.2)$
$\dfrac{115.2}{24} = 4.8$ $x = 4.8$

26. $72 + 6x = 48 + 9x$
$72 + 6x - 6x = 48 + 9x - 6x$
$72 = 48 + 3x$
$72 - 48 = 48 - 48 + 3x$
$24 = 3x$
$\left(\dfrac{1}{3}\right)(24) = \left(\dfrac{1}{3}\right)(3)x$
$\dfrac{24}{3} = x = 8$

27. $4(7x - 3) = 6(3x + 8)$
$28x - 12 = 18x + 48$
$28x - 18x - 12 = 18x - 18x + 48$
$10x - 12 = 48$
$10x - 12 + 12 = 48 + 12$
$10x = 60$
$\left(\dfrac{1}{10}\right)(10)x = \left(\dfrac{1}{10}\right)(60)$
$x = \dfrac{60}{10} = 6$

28. $3(3x - 5) = 4(2x + 9) + 5$
$9x - 15 = 8x + 36 + 5$
$9x - 15 = 8x + 41$
$9x - 8x - 15 = 8x - 8x + 41$
$1x - 15 = 41$

$$1x - 15 + 15 = 41 + 15$$
$$1x = 56$$

29.
$$8(4x - 3) - 12 = 5(4x + 6) + 18$$
$$32x - 24 - 12 = 20x + 30 + 18$$
$$32x - 36 = 20x + 48$$
$$32x - 20x - 36 = 20x - 20x + 48$$
$$12x - 36 = 48$$
$$12x - 36 + 36 = 48 + 36$$
$$12x = 84$$
$$\left(\frac{1}{12}\right)(12)x = \left(\frac{1}{12}\right)(84)$$
$$x = \frac{84}{12} = 7$$

30.
$$\$5,250,000 = \$283x + \$143,650$$
$$\$5,250,000 - \$143,650 = \$283x + \$143,650 - \$143,650$$
$$\$5,106,350 = \$283x$$
$$\left(\frac{1}{\$283}\right)(\$5,106,350) = \left(\frac{1}{\$283}\right)(\$283)x$$
$$\frac{\$5,106,350}{\$283} \approx 18,043.63958 \quad \text{The living area is about } 18,044 \text{ sq ft}$$

31.
$$348 = \frac{5}{9}(F - 32)$$
$$348 = \frac{5}{9}F - \frac{160}{9}$$
$$\frac{3,132}{9} + \frac{160}{9} = \frac{5}{9}F - \frac{160}{9} + \frac{160}{9}$$
$$\frac{3,292}{9} = \frac{5}{9}F$$
$$\left(\frac{9}{5}\right)\frac{3,292}{9} = \left(\frac{9}{5}\right)\left(\frac{5}{9}\right)F$$
$$\frac{3,292}{5} = F = 658.4$$

348 degrees centigrade equal 658.4 degrees Fahrenheit.

32. $C = \dfrac{5}{9}(1{,}979 - 32)$

$C = \dfrac{5}{9}(1{,}947)$

$C = \dfrac{9{,}735}{9} \approx 1{,}081.6666667$

1,979 degrees Fahrenheit is about 1,082 degrees Centigrade.

CHAPTER 2 REVIEW

Section 2.1: Simple Interest

1. $I = PRT = (\$4{,}855)(0.032)\left(\dfrac{7}{12}\right) \approx \$90.62666667 \approx \$90.63$

2. $I = PRT = (\$9{,}700)(0.04625)(2) = \$897.25$

3. $I = PRT = \$470.40 = (\$2{,}400)(R)(4)$
   $\$470.40 = \$9{,}600$
   $\dfrac{1}{\$9{,}600}(\$470.40) = \dfrac{1}{\$9{,}600}(\$9{,}600)R$
   $\dfrac{\$470.40}{\$9{,}600} = 0.049 = R \quad R = 4.9\%$

4. $I = PRT = \$222.75 = \$1{,}650(0.06)T$
   $\$222.75 = \$99T$
   $\dfrac{1}{\$99}(\$222.75) = \dfrac{1}{\$99}(\$99)T$
   $\dfrac{\$222.75}{\$99} = 2.25 = T \quad T = 2.25$ years (2 years, 3 months)

5. $A = P(1 + RT)$
   $A = \$5{,}500\left[1 + (0.06125)\left(\dfrac{42}{12}\right)\right] \approx \$6{,}679.0625 \approx \$6{,}679.06$

6. $A = P(1 + RT)$
   $\$20{,}212 = P(1 + (0.035)(18)) \qquad \$20{,}212 = 1.63P$
   $\dfrac{1}{1.63}(\$20{,}212) = \dfrac{1}{1.63}(1.63)P \qquad \dfrac{\$20{,}212}{1.63} = \$12{,}400 = P$

Section 2.2: Compound Interest

7. $A = P\left(1 + \dfrac{R}{n}\right)^{nT} = \$3,600\left(1 + \dfrac{0.053}{4}\right)^{4\times 4} = \$3,600(1.234428098)$
   $A \approx \$4,443.941153 \approx \$4,443.94$

8. $A = Pe^{RT} = \$3,600\left(e^{0.053\times 4}\right) = \$3,600(1.236147885) \approx \$4,450.132386$
   $A \approx \$4,450.13$

9. $P = \dfrac{A}{\left(1 + \dfrac{R}{n}\right)^{nT}} = \dfrac{\$14,800}{\left(1 + \dfrac{0.07125}{4}\right)^{4\times 8}} = \dfrac{\$14,800}{1.759417745} \approx \$8,411.873781$
   $P \approx \$8,411.87$

10. $P = \dfrac{A}{e^{RT}} = \dfrac{\$7,400}{e^{0.049\times 3}} = \dfrac{\$7,400}{1.158353963} \approx \$6,388.375433$
    $P \approx \$6,388.38$

11. a. $7\dfrac{7}{8}\%$ interest compounded semi-annually.
    $APR = \left(1 + \dfrac{R}{n}\right)^n - 1 = \left(1 + \dfrac{0.07875}{2}\right)^2 - 1 \approx 0.080300391 \approx 8.03\%$

    b. 7.85% interest compounded monthly.
    $APR = \left(1 + \dfrac{R}{n}\right)^n - 1 = \left(1 + \dfrac{0.0785}{12}\right)^{12} - 1 \approx 0.081386867 \approx 8.14\%$

    c. 7.8% interest compounded continuously.
    $APR = e^R - 1 = e^{0.078} - 1 \approx 0.0811226587 \approx 8.11\%$

    The best loan rate is the smallest, $7\dfrac{7}{8}\%$ interest compounded semi-annually.

12. a. $7\dfrac{7}{8}\%$ interest compounded semi-annually.
    $APY = \left(1 + \dfrac{R}{n}\right)^n - 1 = \left(1 + \dfrac{0.07875}{2}\right)^2 - 1 \approx 0.080300391 \approx 8.03\%$

    b. 7.85% interest compounded monthly.
    $APY = \left(1 + \dfrac{R}{n}\right)^n - 1 = \left(1 + \dfrac{0.0785}{12}\right)^{12} - 1 \approx 0.081386867 \approx 8.14\%$

A8

    c. 7.8% interest compounded continuously.
$$APY = e^R - 1 = e^{0.078} - 1 \approx 0.0811226587 \approx 8.11\%$$
    The best savings rate is the largest, 7.85% interest compounded monthly.

Section 2.3: Car Loans

13. $9,999 (cash price) − $2,000 (down payment) = $7,999 (loan)

    a. $7,999(1 + (0.064)(5)) = $10,558.68 $P = \dfrac{\$10{,}558.68}{12(5)} \approx \$175.98$

    b. $P = \dfrac{\left(\dfrac{AR}{n}\right)\left(1+\dfrac{R}{n}\right)^{nT}}{\left(1+\dfrac{R}{n}\right)^{nT}-1} = \dfrac{\left(\dfrac{\$7{,}999(0.064)}{12}\right)\left(1+\dfrac{0.064}{12}\right)^{12\times 5}}{\left(1+\dfrac{0.064}{12}\right)^{12\times 5}-1} \approx \$156.1352204$

    P = $156.14

14. $9,999 (cash price) − $2,000 (down payment) = $7,999 (loan)
$$P = \frac{A}{nT} = \frac{\$7{,}999}{12(5)} = \frac{\$7{,}999}{60} \approx \$133.32$$

15. $22,790 (cash price) − $1,500 (down payment) = $21,290 (loan)
$$P = \frac{A}{nT} = \frac{\$21{,}290}{12(4)} = \frac{\$21{,}290}{48} \approx \$443.54$$

16. $22,790 (cash price) − $1,500 (down payment) = $21,290 (loan)

    a. $21,290(1 + (0.042)(4)) = $24,866.72 $P = \dfrac{\$24{,}866.72}{12(4)} \approx \$518.06$

    b. $P = \dfrac{\left(\dfrac{AR}{n}\right)\left(1+\dfrac{R}{n}\right)^{nT}}{\left(1+\dfrac{R}{n}\right)^{nT}-1} = \dfrac{\left(\dfrac{\$21{,}290(0.042)}{12}\right)\left(1+\dfrac{0.042}{12}\right)^{12\times 4}}{\left(1+\dfrac{0.042}{12}\right)^{12\times 4}-1} \approx \$482.6158134$

    P = $482.62

17. $22,790 (cash price)
$$P = \frac{A}{nT} = \frac{\$22{,}790}{12(4)} = \frac{\$22{,}790}{48} \approx \$474.79$$

18. $64,100 (0.10) = $6,410$ (down payment)
    $64,100$ (cash price) $- $6,410$ (down payment) $= $57,690$ (loan)

   a. $57,590(1 + (0.024(6))) = $65,997.36$ $P = \dfrac{\$65,997.36}{12(6)} = \$916.63$

   b. $P = \dfrac{\left(\dfrac{AR}{n}\right)\left(1+\dfrac{R}{n}\right)^{nT}}{\left(1+\dfrac{R}{n}\right)^{nT}-1} = \dfrac{\left(\dfrac{\$57,690(0.024)}{12}\right)\left(1+\dfrac{0.024}{12}\right)^{12\times 6}}{\left(1+\dfrac{0.024}{12}\right)^{12\times 6}-1} \approx \$861.1236836$

   $P = \$861.12$

19. $8,480 (0.05) = 424 (down payment)
    $8,480$ (cash price) $- 424 (down payment) $= $8,056$ (loan)

   a. $8,056(1 + (0.045)(3)) = $9,143.56$ $P = \dfrac{\$9,143.56}{12(3)} \approx \$253.99$

   b. $P = \dfrac{\left(\dfrac{AR}{n}\right)\left(1+\dfrac{R}{n}\right)^{nT}}{\left(1+\dfrac{R}{n}\right)^{nT}-1} = \dfrac{\left(\dfrac{\$8,056(0.045)}{12}\right)\left(1+\dfrac{0.045}{12}\right)^{12\times 3}}{\left(1+\dfrac{0.045}{12}\right)^{12\times 3}-1} \approx \$239.6412236$

   $P = \$239.64$

20. $18,555 (0.10) = $1,855.50$ (down payment)
    $18,555$ (cash price) $- $1,855.50$ (down payment) $= $16,699.50$ (loan)

   a. $16,699.50(1 + (0.055)(4)) = $20,373.39$ $P = \dfrac{\$20,373.39}{12(4)} \approx \$424.45$

   b. $P = \dfrac{\left(\dfrac{AR}{n}\right)\left(1+\dfrac{R}{n}\right)^{nT}}{\left(1+\dfrac{R}{n}\right)^{nT}-1} = \dfrac{\left(\dfrac{\$16,699.50(0.055)}{12}\right)\left(1+\dfrac{0.055}{12}\right)^{12\times 4}}{\left(1+\dfrac{0.055}{12}\right)^{12\times 4}-1} \approx \$388.3715081$

   $P = \$388.37$

21. $5,880 (0.10) = 588 (down payment)
    $5,880$ (cash price) $- 588 (down payment) $= $5,292$ (loan)

   a. $5,292(1 + (0.0375)(2)) = $5,688.90$ $P = \dfrac{\$5,688.90}{12(2)} \approx \$237.04$

A10

b. $P = \dfrac{\left(\dfrac{AR}{n}\right)\left(1+\dfrac{R}{n}\right)^{nT}}{\left(1+\dfrac{R}{n}\right)^{nT}-1} = \dfrac{\left(\dfrac{\$5,292(0.0375)}{12}\right)\left(1+\dfrac{0.0375}{12}\right)^{12\times 2}}{\left(1+\dfrac{0.0375}{12}\right)^{12\times 2}-1} \approx \$229.2162907$

$P = \$229.22$

Section 2.4: Remaining Balance

22. 5 year loan = 5 ~~years~~ $\left(\dfrac{12 \text{ months}}{\text{~~year~~}}\right) = 60$ months

    2 ~~years~~ $\left(\dfrac{12 \text{ months}}{\text{~~year~~}}\right) + 7$ months $= 24 + 7$ months $= 31$ months

    60 months − 31 months = 29 months remaining on the loan

    a. $B = A - m\left[P - \dfrac{A(R)}{n}\right]$

    $B = \$7,999 - 31\left[\$175.98 - \dfrac{\$7,999(0.064)}{12}\right] \approx \$3,866.12$

    b. $B = P\left(\dfrac{\left(1+\dfrac{R}{n}\right)^x - 1}{\left(\dfrac{R}{n}\right)\left(1+\dfrac{R}{n}\right)^x}\right) = \$156.14\left[\dfrac{\left(1+\dfrac{0.064}{12}\right)^{29} - 1}{\left(\dfrac{0.064}{12}\right)\left(1+\dfrac{0.064}{12}\right)^{29}}\right] \approx \$4,184.96$

    c. $B = x(m) - \left(\dfrac{x(x+1)}{k(k+1)}\right)(k \times m - a) =$

    $29(\$156.14) - \left(\dfrac{29(30)}{60(61)}\right)(60 \times \$156.14 - \$7,999) \approx \$4,202.55$

23. 5 year loan = 5 ~~years~~ $\left(\dfrac{12 \text{ months}}{\text{~~year~~}}\right) = 60$ months

    2 ~~years~~ $\left(\dfrac{12 \text{ months}}{\text{~~year~~}}\right) + 7$ months $= 24 + 7$ months $= 31$ months

    60 months − 31 months = 29 months remaining on the loan
    $B = x(m) = 29 (\$133.32) = \$3,866.28$

24. 4 year loan = 4 ~~years~~ $\left(\dfrac{12 \text{ months}}{\cancel{\text{year}}}\right)$ = 48 months

    1 ~~year~~ $\left(\dfrac{12 \text{ months}}{\cancel{\text{year}}}\right)$ + 2 months = 12 + 2 months = 14 months

    48 months − 14 months = 34 months remaining on the loan
    B = x(m) = 34 ($443.54) = $15,080.36

25. 4 year loan = 4 ~~years~~ $\left(\dfrac{12 \text{ months}}{\cancel{\text{year}}}\right)$ = 48 months

    1 ~~year~~ $\left(\dfrac{12 \text{ months}}{\cancel{\text{year}}}\right)$ + 2 months = 12 + 2 months = 14 months

    48 months − 14 months = 34 months remaining on the loan

    a. $B = A - m\left[P - \dfrac{A(R)}{n}\right]$

    $B = \$21{,}290 - 14\left[\$518.06 - \dfrac{\$21{,}290(0.042)}{12}\right] \approx \$15{,}080.37$

    b. $B = P\left[\dfrac{\left(1+\dfrac{R}{n}\right)^x - 1}{\left(\dfrac{R}{n}\right)\left(1+\dfrac{R}{n}\right)^x}\right] = \$482.62\left[\dfrac{\left(1+\dfrac{0.064}{12}\right)^{34} - 1}{\left(\dfrac{0.042}{12}\right)\left(1+\dfrac{0.042}{12}\right)^{34}}\right] \approx \$15{,}444.91$

    c. $B = x(m) - \left(\dfrac{x(x+1)}{k(k+1)}\right)(k \times m - a) =$

    $B = 34(\$482.62) - \left(\dfrac{34(35)}{48(49)}\right)(48 \times \$482.62 - \$21{,}290) \approx \$15{,}460.03$

26. 4 year loan = 4 ~~years~~ $\left(\dfrac{12 \text{ months}}{\cancel{\text{year}}}\right)$ = 48 months

    1 ~~year~~ $\left(\dfrac{12 \text{ months}}{\cancel{\text{year}}}\right)$ + 2 months = 12 + 2 months = 14 months

    48 months − 14 months = 34 months remaining on the loan
    B = x(m) = 34 ($474.79) = $16,142.86

A12

27. 6 year loan = 6 ~~years~~ $\left(\dfrac{12 \text{ months}}{\cancel{\text{year}}}\right)$ = 72 months

   4 ~~years~~ $\left(\dfrac{12 \text{ months}}{\cancel{\text{year}}}\right)$ + 8 months = 48 + 8 months = 56 months

   72 months − 56 months = 16 months remaining on the loan
   a. The remaining balance is 16($916.63) = $14,666.08
   b. The remaining balance is 16($861.12) = $13,777.92.
   c. The remaining balance is 16($861.12) = $13,777.92.

28. 1 year loan = 3 ~~years~~ $\left(\dfrac{12 \text{ months}}{\cancel{\text{year}}}\right)$ = 36 months

   1 ~~year~~ $\left(\dfrac{12 \text{ months}}{\cancel{\text{year}}}\right)$ + 1 month = 12 + 1 month = 13 months

   36 months − 13 months = 23 months remaining on the loan
   a. The remaining balance is 23($253.99) = $5,841.77.
   b. The remaining balance is 23($239.64) = $5,511.72.
   c. The remaining balance is 23($239.64) = $5,511.72.

29. 4 year loan = 4 ~~years~~ $\left(\dfrac{12 \text{ months}}{\cancel{\text{year}}}\right)$ = 48 months

   3 ~~years~~ $\left(\dfrac{12 \text{ months}}{\cancel{\text{year}}}\right)$ = 36 months

   48 months − 36 months = 12 months remaining on the loan.

   a. $B = A - m\left[P - \dfrac{A(R)}{n}\right]$

   $B = \$16{,}699.50 - 36\left[\$424.45 - \dfrac{\$16{,}699.50(0.055)}{12}\right] \approx \$4{,}174.72$

   b. $B = P\left(\dfrac{\left(1+\dfrac{R}{n}\right)^x - 1}{\left(\dfrac{R}{n}\right)\left(1+\dfrac{R}{n}\right)^x}\right) = \$388.37\left[\dfrac{\left(1+\dfrac{0.055}{12}\right)^{12} - 1}{\left(\dfrac{0.055}{12}\right)\left(1+\dfrac{0.055}{12}\right)^{12}}\right] \approx \$4{,}524.52$

   c. $B = x(m) - \left(\dfrac{x(x+1)}{k(k+1)}\right)(k \times m - a) =$

c. $B = x(m) - \left(\dfrac{x(x+1)}{k(k+1)}\right)(k \times m - a) =$

$B = 12(\$388.37) - \left(\dfrac{12(13)}{48(49)}\right)(48 \times \$388.37 - \$16{,}699.50) \approx \$4{,}531.62$

30. 2 year loan = $2 \, \cancel{\text{years}} \left(\dfrac{12 \text{ months}}{\cancel{\text{year}}}\right) = 24$ months

24 months − 10 months = 14 months remaining on the loan.
a. The remaining balance is 14($237.04) = $3,318.56.
b. The remaining balance is 14($229.22) = $3,209.08.
c. The remaining balance is 14($229.22) = $3,209.08.

CHAPTER 3 REVIEW
Section 3.1: Actual Monthly Payment and Refinancing

1. Down payment = (0.10)($284,900) = $28,490
Loan amount = $284,900 − $28,490 = $256,410

$P = \dfrac{\left(\dfrac{(\$256{,}410)(0.072)}{12}\right)\left(1 + \dfrac{0.072}{12}\right)^{(12 \times 30)}}{\left(1 + \dfrac{0.072}{12}\right)^{(12 \times 30)} - 1} = \dfrac{(\$1{,}538.46)(8.615352653)}{7.615352653}$

$P \approx \$1{,}740.48.$ $\qquad$ PMI = 0.0046($256,410) ≈ $1,179.49.

Escrow = $\dfrac{\$5{,}200 + \$490 + \$1{,}179.49}{12} \approx \$572.46.$

Actual monthly payment = $1,740.48 + $572.46 = $2,312.94

2. Down payment = (0.05)($125,600) = $6,280
Loan amount = $125,600 − $6,280 = $119,320

$P = \dfrac{\left(\dfrac{(\$119{,}320)(0.084)}{12}\right)\left(1 + \dfrac{0.084}{12}\right)^{(12 \times 30)}}{\left(1 + \dfrac{0.084}{12}\right)^{(12 \times 30)} - 1} = \dfrac{(\$835.24)(12.31996288)}{11.31996288}$

$P \approx \$909.0246942.$ Dividing by 2 for bi-weekly, $P = \dfrac{\$909.0246942}{2} \approx \$454.51$

PMI = 0.0038($119,320) ≈ $453.42.
Escrow = $\dfrac{\$4{,}200 + \$350 + \$453.42}{26} \approx \$192.44.$

Actual bi-weekly payment = $454.51 + $192.44 = $646.95.

3. Down Payment = 0.05($200,600.00) = $10,030.00.
   Amount financed = $200,600 − $10,030 = $190,570.
   For the first 7 years.
   a. Interest–only = $\dfrac{0.084}{12}(\$190{,}570) = \$1{,}333.99$.
   b. PMI = 0.0042($190570) ≈ $800.39.
   c. Escrow = $\dfrac{\$4{,}800 + \$400 + \$800.39}{12} \approx \$500.03$.
   d. Actual monthly payment = $1,333.99 + $500.03 = $1,834.02.
   For the remaining 23 years:
   a. $P = \dfrac{\left(\dfrac{(\$190{,}570)(0.084)}{12}\right)\left(1+\dfrac{0.084}{12}\right)^{(12\times 23)}}{\left(1+\dfrac{0.084}{12}\right)^{(12\times 23)} - 1} = \dfrac{(\$1{,}333.99)(6.856995632)}{5.856995632}$

   $P \approx \$1{,}561.75$.
   b. PMI = 0.0042($190570) ≈ $800.39.
   c. Escrow = $\dfrac{\$4{,}800 + \$400 + \$800.39}{12} \approx \$500.03$.
   d. Actual monthly payment = $1,561.75 + $500.03 = $2,061.78.

Exercise 3.2

4A. 30 years = $(30 \text{ years})\left(\dfrac{12 \text{ months}}{1 \text{ year}}\right) = 360$ months

12 years + 2 mos = $(12 \text{ years})\left(\dfrac{12 \text{ mos}}{1 \text{ year}}\right) = 144$ mos + 2 mos = 146 months

360 months − 146 months = 214 months remaining on the loan.

Balance is $A = \$1{,}740.48\left[\dfrac{\left(1+\dfrac{0.072}{12}\right)^{214} - 1}{\left(\dfrac{0.072}{12}\right)\left(1+\dfrac{0.072}{12}\right)^{214}}\right] \approx \$209{,}440.11$

B. Monthly Payment is

$P = \dfrac{\left(\dfrac{(\$209{,}440.11)(0.06125)}{12}\right)\left(1+\dfrac{0.06125}{12}\right)^{(12\times 15)}}{\left(1+\dfrac{0.06125}{12}\right)^{(12\times 15)} - 1} \approx \$1{,}781.55$

C. 214 × $1,740.48 = $372,462.72 to repay the current mortgage.
180 × $1,781.55 = $320,679.00 to repay the new mortgage.
$372,462.72 − $320,679.00 = $51,783.72 saved.

5A. Balance is: $B = \$119{,}320 - \$69.01 \left(\dfrac{\left(1+\dfrac{0.084}{26}\right)^{195} - 1}{\dfrac{0.084}{26}} \right) \approx \$100{,}614.74$.

B. Monthly $P = \dfrac{\left(\dfrac{(\$100{,}614.74)(0.072)}{12}\right)\left(1+\dfrac{0.072}{12}\right)^{(12 \times 15)}}{\left(1+\dfrac{0.072}{12}\right)^{(12 \times 15)} - 1} \approx \$915.64$

Bi-weekly is $916.64 / 2 = $457.82.

C. First months interest is: $\dfrac{\$100{,}614.72(0.072)}{26} \approx \$278.63$

First months principal is $457.82 − $278.63 = $179.19.

To repay 15 year bi-weekly: $x = \dfrac{\log\left(1+\dfrac{\$100{,}614.74(0.072)}{26(\$179.19)}\right)}{\log\left(1+\dfrac{0.072}{26}\right)} \approx 340$

585 − 195 = 390 payments remaining on current mortgage.
390 × $454.51 = $177,258.90 to repay the current mortgage.
340 × $457.82 = $155,658.80 to repay the new mortgage.
$177,258.90 − $155,658.80 = $21,600.10 saved.

6A. The owner had a 7 year of interest–only mortgage. After 9 years 7 months, the owner paid 2 years 7 months on a 23 year amortized mortgage.
23 years = (23 years) (12 months / year) = 276 months
2 years + 7 mos = (2 years) (12 months / year) = (24 + 7) mos = 31 months
276 months − 31 months = 245 months remaining on the loan.

Balance is $A = \$1{,}561.75 \left(\dfrac{\left(1+\dfrac{0.084}{12}\right)^{245} - 1}{\left(\dfrac{0.084}{12}\right)\left(1+\dfrac{0.084}{12}\right)^{245}} \right) \approx \$182{,}715.34$

A16

B. Monthly Payment is

$$P = \frac{\left(\frac{(\$182{,}715.34)(0.07)}{12}\right)\left(1+\frac{0.07}{12}\right)^{(12\times 15)}}{\left(1+\frac{0.07}{12}\right)^{(12\times 15)} - 1} \approx \$1{,}642.30$$

C. 245 × $1,561.75 = $382,628.75 to repay the current mortgage.
180 × $1,642.30 = $295,614.00 to repay the new mortgage.
$382,628.75 − $295,614.00 = $87,014.75 saved.

Exercise 3.3

7.

ITEMIZATION OF SETTLEMENT CHARGES Loan Expenses	
Down Payment	$28,490.00
Loan Application Fee	$ 300.00
Points (1.25 pts.)	$ 3,205.13
Title Search	$ 250.00
Title Insurance	$ 400.00
Recording Fee	$ 50.00
Document Preparation Fee	$ 125.00
Credit Search	$ 50.00
Appraisal Fee	$ 350.00
Survey	$ 275.00
Closing Fee	$ 50.00
Interest (max. 1 month)	$ 1,538.46
Taxes (max. 1 month)	$ 433.33
Insurance	$ 490.00
PMI	$ 1,179.49
Legal (Attorney)	$ 500.00
Total Amount Due on Closing	$37,686.41

8.

ITEMIZATION OF SETTLEMENT CHARGES Loan Expenses	
Down Payment	$ 6,280.00
Loan Application Fee	$ 300.00
Points (0.75 pts.)	$ 894.90
Title Search	$ 250.00
Title Insurance	$ 400.00
Recording Fee	$ 50.00
Document Preparation Fee	$ 125.00

Credit Search	$ 50.00
Appraisal Fee	$ 350.00
Survey	$ 275.00
Closing Fee	$ 50.00
Interest (max. 1 month)	$ 835.24
Taxes (max. 1 month)	$ 350.00
Insurance	$ 350.00
PMI	$ 453.42
Legal (Attorney)	$ 500.00
Total Amount Due on Closing	**$11,513.56**

9.

ITEMIZATION OF SETTLEMENT CHARGES	
Loan Expenses	
Down Payment	$ 10,030.00
Loan Application Fee	$ 350.00
Points (0 pts.)	
Title Search	$ 250.00
Title Insurance	$ 400.00
Recording Fee	$ 50.00
Document Preparation Fee	$ 125.00
Credit Search	$ 50.00
Appraisal Fee	$ 350.00
Survey	$ 275.00
Closing Fee	$ 50.00
Interest (max. 1 month)	$ 1,333.99
Taxes (max. 1 month)	$ 400.00
Insurance	$ 400.00
PMI	$ 800.39
Legal (Attorney)	$ 500.00
Total Amount Due on Closing	**$ 15,364.38**

Section 3.4: Annuities that Grow

10. $P = \dfrac{\dfrac{AR}{n}}{\left(1+\dfrac{R}{n}\right)^{(nT)} - 1} = \dfrac{\dfrac{\$42{,}780(0.0575)}{12}}{\left(1+\dfrac{0.0575}{12}\right)^{(12\times 2)} - 1} = \dfrac{\$204.9875}{0.1215653642} \approx \$1{,}686.23$

11. $P = \dfrac{\dfrac{AR}{n}}{\left(1+\dfrac{R}{n}\right)^{(nT)} - 1} = \dfrac{\dfrac{\$9{,}420(0.065)}{12}}{\left(1+\dfrac{0.065}{12}\right)^{\left(\cancel{12}\times\dfrac{18}{\cancel{12}}\right)} - 1} = \dfrac{\$51.025}{0.1021214212} \approx \$499.65$

12. $A = P\left(\dfrac{\left(1+\dfrac{R}{n}\right)^{(nT)} - 1}{\dfrac{R}{n}}\right) = \$450\left(\dfrac{\left(1+\dfrac{0.057}{4}\right)^{(4\times 45)} - 1}{\dfrac{0.057}{4}}\right) = \dfrac{\$450(11.76741419)}{0.01425}$

$A \approx \$371{,}602.55$

13. $A = \dfrac{P(e^R)(e^{(RT)} - 1)}{e^R - 1} = \dfrac{\$2{,}200(e^{0.063})(e^{(0.063\times 45)} - 1)}{(e^{0.063} - 1)}$

$A = \dfrac{\$2{,}200(1.065026839)(16.0304003)}{0.065026839} \approx \$577{,}610.34$

Section 3.5: Annuities that Decay

14. $A = P\left(\dfrac{\left(1+\dfrac{R}{n}\right)^{(nT)} - 1}{\left(\dfrac{R}{n}\right)\left(1+\dfrac{R}{n}\right)^{(nT)}}\right) = \$6{,}666.67\left(\dfrac{\left(1+\dfrac{0.064}{12}\right)^{(12\times 25)} - 1}{\left(\dfrac{0.064}{12}\right)\left(1+\dfrac{0.064}{12}\right)^{(12\times 25)}}\right)$

$A = \$6{,}666.67\left(\dfrac{3.932019028}{(0.005333333333)(4.932019028)}\right) \approx \$996{,}554.60$

15. $A = P\left(\dfrac{\left(1+\dfrac{R}{n}\right)^{(nT)} - 1}{\left(\dfrac{R}{n}\right)\left(1+\dfrac{R}{n}\right)^{(nT)}}\right) = \$4{,}200\left(\dfrac{\left(1+\dfrac{0.043}{12}\right)^{(12\times 2)} - 1}{\left(\dfrac{0.043}{12}\right)\left(1+\dfrac{0.043}{12}\right)^{(12\times 2)}}\right)$

$A = \$4{,}200\left(\dfrac{0.08963882}{(0.0035833333)(1.08963882)}\right) \approx \$96{,}421.89$

16. $P = \dfrac{\left(\dfrac{AR}{n}\right)\left(1+\dfrac{R}{n}\right)^{nT}}{\left(1+\dfrac{R}{n}\right)^{nT}-1} = \dfrac{\left(\dfrac{(\$790{,}000)(0.062)}{12}\right)\left(1+\dfrac{0.062}{12}\right)^{(12\times 10)}}{\left(1+\dfrac{0.062}{12}\right)^{(12\times 10)}-1}$

$P = \dfrac{(\$4{,}081.666667)(1.855963241)}{0.855963241} \approx \$8{,}850.17$

17. $P = \dfrac{\left(\dfrac{AR}{n}\right)\left(1+\dfrac{R}{n}\right)^{nT}}{\left(1+\dfrac{R}{n}\right)^{nT}-1} = \dfrac{\left(\dfrac{(\$1{,}240{,}000)(0.048)}{12}\right)\left(1+\dfrac{0.048}{12}\right)^{(12\times 20)}}{\left(1+\dfrac{0.048}{12}\right)^{(12\times 20)}-1}$

$P = \dfrac{(\$4{,}960)(2.606700133)}{1.606700133} \approx \$8{,}047.07$

Section 3.6: Annuities

18A. $A = P\left(\dfrac{\left(1+\dfrac{R}{n}\right)^{(nT)}-1}{\left(\dfrac{R}{n}\right)\left(1+\dfrac{R}{n}\right)^{(nT)}}\right) = \$2{,}200\left(\dfrac{\left(1+\dfrac{0.062}{12}\right)^{(12\times 24)}-1}{\left(\dfrac{0.062}{12}\right)\left(1+\dfrac{0.062}{12}\right)^{(12\times 24)}}\right)$

$A = \$2{,}200\left(\dfrac{3.411298936}{(0.0051666667)(4.411298936)}\right) \approx \$329{,}280.13$

B. $P = \dfrac{\dfrac{AR}{n}}{\left(1+\dfrac{R}{n}\right)^{(nT)}-1} = \dfrac{\dfrac{\$329{,}280.13(0.06)}{12}}{\left(1+\dfrac{0.06}{12}\right)^{(12\times 37)}-1} = \dfrac{\$1{,}646.40065}{8.156540484} \approx \$201.85$

C. $\$201.85(12)(37) = \$89{,}621.40$ D. $\$2{,}200(12)(24) = \$633{,}600$

19A. $A = P\left(\dfrac{\left(1+\dfrac{R}{n}\right)^{(nT)}-1}{\left(\dfrac{R}{n}\right)\left(1+\dfrac{R}{n}\right)^{(nT)}}\right) = \$3{,}100\left(\dfrac{\left(1+\dfrac{0.062}{12}\right)^{(12\times 24)}-1}{\left(\dfrac{0.062}{12}\right)\left(1+\dfrac{0.062}{12}\right)^{(12\times 24)}}\right)$

$A = \$3{,}100\left(\dfrac{3.411298936}{(0.0051666667)(4.411298936)}\right) \approx \$463{,}985.64$

A20

B. $P = \dfrac{\dfrac{AR}{n}}{\left(1+\dfrac{R}{n}\right)^{(nT)} - 1} = \dfrac{\dfrac{\$463{,}985.64(0.06)}{12}}{\left(\left(1+\dfrac{0.06}{12}\right)^{(12\times 23)} - 1\right)} = \dfrac{\$2{,}319.9282}{2.961257229} \approx \$783.43$

C. $\$783.43(12)(23) = \$216{,}226.68$ D. $\$3{,}100(12)(24) = \$892{,}800$

20A. $A = P\left(\dfrac{\left(1+\dfrac{R}{n}\right)^{(nT)} - 1}{\dfrac{R}{n}}\right) = \$222\left(\dfrac{\left(1+\dfrac{0.053}{12}\right)^{(12\times 38)} - 1}{\dfrac{0.053}{12}}\right) =$

$\$222\left(\dfrac{6.46007501}{0.0044166667}\right) \approx \$324{,}710.19$

B. $P = \dfrac{\left(\dfrac{AR}{n}\right)\left(1+\dfrac{R}{n}\right)^{nT}}{\left(1+\dfrac{R}{n}\right)^{nT} - 1} = \dfrac{\left(\dfrac{(\$324{,}710.19)(0.067)}{12}\right)\left(1+\dfrac{0.067}{12}\right)^{(12\times 26)}}{\left(1+\dfrac{0.067}{12}\right)^{(12\times 26)} - 1}$

$P = \dfrac{(\$1{,}812.965228)(5.68115719)}{4.68115719} \approx \$2{,}200.26$

C. $\$222(12)(38) = \$101{,}232$ D. $\$2{,}200.26(12)(26) = \$686{,}481.12$

21A. $A = P\left(\dfrac{\left(1+\dfrac{R}{n}\right)^{(nT)} - 1}{\dfrac{R}{n}}\right) = \$222\left(\dfrac{\left(1+\dfrac{0.053}{12}\right)^{(12\times 23)} - 1}{\dfrac{0.053}{12}}\right) =$

$\$222\left(\dfrac{2.374732095}{0.0044166667}\right) \approx \$119{,}363.89$

21B. $P = \dfrac{\left(\dfrac{AR}{n}\right)\left(1+\dfrac{R}{n}\right)^{nT}}{\left(1+\dfrac{R}{n}\right)^{nT}-1} =$

$\dfrac{\left(\dfrac{(\$119{,}363.89)(0.067)}{12}\right)\left(1+\dfrac{0.067}{12}\right)^{(12\times 26)}}{\left(1+\dfrac{0.067}{12}\right)^{(12\times 26)}-1} \approx \$808.82$

21C. $\$222(12)(23) = \$61{,}272$

21D. $\$808.82(12)(26) = \$252{,}351.84$

CHAPTER 4 REVIEW

Section 4.1: Perimeter and Circumference

I. Find the perimeter (circumference) for the following shapes. Round your answers to tenths.

1.

Step	Action
Square the lengths of the two sides that you know. Write the larger value first.	$18^2 = 324$, $13.5^2 = 182.25$ 324, 182.25
Since you are solving for the **HYPOTENUSE, ADD**.	$324 + 182.25 = 506.25$
Square root the result.	$\sqrt{506.25} = 22.5$

P = 18 cm + 13.5 cm + 22.5 cm = 54 cm.

2.

Step	Action
Square the lengths of the two sides that you know. Write the larger value first.	$17^2 = 289$, $43^2 = 1{,}849$ 1,849, 289
Since you are solving for the **LEG, SUBTRACT**.	$1{,}849 - 289 = 1560$
Square root the result.	$\sqrt{1560} \approx 39.49683532 \approx 39.5$

P = 17 m + 43 m + 39.5 m = 99.5 m.

3. P = 18 ft + 10 ft + 18 ft + 10 ft = 56 ft

A22

4.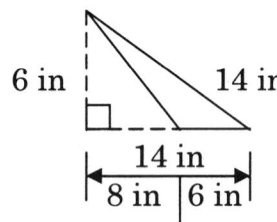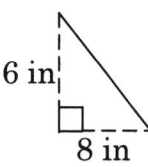

Step	Action
Square the lengths of the two sides that you know. Write the larger value first.	$8^2 = 64$, $6^2 = 36$ 64, 36
Since you are solving for the **HYPOTENUSE, ADD**.	$64 + 36 = 100$
Square root the result.	$\sqrt{100} = 10$

P = 6 in + 10 in + 14 in = 30 in.

5. Circumference = $\pi(18 \text{ ft}) \approx 56.54866776$ ft ≈ 56.5 ft

6. $360° - 79° = 281°$ Circumference = $\dfrac{281°}{360°}(2)(\pi)(24 \text{ m}) \approx 117.7050048$ m

   Perimeter = 117.7050058 m + 24 m + 24 m 165.7050048 m $\approx$ 165.7 m

II. Use Dimensional Analysis to convert the perimeters from Problem I to the specified dimensions.

7. $\dfrac{54 \cancel{\text{cm}}}{1}\left(\dfrac{1 \text{ m}}{100 \cancel{\text{cm}}}\right) = 0.54$ m

8. $\dfrac{99.5 \cancel{\text{m}}}{1}\left(\dfrac{100 \text{ cm}}{1 \cancel{\text{m}}}\right) = 9{,}950$ cm

9. $\dfrac{56 \cancel{\text{ft}}}{1}\left(\dfrac{1 \text{ yard}}{3 \cancel{\text{ft}}}\right) \approx 18.67$ yards

10. $\dfrac{30 \cancel{\text{in}}}{1}\left(\dfrac{1 \text{ ft}}{12 \cancel{\text{in}}}\right) = 2.5$ ft

11. $\dfrac{56.5 \cancel{\text{ft}}}{1}\left(\dfrac{12 \text{ in}}{1 \cancel{\text{ft}}}\right) = 678$ in

12. $\dfrac{165.7 \cancel{\text{m}}}{1}\left(\dfrac{100 \text{ cm}}{1 \cancel{\text{m}}}\right) = 16{,}570$ cm

Section 4.2: Area
I. Find the area for the following shapes. Bump answers up to the next whole number.

13. Area = $\dfrac{13.5 \text{ cm}(18 \text{ cm})}{2} = 121.5$ cm^2. Area = 122 sq cm.

14. Area = $\dfrac{17 \text{ m}(39.5 \text{ m})}{2} \approx 335.75$ m^2. Area = 336 sq m.

15. Area = 18 ft (10 ft) = 180 sq ft.

16. Area = $\dfrac{6 \text{ in}(6 \text{ in})}{2}$ = 18 in²

17. Diameter = 18 ft. Radius = $\dfrac{18 \text{ ft}}{2}$ = 9 ft

    Area = $\pi(9 \text{ ft})^2 \approx 254.4690049$ ft². Area = 255 sq ft.

18. $\dfrac{281°}{360°}(\pi)(24 \text{ m})^2 \approx 1{,}412.460057$ m². Area = 1,413 sq m.

II. Use Dimensional Analysis to convert to the specified dimensions.

19. $\dfrac{122 \, (\cancel{\text{cm}})(\cancel{\text{cm}})}{1}\left(\dfrac{1 \, (\text{in})(\text{in})}{(2.54 \, \cancel{\text{cm}})(3.54 \, \cancel{\text{cm}})}\right) \approx 18.91$ in². Area = 19 sq in.

20. $\dfrac{336 \, (\cancel{\text{m}})(\cancel{\text{m}})}{1}\left(\dfrac{(100 \text{ cm})(100 \text{ cm})}{1 \, (\cancel{\text{m}})(\cancel{\text{m}})}\right) = 3{,}360{,}000$ cm².

21. $\dfrac{180 \, (\cancel{\text{ft}})(\cancel{\text{ft}})}{1}\left(\dfrac{1 \text{ yard}}{(3 \, \cancel{\text{ft}})(3 \, \cancel{\text{ft}})}\right) = 20$ yd²

22. $\dfrac{18 \, (\cancel{\text{in}})(\cancel{\text{in}})}{1}\left(\dfrac{(2.54 \text{ cm})(2.54 \text{ cm})}{(1 \, \cancel{\text{in}})(1 \, \cancel{\text{in}})}\right) \approx 116.1288$ cm². Area = 117 sq cm.

23. $\dfrac{255 \, (\cancel{\text{ft}})(\cancel{\text{ft}})}{1}\left(\dfrac{(12 \text{ in})(12 \text{ in})}{(\cancel{\text{ft}})(\cancel{\text{ft}})}\right) = 36{,}720$ in²

24. $\dfrac{1{,}413 \, (\cancel{\text{m}})(\cancel{\text{m}})}{1}\left(\dfrac{(100 \text{ cm})(100 \text{ cm})}{(\cancel{\text{m}})(\cancel{\text{m}})}\right) = 14{,}130{,}000$ cm²

A24

Section 4.3: Irregular Shapes
I. Find the perimeter and area for the following shapes. Bump answers up to the next whole number.

25.

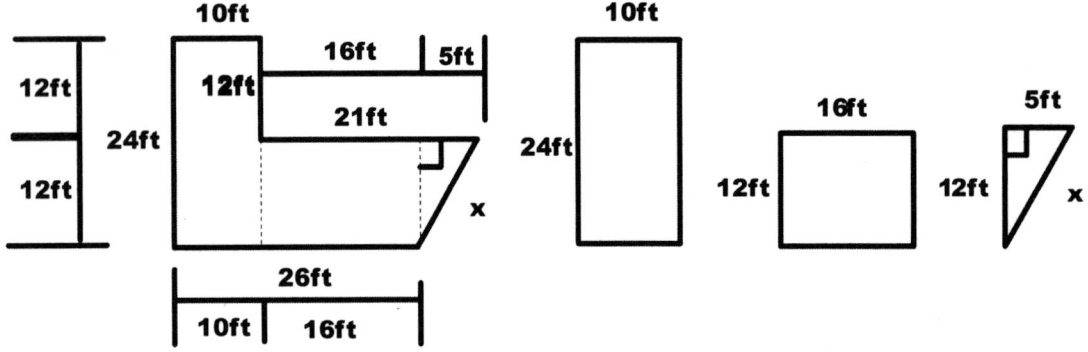

Step	Action
Square the lengths of the two sides that you know. Write the larger value first.	$12^2 = 144, \quad 5^2 = 25$ 144, 25
Since you are solving for the **HYPOTENUSE, ADD**.	$144 + 25 = 169$
Square root the result.	$\sqrt{169} = 13$

Hypotenuse, x = 13 ft
Perimeter = 24 ft + 26 ft + 13 ft + 21 ft + 12 ft + 10 ft = 106 ft.

Area = $10 \text{ ft}(24 \text{ ft}) + 12 \text{ ft}(16 \text{ ft}) + \dfrac{5 \text{ ft}(12 \text{ ft})}{2}$ = (240 + 192 + 30) sq ft

Area = 462 sq ft.

26.

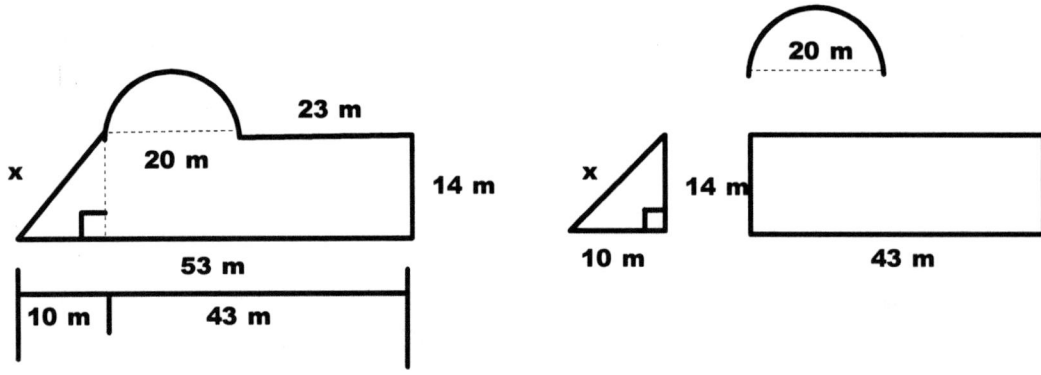

Circumference of the semi-circle = $\dfrac{\pi(20 \text{ m})}{2} \approx 31.4 \text{ m}$.

Step	Action
Square the lengths of the two sides that you know. Write the larger value first.	$14^2 = 196, \quad 10^2 = 100$ 196, 100
Since you are solving for the **HYPOTENUSE, ADD.**	196 + 100 = 296
Square root the result.	$\sqrt{296} \approx 17.2$

Perimeter = 53 m + 14 m + 23 m + 31.4 m + 17.2 m = 138.6 m.
Bump the perimeter up to 139 meters.

Area of the semi-circle = $\dfrac{\pi(10 \text{ m})^2}{2} \approx 157.0796327 \text{ m}^2$. Area ≈ 157.1 sq m.

Area = 157.1 sq m + $\dfrac{14 \text{ m}(10 \text{ m})}{2}$ + 14 m(43 m).

Area = 157.1 sq m + 70 sq m + 602 sq m = 829.1 sq m.
Bump the area to 830 sq m.

27. $360^0 - 55^0 = 305^0$. Circumference = $\dfrac{305^0}{360^0}(2\pi(24 \text{ in})) \approx 127.8 \text{ in}$.

    Perimeter = 127.8 in + 24 in + 24 in = 175.8 in.
    Bump the perimeter up to 176 in.

    Area = $\dfrac{305^0}{360^0}(\pi)(24 \text{ in})^2 \approx 1{,}533.1 \text{ in}^2$. Area = 1,534 sq in.

28.

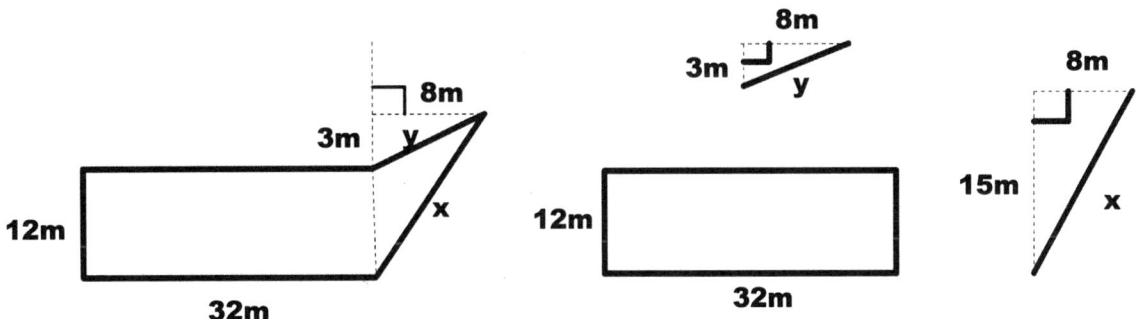

Step	Action
Square the lengths of the two sides that you know. Write the larger value first.	$15^2 = 225, \quad 8^2 = 64$ 225, 64
Since you are solving for the **HYPOTENUSE, ADD.**	225 + 64 = 289
Square root the result.	$\sqrt{289} = 17$

Hypotenuse, x = 17 m

A26

Step	Action
Square the lengths of the two sides that you know. Write the larger value first.	$8^2 = 64$, $3^2 = 9$ 64, 9
Since you are solving for the **HYPOTENUSE, ADD**.	$64 + 9 = 73$
Square root the result.	$\sqrt{73} \approx 8.5$

Perimeter = 12 m + 32 m + 17 m + 8.5 m + 32 m = 101.5 m.
Bump the perimeter up to 102 m.

Area = $(12 \text{ m})(32 \text{ m}) + \dfrac{(8 \text{ m})(12 \text{ m})}{2} = 384$ sq m + 48 sq m = 432 sq m.

II. Use Dimensional Analysis to perform the following conversions.

29. $\dfrac{106 \text{ ft}}{1}\left(\dfrac{1 \text{ yd}}{3 \text{ ft}}\right) \approx 35.3333333$ yd. Bump up to 36 yards.

30. $\dfrac{139 \text{ m}}{1}\left(\dfrac{100 \text{ cm}}{1 \text{ m}}\right) = 13{,}900$ cm.

31. $\dfrac{176 \text{ in}}{1}\left(\dfrac{1 \text{ ft}}{12 \text{ in}}\right) \approx 14.6667$ ft. Bump up to 15 ft.

32. $\dfrac{102 \text{ m}}{1}\left(\dfrac{100 \text{ cm}}{1 \text{ m}}\right) = 10{,}250$ cm.

33. $\dfrac{462 \text{ ft ft}}{1}\left(\dfrac{1 \text{ yd}}{3 \text{ ft}}\right)\left(\dfrac{1 \text{ yd}}{3 \text{ ft}}\right) \approx 51.3333333$ yd^2. Bump up to 52 sq yd.

34. $\dfrac{830 \text{ m m}}{1}\left(\dfrac{100 \text{ cm}}{1 \text{ m}}\right)\left(\dfrac{100 \text{ cm}}{1 \text{ m}}\right) = 8{,}300{,}000$ cm^2.

35. $\dfrac{1{,}534 \text{ in in}}{1}\left(\dfrac{1 \text{ ft}}{12 \text{ in}}\right)\left(\dfrac{1 \text{ ft}}{12 \text{ in}}\right) \approx 10.652778$ ft^2. Bump up to 11 sq ft.

36. $\dfrac{432 \text{ m m}}{1}\left(\dfrac{100 \text{ cm}}{1 \text{ m}}\right)\left(\dfrac{100 \text{ cm}}{1 \text{ m}}\right) = 4{,}320{,}000$ cm^2.

Section 4.4: Volume

I. Find the volume for the following shapes.
Volume = (Area of the base)(Depth)

37. Volume = $(462 \text{ ft}^2)(11 \text{ ft}) = 5{,}082$ ft^3

38. Volume = $(830 \text{ m}^2)(16 \text{ m}) = 13{,}280$ m^3

39. Volume = $(1{,}534 \text{ in}^2)(20 \text{ in}) = 30{,}680$ cu in

40. Volume = $(432 \text{ m}^2)(14 \text{ m}) = 6{,}048$ m^3

II. Use Dimensional Analysis to convert to the specified dimension.

41. $\dfrac{5{,}082 \,\cancel{ft}\,\cancel{ft}\,\cancel{ft}}{1}\left(\dfrac{1 \text{ yd}}{3 \,\cancel{ft}}\right)\left(\dfrac{1 \text{ yd}}{3 \,\cancel{ft}}\right)\left(\dfrac{1 \text{ yd}}{3 \,\cancel{ft}}\right) \approx 188.2 \text{ yd}^3$. Volume = 189 cu yd.

42. $\dfrac{13{,}280 \,\cancel{m}\,\cancel{m}\,\cancel{m}}{1}\left(\dfrac{100 \text{ cm}}{1 \,\cancel{m}}\right)\left(\dfrac{100 \text{ cm}}{1 \,\cancel{m}}\right)\left(\dfrac{100 \text{ cm}}{1 \,\cancel{m}}\right) = 13{,}280{,}000{,}000 \text{ cm}^3$

43. $\dfrac{30{,}680 \,\cancel{in}\,\cancel{in}\,\cancel{in}}{1}\left(\dfrac{1 \text{ ft}}{12 \,\cancel{in}}\right)\left(\dfrac{1 \text{ ft}}{12 \,\cancel{in}}\right)\left(\dfrac{1 \text{ ft}}{12 \,\cancel{in}}\right) \approx 17.755 \text{ ft}^3$.
    Bump up to 18 cu ft.

44. $\dfrac{6{,}048 \,\cancel{m}\,\cancel{m}\,\cancel{m}}{1}\left(\dfrac{100 \text{ cm}}{1 \,\cancel{m}}\right)\left(\dfrac{100 \text{ cm}}{1 \,\cancel{m}}\right)\left(\dfrac{100 \text{ cm}}{1 \,\cancel{m}}\right) = 6{,}048{,}000{,}000 \text{ cm}^3$

III. Automotive conversions

45. $289 \text{ in}^3 = \dfrac{289 \,\cancel{in}\,\cancel{in}\,\cancel{in}}{1}\left(\dfrac{2.54 \text{ cm}}{1 \,\cancel{in}}\right)\left(\dfrac{2.54 \text{ cm}}{1 \,\cancel{in}}\right)\left(\dfrac{2.54 \text{ cm}}{1 \,\cancel{in}}\right) \approx 4{,}735.9 \text{ cc}$

    $V = \dfrac{4{,}735.9 \,\cancel{cc}}{1}\left(\dfrac{1 \text{ liter}}{1{,}000 \,\cancel{cc}}\right) \approx 4.7359 \text{ liters}$

    V = 289 cubic inches is approximately 4.7 liters.

46. $V = 2.8 \text{ liters} = \dfrac{2.8 \text{ liters}}{1}\left(\dfrac{1{,}000 \text{ cc}}{1 \text{ liter}}\right) = \dfrac{2.8 \,\cancel{\text{liters}}}{1}\left(\dfrac{1000 \text{ cc}}{1 \,\cancel{\text{liter}}}\right) = 2{,}800 \text{ cc}$

    $\dfrac{2{,}800\,(\cancel{cm})(\cancel{cm})(\cancel{cm})}{1}\left(\dfrac{1 \text{ in}}{2.54 \,\cancel{cm}}\right)\left(\dfrac{1 \text{ in}}{2.54 \,\cancel{cm}}\right)\left(\dfrac{1 \text{ in}}{2.54 \,\cancel{cm}}\right) \approx 170.9 \text{ in}^3$.

    The volume of the engine is approximately 171 cubic inches.

Section 4.5: Do It Yourself

47. 15 ft   10 ft
     □ 12 ft 5 in   △ 6 ft   □ 6 ft   ◁ 6 ft
                     2.5 ft   2.5 ft

Area of rectangles = $\left(12 + \dfrac{5}{12} \text{ ft}\right)(15 \text{ ft}) = 186.25 \text{ sq ft}$.

$(6 \text{ ft})(10 \text{ ft}) = 60 \text{ sq ft}$

Area of triangle = $\left(\dfrac{1}{2}\right)\left(2+\dfrac{6}{12}\text{ ft}\right)(6\text{ ft}) = 7.5$ sq ft.

$\left(\dfrac{1}{2}\right)\left(2+\dfrac{6}{12}\text{ ft}\right)(6\text{ ft}) = 7.5$ sq ft.

Area of room = 186.25 sq ft + 60 sq ft + 7.5 sq ft + 7.5 sq ft = 261.25 sq ft.

Bump up to 262 sq ft. $\dfrac{262 \text{ sq ft}}{1}\left(\dfrac{1\text{ carton}}{20.67 \text{ sq ft}}\right) \approx 12.7$ cartons

A fraction of a carton cannot be purchased, 13 cartons are needed.
$86.11(13 cartons) = $1,119.43.

48. 11 ft 6 in

13 ft 5 in

Area = $\left(11+\dfrac{6}{12}\text{ft}\right)\left(13+\dfrac{5}{12}\text{ft}\right) \approx 154.29$ sq ft. Bump up to 155 sq ft.

$\dfrac{155 \text{ ft ft}}{1}\left(\dfrac{1\text{ yd}}{3\text{ ft}}\right)\left(\dfrac{1\text{ yd}}{3\text{ ft}}\right) \approx 17.22$ sq yd

A fraction of a square yard cannot be purchased, 18 sq yd are needed.
$21.98(18 sq yards) = $395.64

49. 12 ft 10 in

11 ft 9 in

Area = $\left(12+\dfrac{10}{12}\text{ft}\right)\left(11+\dfrac{9}{12}\text{ft}\right) \approx 150.7916667$ sq ft. Bump up to 151 sq ft.

$\dfrac{151 \text{ ft ft}}{1}\left(\dfrac{12\text{ in}}{1\text{ ft}}\right)\left(\dfrac{12\text{ in}}{1\text{ ft}}\right) = 21{,}744$ sq in.

$\dfrac{21{,}744 \text{ in in}}{1}\left(\dfrac{1\text{ tile}}{64\text{ in in}}\right) \approx 339.75$ tiles. Bump up to 340 tiles.

$\dfrac{340 \text{ tiles}}{1}\left(\dfrac{1\text{ carton}}{24\text{ tiles}}\right) \approx 14.166667$ cartons

A fraction of a carton cannot be purchased, 15 cartons are needed.
15 cartons ($26.88) = $403.20.

50. 12 ft 10 in   12 ft 10 in   11 ft 9 in   11 ft 9 in   12 ft 10 in
☐ 8 ft   ☐ 8 ft   ☐ 8 ft   ☐ 11 ft 9 in   ☐

Area of the two 12 ft 10 in walls = $2\left(12+\frac{10}{12}\text{ft}\right)(8\text{ft}) \approx 205.3$ sq ft

Area of the two 11 ft 9 in walls = $2\left(11+\frac{9}{12}\text{ft}\right)(8\text{ft}) = 188$ sq ft

Area of the ceiling = $\left(12+\frac{10}{12}\text{ft}\right)\left(11+\frac{9}{12}\text{ft}\right) \approx 150.8$ sq ft

Total area = 205.3 + 188 + 150.8 sq ft = 544.1 sq ft. Bump up to 545 sq ft.

$\frac{545\cancel{\text{ft ft}}}{1}\left(\frac{1 \text{ gallon}}{320 \cancel{\text{ft ft}}}\right) \approx 1.703$ gallons. A fraction of a gallon cannot be

purchased, 2 gallons are needed. 2 gallons ($19.99) = $39.98.

51.   12 ft

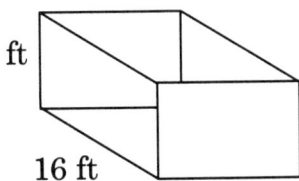

Volume = (12 ft)(16 ft)(8 ft) = 1,536 cubic feet. The AP500 model $169.99.

52. 11 ft 6 in

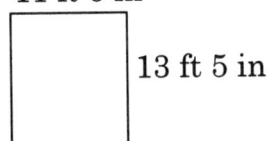

13 ft 5 in

$P = \left(11+\frac{6}{12}\text{ft}\right)+\left(13+\frac{5}{12}\text{ft}\right)+\left(11+\frac{6}{12}\text{ft}\right)+\left(13+\frac{5}{12}\text{ft}\right) \approx 49.8333$ ft

A fraction of a foot cannot be purchased, 50 feet are needed.
50 feet($3.22) = $161.00

53.   15 ft 7 in

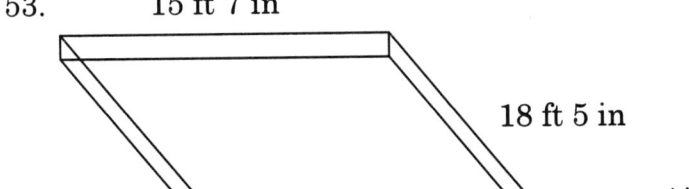

18 ft 5 in

4 in

A30

Volume = $\left(15+\dfrac{7}{12}\text{ ft}\right)\left(18+\dfrac{5}{12}\text{ ft}\right)\left(\dfrac{4}{12}\text{ ft}\right) \approx 95.66435$ cu ft.

Bump volume up to 96 cu ft.

$\dfrac{96\,\cancel{\text{ft}}\,\cancel{\text{ft}}\,\cancel{\text{ft}}}{1}\left(\dfrac{1\text{ yd}}{3\,\cancel{\text{ft}}}\right)\left(\dfrac{1\text{ yd}}{3\,\cancel{\text{ft}}}\right)\left(\dfrac{1\text{ yd}}{3\,\cancel{\text{ft}}}\right) \approx 3.55556$ cu yd

A fraction of a cubic yard cannot be purchased, 4 "yards" are needed.
4 cubic yards ($92.50) = $370.00.

54.   120 ft   160 ft   Area of the lot
                                 Area of rectangle = (120 ft)(300 ft)
   300 ft   300 ft   Area of rectangle = 36,000 sq ft
                                 Area of triangle = $\dfrac{(300\text{ ft})(160\text{ ft})}{2}$
                                 Area of triangle = 24,000 sq ft

Area of the lot = 36,000 sq ft + 24,000 sq ft = 60,000 sq ft.

Other Areas: House is 2,035 sq ft.
   Driveway is a rectangle; Area = (50 ft)(30 ft) = 1,500 sq ft.

Swimming pool is a circle; Area = $\pi(10\text{ ft})^2 \approx 314.2$ sq ft.

Garden is a rectangle; Area = (40 ft)(30 ft) = 1,200 sq ft.

Basketball court is a rectangle; Area = (40 ft)(40 ft) = 1,600 sq ft

Patio is a rectangle; Area = $\left(15+\dfrac{7}{12}\text{ ft}\right)\left(18+\dfrac{5}{12}\text{ ft}\right) \approx 287$ sq ft

Area not seeded = 2,035 sq ft + 1,500 sq ft + 314.2 sq ft + 1,200 sq ft +
        1,600 sq ft + 287 sq ft = 6,936.2 sq ft.
   Bump area up to 6,937 sq ft.

Area to be seeded = 60,000 sq ft – 6,937 sq ft = 53,063 sq ft.

Chapter 5 Review

Section 5.1

1.

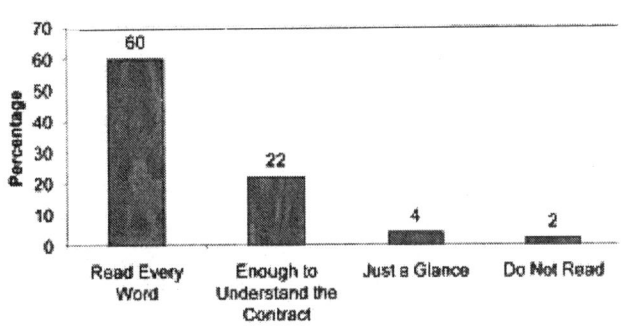

Contract Categories	Percents
Read Every Work	60%
Enough to Understand Contract	22%
Just a Glance	4%
Do Not Read	2%
Total	88%

The sum of the percentages is 88%. Using ± 1% for each category, the interval is 88% to 92%. Since 100% is not contained in this interval, the graph does not satisfy the properties of a graph.

2.

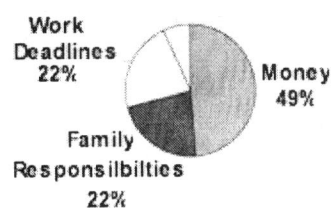

Stress Categories	Percents
Money	49%
Family Responsibilities	22%
Work Deadlines	22%
Commuting	7%
Total	100%

The sum of the percentages is 100%. The graph does satisfy the properties of a graph.

3.

Personal Vehicle of Choice	Frequency	Relative Frequency	Degrees
Car / Station Wagon	1,140	$\frac{1,140}{2,000} = 0.57$	$0.57(360°) \approx 205°$
Pickup Truck	360	$\frac{360}{2,000} = 0.18$	$0.18(360°) \approx 65°$
Sport Utility Vehicle (SUV)	240	$\frac{240}{2,000} = 0.12$	$0.12(360°) \approx 43°$
Van	180	$\frac{180}{2,000} = 0.09$	$0.09(360°) \approx 32°$
Other	80	$\frac{80}{2,000} = 0.04$	$0.04(360°) \approx 14°$
Totals	2,000	1.00	359°

The 359° is an acceptable total. Difference due to rounding.

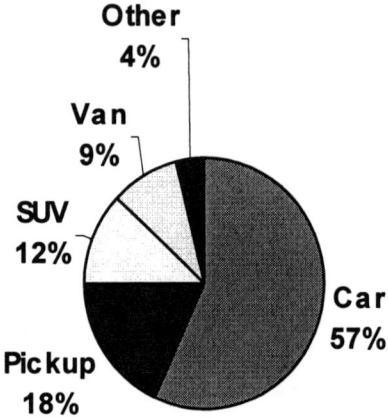

Personal Vehicle of Choice

4. Using the table from problem 3.

Personal Vehicle of Choice

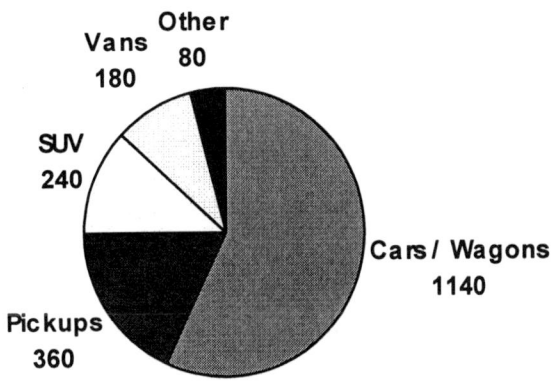

5. Using the table from problem 3.

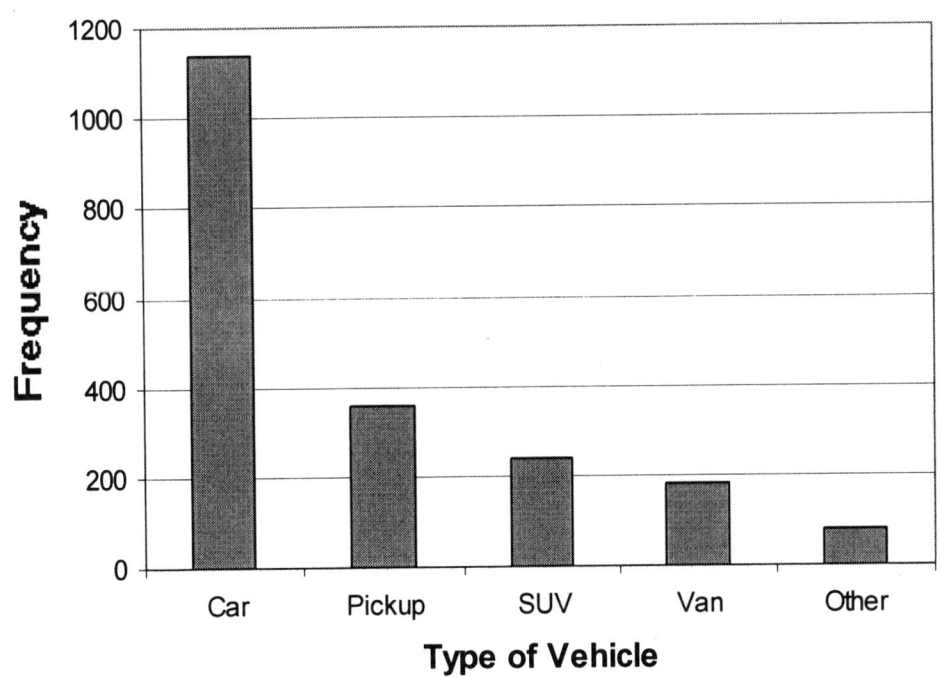

A34

6. Using the table in problem 3.

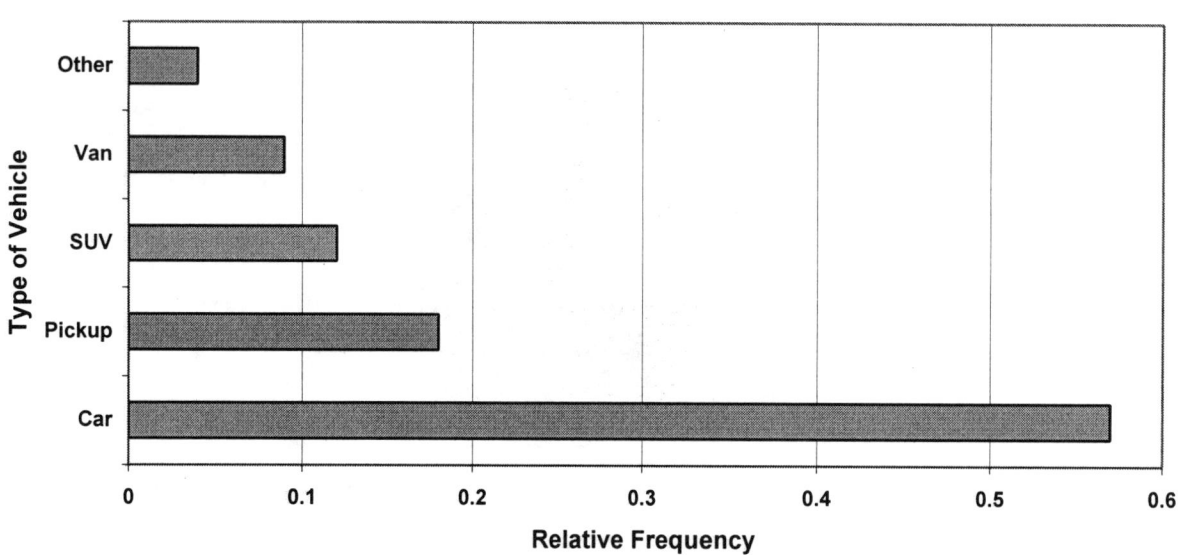

Personal Choice of Vehicle

Section 5.2
7. The sorted data is

$5,699	$5,999	$6,399	$6,499	$6,499	$6,999	$6,999	$7,399
$7,799	$7,899	$7,999	$7,999	$8,499	$8,599	$8,999	$8,999
$9,199	$9,299	$9,499	$9,599	$9,999	$10,199	$10,399	$11,999

Class width $> \dfrac{\$11,999 - \$5,699}{7} = \$900$. Let the class width be $1,000.

Prices (Dollars)	Frequency	Relative Frequency	Class Marks
$5,000 - $6,000	2	$2/24 = 0.0833$	$5,500
$6,000 - $7,000	5	$5/24 = 0.2083$	$6,500
$7,000 - $8,000	5	$5/24 = 0.2083$	$7,500
$8,000 - $9,000	4	$4/24 = 0.1667$	$8,500
$9,000 - $10,000	5	$5/24 = 0.2083$	$9,500
$10,000 - $11,000	2	$2/24 = 0.0833$	$10,500
$11,000 - $12,000	1	$1/24 = 0.0417$	$11,500
Totals	24	0.9999	

The rel. freq. sum of 0.9999 is due to a rounding error.

8. Using the table in problem 7.

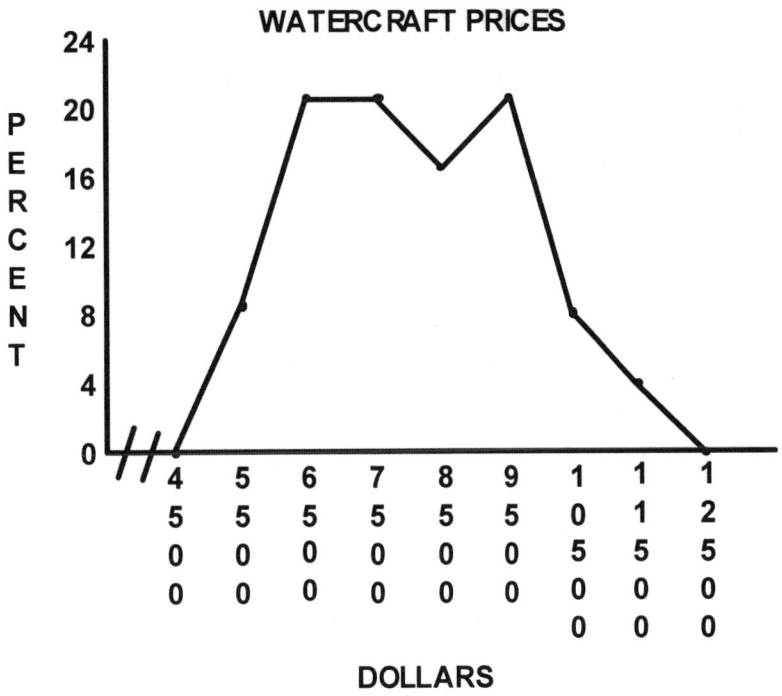

9. Using the table in problem 7.

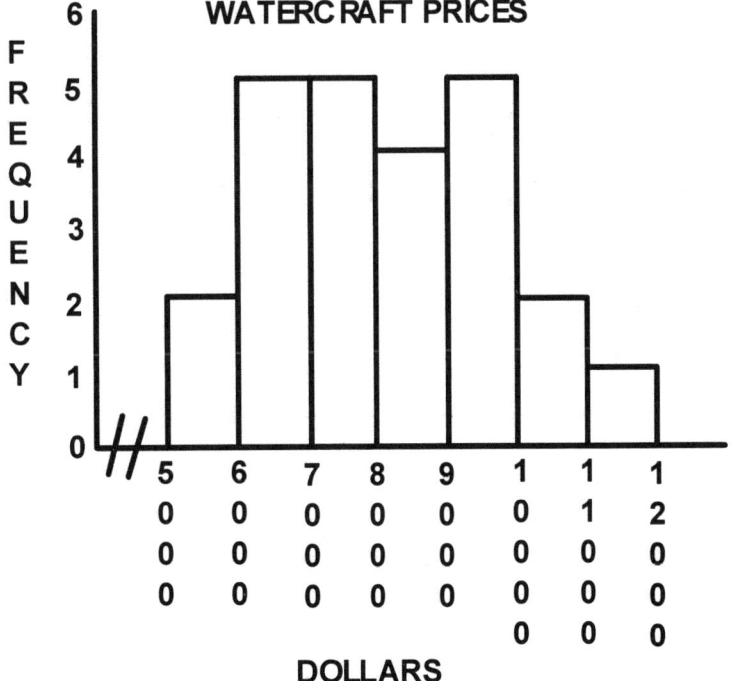

Section 5.3

10A. Using the built in stat function of the calculator, the mean is $8,311.50.

10B. The median is the value at the $\frac{24+1}{2} = 12.5$ position. This is the midpoint of the 12th and 13th positions.

The midpoint is $\frac{\$7,999 + \$8,499}{2} = \$8,249$.

Section 5.4

11A. Using the built in stat function of the calculator, the standard deviation is approximately $1,582.05.

11B. Q1 is the midpoint of the first 12 values. This is the value at the $\frac{12+1}{2} = 6.5$ position. This is the midpoint of the 6th and 7th positions.

Q1 is $\frac{\$6,999 + \$6,999}{2} = \$6,999$.

11C. Q3 is the midpoint of the second 12 values. This is the value at the $\frac{12+1}{2} = 6.5$ position. This is the midpoint of the 6th and 7th positions.

Q3 is $\frac{\$9,299 + \$9,499}{2} = \$9,399$.

12. The boxplot is:
The whiskers are 1.5 (Q3 − Q1) = 1.5($9,399 − $6,999) = $3,600.
The left whisker is $6,999 - $3,600 = $3,399.
The right whisker is $(9,399 + $3,600 = $12,999.

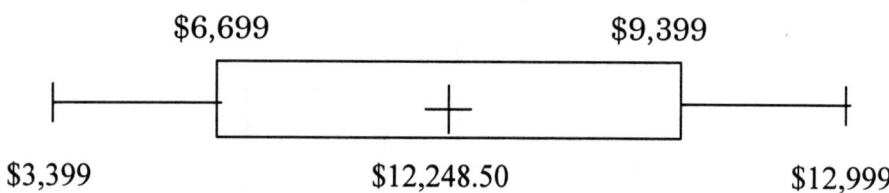

$3,399　　　　　　$6,699　　　　$12,248.50　　　$9,399　　　　$12,999

13. The watercraft data does not have outliers.

Section 5.5

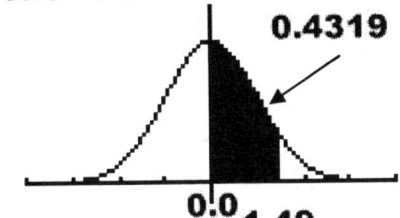
14.
Area = 0.4319

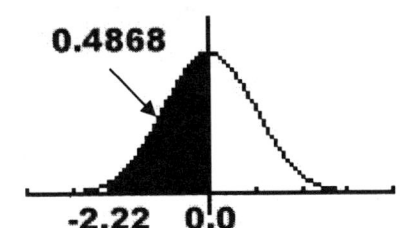
15.
Area = 0.4868

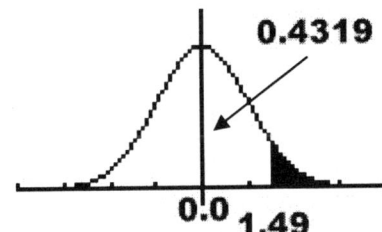
16.
0.5000
− 0.4319
Area = 0.0681

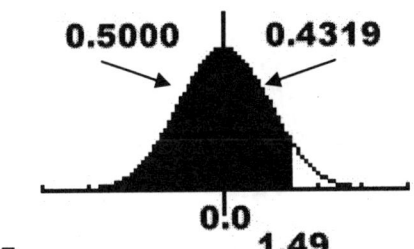
17.
0.5000
+ 0.4319
Area = 0.9319

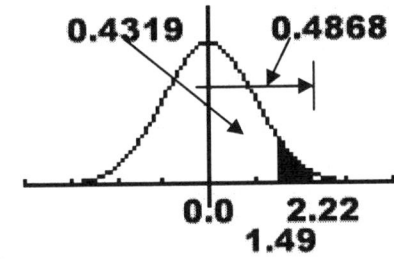
18.
0.4868
− 0.4319
Area = 0.0549

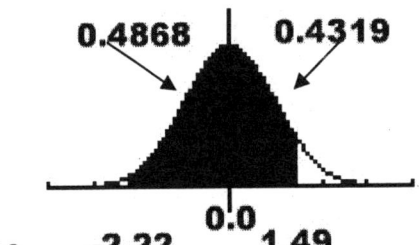
19.
0.4868
+ 0.4319
Area = 0.9187

Section 5.6

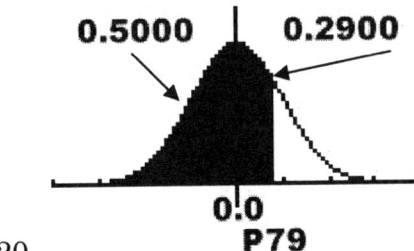
20.

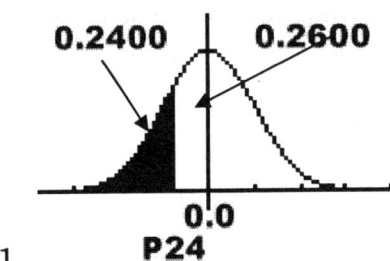
21.

20. 0.2881 **0.2900** 0.2910   21. 0.2580 **0.2600** 0.2611
    ↓   ↓   ↓   ↓
    z = 0.80   z = 0.81   z = 0.70   z = 0.71
    0.2900 is closer to z = 0.81.   0.2600 is closer to z = 0.71.
    $P_{79} = 0.81$.   Since P_{24} is to the left of zero,
                                          it is negative. $P_{24} = -0.71$.

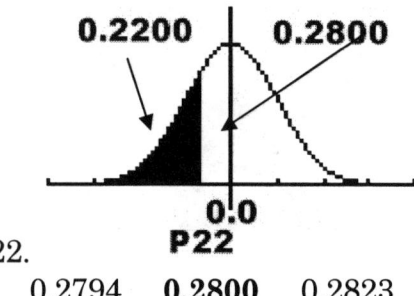
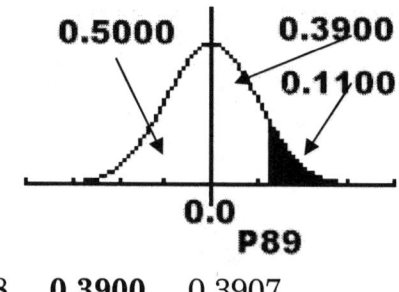

22.
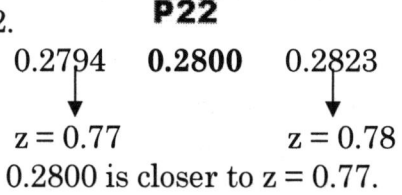

0.2794 **0.2800** 0.2823   0.3888 **0.3900** 0.3907
↓   ↓   ↓   ↓
z = 0.77   z = 0.78   z = 1.22   z = 1.23
0.2800 is closer to z = 0.77.   0.3900 is closer to z = 1.23.
Since P_{22} is to the left of zero, $P_{89} = 1.23$.
z is negative. $P_{22} = -0.77$

The z score marking the lower 22% is –0.77 and the z score marking the upper 11% is 1.23.

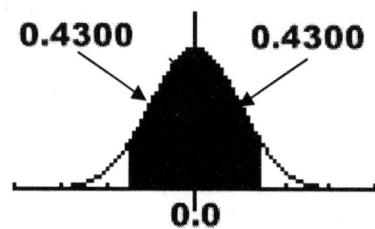

23.
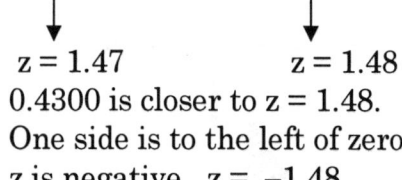

0.4292 **0.4300** 0.4306   0.4292 **0.4300** 0.4306
↓   ↓   ↓   ↓
z = 1.47   z = 1.48   z = 1.47   z = 1.48
0.4300 is closer to z = 1.48.   0.4300 is closer to z = 1.48.
One side is to the left of zero,   z = 1.48.
z is negative. z = –1.48
The z score marking the middle 86% is ± 1.48.

Section 5.7

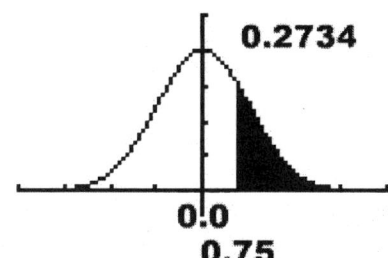

24.

$$z = \frac{x - \mu}{\sigma} = \frac{63 - 60}{4} = \frac{3}{4} = 0.75 \quad P(z > 0.75) = 0.5000$$
$$- \underline{0.2734}$$
$$0.2266$$

P(Battery lasting longer than 63 months) = 0.2266.

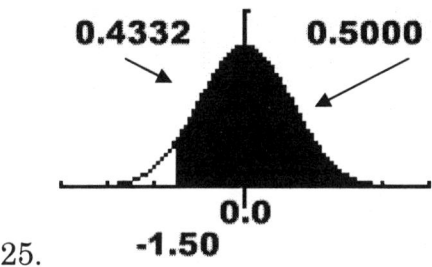

25.

$$z = \frac{x - \mu}{\sigma} = \frac{54 - 60}{4} = \frac{-6}{4} = -1.50 \quad P(z > -1.50) = 0.5000$$
$$+ \underline{0.4332}$$
$$0.9332$$

P(Battery lasting longer than 54 months) = 0.9332.

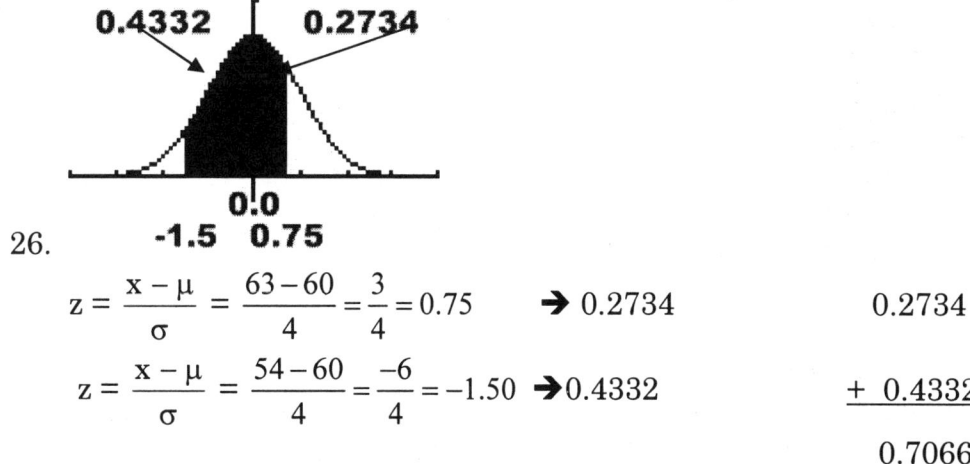

26.

$$z = \frac{x - \mu}{\sigma} = \frac{63 - 60}{4} = \frac{3}{4} = 0.75 \quad \rightarrow 0.2734 \qquad 0.2734$$

$$z = \frac{x - \mu}{\sigma} = \frac{54 - 60}{4} = \frac{-6}{4} = -1.50 \rightarrow 0.4332 \qquad + \underline{0.4332}$$
$$0.7066$$

P(Battery lasting between 54 and 63 months) = 0.7066.

A40

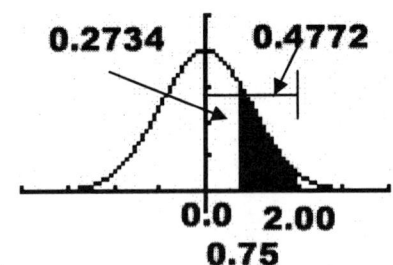

27.
$$z = \frac{x - \mu}{\sigma} = \frac{68 - 60}{4} = \frac{8}{4} = 2.00 \quad \rightarrow 0.4772 \qquad 0.4772$$
$$z = \frac{x - \mu}{\sigma} = \frac{63 - 60}{4} = \frac{3}{4} = 0.75 \quad \rightarrow 0.2734 \qquad \underline{- 0.2734}$$
$$0.2038$$

Section 5.8

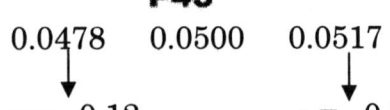

28.
0.0478   0.0500   0.0517

$z = -0.12 \qquad\qquad z = -0.13$

Since 0.0517 is closer to 0.0500, $z = -0.13$.
Substituting these values into the formula for calculating the data value (x) yields: $x = \mu + z(\sigma)$ $x = 1200 + (-0.13)(200) = 1174$

The minimum passing score is 1,174.

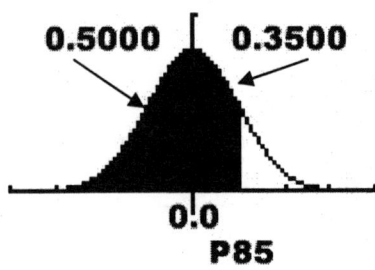

29.
0.3485   0.3500   0.3508

$z = 1.03 \qquad\qquad z = 1.04$

Since 0.3508 is closer to 0.3500, $z = 1.04$.
Substituting these values into the formula for calculating the data value (x) yields: $x = \mu + z(\sigma)$ $x = 600 + (1.04)(102) = 706.08$

The minimum score need to qualify is 706.

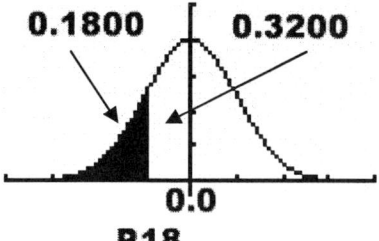

30.

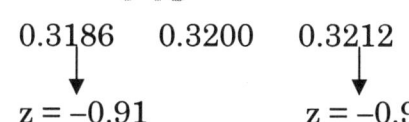

z = −0.91   z = −0.92

Since 0.3212 is closer to 0.3200, z = −0.92.

Substituting these values into the formula for calculating the data value (x) yields: $x = \mu + z(\sigma)$ $x = 128 + (-0.92)(2.4) = 125.792$

Any carton whose weight corresponds to less than 125.792 ounces will be emptied and refilled.

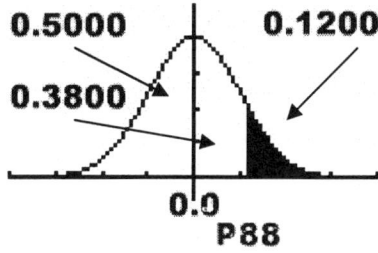

31.

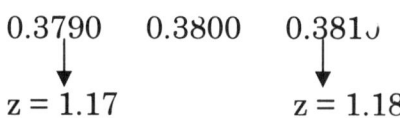

z = 1.17   z = 1.18

Since 0.3800 is in the center of the probabilities, the midpoint of the z scores is calculated. $z = \dfrac{1.17 + 1.18}{2} = \dfrac{2.35}{2} = 1.175$

Substituting these values into the formula for calculating the data value (x) yields: $x = \mu + z(\sigma)$ $x = 128 + (1.175)(2.4) = 130.82$

Any carton whose weight corresponds to more than 130.82 ounces will be emptied and refilled.

Chapter 6 Review
Section 6.1
1.   215 total patients
    − 28 patients experience headaches
    187 patients did not experience headaches

    A. $P(\text{Headache}) = \dfrac{28}{215}$ B. $P(\text{No Headache}) = \dfrac{187}{215}$

2. A. $P(\text{Nausea}) = \dfrac{16}{215}$

   B. $P(\text{No Nausea}) = 1 - P(\text{Nausea}) = 1 - \dfrac{16}{215} = \dfrac{215}{215} - \dfrac{16}{215} = \dfrac{199}{215}$

3.

Cause of Workplace Death	Frequency
Transportation (vehicle accidents, plane crashes)	2,767
Violence (homicides, suicides)	1,318
Hit by equipment, objects	988
Harmful substances, environment	659
Falls	409
Fire, explosion	198
Total	6,339

  A. $P(\text{Violent Death}) = \dfrac{1{,}318}{6{,}339}$

  B.   6,339 total deaths
     − 409 deaths due to falls
     5,930 deaths not due to falls

    $P(\text{Death not caused by a fall}) = \dfrac{5{,}930}{6{,}339}$

  C. $P(\text{Transportation death}) = \dfrac{2{,}767}{6{,}339}$

  D. $P(\text{Death not due to fire, explosion}) = 1 - P(\text{Death due to fire, explosion})$

    $P(\text{Death not due to fire, explosion}) = 1 - \dfrac{198}{6{,}339} = \dfrac{6{,}339}{6{,}339} - \dfrac{198}{6{,}339} = \dfrac{6{,}141}{6{,}339}$

Section 6.2
4. Odds in table are "odds against winning". Losing first and winning second.
   A. Odds against Funny Cide winning are 8 : 1.
      (8 bets are on other horses winning for every 1 bet on Funny Cide winning.)

B. Odds in favor of winning reverses the odds in the table. Winning is first and losing second.
Odds in favor of Ten Most Wanted winning are 1 : 4.

C. P(Hold That Tiger winning) = $\dfrac{\text{winning}}{\text{wining}+\text{lo}\sin g} = \dfrac{1}{1+12} = \dfrac{1}{13}$

D. P(Perfect Drift not winning) = $\dfrac{\text{losing}}{\text{wining}+\text{lo}\sin g} = \dfrac{7}{2+7} = \dfrac{7}{9}$

5. 400 total people surveyed
    − 289 voters in favor of tax increase
    111 voters against the tax increase

A. P(in favor of a tax increase) = $\dfrac{289}{400}$

B. P(against a tax increase) = $\dfrac{111}{400}$

C. Odds in favor of the tax increase, puts in favor first and against second.
Odds in favor of the tax increase are 289 : 111 about 5 : 2.

D. Odds against a tax increase, puts against first and in favor second.
Odds in favor of the tax increase are 111: 289 about 2 : 5.

Section 6.3
6. The random variable x is the prize money minus the raffle ticket price.

Places	Prizes	x	freq	P(x)	x P(x)
First Place	$10,000	$9,980	1	$1/2{,}000 = 0.0005$	$4.99
Second Place	$1,000	$980	1	0.0005	$0.49
Third Place	$750	$730	1	0.0005	$0.365
Fourth Place	$500	$480	1	0.0005	$0.24
Fifth Place	$400	$380	1	0.0005	$0.19
Sixth Place	$300	$280	1	0.0005	$0.14
Seventh Place	$200	$180	1	0.0005	$0.09
Eighth through Twelfth Place	$100	$80	5	$5/2{,}000 = 0.0025$	$0.20
Did not win		− $20	1,988	$1{,}988/2{,}000 = 0.9940$	−$19.88
Totals			2,000	1.0000	−$13.175

A44

A. P(Winning any prize) = $\dfrac{12 \text{ winners}}{2{,}000 \text{ total tickets}}$

B. 2,000 total tickets
    − 5 winners of $100
    1,995 people did not win $100
    Odds against winning $100 is 1,995 : 5 or 399 : 1.

C. 2,000 total tickets
    − 12 winners of any money amount.
    1,988 people did not win any money.
    Odds in favor of losing money is 1,988 : 12 or 497 : 3.

D. P(Winning at least $500) = P($500) + P($750) + P($1,000) + P($10,000)
   P(Winning at least $500) = $\dfrac{4}{2{,}000}$

E. The expected value of this raffle is about −$13.18. This value is the sum of x times P(x) in the above table. It represents the average amount from each $20 ticket that the buyer loses and the charity gains.

Section 6.4
7. The decision tree is:

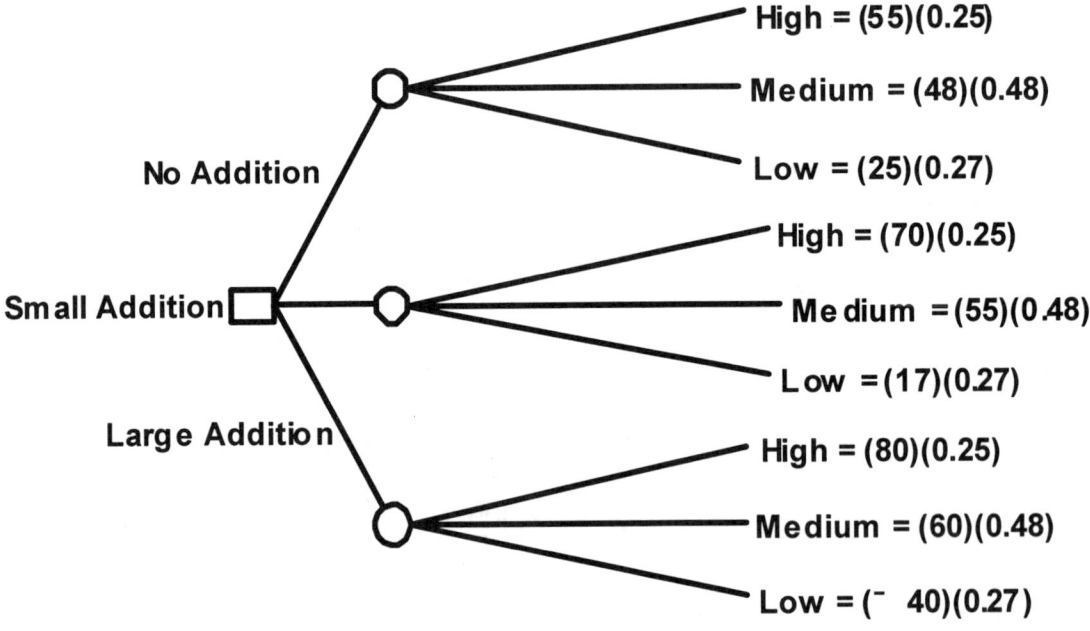

The expected value for each choice is:

No Addition	Small Addition	Large Addition
(55)(0.25) = 13.75	(70)(0.25) = 17.50	(80)(0.25) = 20.00
(48)(0.48) = 23.04	(55)(0.48) = 26.40	(60)(0.48) = 28.80
(32)(0.27) = + 8.64	(17)(0.27) = + 4.59	(−40)(0.27) = + −10.80
45.43	48.49	38.00

Thus, the consulting firm recommends the largest expected profit. The small addition may earn an expected profit of $48.49 thousand ($48,490).

8. The decision tree is:

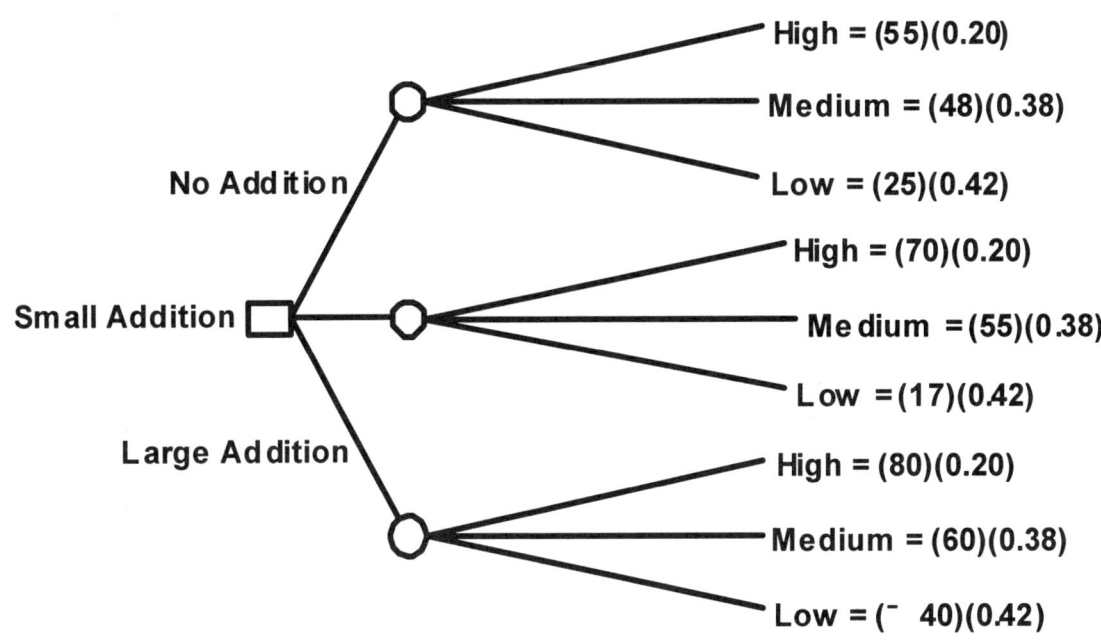

The expected value for each choice is:

No Addition	Small Addition	Large Addition
(55)(0.20) = 11.00	(70)(0.20) = 14.00	(80)(0.20) = 16.00
(48)(0.38) = 18.24	(55)(0.38) = 20.90	(60)(0.38) = 22.80
(32)(0.42) = + 13.44	(17)(0.42) = + 7.14	(−40)(0.42) = + −16.80
42.68	42.04	22.00

Thus, the consulting firm recommends the largest expected profit. The No addition may earn an expected profit of $42.68 thousand ($42,680).

9. In order to maximize profits, the cost is minimized. Thus, the consulting firm recommends the minimum cost. The large addition is recommended by the consulting firm.

Section 6.5

10. The Venn Diagram of problem 1 is:
    b. Label the overlaps first, 10 experience anxiety and dizziness, and 8 experience high blood pressure and dizziness.
    b. Label headaches with 12.
    c. Subtract to account for the other value in the circle.
       Only Anxiety: $18 - 10 = 8$
       Only High Blood Pressure: $15 - 8 = 7$
       Only Dizziness: $24 - 10 - 8 = 6$
    d. Subtract the values from the Venn Diagram to find those patients who did not experience any side effects.
       No side effects: $500 - 8 - 10 - 6 - 8 - 7 - 12 = 449$

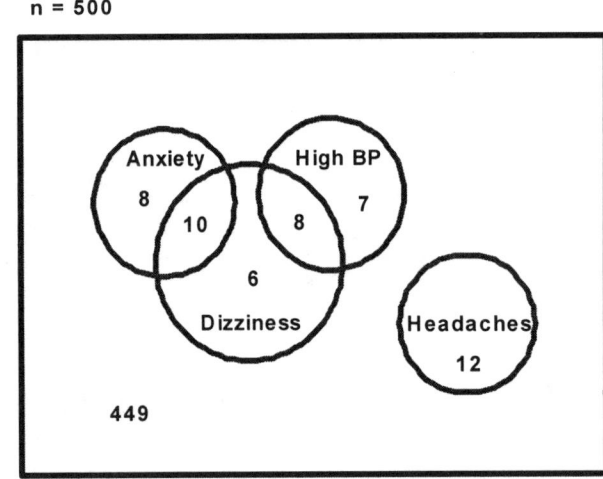

A. P(only headaches) = $\dfrac{12}{500}$

B. P(only dizziness) = $\dfrac{6}{500}$

C. P(only high blood pressure) = $\dfrac{7}{500}$

D. P(only anxiety) = $\dfrac{8}{500}$

E. P(no symptoms) = $\dfrac{449}{500}$

11. The Venn Diagram for Problem 2 is:
    a. Label the major overlap of High Blood Pressure, Dizziness and Headaches = 3.
    b. Subtract 3 to determine other value in the overlaps.
       High blood pressure and headaches; 5 − 3 = 2.
       Headaches and dizziness; 7 − 3 = 4.
       High blood pressure and dizziness; 8 − 3 = 5.
    c. Subtract the values in the circle to get the final value in each circle.
       High blood pressure = 15 − 5 − 3 − 2 = 5.
       Headaches = 12 − 2 − 3 − 4 = 3.
       Dizziness = 24 − 5 − 3 − 4 − 10 = 2.
    d. Label the overlap of dizziness and anxiety = 10.
    e. Subtract 10 to determine the other value in anxiety, 18 − 10 = 8.
    f. Subtract the values from the Venn Diagram to find those patients who did not experience any side effects.
       No side effects: 500 − 5 − 2 − 3 − 4 − 3 − 5 − 2 − 10 − 8 = 458.

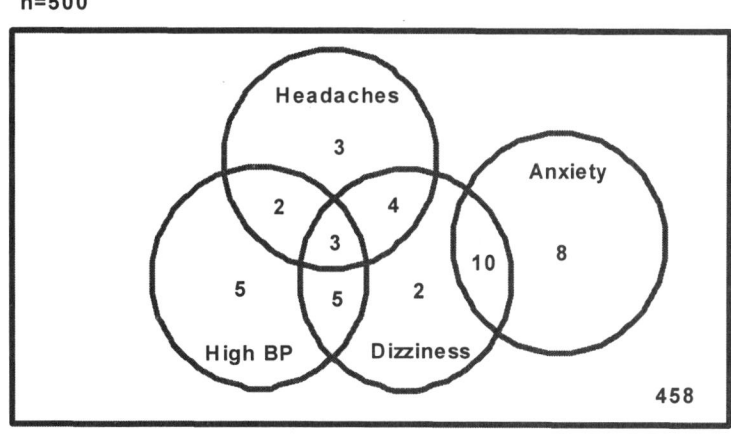

n=500

A. P(only headaches) = $\dfrac{3}{500}$

B. P(only dizziness) = $\dfrac{2}{500}$

C. P(only high blood pressure) = $\dfrac{5}{500}$

D. P(only anxiety) = $\dfrac{8}{500}$

E. P(no symptoms) = $\dfrac{458}{500}$

A48

Section 6.6

12A. P(Headaches OR high blood pressure) = $\dfrac{12+15-0}{500} = \dfrac{27}{500}$

B. P(High blood pressure OR dizziness) = $\dfrac{15+24-8}{500} = \dfrac{31}{500}$

C. P(Headaches OR anxiety) = $\dfrac{12+18-0}{500} = \dfrac{30}{500}$

D. P(Anxiety OR dizziness) = $\dfrac{18+24-10}{500} = \dfrac{32}{500}$

Note: These probabilities can be checked by using the Venn Diagram from Review Chapter 6, Section 6.5, problem 10.

13A. P(Headaches OR high blood pressure) = $\dfrac{12+15-5}{500} = \dfrac{22}{500}$

B. P(High blood pressure OR dizziness) = $\dfrac{15+24-8}{500} = \dfrac{31}{500}$

C. P(Headaches OR anxiety) = $\dfrac{12+18-0}{500} = \dfrac{30}{500}$

D. P(Anxiety OR dizziness) = $\dfrac{18+24-10}{500} = \dfrac{32}{500}$

Note: These probabilities can be checked by using the Venn Diagram from Review Chapter 6, Section 6.5, problem 11.

14.

Reaction to the Medication

Medication	Dizziness	Headache	Swelling	Irregular Heartbeat	Totals
Prescription	18	33	16	10	77
Placebo	18	31	8	9	66
Totals	36	64	24	19	143

A. P(Suffered from Dizziness OR took a Placebo) = $\dfrac{36+66-18}{143} = \dfrac{84}{143}$.

B. P(Took the prescribed medication OR had a headache) = $\dfrac{77+64-33}{143} = \dfrac{108}{143}$.

C. P(Irregular heartbeat OR took the medication) = $\dfrac{19+77-10}{143} = \dfrac{86}{143}$.

D. P(Had a headache OR suffered from swelling) = $\dfrac{64+24-0}{143} = \dfrac{88}{143}$

Section 6.7

15.   32 patients experienced burning sensations
+ 568 patients did not experience burning
600 patients in the study
P(Both patients suffered burning sensations) means
P(Patient 1 AND patient 2 experienced burning).
Since the patient 1 is chosen from the group without replacement, the events are dependent.

A. P(Both patients suffered burning sensations) = $\dfrac{32}{600} \times \dfrac{31}{599} = \dfrac{992}{359,400}$

B. P(Neither patient suffered burning sensations) $\dfrac{568}{600} \times \dfrac{567}{599} = \dfrac{322,056}{359,400}$

16.   24 patients experienced nausea
+ 148 patients did not experience nausea
172 patients in the study
P(Both patients experienced nausea) means
P(Patient 1 AND patient 2 experienced nausea).
Since the patient 1 is chosen from the group without replacement, the events are dependent.

A. P(Both patients experience nausea) = $\dfrac{24}{172} \times \dfrac{23}{171} = \dfrac{552}{29,412}$

B. P(Neither patient experienced nausea) $\dfrac{148}{172} \times \dfrac{147}{171} = \dfrac{21,756}{29,412}$

17. P(Death due to violence) = 0.20
P(Death NOT due to violence) = 1 − P(Death due to violence)
P(Death NOT due to violence) = 1 − 0.20 = 0.80
P(All three deaths are due to violence) means
P(Violent death AND violent death AND violent death).
Since one death has no influence on another death, the events are independent.

A. P(All three deaths are due to violence) =
$(0.20)(0.20)(0.20) = (0.20)^3 = 0.008$

B. P(None of the three deaths are violence) = $(0.80)^3 = 0.512$

18. P(Happy with their job) = 0.33
P(Unhappy with their job) = 1 − P(happy with their job)
P(Unhappy with their job) = 1 − 0.33 = 0.67
P(All four are happy with their jobs) means
P(Happy AND happy AND happy AND happy).
Since one person's satisfaction with their job has no influence on another person's disposition with their job, the events are independent.
A. P(All four are happy with their jobs) =
$(0.33)(0.33)(0.33)(0.33) = (0.33)^4 = 0.01185921 \approx 0.012$
B. P(None of the four are happy with their jobs) =
$(0.67)^4 = 0.20151121 \approx 0.202$

Appendix B

Solution to the Odd Number Problems

Exercise 1.1
Perform the following operations on signed number.

1. $\begin{array}{r}20\\+56\\\hline 76\end{array}$ 3. $\begin{array}{r}20\\+-56\\\hline -36\end{array}$ 5. $\begin{array}{r}20\\-56\\\hline -36\end{array}$ 7. $\begin{array}{r}20\\--56\\\hline 76\end{array}$ 9. $(-934) + 87 = -847$ 11. $(-23) - 45 = -68$

13. $(-861) + (-48) = -909$ 15. $(-3) + (+5) - (-8) = 10$

17. $\$258.31 - \$32.70 - \$124.65 - \$420.88 - \$1{,}257.00 + \$2{,}500.00 = \$923.08$

19. $-1.8 + 0.4 - 0.3 - 0.6 + 1.2 - 0.2 - 0.4 = -1.7$ pounds

21. $(-7)(+8) = -56$ 23. $(-7)(-8) = 56$ 25. $(-13)(9) = -117$

27. $\dfrac{36}{4} = 9$ 29. $\dfrac{24}{-3} = -8$ 31. $\dfrac{54}{-1.6} = -33.75$

33. $(-3)(-2)(+4) = 24$ 35. $(-1.2)(5)(-8) = 48$

37. $\left(\dfrac{15}{16}\right)\left(\dfrac{-48}{125}\right) = \left(\dfrac{\cancel{15}^{\,3}}{\cancel{16}_{\,1}}\right)\left(\dfrac{\cancel{-48}^{\,-3}}{\cancel{125}_{\,25}}\right) = \dfrac{-9}{25}$

39. $(-48)\left(\dfrac{15}{28}\right)\left(\dfrac{49}{-35}\right)(120) = (\cancel{-48}^{-12})\left(\dfrac{\cancel{15}^{\,3}}{\cancel{28}_{\,\cancel{4}_{\,1}}}\right)\left(\dfrac{\cancel{49}^{\,\cancel{7}_{\,1}}}{\cancel{-35}_{\,\cancel{5}_{\,-1}}}\right)(120) = 4{,}320$

Exercise 1.2
Perform the indicated operations. Round to hundredths when necessary.

1. $3 + 4 \times 5 =$
   $3 + 20 = 23$

3. $3 + (-4) \times 5 =$
   $3 - 20 = -17$

5. $24 - 12 \div 4 \times 3 =$
   $24 - 3 \times 3 =$
   $24 - 9 = 15$

7. $(24 - 12) \div (4 \times 3) =$
   $12 \div 12 = 1$

9. $24 - 12 \div 4 \times (-3) =$
   $24 - 3 \times (-3) =$
   $24 - (-9) =$
   $24 + 9 = 33$

B1

11. $(24 - 12) \div (4 \times (-3)) =$
    $12 \div (-12) = -1$

13. $4^3 - 36 \div 2^2 \times 3 =$
    $64 - 36 \div 4 \times 3 =$
    $64 - 9 \times 3 =$
    $64 - 27 = 37$

15. $(4^3 - 36) \div 2^2 \times 3 =$
    $(64 - 36) \div 2^2 \times 3 =$
    $28 \div 2^2 \times 3 =$
    $28 \div 4 \times 3 =$
    $7 \times 3 = 21$

17. $(-4)^3 - 36 \div 2^2 \times 3 =$
    $(-64) - 36 \div 4 \times 3 =$
    $(-64) - 9 \times 3 =$
    $(-64) - 27 = -91$

19. $((-4)^3 - 36) \div 2^2 \times 3 =$
    $(-64 - 36) \div 2^2 \times 3 =$
    $-100 \div 2^2 \times 3 =$
    $-100 \div 4 \times 3 =$
    $-25 \times 3 = -75$

21. $\sqrt{64} \times \sqrt[3]{64} + \sqrt{81} \div \sqrt[4]{81} =$
    $8 \times 4 + 9 \div 3 =$
    $32 + 3 = 35$

23. $7 + 2[50 - 3(4 + 6)] =$
    $7 + 2[50 - 3(10)] =$
    $7 + 2[50 - 30] =$
    $7 + 2[20] =$
    $7 + 40 = 47$

25. $7 + 2[(50 - 3) \times 4 + 6] =$
    $7 + 2[47 \times 4 + 6] =$
    $7 + 2[188 + 6] =$
    $7 + 2[194] =$
    $7 + 388 = 395$

27. $7 + 2[50 - (3 \times 4 + 6)] =$
    $7 + 2[50 - (12 + 6)] =$
    $7 + 2[50 - 18] =$
    $7 + 2[32] =$
    $7 + 64 = 71$

29. $7 + 2[50 - (3 \times (-4) + 6)] =$
    $7 + 2[50 - (-12 + 6)] =$
    $7 + 2[50 - (-6)] =$
    $7 + 2[50 + 6] =$
    $7 + 2[56] = 7 + 112 = 119$

31. $(7 + 2) \times [(50 - 3) \times (-4) + 6] =$
    $9[47 \times (-4) + 6] =$
    $9[-188 + 6] =$
    $9[-182] = -1638$

33. $(7 + 2) \times [50 - (3 \times (-4) + 6)] =$

    $9[50 - (-12 + 6)] =$

    $9[50 - (-6)] =$

    $9[50 + 6] = 9[56] = 504$

35. $\dfrac{(18 - 6) \times 2}{24} =$

    $\dfrac{12 \times 2}{24} =$

    $\dfrac{24}{24} = 1$

37. $\dfrac{100+(4\times 6-4)^3}{2^3+2^2}=$

$\dfrac{100+(24-4)^3}{2^3+2^2}=$

$\dfrac{100+20^3}{2^3+2^2}=$

$\dfrac{100+8000}{8+4}=$

$\dfrac{8100}{12}=675$

39. $\dfrac{(100+4)\times 6-4^3}{2^3+2^2}=$

$\dfrac{104\times 6-4^3}{2^3+2^2}=$

$\dfrac{104\times 6-64}{8+4}=$

$\dfrac{624-64}{12}=\dfrac{560}{12}=46\tfrac{2}{3}$

Exercise 1.3
Solve for x.

1. $\dfrac{x}{12}=\dfrac{2}{8}$ $8x=24$ $\left(\dfrac{1}{\cancel{8}}\right)(\cancel{8})x=\left(\dfrac{1}{\cancel{8}}\right)(\cancel{24})^3$ $x=3$

3. $\dfrac{12}{x}=\dfrac{8}{12}$ $8x=144$ $\left(\dfrac{1}{\cancel{8}}\right)(\cancel{8})x=\left(\dfrac{1}{\cancel{8}}\right)(\cancel{144})^{18}$ $x=18$

5. $\dfrac{5}{9}=\dfrac{x}{12}$ $9x=60$ $\left(\dfrac{1}{\cancel{9}}\right)(\cancel{9})x=\left(\dfrac{1}{\cancel{9}}\right)(\cancel{60})^{20}$ $x=\dfrac{20}{3}=6\tfrac{2}{3}$

7. $\dfrac{5}{8}=\dfrac{12}{x}$ $5x=96$ $\left(\dfrac{1}{\cancel{5}}\right)(\cancel{5})x=\left(\dfrac{1}{5}\right)(96)$ $x=\dfrac{96}{5}=19.2$

9. A. $\dfrac{\text{part}}{\text{whole}}=\dfrac{7}{100}=\dfrac{x}{399.98}$ $100x=2799.86$ $\left(\dfrac{1}{\cancel{100}}\right)(\cancel{100})x=\left(\dfrac{1}{100}\right)(2799.86)$

$$x = \frac{2799.86}{100} = 27.9986 \qquad \text{The sales tax is \$28.00.}$$

B. $399.98 + $28.00 = $427.98

11. 100% — 4% = 96% $\qquad \frac{\text{part}}{\text{whole}} = \frac{96}{100} = \frac{x}{22,167.48} \qquad 100x = 2,128,078.08$

$$\left(\frac{1}{\cancel{100}}\right)(\cancel{100})x = \left(\frac{1}{100}\right)(2128078.08) \qquad x = \frac{2128078.08}{100} = 21280.7808$$

The sales price is $21,280.78.

13. $\frac{\text{part}}{\text{whole}} = \frac{x}{100} = \frac{198}{250} \qquad 250x = 19800 \qquad \left(\frac{1}{\cancel{250}}\right)(\cancel{250})x = \left(\frac{1}{\cancel{250}}\right)(\cancel{19800})$

$$x = \frac{1980}{25} = 79.2 \qquad 79.2\% \text{ have at least one speeding ticket.}$$

15. $\frac{\text{part}}{\text{whole}} = \frac{x}{100} = \frac{18}{400} \qquad 400x = 1800 \qquad \left(\frac{1}{\cancel{400}}\right)(\cancel{400})x = \left(\frac{1}{\cancel{400}}\right)(\cancel{1800})$

$$x = \frac{9}{2} = 4.5 \qquad 4.5\% \text{ of the participants experienced a rash.}$$

17. $\frac{\text{part}}{\text{whole}} = \frac{63}{100} = \frac{252}{x} \qquad 63x = 25200 \qquad \left(\frac{1}{\cancel{63}}\right)(\cancel{63})x = \left(\frac{1}{\cancel{63}}\right)(\cancel{25200})$

x = 400 $\qquad$ Four hundred people were interviewed.

19. $\frac{\text{part}}{\text{whole}} = \frac{65.2}{100} = \frac{163}{x} \qquad 65.2x = 16300 \qquad \left(\frac{1}{\cancel{65.2}}\right)(\cancel{65.2})x = \left(\frac{1}{\cancel{65.2}}\right)(\cancel{16300})$

       x = 250   Two hundred fifty people were interviewed.

Exercise 1.4

1. $\dfrac{\$1 \text{ US}}{\text{foreign equivalent}} = \dfrac{\text{dollar amount (US)}}{\text{foreign equivalent amount}}$ $\dfrac{1}{1.4} = \dfrac{300}{x}$ $1x = 420$

    $300 US is equivalent to 420 Canadian dollars.

3. $\dfrac{\$1 \text{ US}}{\text{foreign equivalent}} = \dfrac{\text{dollar amount (US)}}{\text{foreign equivalent amount}}$ $\dfrac{1}{3.86} = \dfrac{x}{2{,}000}$ $3.86x = 2{,}000$

    $\left(\dfrac{1}{3.86}\right)(3.86)x = \left(\dfrac{1}{3.86}\right)(2{,}000)$ $x = \dfrac{2{,}000}{3.86} \approx 518.134715$

    2,000 Polish zlotys are equivalent to $518.13 US.

5. $\dfrac{\$1 \text{ US}}{\text{foreign equivalent}} = \dfrac{\text{dollar amount (US)}}{\text{foreign equivalent amount}}$ $\dfrac{1}{4.46} = \dfrac{x}{5{,}500}$ $4.46x = 5{,}500$

    $\left(\dfrac{1}{4.46}\right)(4.46)x = \left(\dfrac{1}{4.46}\right)(5{,}500)$ $x = \dfrac{5{,}500}{4.46} \approx 1{,}233.183857$

    5,500 Israeli shekels are equivalent to $1,233.18 US.

7. $\dfrac{\$1 \text{ US}}{\text{foreign equivalent}} = \dfrac{\text{dollar amount (US)}}{\text{foreign equivalent amount}}$ $\dfrac{1}{x} = \dfrac{600}{34{,}600}$ $600x = 34{,}600$

    $\left(\dfrac{1}{600}\right)(600)x = \left(\dfrac{1}{600}\right)(34{,}600)$ $x = \dfrac{34{,}600}{600} \approx 57.66666667$

    $1 US is equivalent to 57.67 Pakistan rupees

9. $\dfrac{\text{in}}{\text{feet}} = \dfrac{1}{4} = \dfrac{4.25}{x}$ $1x = 17$ The length of the living room is 17 feet.

    $\dfrac{\text{in}}{\text{feet}} = \dfrac{1}{4} = \dfrac{3.75}{x}$ $1x = 15$ The width of the living room is 15 feet.

11. $\dfrac{\text{in}}{\text{miles}} = \dfrac{1}{75} = \dfrac{3}{x}$ $1x = 225.$

    It is 225 miles from New York City, NY to Boston, MA.

13. $\dfrac{\text{in}}{\text{feet}} = \dfrac{1}{8} = \dfrac{x}{244}$ $8x = 244$ $\left(\dfrac{1}{\cancel{8}}\right)(\cancel{8})x = \left(\dfrac{1}{\cancel{8}}\right)(\cancel{244}^{\,61})$ $x = \dfrac{61}{2} = 30.5$

The length of the building is 30.5 inches on the blueprint.

15. $\dfrac{\text{in}}{\text{feet}} = \dfrac{1}{4} = \dfrac{x}{52}$ $4x = 52$ $\left(\dfrac{1}{\cancel{4}}\right)(\cancel{4})x = \left(\dfrac{1}{4}\right)(52)$ $x = \dfrac{52}{4} = 13$

The length of the house is 13 inches on the blueprint.

$\dfrac{\text{in}}{\text{feet}} = \dfrac{1}{4} = \dfrac{x}{31}$ $4x = 31$ $\left(\dfrac{1}{\cancel{4}}\right)(\cancel{4})x = \left(\dfrac{1}{4}\right)(31)$ $x = \dfrac{31}{4} = 7.75$

The width of the house is 7.75 inches on the blueprint.

17. $\dfrac{\text{Activity}}{\text{Unit Distance or Unit Time}} = \dfrac{\text{Calories in Food}}{\text{Distance or Time}}$ $\dfrac{\text{calories}}{\text{mile}} = \dfrac{70}{1} = \dfrac{590}{x}$

    $70x = 590$ $\left(\dfrac{1}{\cancel{70}}\right)(\cancel{70}x) = \left(\dfrac{1}{70}\right)(590)$ $x = \dfrac{590}{70} \approx 8.43$

It would take walking 8.43 miles to burn the calories in a McDonald's Big Mac.

19. $\dfrac{\text{Activity}}{\text{Unit Distance or Unit Time}} = \dfrac{\text{Calories in Food}}{\text{Distance or Time}}$ $\dfrac{\text{calories}}{\text{mile}} = \dfrac{225}{1} = \dfrac{390}{x}$

    $225x = 390$ $\left(\dfrac{1}{\cancel{225}}\right)(\cancel{225}x) = \left(\dfrac{1}{225}\right)(390)$ $x = \dfrac{390}{225} \approx 1.73$

It would take power walking 1.73 miles to burn the calories in five Burger King's French toast strips.

21. $\dfrac{\text{Activity}}{\text{Unit Distance or Unit Time}} = \dfrac{\text{Calories in Food}}{\text{Distance or Time}}$ $\dfrac{\text{calories}}{\text{hour}} = \dfrac{493}{1} = \dfrac{415}{x}$

    $493x = 415$ $\left(\dfrac{1}{\cancel{493}}\right)(\cancel{493}x) = \left(\dfrac{1}{493}\right)(415)$ $x = \dfrac{415}{493} \approx 0.84$

It would take 0.84 hours to burn the calories in a 6 inch Wendy's BMT.

23. $\dfrac{\text{Activity}}{\text{Unit Distance or Unit Time}} = \dfrac{\text{Calories in Food}}{\text{Distance or Time}}$ $\dfrac{\text{calories}}{\text{hour}} = \dfrac{493}{1} = \dfrac{730}{x}$

$493x = 730$ $\left(\dfrac{1}{\cancel{493}}\right)(\cancel{493}x) = \left(\dfrac{1}{493}\right)(730)$ $x = \dfrac{730}{493} \approx 1.48$

It would take 1.48 hours to burn the calories in Burger King's enormous omelet sandwich.

25. (FS)(FA) = (OS)(OA), $20(350) = 50x$, $7,000 = 50x$,

$\left(\dfrac{1}{\cancel{50}}\right)(\cancel{7000})^{140} = \left(\dfrac{1}{\cancel{50}}\right)(\cancel{50})^{1}$, $140 = x$.

In order to make 350 ml of 20% solution, 140 ml of 50% solution must be added to 210 ml of distilled water. (Note: 350 − 140 = 210)

27. (FS)(FA) = (OS)(OA), $75(10) = 100x$, $750 = 100x$,

$\left(\dfrac{1}{\cancel{100}}\right)(\cancel{750})^{15}_{2} = \left(\dfrac{1}{\cancel{100}}\right)(\cancel{100})^{1}_{1} x$, $\dfrac{15}{2} = x$, $7.5 = x$.

In order to make 10 ml of 75% solution, 7.5 ml of pure solution must be added to 2.5 ml of distilled water. (Note: 10 − 7.5 = 2.5)

Exercise 1.5
Solve the following equations.

1. $x + 6 = 21$
    $x + 6 - 6 = 21 - 6$
    $x = 15$

3. $17 + x = 11$
    $17 - 17 + x = 11 - 17$
    $x = -6$

5. $x + \dfrac{3}{8} = 10$
    $x + \dfrac{3}{8} - \dfrac{3}{8} = 10 - \dfrac{3}{8}$
    $x = \dfrac{80}{8} - \dfrac{3}{8}$
    $x = \dfrac{77}{8} = 9\dfrac{5}{8}$

7. $12.4 = x - 7.9$
    $12.4 + 7.9 = x - 7.9 + 7.9$
    $20.3 = x$

B8

9. $5x = 120$

$$\left(\frac{1}{\cancel{5}}\right)(\cancel{5})x = \left(\frac{1}{\cancel{5}}\right)(\cancel{120})^{24}$$

$$x = 24$$

11. $144 = 36y$

$$\left(\frac{1}{\cancel{36}}\right)(\cancel{144})^{4} = \left(\frac{1}{\cancel{36}}\right)(\cancel{36})y$$

$$4 = y$$

13. $-x = \frac{43}{8}$

$$(-1)(-x) = (-1)\left(\frac{43}{8}\right)$$

$$x = \frac{-43}{8}$$

15. $3.6z = -25.56$

$$\left(\frac{1}{\cancel{3.6}}\right)(\cancel{3.6})z = \left(\frac{1}{\cancel{3.6}}\right)(\cancel{-25.56})^{-7.1}$$

$$z = -7.1$$

17. $3x + 12 = 42$
$3x + 12 - 12 = 42 - 12$
$3x = 30$

$$\left(\frac{1}{\cancel{3}}\right)(\cancel{3})x = \left(\frac{1}{\cancel{3}}\right)(\cancel{30})^{10}$$

$$x = 10$$

19. $48 - y = 79$
$48 - 48 - y = 79 - 48$
$-y = 31$

$$(-1)(-y) = (-1)(31)$$

$$y = -31$$

21. $5x + \frac{15}{8} = 10$

$$5x + \frac{15}{8} - \frac{15}{8} = \frac{80}{8} - \frac{15}{8}$$

$$5x = \frac{65}{8}$$

$$\left(\frac{1}{5}\right)(5x) = \left(\frac{65}{8}\right)\left(\frac{1}{5}\right) \quad \left(\frac{1}{\cancel{5}}\right)(\cancel{5})x = \left(\frac{1}{\cancel{5}}\right)\left(\frac{\cancel{65}^{13}}{8}\right) \quad x = \frac{13}{8} = 1\frac{5}{8}$$

23.
$$4.2b + 9.4 = 25.36$$
$$4.2b + 9.4 - 9.4 = 25.36 - 9.4$$
$$4.2b = 15.96$$
$$\frac{4.2b}{4.2} = \frac{15.96}{4.2}$$
$$\left(\frac{\cancel{4.2}^{1}}{\cancel{4.2}_{1}}\right)b = \left(\frac{\cancel{15.96}^{3.8}}{\cancel{4.2}}\right)$$
$$b = 3.8$$

25.
$$4(x + 7) = 44$$
$$4x + 28 = 44$$
$$4x + 28 - 28 = 44 - 28$$
$$4x = 16$$
$$\left(\frac{\cancel{4}^{1}}{\cancel{4}_{1}}\right)x = \left(\frac{\cancel{16}^{4}}{\cancel{4}_{1}}\right)$$
$$x = 4$$

27.
$$106 = -2(4x - 5)$$
$$106 = -8x + 10$$
$$106 - 10 = -8x + 10 - 10$$
$$96 = -8x$$
$$\frac{96}{-8} = \frac{-8x}{-8}$$
$$\left(\frac{\cancel{96}^{-12}}{\cancel{-8}}\right) = \left(\frac{\cancel{-8}^{1}}{\cancel{-8}_{1}}\right)x$$
$$-12 = x$$

29.
$$\frac{5}{8}(16x - 8) = 18$$
$$10x - 5 = 18$$
$$10x - 5 + 5 = 18 + 5$$
$$10x = 23$$
$$\frac{10x}{10} = \frac{23}{10}$$
$$\left(\frac{\cancel{10}^{1}}{\cancel{10}_{1}}\right)x = \left(\frac{23}{10}\right)$$
$$x = 2.3$$

31.
$$7.2(3.6z + 4) = 41.76$$
$$25.92z + 28.8 = 41.76$$
$$25.92z + 28.8 - 28.8 = 41.76 - 28.8$$
$$25.92z = 12.96$$
$$\frac{25.92z}{25.92} = \frac{12.96}{25.92}$$
$$\left(\frac{\cancel{25.92}^{1}}{\cancel{25.92}_{1}}\right)z = \left(\frac{12.96}{25.92}\right)$$
$$z = 0.5$$

33.
$$17w + 24 = 15w - 48$$
$$17w - 15w + 24 = 15w - 15w - 48$$
$$2w + 24 = -48$$
$$2w + 24 - 24 = -48 - 24$$
$$2w = -72$$
$$\left(\frac{\cancel{2}^{1}}{\cancel{2}_{1}}\right)w = \left(\frac{\cancel{-72}^{-36}}{\cancel{2}_{1}}\right)$$
$$w = -36$$

35. $\quad 4(2x + 6) = 6x + 6$
$8x + 24 = 6x + 6$
$8x - 6x + 24 = 6x - 6x + 6$
$2x + 24 = 6$
$2x + 24 - 24 = 6 - 24$
$2x = -18$

$$\frac{2x}{2} = \frac{-18}{2} \qquad \left(\frac{\cancel{2}}{\cancel{2}}\right)x = \left(\frac{\cancel{-18}}{\cancel{2}}\right) \qquad x = -9$$

(with 1, 1 under left side and -9, 1 under right)

37. $\quad -2(8x - 5) = 3(-5x + 11)$
$-16x + 10 = -15x + 33$
$-16x + 16x + 10 = -15x + 16x + 33$
$10 = 1x + 33$
$10 - 33 = 1x + 33 - 33$
$-23 = x$

39. $-3(6x - 12) + 20 = 5(4x - 2) + 47$
$-18x + 36 + 20 = 20x - 10 + 47$
$-18x + 56 = 20x + 37$
$-18x + 18x + 56 = 20x + 18x + 37$
$56 = 38x + 37$
$56 - 37 = 38x + 37 - 37$
$19 = 38x$

$$\frac{19}{38} = \frac{38x}{38} \qquad \left(\frac{\cancel{19}}{\cancel{38}}\right) = \left(\frac{\cancel{38}}{\cancel{38}}\right)x \qquad 0.5 = x$$

41. $\qquad 250{,}000 = 130.80A + 24{,}955.70$
$250{,}000 - 24{,}955.70 = 130.80A + 24{,}955.70 - 24{,}955.70$
$225{,}044.30 = 130.80A$

$$\left(\frac{1}{130.80}\right)(225{,}044.30) = \left(\frac{1}{130.80}\right)(130.80)A$$

$$\frac{225{,}044.30}{130.80} \approx 1{,}720.52 \approx A$$

The living area of the house is approximately 1,721 square feet.

43.
$$500{,}000 = 288.50\text{A} + 49{,}790$$
$$500{,}000 - 49{,}790 = 288.50\text{A} + 49{,}790 - 49{,}790$$
$$450{,}210 = 288.50\text{A}$$

$$\left(\frac{1}{288.50}\right)(450{,}210) = \left(\frac{1}{\cancel{288.50}}\right)(\overset{1}{\cancel{288.50}})\text{A}$$

$$\frac{450{,}210}{288.50} \approx 1{,}560.52 \approx \text{A}$$

The living area of the house is approximately 1,561 square feet.

45.
$$1{,}068 = \frac{5}{9}(\text{F} - 32)$$
$$1{,}068 = \frac{5}{9}\text{F} - \frac{160}{9}$$
$$1{,}068 + \frac{160}{9} = \frac{5}{9}\text{F} - \frac{160}{9} + \frac{160}{9}$$
$$\frac{9{,}612}{9} + \frac{160}{9} = \frac{5}{9}\text{F}$$

$$\frac{9{,}772}{9} = \frac{5}{9}\text{F} \qquad \left(\frac{9}{5}\right)\left(\frac{9{,}772}{9}\right) = \left(\frac{9}{5}\right)\left(\frac{5}{9}\right)\text{F} \qquad \left(\frac{\cancel{9}}{5}\right)\left(\frac{9{,}772}{\cancel{9}}\right) = \left(\frac{\cancel{9}}{\cancel{5}}\right)\left(\frac{\cancel{5}}{\cancel{9}}\right)\text{F}$$

$$\frac{9{,}772}{5} = 1{,}954.4 = \text{F} \qquad \text{The melting point of gold is } 1{,}954.4\ {}^\circ\text{F}.$$

47.
$$-183 = \frac{5}{9}(\text{F} - 32)$$
$$-183 = \frac{5}{9}\text{F} - \frac{160}{9}$$
$$-183 + \frac{160}{9} = \frac{5}{9}\text{F} - \frac{160}{9} + \frac{160}{9}$$
$$\frac{-1{,}647}{9} + \frac{160}{9} = \frac{5}{9}\text{F}$$
$$\frac{-1{,}487}{9} = \frac{5}{9}\text{F} \qquad\qquad \left(\frac{9}{5}\right)\left(\frac{-1{,}487}{9}\right) = \left(\frac{9}{5}\right)\left(\frac{5}{9}\right)\text{F}$$

B11

$$\left(\frac{1}{\cancel{5}}\right)\left(\frac{-1,487}{\cancel{9}}\right)^{1} = \left(\frac{\cancel{9}}{\cancel{5}}\right)^{1}\left(\frac{\cancel{5}}{\cancel{9}}\right)^{1} F \qquad\qquad \frac{-1,487}{5} = -297.4 = F$$

The boiling point of liquid oxygen is $-297.4\ ^\circ F$.

49. $$1,538 = \frac{5}{9}(F - 32)$$

$$1,538 = \frac{5}{9}F - \frac{160}{9}$$

$$1,538 + \frac{160}{9} = \frac{5}{9}F - \frac{160}{9} + \frac{160}{9}$$

$$\frac{13,842}{9} + \frac{160}{9} = \frac{5}{9}F$$

$$\frac{14,002}{9} = \frac{5}{9}F \qquad\qquad \left(\frac{9}{5}\right)\left(\frac{14,002}{9}\right) = \left(\frac{9}{5}\right)\left(\frac{5}{9}\right)F$$

$$\left(\frac{1}{\cancel{5}}\right)\left(\frac{14,002}{\cancel{9}}\right)^{1} = \left(\frac{\cancel{9}}{\cancel{5}}\right)^{1}\left(\frac{\cancel{5}}{\cancel{9}}\right)^{1} F \qquad\qquad \frac{14.002}{5} = 2,800.4 = F$$

The melting point of iron is $2,800.4\ ^\circ F$.

51. $C = \frac{5}{9}(1,761 - 32) \qquad C = \frac{5}{9}(1,729) \qquad C \approx 960.555556$

The melting point of Silver is about 960.6 degrees Centigrade

53. $C = \frac{5}{9}(787 - 32) \qquad C = \frac{5}{9}(755) \qquad C \approx 419.4444444$

The melting point of Zinc is about 419.4 degrees Centigrade.

55. $C = \frac{5}{9}(1,220 - 32) \qquad C = \frac{5}{9}(1,188) \qquad C = 660$

The melting point of Aluminum is 660 degrees Centigrade.

Exercise 2.1

1. $I = PRT \quad I = \$2,400(0.055)(1) = \$132$

3. $I = PRT \quad I = (\$2,100)(0.06875)\left(\dfrac{17}{12}\right) = \$204.53125 \qquad I \approx \$204.53$

5. $I = PRT \quad \$25.50 = \$1,800(R)(0.25)$
$I = \$25.50 = 450R$

$$\left(\dfrac{1}{\cancel{\$450}}\right)(\cancel{\$}25.50) = \left(\dfrac{1}{\cancel{\$450}}\right)(\cancel{\$450})R$$

$$\dfrac{25.50}{450} = R \approx 0.566666667 \qquad R \approx 5.7\%$$

7. $I = PRT \quad \$660 = \$2,000(0.06)(T)$
$\$660 = \$120T$

$$\left(\dfrac{1}{\cancel{\$120}}\right)(\cancel{\$660}) = \left(\dfrac{1}{\cancel{\$120}}\right)(\cancel{\$120})T$$

$$\dfrac{11}{2} = 5.5 \text{ years} = T$$

9. $I = PRT \quad \$234 = P(0.065)\left(\dfrac{8}{12}\right)$

$$\$234 = \dfrac{0.52}{12}P$$

$$\left(\dfrac{12}{\cancel{0.52}}\right)(\cancel{\$234}) = \left(\dfrac{\cancel{12}}{\cancel{0.52}}\right)\left(\dfrac{\cancel{0.52}}{\cancel{12}}\right)P \qquad \$5,400 = P$$

11. $A = P(1 + RT) \quad A = \$12,000(1 + (0.0525)(2)) \qquad A = \$13,260$

13. $A = P(1 + RT) \quad A = \$4,800\left(1 + (0.06625)\left(\dfrac{42}{12}\right)\right) \qquad A = \$5,913.00$

15. $A = P(1 + RT)$ $\$2{,}760 = \$2{,}400(1 + R(3))$
$$\$2{,}760 = \$2{,}400 + 7{,}200\,R$$
$$\$2{,}760 - \$2{,}400 = \$2{,}400 + 7{,}200\,R - \$2{,}400$$
$$\$360 = 7{,}200\,R$$

$$\left(\frac{1}{\cancel{7{,}200}}\right)\!\overset{1}{\cancel{(360)}} = \left(\frac{1}{\cancel{7{,}200}}\right)\!\overset{1}{\cancel{(7{,}200)}}R$$
$$201$$

$$\frac{1}{20} = 0.05 = R \qquad R = 5\%$$

17. $A = P(1 + RT)$ $\$2{,}760 = \$2{,}400\bigl(1 + 0.05(T)\bigr)$
$$\$2{,}760 = \$2{,}400 + 120T$$
$$\$2{,}760 - \$2{,}400 = \$2{,}400 + 120\,T - \$2{,}400$$
$$\$360 = 120\,T$$

$$\left(\frac{1}{\cancel{120}}\right)\!\overset{3}{\cancel{(360)}} = \left(\frac{1}{\cancel{120}}\right)\!\overset{1}{\cancel{(120)}}T$$
$$11$$

$$3 = T$$

$$(3\;\cancel{\text{Years}})\left(\frac{12\,\text{months}}{1\,\cancel{\text{year}}}\right) = 36\text{ months}$$

19. $P = \dfrac{A}{1+RT}$ $P = \dfrac{\$9{,}120}{1+(0.035)(4)} = \dfrac{\$9{,}120}{1.14} = \$8{,}000$

Exercise 2.2

1. $\$2{,}400\left(1 + \dfrac{0.051}{1}\right)^{(1\times 3)} \approx \$2{,}786.25$

3. $\$5{,}000\left(1 + \dfrac{0.066}{12}\right)^{(12\times 18)} \approx \$16{,}349.26$

5. $\$5{,}000(e^{(0.06375\times 2)}) \approx \$5{,}679.92$

7. $5{,}000\left(1+\dfrac{0.06375}{52}\right)^{(52\times 2)} \approx \$5{,}679.48$

9. $\dfrac{\$5{,}400}{\left(1+\dfrac{0.042}{2}\right)^{(2\times 1.5)}} \approx \$5{,}073.60$

11. $\dfrac{\$4{,}400}{e^{(0.048\times 4)}} \approx \$3{,}631.35$

13. $\left(1+\dfrac{0.0625}{4}\right)^{4} - 1 \approx 0.0639801621 \approx 6.4\%$

15. $\left(1+\dfrac{0.06625}{2}\right)^{2} - 1 \approx 0.067347266 \approx 6.7\%$

17A. $\left(1+\dfrac{0.04625}{1}\right)^{1} - 1 = 0.04625$

  B. $\left(1+\dfrac{0462}{52}\right)^{52} - 1 \approx 0.0472623658$

  C. $\left(1+\dfrac{0.0461}{365}\right)^{365} - 1 \approx 0.0471760753$

  D. $e^{0.046} - 1 \approx 0.047074411$

The most profitable APY for the savings account is 0.0472623658 derived from 4.62% compounded weekly.

19A. $\left(1+\dfrac{0.04375}{2}\right)^{2} - 1 \approx 0.044228516$

  B. $\left(1+\dfrac{0.0437}{4}\right)^{4} - 1 \approx 0.0444213638$

  C. $\left(1+\dfrac{0.0435}{12}\right)^{12} - 1 \approx 0.0443778469$

  D. $e^{0.0432} - 1 \approx 0.0441467033$

The most beneficial APR for the borrower is 0.0441467033 derived from 4.32% compounded continuously.

Exercise 2.3

1. Loan amount = $26,999 - $3,000 = $23,999

   A. $23,999(1 + (0.09)(5)) = \$34,798.55 \qquad P = \dfrac{\$34,798.55}{12(5)} \approx \$579.98$

   B. $P = \dfrac{\left(\dfrac{(\$23,999)(0.09)}{12}\right)\left(1+\dfrac{0.09}{12}\right)^{(12\times 5)}}{\left(1+\dfrac{0.09}{12}\right)^{(12\times 5)} - 1} = \dfrac{(\$179.9925)(1.565681027)}{0.565681027}$

   $P \approx \$498.18$

3. Loan amount = $19,790 - $2,000 = $17,790

   $P = \dfrac{\$17,790}{12(3)} = \dfrac{\$17,790}{36} = \$494.17$

5. Down payment = (0.10)($46,700) = $4,670
   Loan amount = $46,700 - $4,670 = $42,030

   A. $42,030(1 + (0.092)(6)) = \$65,230.56 \qquad P = \dfrac{\$65,230.56}{72} = \$905.98$

   B. $P = \dfrac{\left(\dfrac{(\$42,030)(0.092)}{12}\right)\left(1+\dfrac{0.092}{12}\right)^{(12\times 6)}}{\left(1+\dfrac{0.092}{12}\right)^{(12\times 6)} - 1} = \dfrac{(\$322.23)(1.733070608)}{0.733070608}$

   $P \approx \$761.79$

7. 0% Down means no down payment. Loan amount = $18,400

   $P = \dfrac{\$18,400}{4(12)} = \dfrac{\$18,400}{48} \approx \$383.33$

9. Loan amount = $12,480 - $500 = $11,980

   A. $11,980(1 + (0.068)(3)) = \$14,423.92 \qquad P = \dfrac{\$14,423.92}{36} \approx \$400.66$

   B. $P = \dfrac{\left(\dfrac{(\$11,980)(0.068)}{12}\right)\left(1+\dfrac{0.068}{12}\right)^{(12\times 3)}}{\left(1+\dfrac{0.068}{12}\right)^{(12\times 3)} - 1} = \dfrac{(\$67.88666667)(1.225592223)}{0.225592223}$

   $P \approx \$368.81$

11. Loan amount = $10,500 − $500 = $10,000
$$P = \frac{\$10,000}{24} \approx \$416.67$$

13. Down Payment = (0.10)($22,680) = $2,268
    Loan Amount = $22,680 − $2,268 = $20,412

    A. $20,412(1 + (0.0575)(4)) = \$25,106.76 \qquad P = \frac{\$25,106.76}{48} \approx \$523.06$

    B. $P = \dfrac{\left(\dfrac{(\$20,412)(0.0575)}{12}\right)\left(1+\dfrac{0.0575}{12}\right)^{(12\times 4)}}{\left(1+\dfrac{0.0575}{12}\right)^{(12\times 4)}-1} = \dfrac{(\$97.8075)(1.257908866)}{(0.257908866)}$

    $P \approx \$477.04$

15. $P = \dfrac{\$16,200}{12(5)} = \dfrac{\$16,200}{60} = \$270.00$

17. Down Payment = (0.15)($170,000) = $25,500
    Loan Amount = $170,000 − $25,500 = $144,500

    A. $\$144,500(1 + (0.07)(15)) = \$296,225 \qquad P = \dfrac{\$296,225}{180} \approx \$1,645.69$

    B. $P = \dfrac{\left(\dfrac{(\$144,500)(0.07)}{12}\right)\left(1+\dfrac{0.07}{12}\right)^{(12\times 15)}}{\left(1+\dfrac{0.07}{12}\right)^{(12\times 15)}-1} = \dfrac{(\$842.9166667)(2.848946731)}{1.848946731}$

    $P \approx \$1,298.81$

Exercise 2.4

1. 5 years = $(5 \text{ years})\left(\dfrac{12 \text{ months}}{1 \text{ year}}\right) = 60$ months

   2 years and 4 months = $(2 \text{ years})\left(\dfrac{12 \text{ months}}{1 \text{ year}}\right) = (24 + 4)$ months = 28 months

   A. $B = \$23,999 - 28\left[\$579.98 - \dfrac{\$23,999(0.09)}{12}\right] = \$12,799.35$

   B. 60 payments − 28 payment made = 32 payments left.
      32($579.98) = $18,559.36

B18

3. 5 years = $(5 \text{ years})\left(\dfrac{12 \text{ months}}{1 \text{ year}}\right)$ = 60 months

2 years and 4 months = $(2 \text{ years})\left(\dfrac{12 \text{ months}}{1 \text{ year}}\right)$ = (24 + 4) months = 28 months

60 months − 28 months = 32 months remaining on the loan.
B = 32($399.98) = $12,799.36

5. 3 years = $(3 \text{ years})\left(\dfrac{12 \text{ months}}{1 \text{ year}}\right)$ = 36 months

1 year and 9 months = $(1 \text{ year})\left(\dfrac{12 \text{ months}}{1 \text{ year}}\right)$ = (12 + 9) months = 21 months

36 months − 21 months = 15 months remaining on the loan.

a. B = $563.65 $\left(\dfrac{\left(1+\dfrac{0.0875}{12}\right)^{15}-1}{\left(\dfrac{0.0875}{12}\right)\left(1+\dfrac{0.0875}{12}\right)^{15}}\right)$ $563.65\left(\dfrac{0.115138001}{(0.0072916667)(1.115138001)}\right)$

B ≈ $7,981.28416   The remaining balance is $7,981.28.

b. B = $15(\$563.65) - \left(\dfrac{15(16)}{36(37)}\right)(36 \times \$563.65 - \$17{,}790)$

B = $8,454.75 − $\dfrac{240}{1{,}332}$($2,501.40) ≈ $8,004.047297

The remaining balance is $8,004.05.

c. 15($563.65) = $8,454.75   The remaining balance is $8,454.75.

7. 6 years = $(6 \text{ years})\left(\dfrac{12 \text{ months}}{1 \text{ year}}\right)$ = 72 months

4 years and 8 months = $(4 \text{ years})\left(\dfrac{12 \text{ months}}{1 \text{ year}}\right)$ = (48 + 8) months = 56 months

72 months − 56 months = 16 months remaining on the loan.

a. B = $761.79 $\left(\dfrac{\left(1+\dfrac{0.092}{12}\right)^{16}-1}{\left(\dfrac{0.092}{12}\right)\left(1+\dfrac{0.092}{12}\right)^{16}}\right)$ = $761.79\left(\dfrac{0.129978758}{(0.0076666667)(1.129978758)}\right)$

B ≈ $11,429.5936   The remaining balance is $11,429.59.

b. $B = 16(\$761.79) - \left(\dfrac{16(17)}{72(73)}\right)(72 \times \$761.79 - \$42{,}030)$

$B = \$12{,}188.64 - \dfrac{272}{5{,}256}(\$12{,}818.88) \approx \$11{,}525.25808$

The remaining balance is $11,525.26.

c. $16(\$761.79) = \$12{,}188.64$ The remaining balance is $12,188.64.

9. 6 years $= (6 \text{ years})\left(\dfrac{12 \text{ months}}{1 \text{ year}}\right) = 72$ months

    4 years and 8 months $= (4 \text{ years})\left(\dfrac{12 \text{ months}}{1 \text{ year}}\right) = (48 + 8)$ months $= 56$ months

  A. $B = \$42{,}030 - 56\left[\$905.98 - \dfrac{\$42{,}030(0.092)}{12}\right] = \$9{,}340.00$

  B. 72 months − 56 months = 16 months remaining on the loan.
     16($905.98) = $14,495.68.

11. 4 years $= (4 \text{ years})\left(\dfrac{12 \text{ months}}{1 \text{ year}}\right) = 48$ months

     3 years $= (3 \text{ years})\left(\dfrac{12 \text{ months}}{1 \text{ year}}\right) = 36$ months

     48 months − 36 months = 12 months remaining on the loan.

  a. $B = \$425.92\left(\dfrac{\left(1+\dfrac{0.079}{12}\right)^{12} - 1}{\left(\dfrac{0.079}{12}\right)\left(1+\dfrac{0.079}{12}\right)^{12}}\right) = \$425.92\left(\dfrac{0.081924169}{(0.0065833333)(1.081924169)}\right)$

     $B \approx \$4{,}898.886884$ The remaining balance is $4,898.89.

  b. $B = 12(\$425.92) - \left(\dfrac{12(13)}{48(49)}\right)(48 \times \$425.92 - \$17{,}480)$

     $B = \$5{,}111.04 - \dfrac{156}{2{,}352}(\$2{,}964.16) \approx \$4{,}914.437551$

     The remaining balance is $4,914.44.

  c. $12(\$425.92) = \$5{,}111.04$ The remaining balance is $5,111.04.

13. 3 years = $(3 \cancel{\text{years}})\left(\dfrac{12 \text{ months}}{1 \cancel{\text{year}}}\right)$ = 36 months

   1 year and 6 months = $(1 \cancel{\text{year}})\left(\dfrac{12 \text{ months}}{1 \cancel{\text{year}}}\right)$ = (12 + 6) months = 18 months

   36 months − 18 months = 18 months remaining on the loan.

   a. B = 368.81\left(\dfrac{\left(1+\dfrac{0.068}{12}\right)^{18}-1}{\left(\dfrac{0.068}{12}\right)\left(1+\dfrac{0.068}{12}\right)^{18}}\right)$ = 368.81\left(\dfrac{0.107064688}{(0.0056666667)(1.107064688)}\right)$

   B ≈ $6,294.312164   The remaining balance is $6,294.31.

   b. B = 18($368.81) − $\left(\dfrac{18(19)}{36(37)}\right)$(36 × $368.81 − $11,980)

   B = $6,638.58 − $\dfrac{342}{1,332}$($1,297.16) ≈ $6,305.525405

   The remaining balance is $6,305.53.

   c. 18($368.81) = $6,638.58   The remaining balance is $6,638.58.

15. 3 years = $(3 \cancel{\text{years}})\left(\dfrac{12 \text{ months}}{1 \cancel{\text{year}}}\right)$ = 36 months

    1 year and 6 months = $(1 \cancel{\text{year}})\left(\dfrac{12 \text{ months}}{1 \cancel{\text{year}}}\right)$ = (12 + 6) months = 18 months

    A. B = $11,980 − 18$\left[\$400.66 - \dfrac{\$11,980(0.068)}{12}\right]$ = $5,990.08

    B. 36 months − 18 months = 18 months remaining on the loan.
       18($400.66) = $7,211.88   The remaining balance is $7,211.88.

17. 4 years = $(4 \cancel{\text{years}})\left(\dfrac{12 \text{ months}}{1 \cancel{\text{year}}}\right)$ = 48 months

    2 years and 11 months = $(2 \cancel{\text{years}})\left(\dfrac{12 \text{ months}}{1 \cancel{\text{year}}}\right)$ = (24 + 11) months = 35 months

    A. B = $20,412 − 35$\left[\$516.14 - \dfrac{\$20,412(0.0575)}{12}\right]$ ≈ $5,770.36

    B. 48 months − 35 months = 13 months remaining on the loan.
       13($516.14) = $6,709.82

19. $5 \text{ years} = (5 \text{ years})\left(\dfrac{12 \text{ months}}{1 \text{ year}}\right) = 60 \text{ months}$

$4 \text{ years} = (4 \text{ years})\left(\dfrac{12 \text{ months}}{1 \text{ year}}\right) = 48 \text{ months}$

60 months − 48 months = 12 months remaining on the loan.
12($270.00) = $3,240.00

21. $5 \text{ years} = (5 \text{ years})\left(\dfrac{12 \text{ months}}{1 \text{ year}}\right) = 60 \text{ months}$

$4 \text{ years} = (4 \text{ years})\left(\dfrac{12 \text{ months}}{1 \text{ year}}\right) = 48 \text{ months}$

A. $B = \$15{,}390 - 48\left[\$297.17 - \dfrac{\$15{,}390(0.0575)}{12}\right] = \$4{,}665.54$

B. 60 months − 48 months = 12 months remaining on the loan.
12($297.17) = $3,566.04

Exercise 3.1

1. Down payment = (0.05)($96,000) = $4,800
Loan amount = $96,000 − $4,800 = $91,200

$$P = \dfrac{\left(\dfrac{(\$91{,}200)(0.079)}{12}\right)\left(1+\dfrac{0.079}{12}\right)^{(12\times 30)}}{\left(1+\dfrac{0.079}{12}\right)^{(12\times 30)} - 1} = \dfrac{(\$600.40)(10.61462557)}{9.61462557}$$

P ≈ $662.85.

....PMI = 0.0032($91,200) = $291.84.

$\text{Escrow} = \dfrac{\text{Taxes} + \text{Insurance} + \text{PMI}}{12} = \dfrac{\$2{,}100 + \$230 + \$291.84}{12} \approx \$218.49$

Actual monthly payment = $662.85 + $218.49 = $881.34.

3. Down payment = (0.10)($249,000) = $24,900
Loan amount = $249,000 − $24,900 = $224,100

$$P = \dfrac{\left(\dfrac{(\$224{,}100)(0.084)}{12}\right)\left(1+\dfrac{0.084}{12}\right)^{(12\times 30)}}{\left(1+\dfrac{0.084}{12}\right)^{(12\times 30)} - 1} = \dfrac{(\$1{,}568.70)(12.31996288)}{11.31996288}$$

P ≈ $1,707.28

B22

$$\text{PMI} = 0.0056(\$224{,}100) = \$1{,}254.96$$
$$\text{Escrow} = \frac{\text{Taxes} + \text{Insurance} + \text{PMI}}{12} = \frac{\$6{,}200 + \$550 + \$1{,}254.96}{12} = \$667.08.$$
Actual monthly payment = $\$1{,}707.28 + \$667.08 = \$2{,}374.36$.

5. Down payment = $(0.05)(\$96{,}000) = \$4{,}800$
Loan amount = $\$96{,}000 - \$4{,}800 = \$91{,}200$

$$P = \frac{\left(\dfrac{(\$91{,}200)(0.079)}{12}\right)\left(1 + \dfrac{0.079}{12}\right)^{(12 \times 30)}}{\left(1 + \dfrac{0.079}{12}\right)^{(12 \times 30)} - 1} = \frac{(\$600.40)(10.61462557)}{9.614625568}$$

$P \approx \$662.8465296$. Then dividing by 2 for bi-weekly, $\dfrac{P}{2} = \dfrac{\$662.8465296}{2} \approx \$331.42$

$\text{PMI} = 0.0032(\$91{,}200) = \$291.84$.
$$\text{Escrow} = \frac{\text{Taxes} + \text{Insurance} + \text{PMI}}{26} = \frac{\$2{,}100 + \$230 + \$291.84}{26} \approx \$100.84.$$
Actual monthly payment = $\$331.42 + \$100.84 = \$432.26$.

7. Down payment = $(0.10)(\$249{,}000) = \$24{,}900$
Loan amount = $\$249{,}000 - \$24{,}900 = \$224{,}100$

$$P = \frac{\left(\dfrac{(\$224{,}100)(0.084)}{12}\right)\left(1 + \dfrac{0.084}{12}\right)^{(12 \times 30)}}{\left(1 + \dfrac{0.084}{12}\right)^{(12 \times 30)} - 1} = \frac{(\$1{,}568.70)(12.31996288)}{11.31996288}$$

$P \approx \$1{,}707.278193$. Dividing by 2 for bi-weekly, $\dfrac{P}{2} = \dfrac{\$1{,}707.278193}{2} \approx \$853.64$

$\text{PMI} = 0.0056(\$224{,}100) = \$1{,}254.96$
$$\text{Escrow} = \frac{\text{Taxes} + \text{Insurance} + \text{PMI}}{26} = \frac{\$6{,}200 + \$550 + \$1{,}254.96}{26} = \$307.88.$$
Actual monthly payment = $\$853.64 + \$307.88 = \$1{,}161.52$.

9. Down payment = $(0.05)(\$96{,}000) = \$4{,}800$
Loan amount = $\$96{,}000 - \$4{,}800 = \$91{,}200$
For the first 5 years: a. Interest–only = $\dfrac{0.079}{12}(\$91{,}200) = \$600.40$.
b. $\text{PMI} = 0.0032(\$91{,}200) = \$291.84$.
c. $\text{Escrow} = \dfrac{\text{Taxes} + \text{Insurance} + \text{PMI}}{12} = \dfrac{\$2{,}100 + \$230 + \$291.84}{12} \approx \$218.49$.

d. Monthly payment A = $600.40 + $218.49 = $818.89.
For the remaining 25 years

a. $P = \dfrac{\left(\dfrac{(\$91,200)(0.079)}{12}\right)\left(1+\dfrac{0.079}{12}\right)^{(12\times 25)}}{\left(1+\dfrac{0.079}{12}\right)^{(12\times 25)} - 1} = \dfrac{(\$600.40)(7.160124367)}{6.160124367}$

$P \approx \$697.87.$

b. PMI = 0.0032($91,200) = $291.84.

c. Escrow = $\dfrac{\text{Taxes} + \text{Insurance} + \text{PMI}}{12} = \dfrac{\$2,100 + \$230 + \$291.84}{12} \approx \$218.49$

d. Actual monthly payment = $697.87 + $218.49 = $916.36.

11. Down payment = (0.10)($249,000) = $24,900
    Loan amount = $249,000 − $24,900 = $224,100
    For the first 7 years:

    a. Interest–only = $\dfrac{0.084}{12}(\$224,100) = \$1,568.70$.

    b. PMI = 0.0056($224,100) = $1,254.96

    c. Escrow = $\dfrac{\text{Taxes} + \text{Insurance} + \text{PMI}}{12} = \dfrac{\$6,200 + \$550 + \$1,254.96}{12} = \$667.08.$

    d. Actual monthly payment = $1,568.70 + $667.08 = $2,235.78.
    For the remaining 23 years

    a. $P = \dfrac{\left(\dfrac{(\$224,100)(0.084)}{12}\right)\left(1+\dfrac{0.084}{12}\right)^{(12\times 23)}}{\left(1+\dfrac{0.084}{12}\right)^{(12\times 23)} - 1} = \dfrac{(\$1,568.70)(6.856995632)}{5.856995632}$

    $P \approx \$1,836.53.$

    b. PMI = 0.0056($224,100) = $1,254.96

    c. Escrow = $\dfrac{\text{Taxes} + \text{Insurance} + \text{PMI}}{12} = \dfrac{\$6,200 + \$550 + \$1,254.96}{12} = \$667.08.$

    d. Actual monthly payment = $1,836.53 + $667.08 = $2,503.61.

Exercise 3.2

1A. 30 years (12 months / year) = 360 months
    10 years 9 mos = (10 years) (12 months / year) + 9 mos = 129 months
    360 months − 129 months = 231 months remaining on the loan.

B24

Balance is $A = \$662.85 \left(\dfrac{\left(1 + \dfrac{0.079}{12}\right)^{231} - 1}{\left(\dfrac{0.079}{12}\right)\left(1 + \dfrac{0.079}{12}\right)^{231}} \right) \approx \$78{,}571.40$

B. Monthly Payment is $P = \dfrac{\left(\dfrac{(\$78{,}571.40)(0.066)}{12}\right)\left(1 + \dfrac{0.066}{12}\right)^{(12 \times 15)}}{\left(1 + \dfrac{0.066}{12}\right)^{(12 \times 15)} - 1} \approx \$688.77$

C. 231 × $662.85 = $153,118.35 to repay the current mortgage.
180 × $688.77 = $123,978.60 to repay the new mortgage.
$153,118.35 − $123,978.60 = $29,139.75 saved.

3A. 30 years (12 months / year) = 360 months
8 years (12 months / year) = 96 months
360 months − 96 months = 264 months remaining on the loan.

Balance is $A = \$2{,}094.22 \left(\dfrac{\left(1 + \dfrac{0.0775}{12}\right)^{264} - 1}{\left(\dfrac{0.0775}{12}\right)\left(1 + \dfrac{0.0775}{12}\right)^{264}} \right) \approx \$264{,}999.64$

B. Monthly Payment is $P = \dfrac{\left(\dfrac{(\$264{,}999.64)(0.063)}{12}\right)\left(1 + \dfrac{0.063}{12}\right)^{(12 \times 10)}}{\left(1 + \dfrac{0.063}{12}\right)^{(12 \times 10)} - 1} \approx \$2{,}982.12$

C. 264 × $2,094.22 = $552,874.08 to repay the current mortgage.
120 × $2,982.12 = $357,854.40 to repay the new mortgage.
$552,874.08 − $357,854.40 = $195,019.68 saved.

5A. Balance is: $B = \$91{,}200 - \$54.31 \left(\dfrac{\left(1 + \dfrac{0.079}{26}\right)^{182} - 1}{\dfrac{0.079}{26}} \right) \approx \$78{,}026.67$.

B. Monthly $P = \dfrac{\left(\dfrac{(\$78{,}026.67)(0.066)}{12}\right)\left(1 + \dfrac{0.066}{12}\right)^{(12 \times 15)}}{\left(1 + \dfrac{0.066}{12}\right)^{(12 \times 15)} - 1} \approx \$683.99$

Bi-weekly is $683.99 / 2 = $342.00.

C. First months interest is: $\dfrac{\$78{,}026.67(0.066)}{26} \approx \$198.07$

First months principal is $342.00 − $198.07 = $143.93.

Payments to repay 15 year bi-weekly: $x = \dfrac{\log\left(1 + \dfrac{\$78{,}026.67(0.066)}{26(\$143.93)}\right)}{\log\left(1 + \dfrac{0.066}{26}\right)} \approx 342$

597 − 182 = 415 payments remaining on current mortgage.
415 × $331.42 = $137,539.30 to repay the current mortgage.
342 × $342.00 = $116,964.00 to repay the new mortgage.
$137,539.30 − $116,964.00 = $20,575.30 saved.

7A. Balance is: $B = \$224{,}100 - \$129.62\left(\dfrac{\left(1 + \dfrac{0.084}{26}\right)^{330} - 1}{\dfrac{0.084}{26}}\right) \approx \$147{,}903.14$.

B. Monthly $P = \dfrac{\left(\dfrac{(\$147{,}903.14)(0.072)}{12}\right)\left(1 + \dfrac{0.072}{12}\right)^{(12 \times 10)}}{\left(1 + \dfrac{0.072}{12}\right)^{(12 \times 10)} - 1} \approx \$1{,}732.57$

Bi-weekly is $1,732.57 / 2 = $866.29.

C. First months interest is: $\dfrac{\$147{,}903.14(0.072)}{26} \approx \$409.58$.

First months principal is $866.29 − $409.58 = $456.71.

Payments to repay 10 year bi-weekly: $x = \dfrac{\log\left(1 + \dfrac{\$147{,}903.14(0.072)}{26(\$456.51)}\right)}{\log\left(1 + \dfrac{0.072}{26}\right)} \approx 232$.

585 − 390 = 255 payments remaining on current mortgage.
255 × $853.64 = $217,678.20 to repay the current mortgage.
232 × $866.29 = $200,979.28 to repay the new mortgage.
$217,678.20 − $200,979.28 = $16,698.92 saved.

9A. The interest–only period was for 5 years. After 7 years 8 months the mortgage was amortized for 2 years 8 months of the 25 year amortization period.

25 years (12 months / year) = 300 months
2 years 8 mos = (2 years) (12 months / year) + 8 mos = (24 + 8) mos = 32 months
300 months − 32 months = 268 months remaining on the loan.

Balance is $A = \$697.87 \left(\dfrac{\left(1+\dfrac{0.079}{12}\right)^{268} - 1}{\left(\dfrac{0.079}{12}\right)\left(1+\dfrac{0.079}{12}\right)^{268}} \right) \approx \$87{,}741.41$

B. Monthly Payment is $P = \dfrac{\left(\dfrac{(\$87{,}741.41)(0.063)}{12}\right)\left(1+\dfrac{0.063}{12}\right)^{(12\times 15)}}{\left(1+\dfrac{0.063}{12}\right)^{(12\times 15)} - 1} \approx \$754.71$

C. 268 × $697.87 = $187,029.16 to repay the current mortgage.
180 × $754.71 = $135,847.80 to repay the new mortgage.
$187,029.16 − $135,847.80 = $51,181.36 saved.

11A. The interest–only period was for 7 years. After 5 years this leaves 2 years on the interest–only period and 23 years on the amortization period.
Down payment = (0.10)($249,000) = $24,900.
Loan amount = $249,000 − $24,900 = $224,100.
After 5 years of interest–only payments, the remaining balance is $224,100.

B. Monthly Payment is $P = \dfrac{\left(\dfrac{(\$224{,}100)(0.071)}{12}\right)\left(1+\dfrac{0.071}{12}\right)^{(12\times 15)}}{\left(1+\dfrac{0.071}{12}\right)^{(12\times 15)} - 1} \approx \$2{,}026.82$

C. To repay the current mortgage:
1. 2 years of interest–only payments,

$2 \text{ years} = (2 \text{ years})\left(\dfrac{12 \text{ months}}{1 \text{ year}}\right) = 24 \text{ months}$

24 × $1,568.70 = $37,648.80 and
2. 23 years of amortized payments,

$23 \text{ years} = (23 \text{ years})\left(\dfrac{12 \text{ months}}{1 \text{ year}}\right) = 276 \text{ months}$

276 × $1,836.53 = $506,882.28.
It costs $37,648.80 + $506,882.28 = $544,531.08 to repay the current mortgage.
To repay the new mortgage, 180 × $2,026.82 = $364,827.60.
$544,531.08 − $364,827.60 = $179,703.48 saved.

Exercise 3.3

1.

ITEMIZATION OF SETTLEMENT CHARGES	
Down Payment	$ 4,800.00
Loan Application Fee	$ 300.00
Points (3 pts.)	$ 2,736.00
Title Search	$ 250.00
Title Insurance	$ 400.00
Recording Fee	$ 50.00
Document Preparation Fee	$ 125.00
Credit Search	$ 50.00
Appraisal Fee	$ 350.00
Survey	$ 275.00
Closing Fee	$ 50.00
Interest (max. 1 month)	$ 570.00
Taxes (max. 1 month)	$ 175.00
Insurance	$ 230.00
PMI	$ 291.84
Legal (Attorney)	$ 500.00
Total Amount Due on Closing	$11,152.84

3.

ITEMIZATION OF SETTLEMENT CHARGES	
Down Payment	$32,480.00
Loan Application Fee	$ 325.00
Points (pts.)	$
Title Search	$ 250.00
Title Insurance	$ 400.00
Recording Fee	$ 50.00
Document Preparation Fee	$ 125.00
Credit Search	$ 50.00
Appraisal Fee	$ 350.00
Survey	$ 275.00
Closing Fee	$ 50.00
Interest (max. 1 month)	$ 1,887.90
Taxes (max. 1 month)	$ 405.83
Insurance	$ 409.00
PMI	$ 1,753.92
Legal (Attorney)	$ 500.00
Total Amount Due on Closing	$39,311.65

B28

5.

ITEMIZATION OF SETTLEMENT CHARGES	
Down Payment	$24,900.00
Loan Application Fee	$ 500.00
Points (pts.)	$
Title Search	$ 250.00
Title Insurance	$ 400.00
Recording Fee	$ 50.00
Document Preparation Fee	$ 125.00
Credit Search	$ 50.00
Appraisal Fee	$ 350.00
Survey	$ 275.00
Closing Fee	$ 50.00
Interest (max. 1 month)	$ 1,568.70
Taxes (max. 1 month)	$ 516.67
Insurance	$ 550.00
PMI	$ 1,254.96
Legal (Attorney)	$ 500.00
Total Amount Due on Closing	$31,340.33

7.

ITEMIZATION OF SETTLEMENT CHARGES	
Down Payment	$24,740.00
Loan Application Fee	$ 300.00
Points (2.5 pts.)	$11,751.50
Title Search	$ 250.00
Title Insurance	$ 400.00
Recording Fee	$ 50.00
Document Preparation Fee	$ 125.00
Credit Search	$ 50.00
Appraisal Fee	$ 350.00
Survey	$ 275.00
Closing Fee	$ 50.00
Interest (max. 1 month)	$ 2,663.67
Taxes (max. 1 month)	$ 903.33
Insurance	$ 1,086.00
PMI	$ 3,008.38
Legal (Attorney)	$ 500.00
Total Amount Due on Closing	$46,502.88

Exercise 3.4

1. $P = \dfrac{\dfrac{AR}{n}}{\left(1+\dfrac{R}{n}\right)^{(nT)} - 1} = \dfrac{\dfrac{\$9{,}000(0.065)}{12}}{\left(1+\dfrac{0.065}{12}\right)^{\left(12\times\frac{18}{12}\right)} - 1} = \dfrac{\$48.75}{0.1021214212} \approx \$477.37$

3. $P = \dfrac{\dfrac{AR}{n}}{\left(1+\dfrac{R}{n}\right)^{(nT)} - 1} = \dfrac{\dfrac{\$38{,}000(0.068)}{12}}{\left(1+\dfrac{0.068}{12}\right)^{\left(12\times\frac{18}{12}\right)} - 1} = \dfrac{\$215.3333333}{0.1070646877} \approx \$2{,}011.25$

5. $P = \dfrac{\dfrac{AR}{n}}{\left(1+\dfrac{R}{n}\right)^{(nT)} - 1} = \dfrac{\dfrac{\$50{,}000(0.079)}{12}}{\left(1+\dfrac{0.079}{12}\right)^{(12\times 6)} - 1} = \dfrac{\$329.1666667}{0.603913474} \approx \$545.06$

7. $P = \dfrac{\dfrac{AR}{n}}{\left(1+\dfrac{R}{n}\right)^{(nT)} - 1} = \dfrac{\dfrac{\$12{,}000{,}000(0.068)}{4}}{\left(1+\dfrac{0.068}{4}\right)^{(4\times 6)} - 1} = \dfrac{\$204{,}000}{0.4986591519} \approx \$409{,}097.07$

9. $P = \dfrac{\dfrac{AR}{n}}{\left(1+\dfrac{R}{n}\right)^{(nT)} - 1} = \dfrac{\dfrac{\$50{,}000{,}000(0.072)}{2}}{\left(1+\dfrac{0.072}{2}\right)^{(2\times 10)} - 1} = \dfrac{\$1{,}800{,}000}{1.028593867} \approx \$1{,}749{,}961.82$

11. $A = P\left(\dfrac{\left(1+\dfrac{R}{n}\right)^{(nT)} - 1}{\dfrac{R}{n}}\right) = \$500\left(\dfrac{\left(1+\dfrac{0.05}{4}\right)^{(4\times 45)} - 1}{\dfrac{0.05}{4}}\right) = \dfrac{\$500(8.356334493)}{0.0125}$

$A \approx \$334{,}253.38$

B30

13. $A = \dfrac{P(e^R)(e^{(RT)} - 1)}{e^R - 1} = \dfrac{\$2{,}000(e^{0.05})(e^{(0.05 \times 45)} - 1)}{(e^{0.05} - 1)}$

$A = \dfrac{\$2{,}000(1.051271096)(8.487735836)}{0.0512710964} \approx \$348{,}067.90$

15. $A = P\left(\dfrac{\left(1 + \dfrac{R}{n}\right)^{(nT)} - 1}{\dfrac{R}{n}}\right) = \$300\left(\dfrac{\left(1 + \dfrac{0.054}{12}\right)^{(12 \times 38)} - 1}{\dfrac{0.054}{12}}\right) = \dfrac{\$300(6.747706038)}{0.0045}$

$A \approx \$449{,}847.07$

17. $A = P\left(\dfrac{\left(1 + \dfrac{R}{n}\right)^{(nT)} - 1}{\dfrac{R}{n}}\right) = \$2{,}000\left(\dfrac{\left(1 + \dfrac{0.064}{1}\right)^{(1 \times 22)} - 1}{\dfrac{0.064}{1}}\right) = \dfrac{\$2{,}000(2.914856208)}{0.064}$

$A \approx \$91{,}089.26$

19. $A = P\left(\dfrac{\left(1 + \dfrac{R}{n}\right)^{(nT)} - 1}{\dfrac{R}{n}}\right) = \$200{,}000\left(\dfrac{\left(1 + \dfrac{0.064}{4}\right)^{(4 \times 10)} - 1}{\dfrac{0.064}{4}}\right) =$

$A = \dfrac{\$200{,}000(0.8868975384)}{0.016} \approx \$11{,}086{,}219.23$

Exercise 3.5

1. $A = P\left(\dfrac{\left(1 + \dfrac{R}{n}\right)^{(nT)} - 1}{\left(\dfrac{R}{n}\right)\left(1 + \dfrac{R}{n}\right)^{(nT)}}\right) = \$20{,}000\left(\dfrac{\left(1 + \dfrac{0.06}{1}\right)^{(1 \times 4)} - 1}{\left(\dfrac{0.06}{1}\right)\left(1 + \dfrac{0.06}{1}\right)^{(1 \times 4)}}\right)$

$A = \$20{,}000\left(\dfrac{0.26247696}{(0.06)(1.26247696)}\right) \approx \$69{,}302.11$

3. $A = P\left(\dfrac{\left(1+\dfrac{R}{n}\right)^{(nT)} - 1}{\left(\dfrac{R}{n}\right)\left(1+\dfrac{R}{n}\right)^{(nT)}}\right) = \$3{,}000\left(\dfrac{\left(1+\dfrac{0.043}{12}\right)^{(12\times 2)} - 1}{\left(\dfrac{0.043}{12}\right)\left(1+\dfrac{0.043}{12}\right)^{(12\times 2)}}\right)$

$A = \$3{,}000\left(\dfrac{0.08963882}{(0.035833333)(1.08963882)}\right) \approx \$68{,}872.78$

5. $A = P\left(\dfrac{\left(1+\dfrac{R}{n}\right)^{(nT)} - 1}{\left(\dfrac{R}{n}\right)\left(1+\dfrac{R}{n}\right)^{(nT)}}\right) = \$2{,}000\left(\dfrac{\left(1+\dfrac{0.038}{12}\right)^{(12\times 3)} - 1}{\left(\dfrac{0.038}{12}\right)\left(1+\dfrac{0.038}{12}\right)^{(12\times 3)}}\right)$

$A = \$2{,}000\left(\dfrac{0.1205502734}{(0.031666667)(1.120550273)}\right) \approx \$67{,}946.09$

7A. $P = \dfrac{\left(\dfrac{AR}{n}\right)\left(1+\dfrac{R}{n}\right)^{nT}}{\left(1+\dfrac{R}{n}\right)^{nT} - 1} = \dfrac{\left(\dfrac{\$100{,}000(0.058)}{4}\right)\left(1+\dfrac{0.058}{4}\right)^{(4\times 10)}}{\left(1+\dfrac{0.058}{4}\right)^{(4\times 10)} - 1}$

$P = \dfrac{(\$1{,}450)(1.778615427)}{0.778615427} \approx \$3{,}312.28$ per quarter

B. $\dfrac{\$3{,}312.28}{\cancel{\text{quarter}}}\left(\dfrac{4\ \cancel{\text{quarter}}}{1\ \cancel{\text{year}}}\right)(10\ \cancel{\text{year}}) = \$132{,}491.20$

9A. $P = \dfrac{\left(\dfrac{AR}{n}\right)\left(1+\dfrac{R}{n}\right)^{nT}}{\left(1+\dfrac{R}{n}\right)^{nT} - 1} = \dfrac{\left(\dfrac{\$350{,}000(0.052)}{1}\right)\left(1+\dfrac{0.052}{1}\right)^{(1\times 4)}}{\left(1+\dfrac{0.052}{1}\right)^{(1\times 4)} - 1}$

$P = \dfrac{(\$18{,}200)(1.224793744)}{0.224793744} \approx \$99{,}163.11$ per year

B. $\dfrac{\$99{,}163.11}{\cancel{\text{year}}}(4\ \cancel{\text{years}}) = \$396{,}652.44$

Exercise 3.6

1A. $A = P\left(\dfrac{\left(1+\dfrac{R}{n}\right)^{(nT)} - 1}{\left(\dfrac{R}{n}\right)\left(1+\dfrac{R}{n}\right)^{(nT)}}\right) = \$1,500\left(\dfrac{\left(1+\dfrac{0.062}{12}\right)^{(12\times 24)} - 1}{\left(\dfrac{0.062}{12}\right)\left(1+\dfrac{0.062}{12}\right)^{(12\times 24)}}\right)$

$A = \$1,500\left(\dfrac{3.411298936}{(0.0051666667)(4.411298936)}\right) \approx \$224,509.18$

B. $P = \dfrac{\dfrac{AR}{n}}{\left(1+\dfrac{R}{n}\right)^{(nT)} - 1} = \dfrac{\dfrac{\$224,509.18(0.06)}{12}}{\left(1+\dfrac{0.06}{12}\right)^{(12\times 40)} - 1} = \dfrac{\$1,122.5459}{9.957453672} \approx \$112.73$

C. $\$112.73(12)(40) = \$54,110.40$
D. $\$1,500(12)(24) = \$432,000.00$

3A. $A = P\left(\dfrac{\left(1+\dfrac{R}{n}\right)^{(nT)} - 1}{\left(\dfrac{R}{n}\right)\left(1+\dfrac{R}{n}\right)^{(nT)}}\right) = \$1,500\left(\dfrac{\left(1+\dfrac{0.062}{12}\right)^{(12\times 24)} - 1}{\left(\dfrac{0.062}{12}\right)\left(1+\dfrac{0.062}{12}\right)^{(12\times 24)}}\right)$

$A = \$1,500\left(\dfrac{3.411298936}{(0.0051666667)(4.411298936)}\right) \approx \$224,509.1875$

B. $P = \dfrac{\dfrac{AR}{n}}{\left(1+\dfrac{R}{n}\right)^{(nT)} - 1} = \dfrac{\dfrac{\$224,509.18(0.06)}{12}}{\left(1+\dfrac{0.06}{12}\right)^{(12\times 24)} - 1} = \dfrac{\$1,122.5459}{3.205578908} \approx \$350.19$

C. $\$350.19(12)(24) = \$100,854.72$
D. $\$1,500(12)(24) = \$432,000.00$

5A. $A = P\left(\dfrac{\left(1+\dfrac{R}{n}\right)^{(nT)} - 1}{\left(\dfrac{R}{n}\right)\left(1+\dfrac{R}{n}\right)^{(nT)}}\right) = \$2,000\left(\dfrac{\left(1+\dfrac{0.062}{12}\right)^{(12\times 24)} - 1}{\left(\dfrac{0.062}{12}\right)\left(1+\dfrac{0.062}{12}\right)^{(12\times 24)}}\right)$

$A = \$2,000\left(\dfrac{3.411298936}{(0.0051666667)(4.411298936)}\right) \approx \$299,345.57$

B. $P = \dfrac{\dfrac{AR}{n}}{\left(1+\dfrac{R}{n}\right)^{(nT)} - 1} = \dfrac{\dfrac{\$299{,}345.57(0.06)}{12}}{\left(1+\dfrac{0.06}{12}\right)^{(12\times 40)} - 1} = \dfrac{\$1{,}496.72785}{9.957453672} \approx \$150.31$

C. $\$150.31(12)(40) = \$72{,}148.80$
D. $\$2{,}000(12)(24) = \$576{,}000.00$

7A. $A = P\left(\dfrac{\left(1+\dfrac{R}{n}\right)^{(nT)} - 1}{\dfrac{R}{n}}\right) = \$175\left(\dfrac{\left(1+\dfrac{0.048}{12}\right)^{(12\times 40)} - 1}{\dfrac{0.048}{12}}\right) = \$175\left(\dfrac{5.794885585}{0.004}\right)$

$A \approx \$253{,}526.24$

B. $P = \dfrac{\left(\dfrac{AR}{n}\right)\left(1+\dfrac{R}{n}\right)^{nT}}{\left(1+\dfrac{R}{n}\right)^{nT} - 1} = \dfrac{\left(\dfrac{\$253{,}526.24(0.058)}{12}\right)\left(1+\dfrac{0.058}{12}\right)^{(12\times 20)}}{\left(1+\dfrac{0.058}{12}\right)^{(12\times 20)} - 1}$

$P = \dfrac{(\$1{,}225.376827)(3.181031983)}{2.181031983} \approx \$1{,}787.21$

C. $(\$175.00)(12)(40) = \$84{,}000.00$
D. $(\$1{,}787.21)(12)(20) = \$428{,}930.40$

9A. $A = P\left(\dfrac{\left(1+\dfrac{R}{n}\right)^{(nT)} - 1}{\dfrac{R}{n}}\right) = \$175\left(\dfrac{\left(1+\dfrac{0.048}{12}\right)^{(12\times 21)} - 1}{\dfrac{0.048}{12}}\right) = \$175\left(\dfrac{1.73461145}{0.004}\right)$

$A \approx \$75{,}889.25$

B. $P = \dfrac{\left(\dfrac{AR}{n}\right)\left(1+\dfrac{R}{n}\right)^{nT}}{\left(1+\dfrac{R}{n}\right)^{nT} - 1} = \dfrac{\left(\dfrac{\$75{,}889.25(0.058)}{12}\right)\left(1+\dfrac{0.058}{12}\right)^{(12\times 20)}}{\left(1+\dfrac{0.058}{12}\right)^{(12\times 20)} - 1}$

$P = \dfrac{(\$366.7980417)(3.181031983)}{2.181031983} \approx \$534.97$

C. $(\$175)(12)(21) = \$44{,}100.00$
D. $(\$534.97)(12)(20) = \$128{,}392.80$

B34

Exercise 4.1

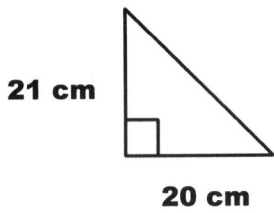

1. $21^2 + 20^2 = 441 + 400 = 841$
   $\sqrt{849} = 29$
   P = 21 + 20 + 29 = 70
   The perimeter is 70 centimeters.

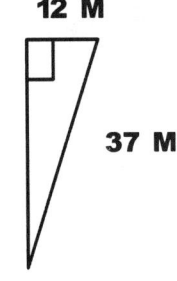

3. $37^2 - 12^2 = 1{,}369 - 144 = 1{,}225$
   $\sqrt{1{,}225} = 35$
   P = 12 + 35 + 37 = 84
   The perimeter is 84 meters.

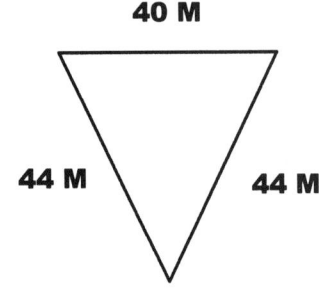

5. $27.5^2 - 22^2 = 272.25$
   $\sqrt{2172.5} = 16.5$
   P = 16.5 + 22 + 27.5 = 66
   The perimeter is 66 yards.

7. P = 40 + 44 + 44 = 128
   The perimeter is 128 meters.

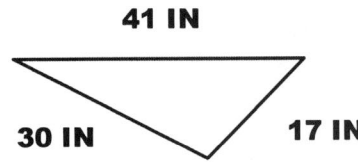

9. P = 17 + 30 + 41 = 88
   The perimeter is 88 inches.

10 YD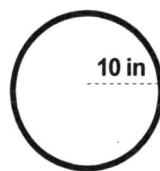
30 YD

11. P = 10 + 30 + 10 + 30 = 80
    The perimeter is 80 yards.

48 MM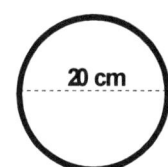

13. P = 48 + 48 + 48 + 48 = 192
    The perimeter is 192 millimeters.

10 in

15. C = 2πr = 2π(10) ≈ 62.83185307
    The circumference is 62.8 inches.

20 cm

17. C = πd = π(20) ≈ 62.83185307
    The circumference is 62.8 cm.

130° 48 M

19. Sector =
    $\frac{130^0}{360^0}(2\pi r) = \frac{130^0}{360^0}(2)(\pi)(48)$
    Sector ≈ 108.9085453
    P ≈ 108.9 + 48 + 48 = 204.9
    The perimeter is 204.9 meters.

II. Convert the perimeters from Problem I to the specified dimensions.

1. $(70 \text{ centimeters})\left(\frac{1 \text{ meter}}{100 \text{ centimeters}}\right) = 0.7$ meters.

3. $(84 \text{ meters})\left(\frac{100 \text{ centimeters}}{1 \text{ meters}}\right) = 8,400$ centimeters.

5. $(66 \text{ yards})\left(\frac{3 \text{ feet}}{1 \text{ yard}}\right) = 198$ feet.

B35

B36

7. $\left(128 \cancel{\text{meters}}\right)\left(\dfrac{100 \text{ centimeters}}{1 \cancel{\text{meters}}}\right) = 12{,}800$ centimeters.

9. $\left(88 \cancel{\text{inches}}\right)\left(\dfrac{1 \text{ yard}}{36 \cancel{\text{inches}}}\right) \approx 2.444$ yards

11. $\left(80 \cancel{\text{yards}}\right)\left(\dfrac{3 \text{ feet}}{1 \cancel{\text{yard}}}\right) = 240$ feet.

13. $\left(192 \cancel{\text{millimeters}}\right)\left(\dfrac{1 \text{ inch}}{25.4 \cancel{\text{millimeters}}}\right) \approx 7.559$ inches.

15. $\left(62.8 \cancel{\text{inches}}\right)\left(\dfrac{1 \text{ yard}}{36 \cancel{\text{inches}}}\right) \approx 1.744$ yards.

17. $\left(62.8 \cancel{\text{centimeters}}\right)\left(\dfrac{1 \text{ meter}}{100 \cancel{\text{centimeters}}}\right) = 0.628$ meters.

19. $\left(204.9 \cancel{\text{meters}}\right)\left(\dfrac{3.28 \text{ feet}}{1 \cancel{\text{meters}}}\right) \approx 672.072$ feet.

Exercise 4.2

I. Find the area for the following figures. Take the result to the next whole number.

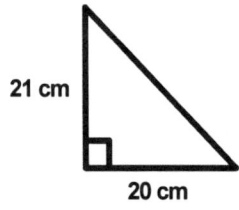
21 cm
20 cm

1. Area = 0.5(base)(height)
   Area = (0.5)(20 cm)(21 cm) = 210 square centimeters.

12 m
37 m

3. $37^2 - 12^2 = 1{,}369 - 144 = 1{,}225$
   $\sqrt{1{,}225} = 35$ is the height
   Area = 0.5(base)(height)
   Area = 0.5 (12 m)(35 m) = 210 square meters.

B37

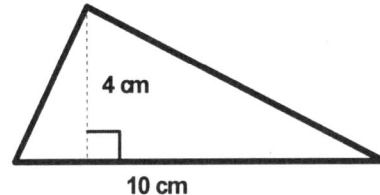

5. Area = 0.5(base)(height)
   Area = 0.5(10 cm)(4 cm)
   Area = 20 square cm.

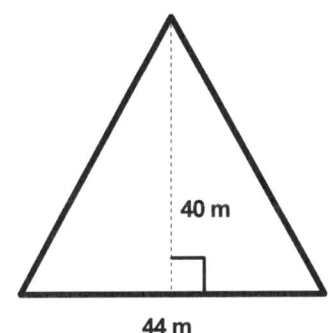

7. Area = 0.5(base)(height)
   Area = 0.5(44 m)(40 m)
   Area = 880 square meters.

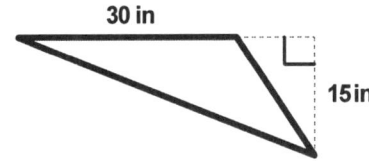

9. Area = 0.5(base)(height)
   Area = 0.5(30 in)(15 in)
   Area = 225 square inches.

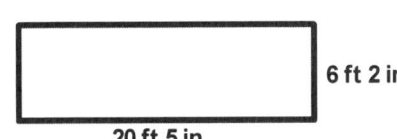

11. Area = (length)(width)
    Area = $\left(20+\dfrac{5}{12}\right)\left(6+\dfrac{2}{12}\right)$
    Area ≈ 125.9027778
    Area is approximately 126 sq ft.

13. Area = (length)(width)
    Area = $\left(4+\dfrac{1}{12}\right)\left(4+\dfrac{1}{12}\right)$
    Area ≈ 16.67361111
    Area is approximately 17 sq ft.

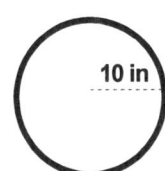

15. Area = $\pi r^2 = \pi(10^2)$
    Area ≈ 314.1592654 sq in
    Area is approximately 315 sq in.

B38

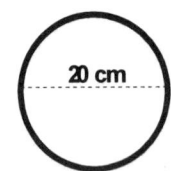

17. Radius = $\dfrac{\text{Diameter}}{2} = \dfrac{10}{2} = 10$ cm

Area = $\pi r^2 = \pi(10^2)$

Area ≈ 314.1592654 sq cm
Area is approximately 315 sq cm.

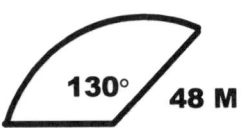

19. Area = $\dfrac{130^0}{360^0}\pi(48^2)$

Area ≈ 2,613.805088 sq m
Area is approximately 2,614 sq m.

II. Use Dimensional Analysis to convert the areas from Problem I to the specified dimensions.

1. $210(\cancel{\text{cm}})(\cancel{\text{cm}})\left(\dfrac{1 \text{ inch}}{2.54 \cancel{\text{cm}}}\right)\left(\dfrac{1 \text{ inch}}{2.54 \cancel{\text{cm}}}\right) \approx 32.55$ sq in.

   Area is bumped up to 33 sq in.

3. $210(\cancel{\text{m}})(\cancel{\text{m}})\left(\dfrac{1.09 \text{ yd}}{1 \cancel{\text{m}}}\right)\left(\dfrac{1.09 \text{ yd}}{1 \cancel{\text{m}}}\right) = 249.501$ square yards.

   Area is bumped up to 250 square yards.

5. $20(\cancel{\text{cm}})(\cancel{\text{cm}})\left(\dfrac{1 \text{ inch}}{2.54 \cancel{\text{cm}}}\right)\left(\dfrac{1 \text{ inch}}{2.54 \cancel{\text{cm}}}\right) \approx 3.10$ square inches.

   Area is bumped up to 4 square inches.

7. $880(\cancel{\text{m}})(\cancel{\text{m}})\left(\dfrac{100 \text{ cm}}{1 \cancel{\text{m}}}\right)\left(\dfrac{100 \text{ cm}}{1 \cancel{\text{m}}}\right) = 8{,}800{,}000$ square centimeters.

9. $225(\cancel{\text{in}})(\cancel{\text{in}})\left(\dfrac{25.4 \text{ mm}}{1 \cancel{\text{in}}}\right)\left(\dfrac{25.4 \text{ mm}}{1 \cancel{\text{in}}}\right) = 145{,}161$ sq mm.

11. $126(\cancel{\text{ft}})(\cancel{\text{ft}})\left(\dfrac{1 \text{ yard}}{3 \cancel{\text{ft}}}\right)\left(\dfrac{1 \text{ yard}}{3 \cancel{\text{ft}}}\right) = 14.0$ sq yd.

13. $17(\cancel{\text{ft}})(\cancel{\text{ft}})\left(\dfrac{12 \text{ inches}}{1 \cancel{\text{ft}}}\right)\left(\dfrac{12 \text{ inches}}{1 \cancel{\text{ft}}}\right) \approx 2{,}448$ square inches.

15. $315(\cancel{in})(\cancel{in})\left(\dfrac{1 \text{ foot}}{12 \cancel{in}}\right)\left(\dfrac{1 \text{ foot}}{12 \cancel{in}}\right) \approx 2.1875$ sq ft .

Area is bumped up to 3 square feet.

17. $315(\cancel{cm})(\cancel{cm})\left(\dfrac{1 \text{ meter}}{100 \cancel{cm}}\right)\left(\dfrac{1 \text{ meter}}{100 \cancel{cm}}\right) = 0.0315$ sq m .

Area is bumped up to 1 square meter.

19. $2,614(\cancel{m})(\cancel{m})\left(\dfrac{1,000 \text{ mm}}{1 \cancel{m}}\right)\left(\dfrac{1,000 \text{ mm}}{1 \cancel{m}}\right) = 2,614,000,000$ sq mm .

Exercises: 4.3
I. Find the perimeters and area for the following shapes. (Round answers to tenths.)
1. This figure can be partitioned into

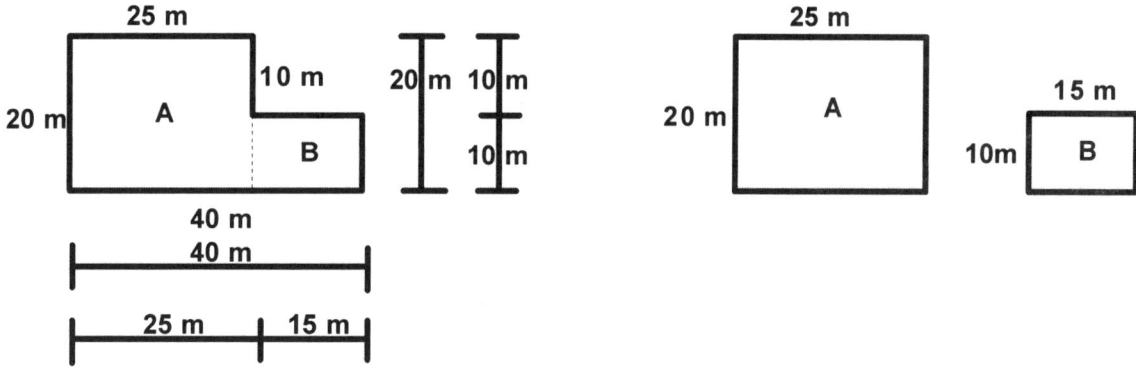

Perimeter = 25 m + 20 m + 40 m + 10 m + 15 m + 10 m = 120 m
Area = (25 m)(20 m) + (15 m)(10 m) = (500 + 150) sq m = 650 square meters

B40

3. This figure can be partitioned into

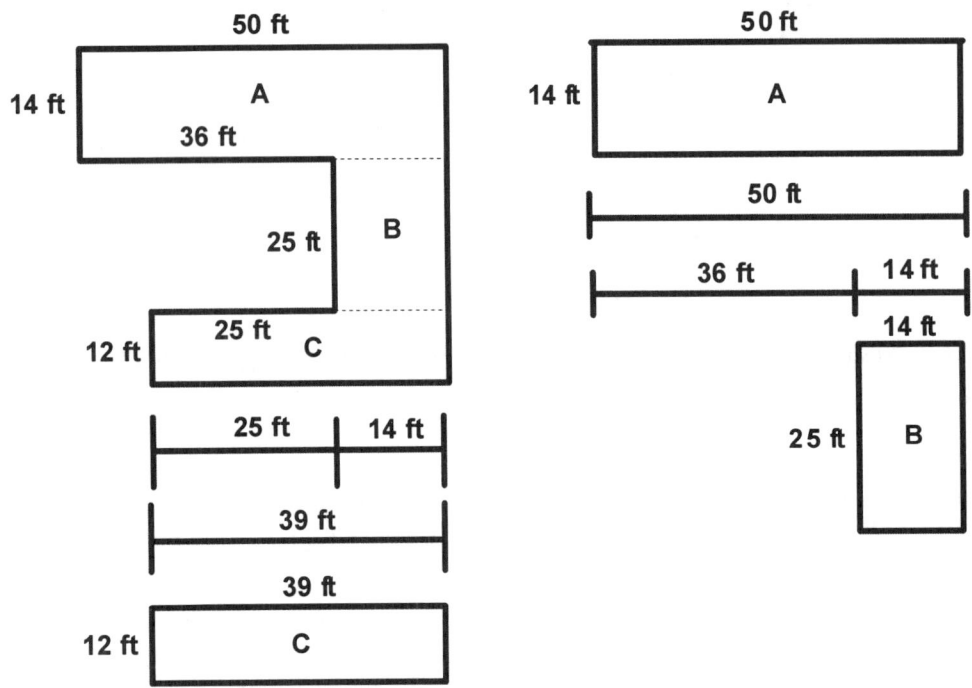

P = 50 ft + 14 ft + 36 ft + 25 ft + 25 ft + 12 ft + 39 ft + (12 + 25 + 14)ft = 252 feet
A = (14 ft)(50 ft) + (25 ft)(14 ft) + (12 ft)(39 ft) = 500 sq ft + 350 sq ft + 468 sq ft
A = 1,518 square feet

5. This figure can be partitioned into

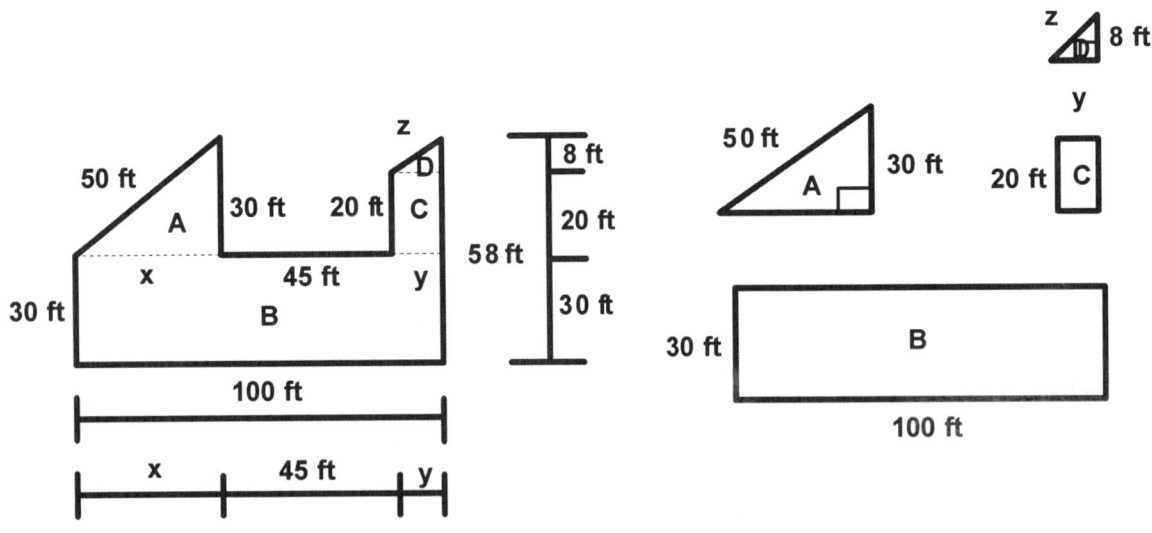

For x, the leg of right triangle A: $50^2 - 30^2 = 2,500 - 900 = 1,600$
$\sqrt{1,600} = 40 \quad x = 40$

Then y = 100 − (40 + 45) = 15.

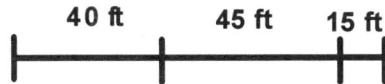

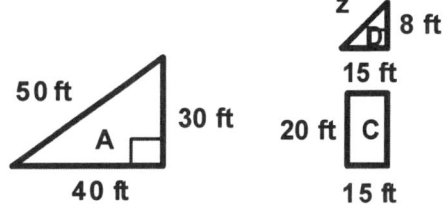

This produces the following parts:
   For z, the hypotenuse of the right triangle: $15^2 + 8^2 = 225 + 64 = 289$
   $$\sqrt{289} = 17$$
P = (50 + 30 + 100 + 58 + 17 + 20 + 45 + 30) ft = 350 ft
A = 0.5(30 ft)(40 ft) + (30 ft)(100 ft) + (15 ft)(20 ft) + 0.5(15 ft)(8 ft)
A = 600 sq ft + 3,000 sq ft + 300 sq ft + 60 sq ft = 3,960 square feet

7.   This figure can be partitioned into

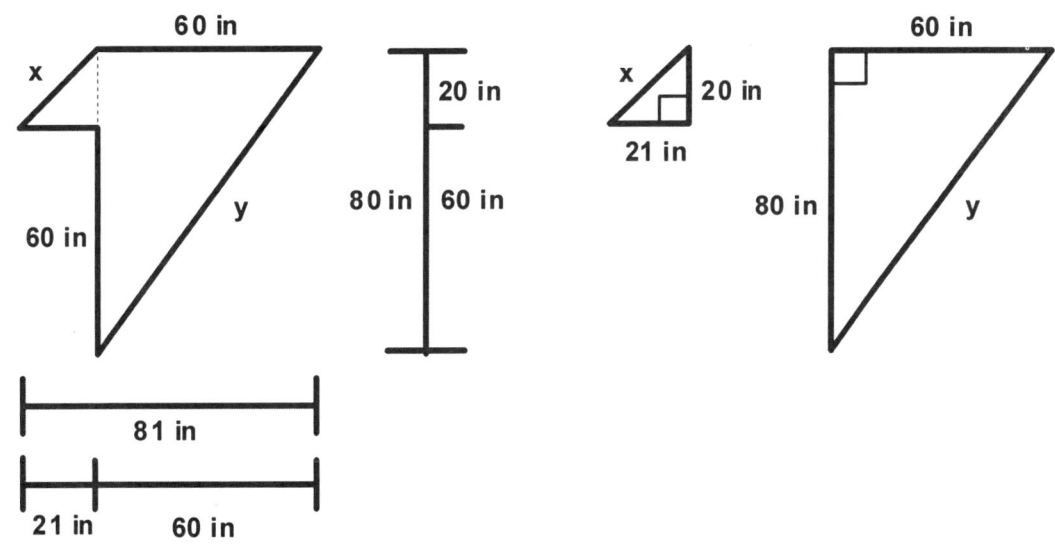

For x, the hypotenuse: $21^2 + 20^2 = 441 + 400 = 841$ $\sqrt{841} = 29$
For y, the hypotenuse: $80^2 + 60^2 = 6,400 + 3,600 = 10,000$ $\sqrt{10,000} = 100$
P = (29 + 21 + 60 + 100 + 60) in = 270 inches
A = 0.5(21 in)(20 in) + 0.5(60 in)(80 in) = 210 sq in + 2,400 sq in = 2,610 sq in.

B42

9.   This figure can be partitioned into

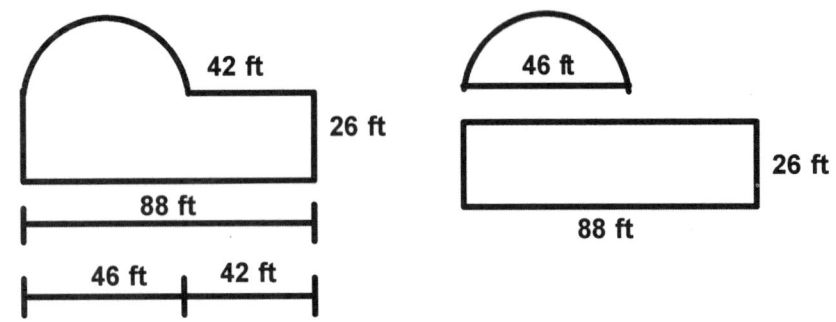

$P = \frac{1}{2}(\pi)(46)$ ft + 26 ft + 88 ft + 26 ft + 42 ft

P = 72.25663103 ft + 26 ft + 88 ft + 26 ft + 42 ft ≈ 254.25663103 ft ≈ 254.3 feet
The perimeter is bumped up to 255 feet.
$A = (26\text{ ft})(88\text{ ft}) + 0.5(\pi)(23)^2$ = 2,288 sq ft + 830.9512569 ≈ 3,118.9512569
The area is bumped up to 3,119 square feet

11.   This figure can be partitioned into

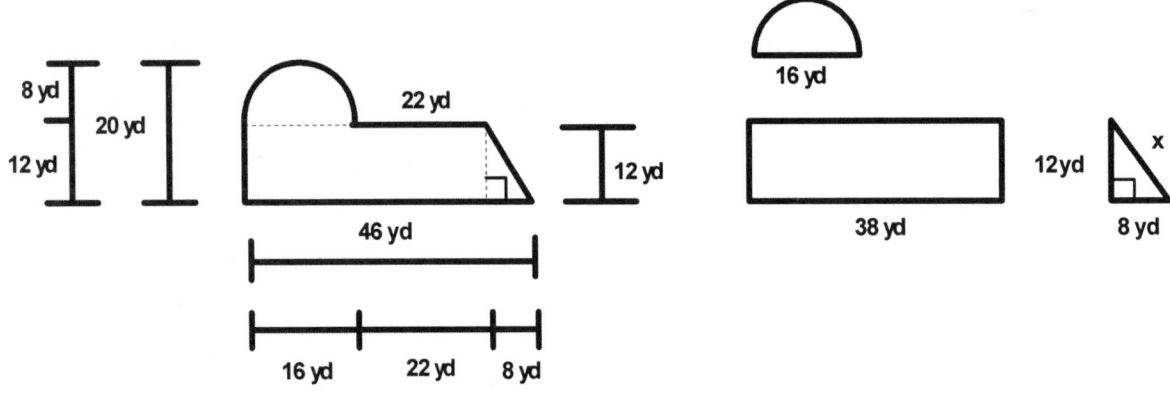

The radius of the circle is 8 yards. This makes the diameter of the circle 176 yards.
For x, the hypotenuse of the right triangle, $12^2 + 8^2 = 144 + 64 = 208$
$$\sqrt{208} \approx 14.4222051 \approx 14.4$$
P = 0.5(π)(16) yd + 12 yd + 46 yd + 14.4 yd + 22 yd
P ≈ 25.13274123 yd + 12 yd + 46 yd + 14.4 yd + 22 yd ≈ 119.53274123 yd
The perimeter is bumped up to 120 yards.
$A = 0.5(\pi)(8\text{ yd})^2 + (12\text{ yd})(38\text{ yd}) + 0.5(12\text{ yd})(8\text{ yd})$

A = 100.5309649 sq yd + 456 sq yd + 48 sq yd ≈ 604.5309649 sq yd.
The area is bumped up to 605 square yards.

II. Use Dimensional Analysis to convert the perimeters and areas from Problem I to the specified dimensions.

1. $P = 120 \text{ m} = 120 \cancel{\text{m}} \left(\dfrac{1.09 \text{ yd}}{1 \cancel{\text{m}}} \right) \approx 130.8 \text{ yards}.$

    The perimeter is bumped up to 131 yards.

    $A = 650 \text{m}^2 = 650 (\cancel{\text{m}})(\cancel{\text{m}}) \left(\dfrac{1.09 \text{ yd}}{1 \cancel{\text{m}}} \right) \left(\dfrac{1.09 \text{ yd}}{1 \cancel{\text{m}}} \right) \approx 772.265 \text{ square yards}.$

    The area is bumped up to 773 square yards.

3. $P = 252 \text{ ft} = 252 \cancel{\text{ft}} \left(\dfrac{1 \text{ yard}}{3 \cancel{\text{ft}}} \right) = 84 \text{ yards}.$

    $A = 1{,}518 \text{ ft}^2 = 1{,}518 (\cancel{\text{ft}})(\cancel{\text{ft}}) \left(\dfrac{1 \text{ yd}}{3 \cancel{\text{ft}}} \right) \left(\dfrac{1 \text{ yd}}{3 \cancel{\text{ft}}} \right) \approx 168.7 \text{ sq yd}.$

    The area is bumped up to 169 square yards.

5. $P = 350 \text{ ft} = 350 \cancel{\text{ft}} \left(\dfrac{1 \text{ yard}}{3 \cancel{\text{feet}}} \right) \approx 116.7 \text{ yards}.$

    The perimeter is bumped up to 117 yards.

    $A = 3{,}960 \text{ ft}^2 = 3{,}960 (\cancel{\text{ft}})(\cancel{\text{ft}}) \left(\dfrac{1 \text{ yard}}{3 \cancel{\text{ft}}} \right) \left(\dfrac{1 \text{ yard}}{3 \cancel{\text{ft}}} \right) = 440 \text{ square yards}.$

7. $P = 270 \text{ in} = 270 \cancel{\text{in}} \left(\dfrac{1 \text{ yard}}{36 \cancel{\text{inches}}} \right) = 7.5 \text{ yards}.$

    The perimeter is bumped up to 8 yards.

    $A = 2{,}610 \text{ in}^2 = 2{,}610 (\cancel{\text{in}})(\cancel{\text{in}}) \left(\dfrac{1 \text{ yd}}{36 \cancel{\text{in}}} \right) \left(\dfrac{1 \text{ yd}}{36 \cancel{\text{in}}} \right) \approx 2.0 \text{ sq yd}.$

    The area is bumped up to 3 square yards.

9. $P = 255 \text{ ft} = 255 \cancel{\text{ft}} \left(\dfrac{1 \text{ yard}}{3 \cancel{\text{feet}}} \right) = 85 \text{ yards}.$

    $A = 3{,}119 \text{ ft}^2 = 3{,}119 (\cancel{\text{ft}})(\cancel{\text{ft}}) \left(\dfrac{1 \text{ yd}}{3 \cancel{\text{ft}}} \right) \left(\dfrac{1 \text{ yd}}{3 \cancel{\text{ft}}} \right) \approx 346.6 \text{ sq yd}.$

    The area is bumped up to 347 square yards.

B44

11. $P = 120 \text{ yd} = 120 \cancel{\text{yd}} \left(\dfrac{3 \text{ feet}}{1 \cancel{\text{yd}}} \right) = 360 \text{ feet}.$

$A = 605 \text{ yd}^2 = 605 (\cancel{\text{yd}})(\cancel{\text{yd}}) \left(\dfrac{3 \text{ ft}}{1 \cancel{\text{yd}}} \right) \left(\dfrac{3 \text{ ft}}{1 \cancel{\text{yd}}} \right) = 5{,}445 \text{ square feet}.$

Exercise 4.4

I. Find the volume (cubic feet) of these three-dimensional solids based on the figures in Exercise 4.3. Round the answers to tenths.

1. Area of base, from previous section = 650 square meters
Volume = (Area of the base) × (height) = (650 sq m) × (10 m) = 6,500 cubic meters.

3. Area of base, from previous section = 1,518 square feet
Volume = (Area of the base) × (height) = $(1{,}518 \text{ sq ft}) \times \left(7 + \dfrac{3}{12} \right) \text{ft} = 11{,}005.5 \text{ cu ft}.$
The volume is bumped up to 11,006 cubic feet.

5. Area of base, from previous section = 3,960 square feet
Volume = (Area of the base) × (height) = $(3{,}960 \text{ sq ft}) \times \left(3 + \dfrac{8}{12} \right) \text{ft} = 14{,}520 \text{ cu ft}.$

7. Area of base, from previous section = 2,610 square inches
Volume = (Area of the base) × (height) = (2,610 sq in) × (25 in) = 65,250 cu in.

9. Area of base, from previous section = 3,119 square feet
Volume = (Area of the base) × (height) = $(3{,}119 \text{ sq ft}) \times \left(17 + \dfrac{9}{12} \right) \text{ft} =$
55,362.25 cubic feet. The volume is bumped up to 55,363 cubic feet.

11. Area of base, from previous section = 604.5 square yards
Volume = (Area of the base) × (height) = (604.5 sq yd) × (8 yd) = 4,836 cu yd.

II. Use Dimensional Analysis to convert the volumes from Problem I to the specified dimension. Round the answer to tenths.

1. $6{,}500 (\cancel{\text{m}})(\cancel{\text{m}})(\cancel{\text{m}}) \left(\dfrac{3.28 \text{ ft}}{1 \cancel{\text{m}}} \right) \left(\dfrac{3.28 \text{ ft}}{1 \cancel{\text{m}}} \right) \left(\dfrac{3.28 \text{ ft}}{1 \cancel{\text{m}}} \right) \approx 229{,}369.088 \text{ cubic feet}.$
Bumping up to the next whole number, 6,500 cubic meters is 229,370 cubic feet.

3. $\left(\dfrac{11{,}006 \cancel{ft^3}}{1}\right)\left(\dfrac{7.48 \text{ gallons}}{\cancel{ft^3}}\right) \approx 82{,}324.88$ gallons.

Bumping up to the next whole number, 11,006 cubic feet is 82,325 gallons.

5. $14{,}520(\cancel{ft})(\cancel{ft})(\cancel{ft})\left(\dfrac{1 \text{ yd}}{3 \cancel{ft}}\right)\left(\dfrac{1 \text{ yd}}{3 \cancel{ft}}\right)\left(\dfrac{1 \text{ yd}}{3 \cancel{ft}}\right) = \dfrac{14{,}520}{27} \text{yd}^3 \approx 537.7777778$ cubic yards.

Bumping up to the next whole number, 14,520 cu ft is 538 cu yd.

7. $65{,}250(\cancel{in})(\cancel{in})(\cancel{in})\left(\dfrac{1 \text{ ft}}{12 \cancel{in}}\right)\left(\dfrac{1 \text{ ft}}{12 \cancel{in}}\right)\left(\dfrac{1 \text{ ft}}{12 \cancel{in}}\right) = \dfrac{65{,}250}{1{,}728} \text{ft}^3 \approx 37.76041667$ cu ft.

Bumping up to the next whole number, 65,250 cu in is 38 cu ft.

9. $55{,}363(\cancel{ft})(\cancel{ft})(\cancel{ft})\left(\dfrac{1 \text{ yd}}{3 \cancel{ft}}\right)\left(\dfrac{1 \text{ yd}}{3 \cancel{ft}}\right)\left(\dfrac{1 \text{ yd}}{3 \cancel{ft}}\right) = \dfrac{55{,}363}{27} \text{yd}^3 \approx 2050.481481$ cu yd.

Bumping up to the next whole number, 55,363 cu ft is 2,051 cu yd.

11. $4{,}836(\cancel{yd})(\cancel{yd})(\cancel{yd})\left(\dfrac{3 \text{ ft}}{1 \cancel{yd}}\right)\left(\dfrac{3 \text{ ft}}{1 \cancel{yd}}\right)\left(\dfrac{3 \text{ ft}}{1 \cancel{yd}}\right) = 130{,}572$ cubic feet.

III. Automotive conversions

1. $V = 327$ cu in $= \dfrac{327 \cancel{in}\,\cancel{in}\,\cancel{in}}{1}\left(\dfrac{2.54 \text{ cm}}{1 \cancel{in}}\right)\left(\dfrac{2.54 \text{ cm}}{1 \cancel{in}}\right)\left(\dfrac{2.54 \text{ cm}}{1 \cancel{in}}\right) \approx 5{,}358.8$ cc

$V = \dfrac{5{,}358.8 \cancel{cc}}{1}\left(\dfrac{1 \text{ liter}}{1{,}000 \cancel{cc}}\right) \approx 5.3588$ liters

The volume of a 327 cubic inch engine is approximately 5.4 liters.

3. $V = 454$ cu in $= \dfrac{454 \cancel{in}\,\cancel{in}\,\cancel{in}}{1}\left(\dfrac{2.54 \text{ cm}}{1 \cancel{in}}\right)\left(\dfrac{2.54 \text{ cm}}{1 \cancel{in}}\right)\left(\dfrac{2.54 \text{ cm}}{1 \cancel{in}}\right) \approx 7{,}439.7$ cc

$V = \dfrac{7{,}437.7 \cancel{cc}}{1}\left(\dfrac{1 \text{ liter}}{1{,}000 \cancel{cc}}\right) \approx 7.4377$ liters.

The volume of a 454 cubic inch engine is approximately 7.4 liters.

B46

5. $V = 3.8 \text{ liters} = \dfrac{3.8 \text{ liters}}{1}\left(\dfrac{1,000 \text{ cc}}{1 \text{ liter}}\right) = \dfrac{3.8 \text{ liters}}{1}\left(\dfrac{1000 \text{ cc}}{1 \text{ liter}}\right) = 3,800 \text{ cc}$

$V = \dfrac{3,800(\text{cm})(\text{cm})(\text{cm})}{1}\left(\dfrac{1 \text{ in}}{2.54 \text{ cm}}\right)\left(\dfrac{1 \text{ in}}{2.54 \text{ cm}}\right)\left(\dfrac{1 \text{ in}}{2.54 \text{ cm}}\right) \approx 231.89 \text{ cu in.}$

The volume of the engine is approximately 231.9 cubic inches.

Exercise 4.5

1. The Dining Room is 13 ft 6 in by 11 ft.

   Area is $\left(13 + \dfrac{6}{12} \text{ ft}\right)(11 \text{ ft}) = 148.5$ square feet

   Bumping up the area of the dining room is 149 sq ft.

   Cartons needed is $149\,(\text{ft})(\text{ft})\left(\dfrac{\text{carton}}{12.94\,(\text{ft})(\text{ft})}\right) \approx 11.51468315$ cartons.

   Since you cannot purchase a fraction of a carton, 12 cartons are needed.

   Price $= \dfrac{\$37.39}{\text{carton}}(12 \text{ cartons}) = \$448.68$

3. The Family Room is 22 ft 4 in by 15 ft.

   The area of two walls is $(2)\left(22 + \dfrac{4}{12} \text{ ft}\right)(8 \text{ ft}) \approx 357.3$ sq ft

   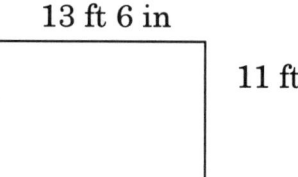

   The area of two walls is $(2)(15 \text{ ft})(8 \text{ ft}) = 240$ sq ft

   Total area to be wallpapered is 357.3 sq ft + 240 sq ft = 597.3 sq ft.
   Bump the area up to 598 square feet.

   $\left(\dfrac{598 \text{ sq ft}}{1}\right)\left(\dfrac{1 \text{ double bolt}}{56.37 \text{ sq ft}}\right) \approx 10.6$ double bolts

   A fraction of double bolts cannot be purchased. Bumping up the number of bolts to 11 double bolts.

   $(11 \text{ double bolts})\left(\dfrac{\$14.96}{\text{double bolt}}\right) = \$164.56$

   It will cost \$164.56 to wallpaper the family room.

B47

5. The Kitchen is 14 ft by 10 ft.
   Area is (14 ft)(10 ft) = 140 square feet

   14 ft
   10 ft

   Convert to square yards

   $$140(\cancel{ft})(\cancel{ft})\left(\frac{yd}{3(\cancel{ft})}\right)\left(\frac{yd}{3(\cancel{ft})}\right) = \frac{140}{9} \text{ sq yd} \approx 15.55555556 \text{ square yards.}$$

   Since you cannot purchase a fraction of a square yard, 16 square yards are needed.

   $$\text{Price} = 16(\cancel{yd})(\cancel{yd})\left(\frac{\$9.66}{(\cancel{yd})(\cancel{yd})}\right) = \$154.56$$

7. The Study is 11 ft by 11 ft. 6 in

   11 ft
   11 ft 6 in

   $$P = 11 \text{ ft} + \left(11 + \frac{6}{12} \text{ ft}\right) + 11 \text{ ft} + \left(11 + \frac{6}{12} \text{ ft}\right) = 45 \text{ feet}$$

   Number of panels needed is $45 \, (\cancel{ft}) \left(\frac{\text{panel}}{4 \, (\cancel{ft})}\right) = 11.25$ panels.

   Since you cannot purchase a fraction of a panel, 12 panels are needed.

   $$\text{Price} = \frac{\$19.99}{\cancel{\text{panel}}} (12 \, \cancel{\text{panels}}) = \$239.88$$

9. Area of the lot is

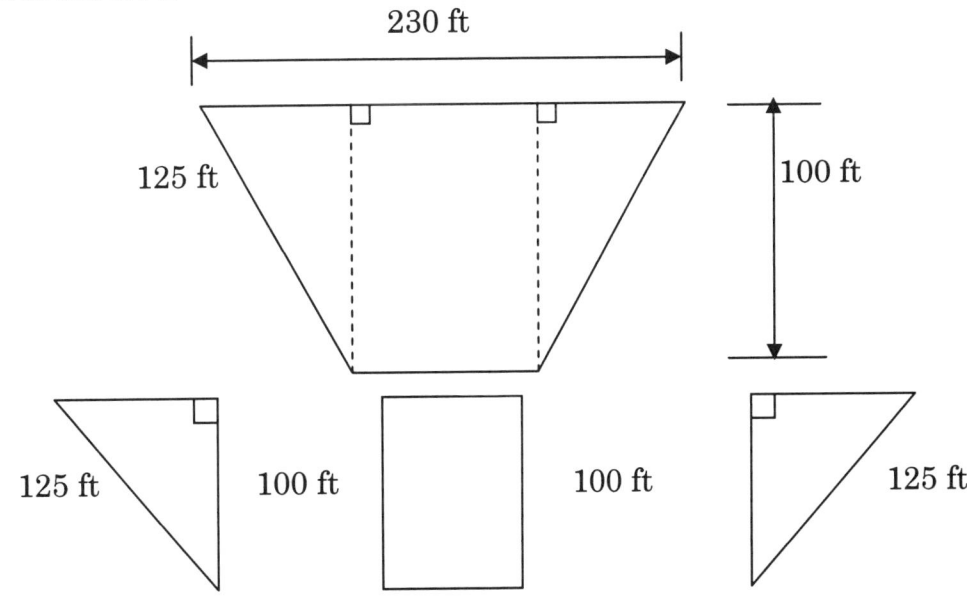

Solving for the third side of the triangle
$125^2 - 100^2 = 5{,}625$ $\sqrt{5{,}625} = 75$

Calculating the distance across the top of the lot.

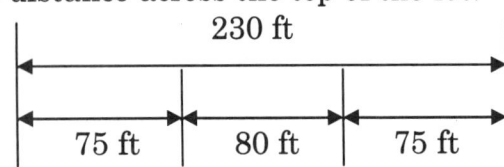

The areas are:

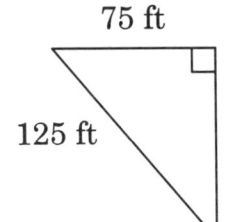 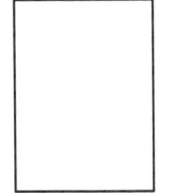 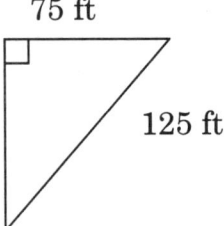

The area of the lot is (0.5)(100 ft)(75 ft) + (80 ft)(100 ft) + (0.5)(100 ft)(75 ft)
Area of lot is 15,500 square feet.

The house is 32 ft by 64 ft.
Area is (32 ft)(64 ft) = 2,048 sq ft

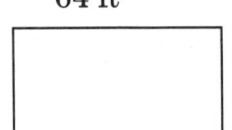

The driveway is 30 ft by 20 ft
The area is (30 ft)(20 ft) = 600 square feet

The swimming pool is 18 ft across
Diameter is 18 ft. Radius is $\dfrac{18}{2}$ ft = 9 ft

Area is $\pi(9\text{ ft})^2 \approx 254.5$ square feet.

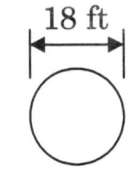

The garden is 24 ft by 32 ft.
The area is (24 ft)(32 ft) = 768 sq ft

Unsodded area is 2,048 sq ft + 600 sq ft + 254.5 sq ft + 768 sq ft = 3,670.5 sq ft.
Bump up to area to 3,671 square feet.
Area to be sodded is 15,500 sq ft − 3,671 sq ft = 11,829 sq ft.

A roll of sod is 6 ft by 18 inches, Area is $6 \text{ ft}\left(\frac{18}{12} \text{ ft}\right) = 9$ square feet

Number of rolls of sod needed $11,829 \text{ (ft)(ft)}\left(\frac{\text{roll}}{9 \text{ ft ft}}\right) \approx 1,314.333333$

Since you cannot purchase a fraction of a roll, 1315 rolls are needed.

Cost $= \left(\frac{\$1.19}{\text{roll}}\right) 1,315 \text{ rolls} = \$1,564.85$

11. The roof is partitioned into six sections.
    Two sections of 38 ft 10 in by 15 ft 3 in

    Area $= (2)\left(38 + \frac{10}{12} \text{ ft}\right)\left(15 + \frac{3}{12} \text{ ft}\right) \approx 1,184.4$ sq ft.

    Two sections of 26 ft 6 in by 14 ft 5 in

    Area $= (2)\left(26 + \frac{6}{12} \text{ ft}\right)\left(14 + \frac{5}{12} \text{ ft}\right) \approx 764.1$ sq ft.

    Two sections of 17 ft 11 in by 16 ft

    Area $= (2)\left(17 + \frac{11}{12} \text{ ft}\right)(16 \text{ ft}) \approx 573.3$ sq ft.

    Total area = 1,184.4 sq ft + 764.1 sq ft + 573.3 sq ft = 2,521.8
    Bump the area up to 2,522 sq ft.

    $\dfrac{2,522 \text{ sq ft}}{1}\left(\dfrac{1 \text{ bundle}}{33\frac{1}{3} \text{ sq ft}}\right) \approx 75.66$ bundles.

    Since you cannot purchase a fraction of a bundle, 76 bundles are needed.
    Price = 76 ($12.79) = $972.04

B50

Exercise 5.1

1. The sum of the percentages is 93%. Using ± 1% for each category, the interval is 90% to 96%. Since 100% is not contained in this interval, the graph is **not** properly constructed.

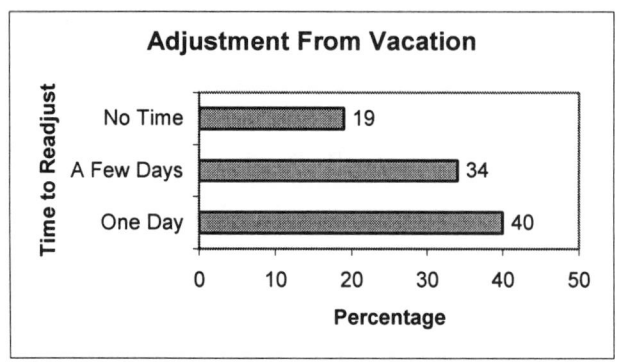

3. a. "How Often We Check e-mail" is a definitive title.
   b. The numeric values are labeled as percents. The sum of the percentages is 100%. This sum is appropriate.
   c. Each slice has a definitive label.
   d. The chart's slices complete 360 degrees. It's a complete circle.)
   The graph is properly constructed.

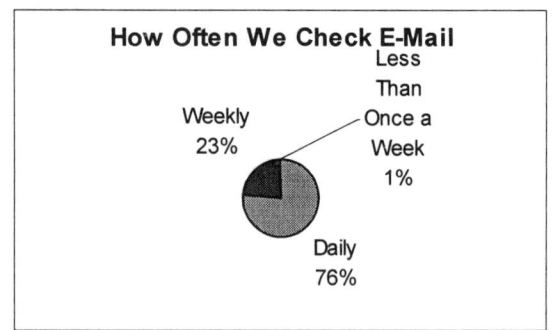

5.

Job Candidate's Mistake	Frequency
Interview	448
Resume	294
Cover Letter	126
Reference Checks	126
Interview Follow-up	98
Screening Call	84
Other	28
Don't Know	196

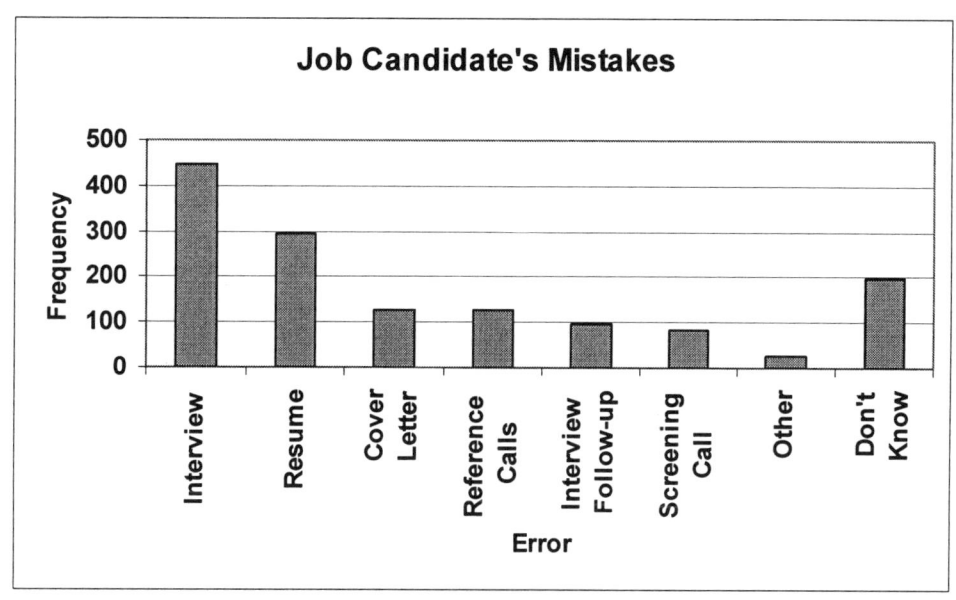

7.

Distribution	Cents / Dollar	Degrees
Administrative Costs	$0.128	$0.128(360^0) \approx 46^0$
Nurturing Children	$0.253	$0.253(360^0) \approx 91^0$
Develop Self-Sufficiency	$0.174	$0.174(360^0) \approx 63^0$
Health & Wellness	$0.140	$0.14(360^0) \approx 50^0$
Strengthen Families	$0.192	$0.192(360^0) \approx 69^0$
Strong Communities	$0.113	$0.113(360^0) \approx 41^0$

United Way Distributions

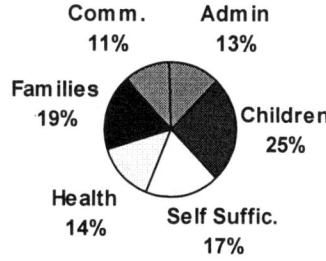

B52

9.

Who's Watching the Kids?	Frequency	Relative Frequency
Center based (child care center)	3,480	3,480/12,000 = 0.29
Family child care homes	1,560	1,560/12,000 = 0.13
Babysitter	600	600/12,000 = 0.05
Care of relative	3,120	3,120/12,000 = 0.26
Parental care	3,240	3,240/12,000 = 0.27
Totals	12,000	12,000/12,000 = 1.00

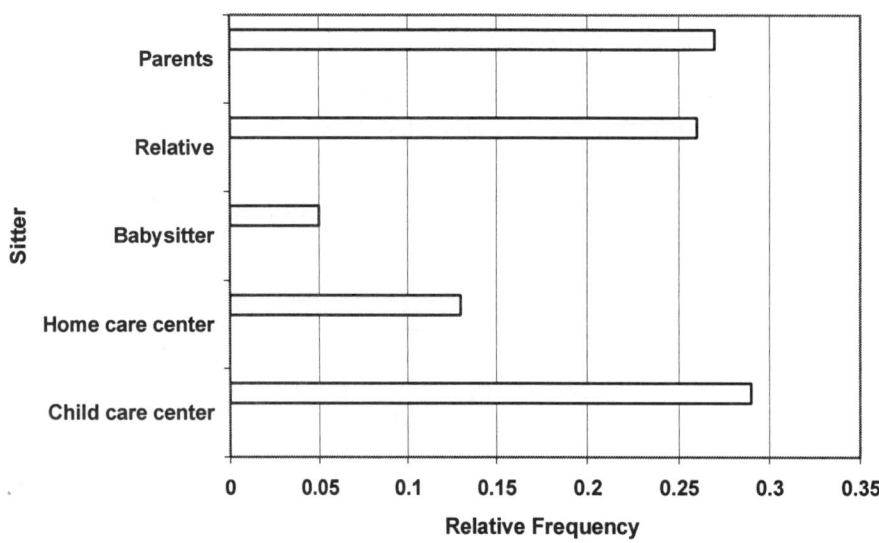

11.

Response	Frequency	Rel Freq
Happy with Career	112	112/336 = 0.33
Enjoy job, but its not my career choice	64	64/336 = 0.19
Job is OK, but it is not my career choice	64	64/336 = 0.19
Don't like my job, but it is my career path	20	20/336 = 0.06
My job just pays my expenses	76	76/336 = 0.23
Totals	336	336/336 = 1.00

Job Satisfaction (Workers in Their 20's)

13.

Distribution	Amount	Rel. Freq.
Research	$1.35	$1.35/$5 = 0.27
Patient Services	$1.05	$1.05/$5 = 0.21
Public Education	$0.90	$0.90/$5 = 0.18
Fund Raising	$0.90	$0.90/$5 = 0.18
Prof. Ed.	$0.45	$0.45/$5 = 0.09
Management	$0.35	$0.35/$5 = 0.07
Total	$5.00	1.00

$5 Contribution to the American Cancer Society

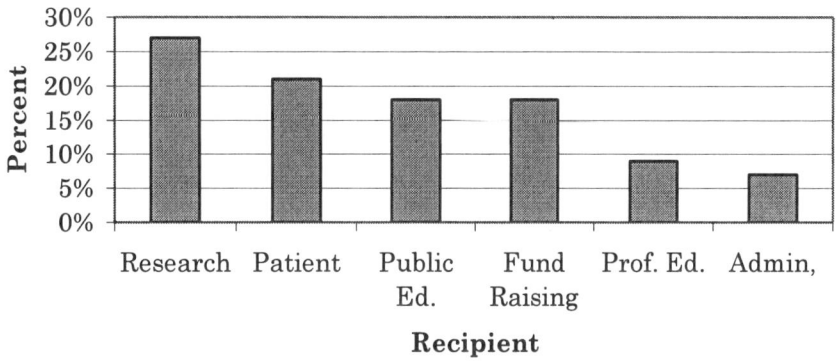

B54

15.

Flavor	Freq	Rel Freq	Degrees
Cherry	200	200/2,000 = 0.10	$0.10(360°) = 36°$
Lemon meringue	220	220/2,000 = 0.11	$0.11(360°) \approx 40°$
Sweet Potato	170	170/2,000 = 0.085	$0.085(360°) \approx 31°$
Pumpkin	170	170/2,000 = 0.085	$0.085(360°) \approx 31°$
Apple	500	500/2,000 = 0.25	$0.25(360°) = 90°$
Chocolate	280	280/2,000 = 0.14	$0.14(360°) \approx 50°$
Other	460	460/2,000 = 0.23	$0.23(360°) \approx 83°$
Total	2,000	2,000/2,000 = 1.000	361° (rounding)

Favorite Pie Flavors

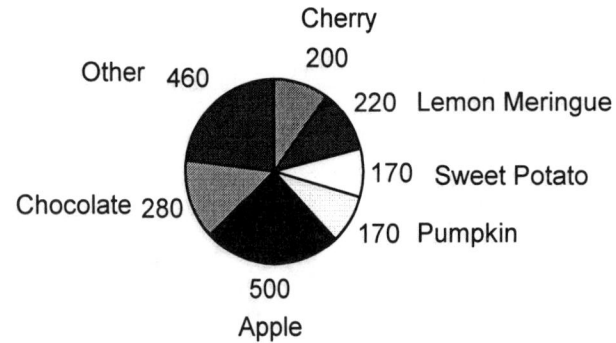

Exercise 5.2

1. One possible way to arrange this data is:

   Class width $> \dfrac{\$82{,}550 - \$12{,}255}{7} \approx \$10{,}042.14$. Let the class width be $11,000.

Dollars Thousands	Freq.
12 – 23	17
23 – 34	8
34 – 45	7
45 – 56	3
56 – 67	3
67 – 78	1
78 – 89	1
Totals	40

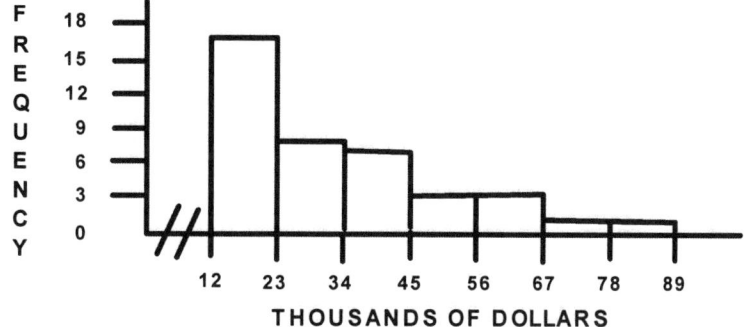

TWO-DOOR SPORT COUPE PRICES

3. One possible way to arrange this data is:

Class width $> \dfrac{\$83{,}820 - \$19{,}735}{6} \approx \$10{,}680.83$. Let the class width be $11,000.

Dollars (Thousands)	Freq.	Rel. Freq.
19 – 30	13	$13/40 \approx 0.33$
30 – 41	7	$7/40 \approx 0.18$
41 – 52	13	$13/40 \approx 0.33$
52 – 63	2	$2/40 \approx 0.05$
63 – 74	1	$1/40 \approx 0.03$
74 – 85	4	$4/40 \approx 0.10$
Totals	40	1.02*

* Rounding error

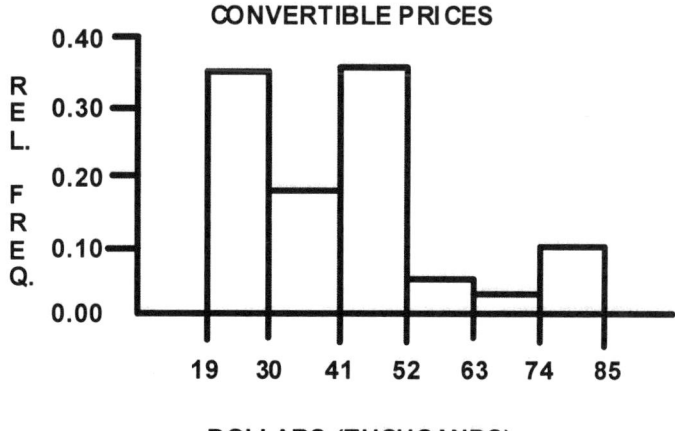

CONVERTIBLE PRICES

B56

5. One possible way to arrange this data is:

Class width $> \dfrac{\$367{,}000 - \$55{,}000}{7} \approx \$44{,}571.43$. Let the class width be $50,000.

Dollars (Thousands)	Freq.	Class Mark
20 – 70	3	45
70 – 120	9	95
120 – 170	11	145
170 – 220	6	195
220 – 270	5	245
270 – 320	3	295
320 – 370	2	345
Totals	39	

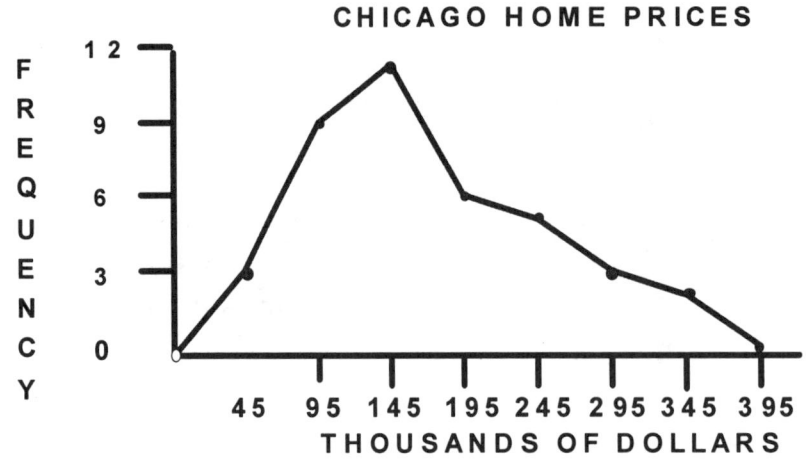

7. One possible way to arrange this data is:

Class width $> \dfrac{\$268{,}650 - \$74{,}900}{8} \approx \$24{,}218.75$. Let the class width be $30,000.

Dollars Thousands	Freq.	Class Mark
50 – 80	2	65
80 – 110	13	95
110 – 140	11	125
140 – 170	8	155
170 – 200	1	185
200 – 230	2	215
230 – 260	1	245
260 – 290	2	275
Totals	40	

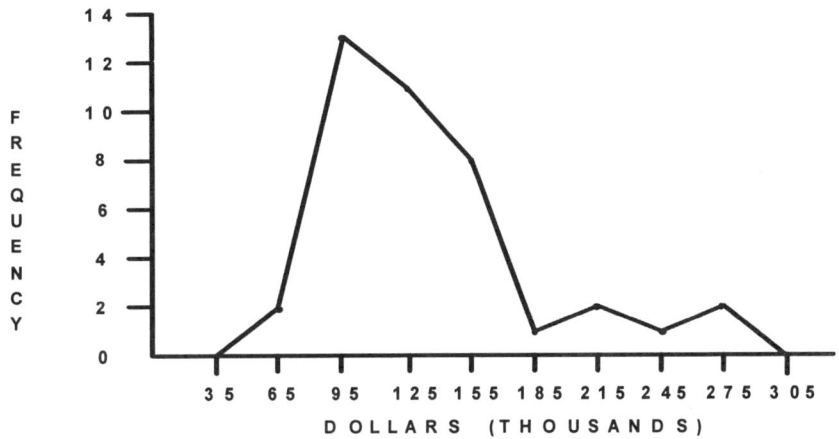

9. One possible way to arrange this data is:

Class width $> \dfrac{\$19{,}200 - \$6{,}999}{7} = \$1{,}743$. Let the class width be $2,000.

$ Thousands	Freq.	Rel. Freq.	Class Mark
6 – 8	6	$6/29 \approx 0.21$	7
8 – 10	7	$7/29 \approx 0.24$	9
10 – 12	6	$6/29 \approx 0.21$	11
12 – 14	1	$1/29 \approx 0.03$	13
14 – 16	4	$4/29 \approx 0.14$	15
16 – 18	3	$3/29 \approx 0.10$	17
18 – 20	2	$2/29 \approx 0.07$	19
Totals	29	1.00	

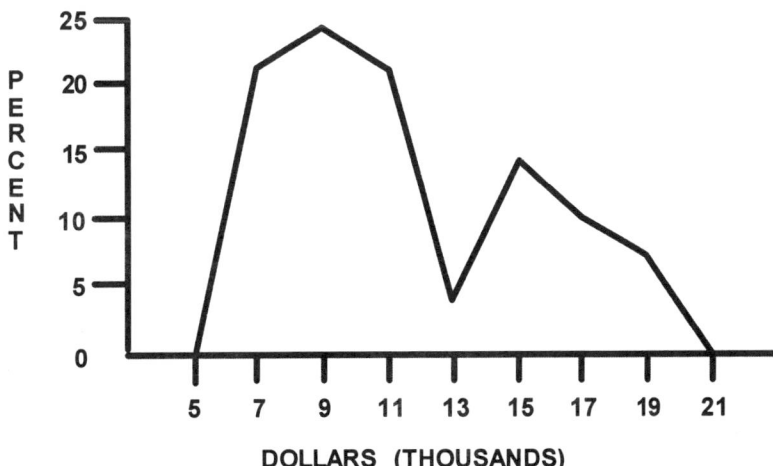

B58

Exercise 5.3
1. The mean value from the calculator is 32889.375.
   The mean price of a Sport Coupe is approximately $32,889.38.

3. The sorted data
   $19,735 $19,990 $20,080 $21,245 $22,015 $22,255 $23,300 $23,995
   $24,140 $25,760 $25,915 $26,710 $28,365 $31,500 $31,870 $32,000
   $32,155 $37,120 $37,470 **$39,450** **$41,000** $41,430 $42,995 $43,270
   $43,395 $43,500 $44,640 $44,995 $45,320 $45,500 $46,500 $48,100
   $49,930 $53,140 $55,600 $71,200 $74,970 $78,390 $80,400 $83,820

   There are 40 data values. $\frac{n+1}{2} = \frac{40+1}{2} = \frac{41}{2} = 20.5$.

   The median is half way between the 20th and 21st values.

   The median for a convertible is $\frac{\$39,450 + \$41,000}{2} = \frac{\$80,450}{2} = \$40,225$

5. The mean value from the calculator is 165,900.
   The mean price for a home in Chicago is $165,900.

7. The mean value from the calculator is 133011.625
   The mean price for a townhouse in a Chicago suburb is approximately $133,011.63.

9. The sorted data is:
   $ 6,999 $ 7,399 $ 7,399 $ 7,895 $ 7,899 $ 7,999 $ 8,199 $ 8,399
   $ 8,595 $ 8,990 $ 8,995 $ 9,699 $ 9,999 $10,449   **$10,599** $10,899
   $11,770 $11,999 $11,999 $12,299 $14,399 $14,500 $15,000 $15,600
   $16,190 $16,700 $16,990 $18,200 $19,200

   There are 29 data values. $\frac{n+1}{2} = \frac{29+1}{2} = \frac{30}{2} = 15$.

   The 15th value is $10,599.
   The median price for a new 1100 cc to 1200 cc motorcycle is $10,599.

Exercises 5.4
1. The standard deviation from the calculator is 17508.98
   The standard deviation for a sport coupe is approximately $17,508.98

3. The sorted data

$19,735 $19,990 $20,080 $21,245 $22,015 $22,255 $23,300 $23,995
$24,140 $25,760 $25,915 $26,710 $28,365 $31,500 $31,870 $32,000
$32,155 $37,120 $37,470 $39,450 $41,000 $41,430 $42,995 $43,270
$43,395 $43,500 $44,640 $44,995 $45,320 $45,500 $46,500 $48,100
$49,930 $53,140 $55,600 $71,200 $74,970 $78,390 $80,400 $83,820

The median is the value at the $\frac{n+1}{2}$ location, $\frac{40+1}{2} = \frac{41}{2} = 20.5$ value.

This is the midpoint of the 20th and 21st values. The median is
$\frac{\$39,450 + \$41,000}{2} = \frac{\$80,450}{2} = \$40,225$.

The first quartile is the median of the first 20 values, $\frac{20+1}{2} = \frac{21}{2} = 10.5$ value.

This is the midpoint of the 10th and 11th values.

The first quartile is $\frac{\$25,760 + \$25,915}{2} = \frac{\$51,675}{2} = \$25,837.50$.

The third quartile is the median of the second 20 values $\frac{20+1}{2} = \frac{21}{2} = 10.5$ value.

This is the midpoint of the 10th and 11th values for the second set of 20 values.

The third quartile is $\frac{\$45,500 + \$46,500}{2} = \frac{\$92,000}{2} = \$46,000$.

IQR = Third Quartile – First Quartile = $46,000.00 – $25,837.50 = $20,162.50.
1.5(IQR) = 1.5($20,162.50) = $30,243.75.
Left Whisker = First Quartile – 1.5(IQR) = $25,837.50 – $30,243.75 = $0.00.
Note: The price of a convertible cannot be negative, the left whisker is zero.
Right Whisker = Third Quartile + 1.5(IQR) = $46,000.00 + $30,243.75
Right Whisker = $76,243.75.

The boxplot is:

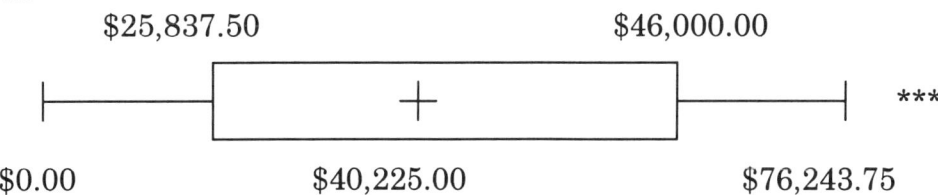

Note: The three asterisks represent the three prices ($78,390 $80,400 and $83,820) of convertibles that are unusually high compared to the other prices in the data set.

B60

5. The standard deviation from the calculator is 78297.36186.
   The standard deviation for a home in Chicago is approximately $78,297.36.

7. The sorted data is:

   $ 74,900 $ 79,900 $ 80,000 $ 84,000 $ 89,500 $ 91,000 $ 92,500
   $ 94,000 $ 95,000 $ 97,000 $ 99,500 $101,000 $101,500 $103,990
   $108,000 $110,000 $113,700 $114,475 $115,000 $117,500 $118,000
   $124,700 $125,250 $127,500 $135,000 $139,000 $142,000 $145,000
   $146,000 $149,000 $149,900 $159,000 $161,500 $162,000 $178,500
   $205,000 $224,000 $235,000 $263,000 $268,650

   There are 40 data values. $\dfrac{n+1}{2} = \dfrac{40+1}{2} = \dfrac{41}{2} = 20.5$.

   The median is half way between the 20th and 21st values.

   The median for a townhouse is $\dfrac{\$117,500 + \$118,000}{2} = \dfrac{\$235,500}{2} = \$117,750$.

   The first quartile is the median of the first 20 values, $\dfrac{40+1}{2} = \dfrac{21}{2} = 10.5$ value.

   This is the midpoint of the 10th and 11th values.

   The first quartile is $\dfrac{\$97,000 + \$99,500}{2} = \dfrac{\$196,500}{2} = \$98,250$.

   The third quartile is the median of the second 20 values $\dfrac{40+1}{2} = \dfrac{21}{2} = 10.5$ value.

   This is the midpoint of the 10th and 11th values for the second set of 20 values.

   The third quartile is $\dfrac{\$149,000 + \$149,900}{2} = \dfrac{\$298,900}{2} = \$149,450$.

   IQR = Third Quartile − First Quartile = $149,450 − $98,250 = $51,200.
   1.5(IQR) = 1.5($51,200) = $76,800.
   Left Whisker = First Quartile − 1.5(IQR) = $98,250 − $76,800 = $21,450.
   Right Whisker = Third Quartile + 1.5(IQR) = $149,450 + $76,800 = $226,250.

   The boxplot is:

   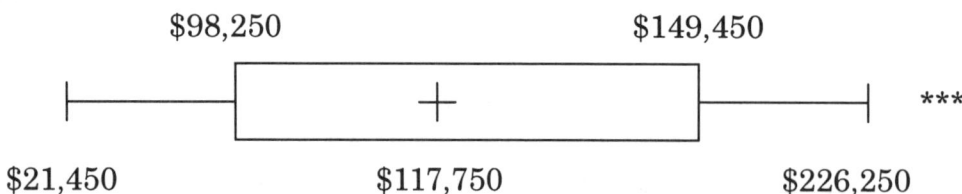

   Note: The three asterisks represent the three prices ($235,000 $263,000 and $268,650) of townhouses in suburban Chicago that are unusually high compared to the other prices in the data set.

9. The sorted data is:

$ 6,999	$ 7,399	$ 7,399	$ 7,895	$ 7,899	$ 7,999	$ 8,199
$ 8,399	$ 8,595	$ 8,990	$ 8,995	$ 9,699	$ 9,999	$10,449
$10,599	$10,899	$11,770	$11,999	$11,999	$12,299	$14,399
$14,500	$15,000	$15,600	$16,190	$16,700	$16,990	$18,200
$19,200

The median is the value at the $\frac{n+1}{2}$ location, $\frac{29+1}{2} = \frac{30}{2} = 15^{th}$ value.

The median is $10,599.

The first quartile is the median of the first 14 values, $\frac{14+1}{2} = \frac{15}{2} = 7.5$ value.

This is the midpoint of the 7th and 8th values. The first quartile is
$\frac{\$8,199 + \$8,399}{2} = \frac{\$16,598}{2} = \$8,299$. The first quartile is $8,299.

The third quartile is the median of the second 14 values, $\frac{14+1}{2} = \frac{15}{2} = 7.5$ value.

This is the midpoint of the 7th and 8th values for the second set of 14 values.
$\frac{\$14,500 + \$15,000}{2} = \frac{\$29,500}{2} = \$14,750$. The third quartile is $14,750.

IQR = Third Quartile − First Quartile = $14750 − $8,299 = $6,451.
1.5(IQR) = 1.5($6,451) = $9,676.50.
Left Whisker = First Quartile − 1.5(IQR) = $8,299 − $9,676.50 = $0.
Note: Since the price of a motorcycle cannot be negative, the left whisker is zero.
Right Whisker = Third Quartile + 1.5(IQR) = $14,750 + $9,676.50 = $24,426.50.
The boxplot is:

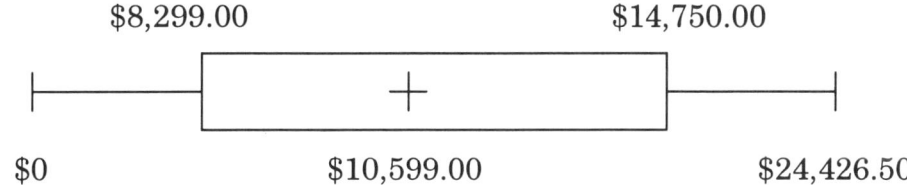

There are no outliers for this data set.

Exercise 5.5

1.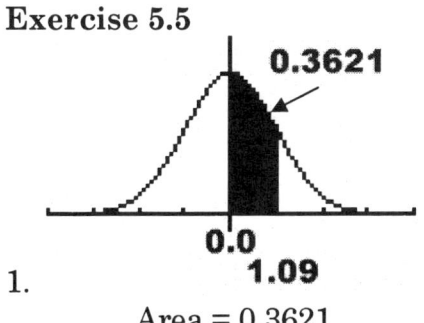
Area = 0.3621

3.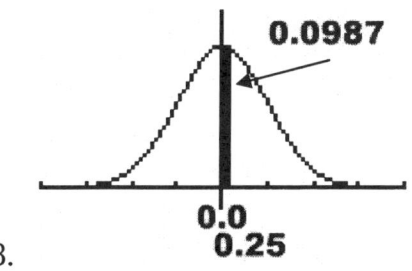
Area = 0.0987

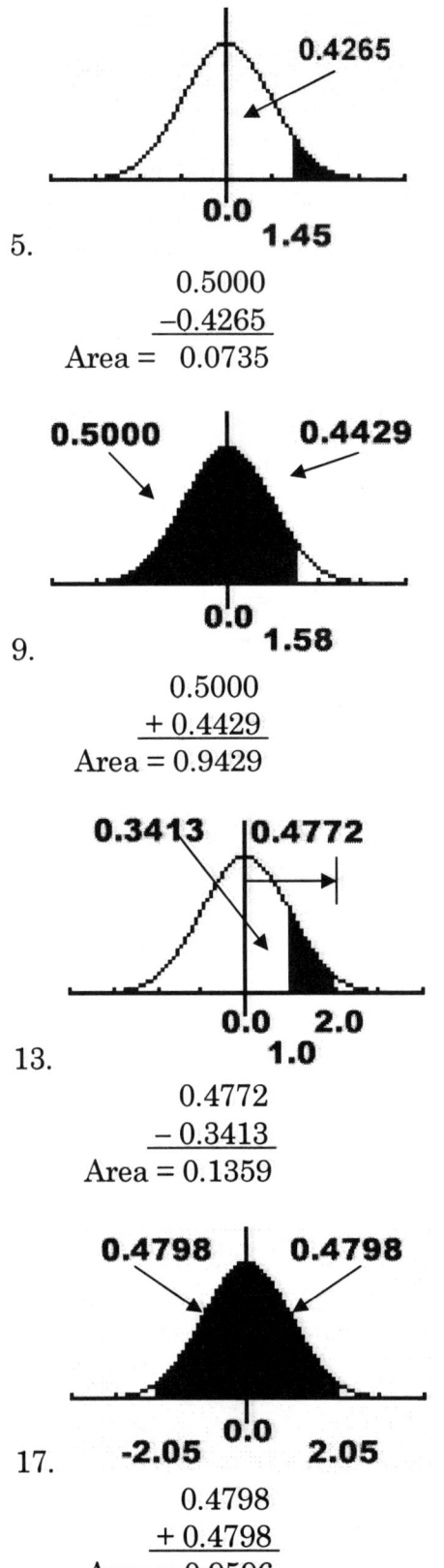
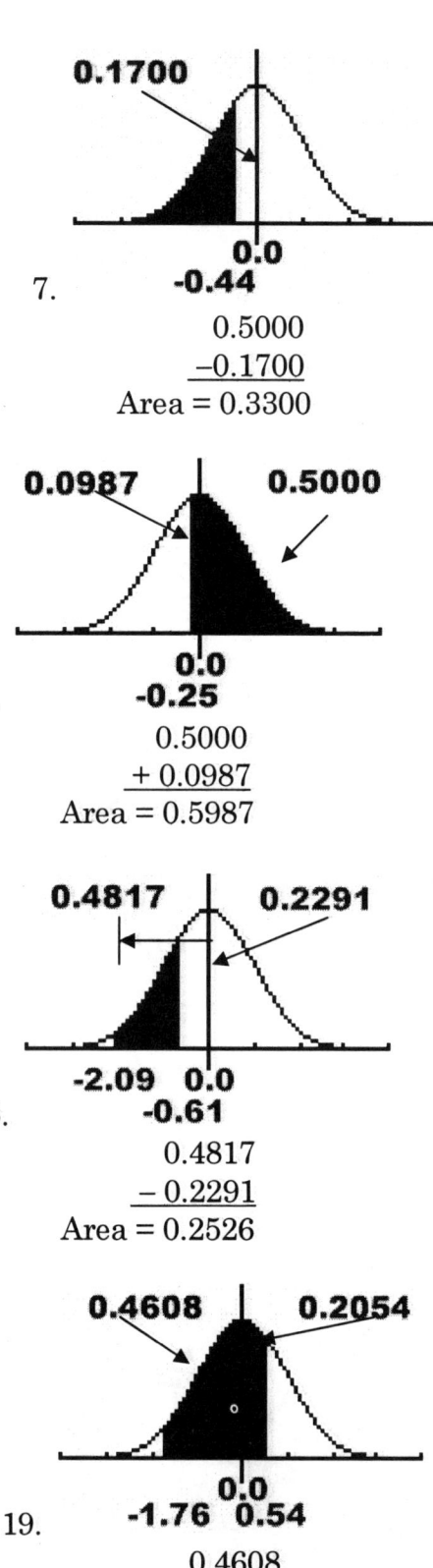

Exercise 5.6

1.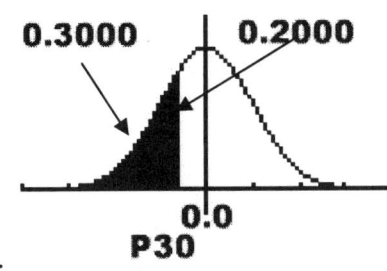

0.1985   **0.2000**   0.2019
  ↓   ↓

$z = 0.52$ $z = 0.53$
0.2000 is closer to $z = 0.52$.
Since P_{30} is to the left of zero,
it is negative. $P_{30} = -0.52$.

3.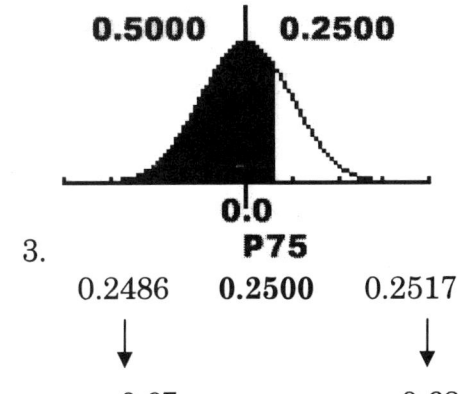

0.2486   **0.2500**   0.2517
  ↓   ↓

$z = 0.67$ $z = 0.68$
0.2500 is closer to $z = 0.67$.
$P_{75} = 0.67$.

5.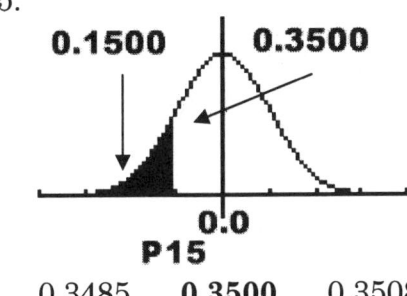

0.3485   **0.3500**   0.3508
  ↓   ↓

$z = 1.03$ $z = 1.04$
0.3500 is closer to $z = 1.04$.
Since P_{15} is to the left of zero, $P_{15} = -1.04$ and $P_{85} = 1.04$.
The z scores marking the upper and lower 15 percentiles are ± 1.04

7.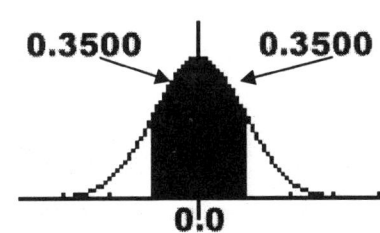

0.3485   **0.3500**   0.3508
  ↓   ↓

$z = 1.03$ $z = 1.04$

B64

0.3500 is closer to z = 1.04.
Since one z-score is to the left of zero, z is negative, z = –1.04.
The z scores that correspond to the middle 70% are ± 1.04.

9.

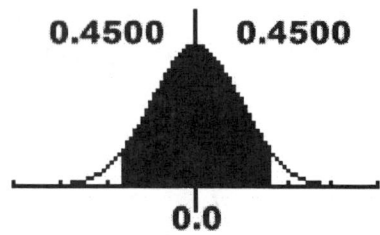

0.4495   **0.4500**   0.4505
  ↓   ↓
z = 1.64   z = 1.65

0.4500 is centered between 0.4495 and 0.4505.

The midpoint the z scores is used $z = \dfrac{1.64 + 1.65}{2} = 1.645$.

Since one z-score is to the left of zero, z is negative, z = –1.6454.
The z scores that correspond to the middle 90% are ± 1.645.

Exercise 5.7

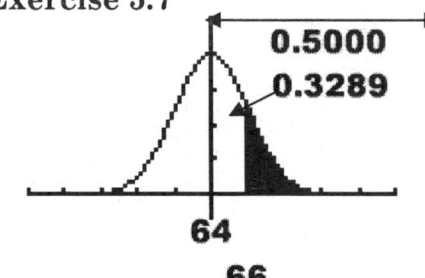

1.
$z = \dfrac{x - \mu}{\sigma} = \dfrac{66 - 64}{2.1} \approx 0.95 \to 0.3289$

$z = \dfrac{x - \mu}{\sigma} = \dfrac{66 - 64}{2.1} \approx 0.95 \to 0.3289$

0.5000
− 0.3289
0.1711

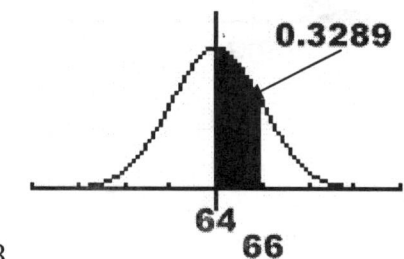

3.
$z = \dfrac{x - \mu}{\sigma} = \dfrac{66 - 64}{2.1} \approx 0.95 \to 0.3289$

$z = \dfrac{64 - 64}{2.1} = 0 \to 0.0000$

0.3289
− 0.0000
0.3289

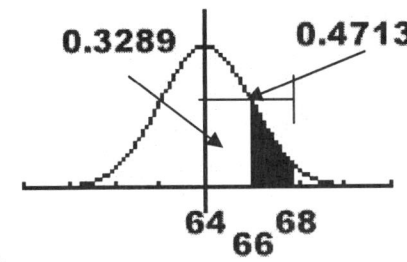

5.
$$z = \frac{x - \mu}{\sigma} = \frac{68 - 64}{2.1} \approx 1.99 \to 0.4713$$
$$z = \frac{x - \mu}{\sigma} = \frac{66 - 64}{2.1} \approx 0.95 \to 0.3289$$

0.4713
− 0.3289
0.1424

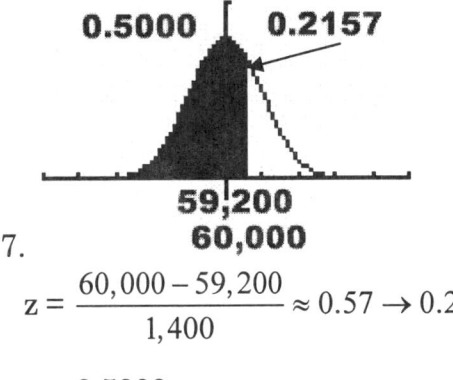

7.
$$z = \frac{60,000 - 59,200}{1,400} \approx 0.57 \to 0.2157$$

0.5000
+ 0.2157
0.7157

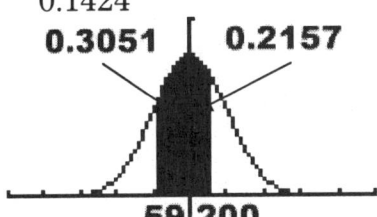

9.
$$z = \frac{58,000 - 59,200}{1,400} \approx -0.86 \to 0.3051$$
$$z = \frac{60,000 - 59,200}{1,400} \approx 0.57 \to 0.2157$$

0.3051
+ 0.2157
0.5208

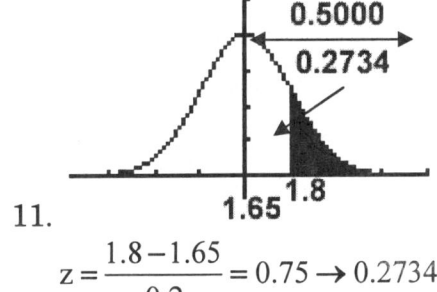

11.
$$z = \frac{1.8 - 1.65}{0.2} = 0.75 \to 0.2734$$

0.5000
− 0.2734
0.2266

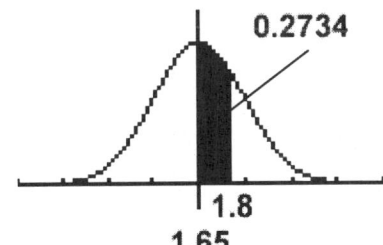

13.
$$z = \frac{1.8 - 1.65}{0.2} = 0.75 \to 0.2734$$

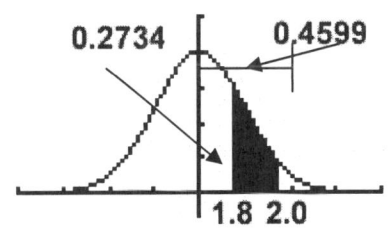

15.
$$z = \frac{2.0 - 1.65}{0.2} = 1.75 \to 0.4599$$
$$z = \frac{1.8 - 1.65}{0.2} = 0.75 \to 0.2734$$
0.4599 − 0.2734 = 0.1865

B66

Section 5.8

1.
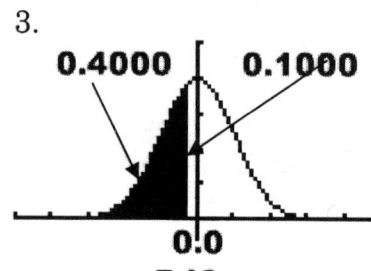

0.3485   **0.3500**   0.3508
↓   ↓
$z = 1.03$ $z = 1.04$
0.3500 is closer to $z = 1.04$.
Since P_{15} is to the left of zero, it is negative.
$P_{15} = -1.04$.
$x = \mu + z\sigma$
$x = 1.65 + (-1.04)(0.2) = 1.442$.

3.
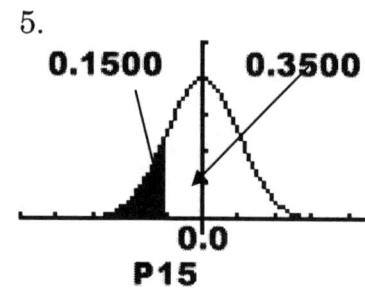

0.3997   **0.4000**   0.4015
↓   ↓
$z = 1.28$ $z = 1.29$
0.4000 is closer to $z = 1.28$.
Since P_{40} is to the left of zero, is negative. $P_{40} = -1.28$.
$x = \mu + z\sigma$
$x = 900 + (-1.28)(66) = 815.52$.

5.
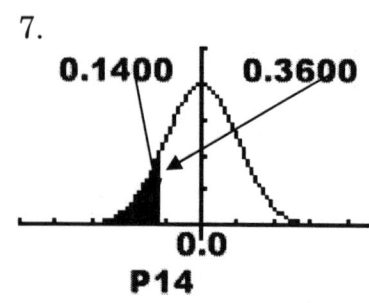

0.3485   **0.3500**   0.3508
↓   ↓
$z = 1.03$ $z = 1.04$
0.3500 is closer to $z = 1.04$.
Since P_{15} is to the left of zero, it is negative. $P_{15} = -1.04$.
$x = \mu + z\sigma$
$x = 64 + (-1.04)(2.1) = 61.816$.

7.

0.3599   **0.3600**   0.3621
↓   ↓
$z = 1.08$ $z = 1.09$
0.3600 is closer to $z = 1.08$.
Since it is to the left of zero, z is negative. $P_{14} = -1.08$.
$x = \mu + z\sigma$
$x = 59{,}200 + (-1.08)(1{,}400) = 57{,}588$.

9.

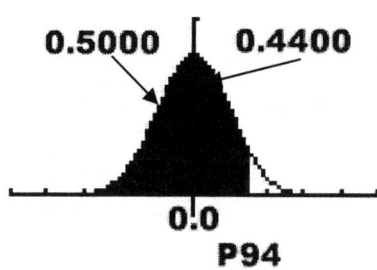

0.4394 **0.4400** 0.4406

z = 1.55   z = 1.56

0.4400 is the midpoint of = 1.55 and z = 1.56.

$z = \dfrac{1.55 + 1.56}{2} = 1.555$. $P_{94} = 1.55$.

$x = \mu + z\sigma$ $x = 7.3 + (1.555)(2.3) = 10.8765$.

Exercise 6.1

1A. $\dfrac{6}{36} \approx 0.167$ B. $\dfrac{2}{36} \approx 0.056$ C. $1 - \dfrac{6}{36} = \dfrac{30}{36} \approx 0.833$

D. $1 - \dfrac{2}{36} = \dfrac{34}{36} \approx 0.944$ E. $\dfrac{0}{36} = 0$ F. 1

3. $\dfrac{116}{1,630} \approx 0.071$

5. 1,630 patients in the sample space: $\dfrac{32}{1,630} \approx 0.020$

7. 76,000 accidental deaths in the sample space.
   A. $\dfrac{4,800}{76,000} \approx 0.063$ B. $1 - \dfrac{1,600}{76,000} = \dfrac{74,400}{76,000} \approx 0.979$ C. $\dfrac{6,500}{76,000} \approx 0.086$

   D. $1 - \dfrac{42,000}{76,000} = \dfrac{34,000}{76,000} \approx 0.447$

9. 26,138,570 crimes in the sample space.
   A. $\dfrac{23,760}{26,138,570} \approx 0.001$ B. $1 - \dfrac{672,480}{26,138,570} = \dfrac{25,466090}{26,138,570} \approx 0.974$

   C. $\dfrac{12,505,900}{26,138,570} \approx 0.478$ D. $1 - \dfrac{7,915,200}{26,138,570} = \dfrac{18,223,370}{26,138,570} \approx 0.697$

11. 1,998 families in the sample space.

   A. $\dfrac{235}{1,998} \approx 0.118$ B. $1 - \dfrac{235}{1,998} \approx 0.882$ C. $\dfrac{523}{1,998} \approx 0.262$

   D. $\dfrac{1,475}{1,998} \approx 0.738$ E. 0   F. 1

13. 3,200 people in the sample space.

   A. $\dfrac{325}{3,200} \approx 0.102$ B. $\dfrac{289}{3,200} \approx 0.090$ C. $1 - \dfrac{228}{3,200} \approx 0.929$ D. $\dfrac{125}{3,200} \approx 0.039$

15. 5,965 households in the sample space.

   A. $\dfrac{2,412}{5,965} \approx 0.404$ B. $1 - \dfrac{1,716}{5,965} \approx 0.712$ C. 0.000

   D. $1.00 - 0.00 = 1.00$ E. $\dfrac{135}{5,965} \approx 0.023$

17. 1,200 tickets in the sample space (60 winning tickets and 1,140 losing tickets).

   A. $\dfrac{1}{1,200} \approx 0.0008$ B. $1 - \dfrac{1}{1,200} \approx 0.9992$ C. $\dfrac{9}{1,200} = 0.0075$

   D. $1 - \dfrac{7}{1,200} \approx 0.9942$ E. $\dfrac{40}{1,200} \approx 0.033$ F. $\dfrac{1,140}{1,200} = 0.95$

19. 3,300 tickets in the sample space (20 winning tickets and 3,280 losing tickets).

   A. $\dfrac{1}{3,300} \approx 0.000$ B. $1 - \dfrac{10}{3,300} \approx 0.997$ C. 0   D. $\dfrac{3,280}{3,300} \approx 0.994$

   E. $\dfrac{5}{3,300} \approx 0.002$

   Note: The 0.000 is not equivalent to 0. The 0 means never. 0.000 means that due to rounding to three decimal places, the result is 0.000. However, there is a value beyond the third decimal place.

Exercise 6.2

1. Given: Odds against "Silver Charm" winning are 5 : 1,
   5 chances against winning and 1 chance of winning.

   A. Odds in favor of "Silver Charm" winning are 1 : 5

   B. P(Silver Charm losing) = $\dfrac{5}{5+1} = \dfrac{5}{6} \approx 0.833$

   C. P(Silver Charm winning) = $1 - \dfrac{5}{6} = \dfrac{1}{6} \approx 0.167$

3. Given: P(headache after taking medication) = 0.7% = 0.007 = $\frac{7}{1,000}$

   A. P(No headache) = 1 − 0.007 = 0.993
   B. P(Headache) = $\frac{7}{1,000}$. That means there are 7 chances of a headache and (1,000 − 7) = 993 chances of no headache. Odds in favor of a headache is 7 : 993
      Simplifying, odds in favor of a headache are about 1 : 142
   C. Odds of no headache is 993 : 7. Simplifying, odds of no headache are about 142 : 1.
   D. Odds against a headache is the same as odds of no headache, 993 : 7
      Simplifying, odds of no headache are about 142 : 1.
   E. Odds against no headache is the same as odds of a headache, 7 : 993
      Simplifying, odds in favor of a headache are about 1 : 142

5A. P(Green) = $\frac{6}{31}$ ≈ 0.194

   B. There are 6 chances of a green and (31 − 6) = 25 chances of not green.
      Odds against a green candy are 25 : 6
      Simplifying, odds against a green candy are about 4 : 1.
   C. P(Orange) = $\frac{12}{31}$ ≈ 0.387
   D. There are 12 chances of an orange and (31 − 12) = 19 chances of not orange.
      Odds in favor of an orange candy are 12 : 19
      Simplifying, odds in favor of an orange candy are about 2 : 3.

7. Given 250 total votes: 155 were against the tax increase and 95 (250 − 155 = 95) were in favor the tax increase.

   A. P(voting in favor of a tax increase for the library) = $\frac{95}{250}$ = 0.38.
   B. Odds in favor of voting in favor of a tax increase for the library are 95 : 155.
      Simplifying, odds in favor of voting in favor of a tax increase for the library are about 2 : 3.
   C. Odds against voting a tax increase for the library are 155 : 95.
      Simplifying, odds against voting a tax increase for the library are about 3 : 2.

9A. P(Collision with a train) = $\frac{600}{42,000}$ ≈ 0.014

   B. There are 6,200 chances of death by a pedestrian accident and (42,000 − 6,200) = 35,800 chances of death not by a pedestrian accident.
      Odds against a pedestrian accident are 35,800 : 6,200.
      Simplifying, odds against a pedestrian accident are about 6 : 1.

C. P(Collision with another motor vehicle) = $\dfrac{17,900}{42,000} \approx 0.426$

D. There are 11,900 chances of death due to a collision with a fixed object and (42,000 − 11,900) = 30,100 chances of death by not colliding with a fixed object. Odds in favor of death due to a collision with a fixed object are 11,900 : 30,100. Odds in favor of death due to a collision with a fixed object are about 2 : 5.

Exercise 6.3

x	Freq.	P(x)	x • P(x)
0	523	0.262	0.000
1	434	0.217	0.217
2	658	0.329	0.658
3	235	0.118	0.354
4	117	0.059	0.236
5	21	0.011	0.055
6	8	0.004	0.024
7	0	0.000	0.000
8	2	0.001	0.008
Totals	1998	1.001*	1.552

* Note: Error due to Rounding

1A. P(3) + P(4) + P(5) + P(6) + P(7) + P(8) = 0.192
1B. P(0) + P(1) + P(2) + P(3) = 0.926
1C. P(4) + P(5) + P(6) + P(7) + P(8) = 0.075
1D. P(0) + P(1) + P(2) + P(3) = 0.926
1E. E = 1.552   There are between 1 to 2 children living with their parents.

x	Freq.	P(x)	x • P(x)
0	125	0.039	0.000
1	446	0.139	0.139
2	365	0.114	0.228
3	442	0.138	0.414
4	276	0.086	0.344
5	228	0.071	0.355
6	325	0.102	0.612
7	409	0.128	0.896
8	289	0.090	0.720
9	126	0.039	0.351
10	56	0.018	0.180
11	87	0.027	0.297
12	26	0.008	0.096
Totals	3200	0.999*	4.632

* Note: Error due to Rounding

3A. P(6) + P(7) + P(8) + P(9) + P(10) + P(11) + P(12) = 0.412
3B. P(0) + P(1) + P(2) + P(3) + P(4) + P(5) + P(6) + P(7) + P(8) = 0.908
3C. P(6) + P(7) + P(8) + P(9) + P(10) + P(11) + P(12) = 0.412
3D. P(0) + P(1) + P(2) + P(3) + P(4) + P(5) + P(6) + P(7) = 0.818
3E. E = 4.632 The average number of movies seen by a person is between 4 and 5.

x	Freq	P(x)	x • P(x)
0	135	0.023	0.000
1	878	0.147	0.147
2	2,412	0.404	0.808
3	1,716	0.288	0.864
4	623	0.104	0.416
5	201	0.034	0.170
Totals	5,965	1.000	2.405

5A. P(2) + P(3) + P(4) + P(5) = 0.830   5B. P(0) + P(1) + P(2) + P(3) = 0.862
5C. P(3) + P(4) + P(5) = 0.426   5D. P(0) + P(1) + P(2) = 0.574
5E. E = 2.405 The average number of motor vehicles per household is between 2 and 3.

Exercise 6.4

1.

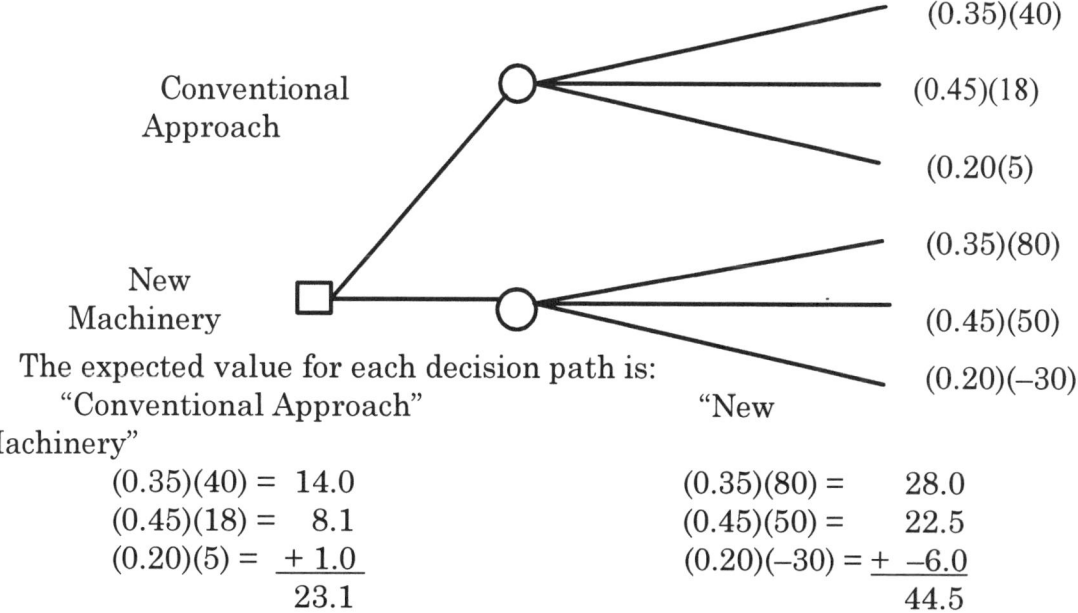

The expected value for each decision path is:
  "Conventional Approach"   "New Machinery"

  (0.35)(40) =   14.0   (0.35)(80) =   28.0
  (0.45)(18) =   8.1   (0.45)(50) =   22.5
  (0.20)(5) = + 1.0   (0.20)(−30) = + −6.0
                 23.1   44.5

The recommendation is to purchase "New Machinery". The expected profit is 44.5 million dollars ($44,500,000).

3.

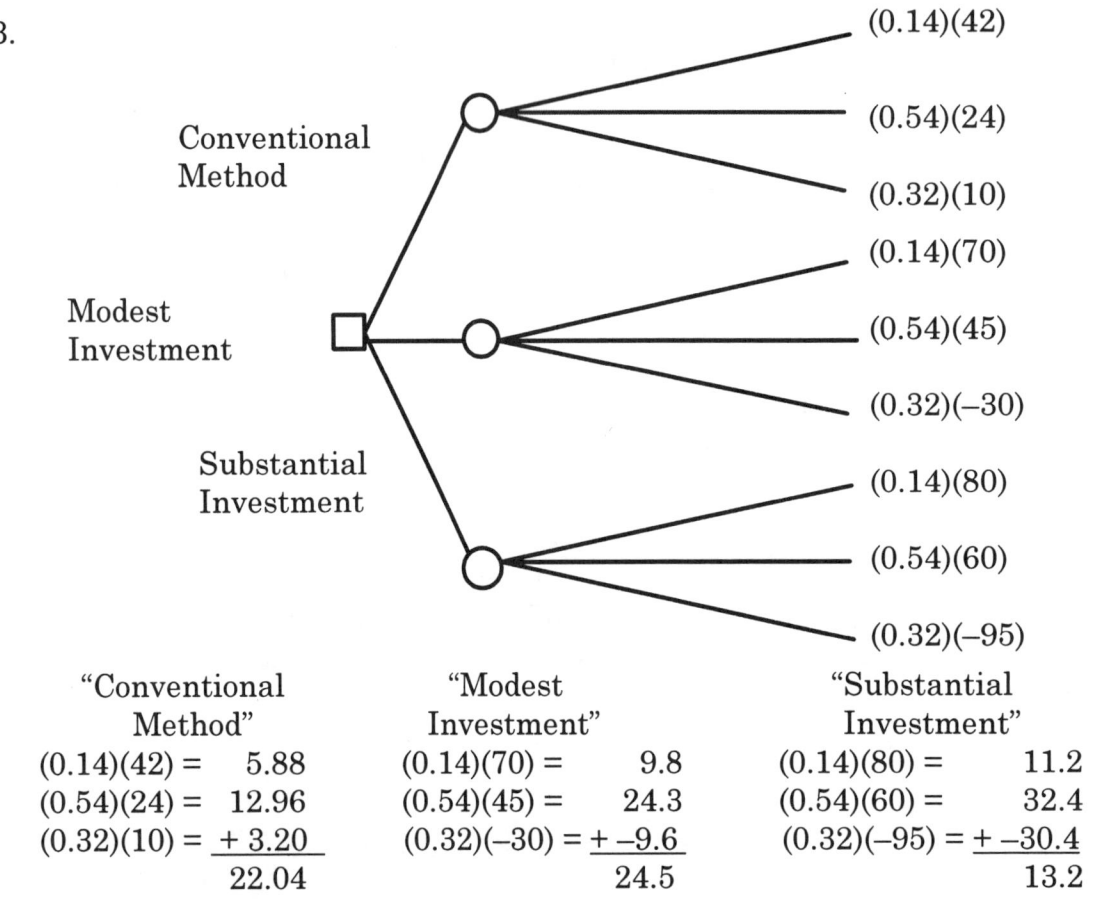

"Conventional   "Modest   "Substantial
  Method"   Investment"   Investment"
(0.14)(42) = 5.88   (0.14)(70) =   9.8   (0.14)(80) =   11.2
(0.54)(24) = 12.96   (0.54)(45) = 24.3   (0.54)(60) =   32.4
(0.32)(10) = + 3.20 (0.32)(–30) = + –9.6 (0.32)(–95) = + –30.4
              22.04   24.5   13.2

The recommendation is to make a "Modest Investment in New Machinery". The expected profit is 24.5 million dollars ($24,500,000).

5.

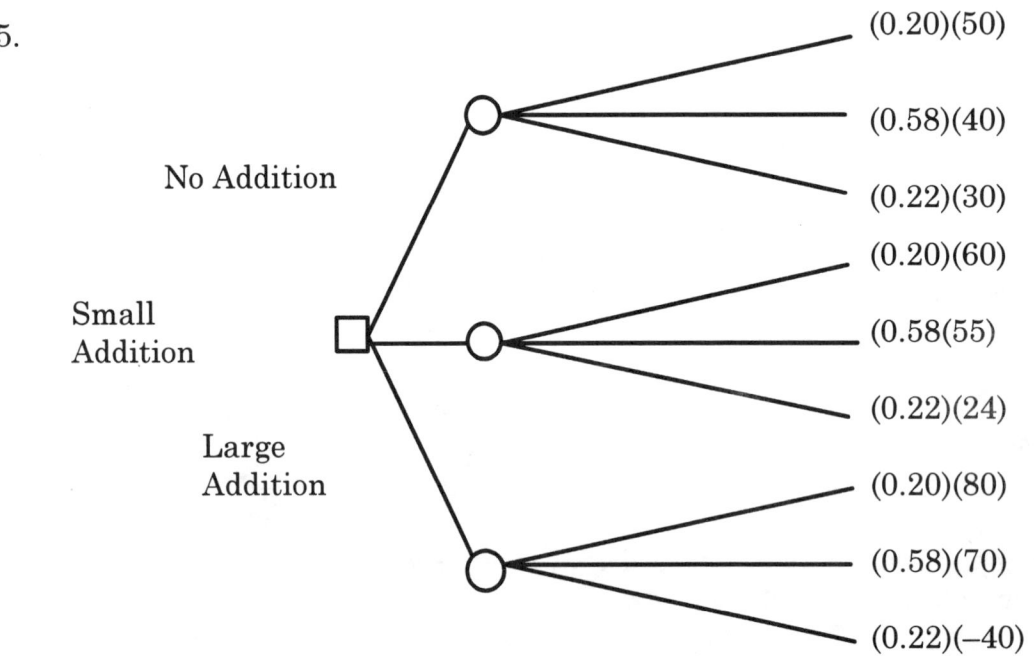

The expected value for each decision path is:

"No Addition"	"Small Addition"	"Huge Addition"
(0.20)(50) = 10.0	(0.20)(60) = 12.00	(0.20)(80) = 16.0
(0.58)(40) = 23.2	(0.58)(55) = 31.90	(0.58)(70) = 40.6
(0.22)(30) = + 6.6	(0.22)(24) = + 5.28	(0.22)(–40) = + – 8.8
39.8	49.18	47.8

The recommendation is to build a "Small Addition". The expected profit is 49.18 million dollars ($49,180,000).

7.

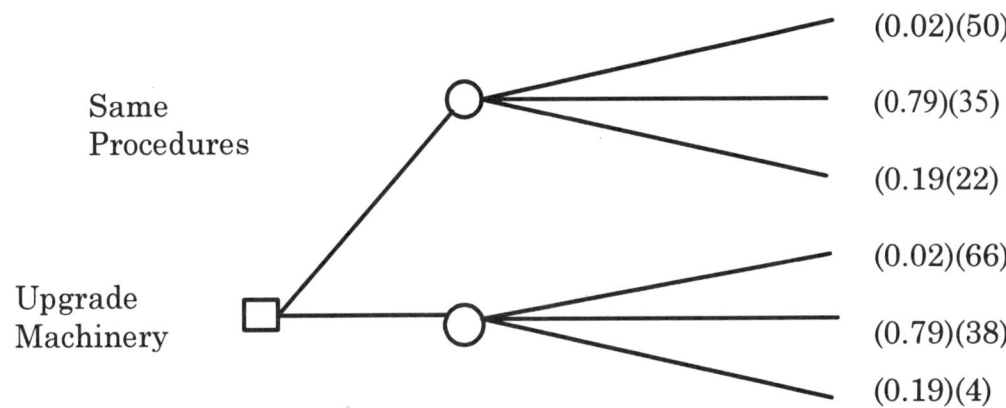

The expected value for each decision path is:

"Same Procedure"	"Upgrade Machinery"
(0.02)(50) = 1.00	(0.02)(66) = 1.32
(0.79)(35) = 27.65	(0.79)(38) = 30.02
(0.19)(22) = + 4.18	(0.19)(4) = + 0.76
32.83	32.10

The recommendation is to "Upgrade Machinery". The expected cost is 32.1 million dollars ($32,100,000).

B74

9.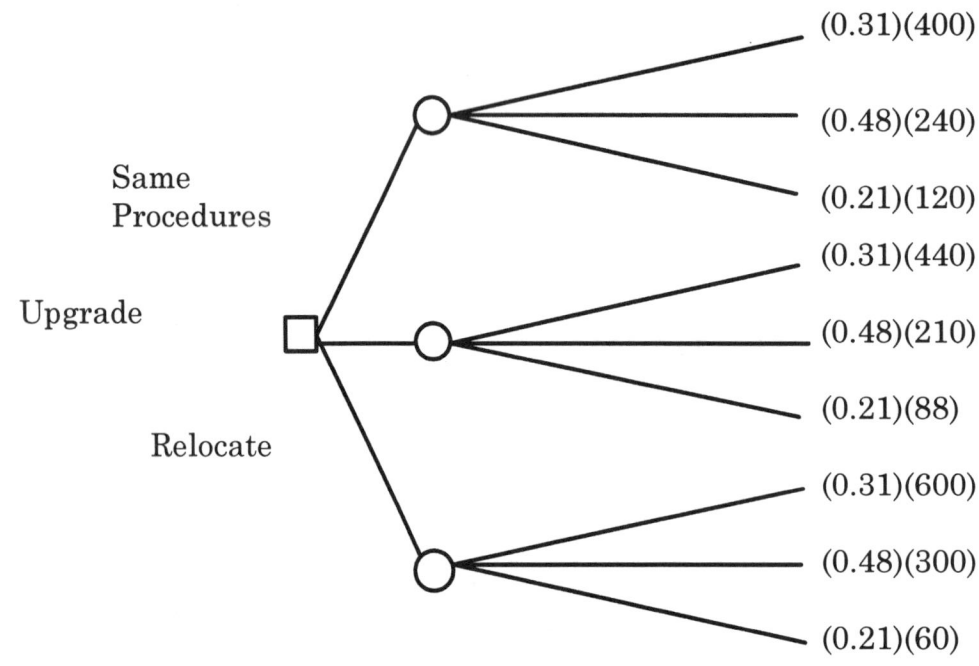

The expected value for each decision path is:

"Same Procedure"	"Upgrade Machinery"	"Relocate"
(0.31)(400) = 124.0	(0.31)(440) = 136.40	(0.31)(600) = 186.0
(0.48)(240) = 115.2	(0.48)(210) = 100.80	(0.48)(300) = 144.0
(0.21)(120) = + 25.2	(0.21)(88) = + 18.48	(0.21)(60) = + 12.6
264.4	255.68	342.6

The recommendation is to "Upgrade Machinery". The expected cost is 255.68 million dollars ($255,680,000).

Exercise 6.5

1. n = 200

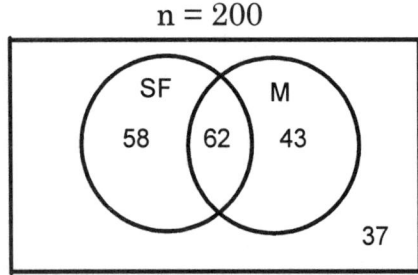

SF = Science Fiction
M = Movie of the Week

A. P(Only Sci-Fi) = $\frac{58}{200}$ = 29%

B. P(Only Movie of the Week) = $\frac{43}{200}$ = 21.5%

C. P(Neither Sci-Fi or Movie) = $\frac{37}{200}$ = 18.5%

B75

Exercise 6.5

3. n = 607

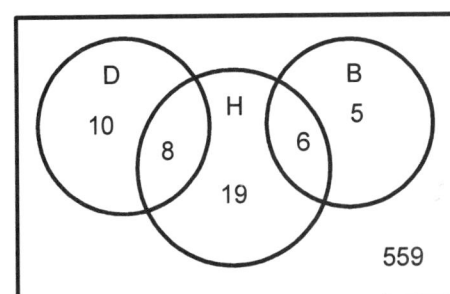

H = Headaches
D = Dizziness
B = Breathing difficulties

A. P(Only headaches) = $\dfrac{19}{607} \approx 3.1\%$

B. P(Only dizziness) = $\dfrac{10}{607} \approx 1.6\%$

C. P(Only breathing difficulties) = $\dfrac{5}{607} \approx 0.8\%$

D. P(none of these symptoms) = $\dfrac{559}{607} \approx 92.1\%$

5. n = 120

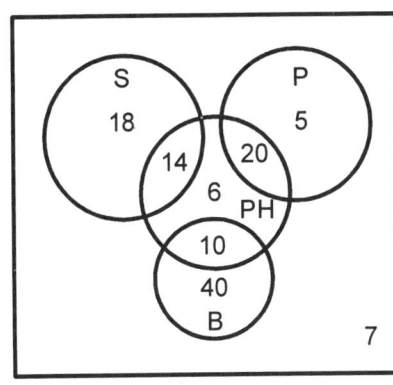

S = Salad Bar
P = Indoor Play Area
PH = Indoor Public Phones
B = Open for Breakfast

A. P(Only an indoor phone) = $\dfrac{6}{120} = 5\%$

B. P(Only a salad bar) = $\dfrac{18}{120} = 15\%$

C. P(Only open for breakfast) = $\dfrac{40}{120} \approx 33.3\%$

D. P(Only an indoor play area) = $\dfrac{5}{120} = 4.2\%$

E. P(None of these four categories) = $\dfrac{7}{120} \approx 5.8\%$

Exercise 6.6

1. n = 200

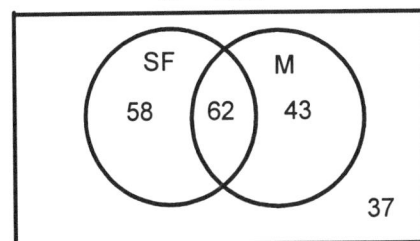

SF = Science Fiction
M = Movie of the Week

A. $\dfrac{120}{200} = 0.60$

B. $\dfrac{105}{200} = 0.525$

C. $\dfrac{58+62+43}{200} = \dfrac{163}{200} = 0.815$ or $\dfrac{120+105-62}{200} = \dfrac{163}{200} = 0.815$

D. $\dfrac{37}{200} = 0.185$

3.   n = 607

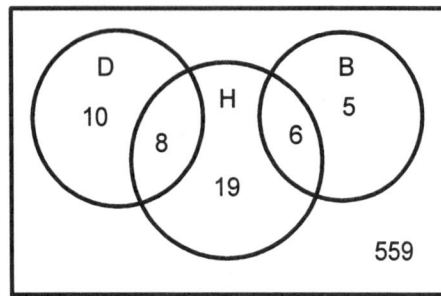

H = Headaches
D = Dizziness
B = Breathing difficulties

A. $\dfrac{19}{607} \approx 0.031$

B. $\dfrac{10}{607} \approx 0.016$

C. $\dfrac{5}{607} \approx 0.008$

D. $\dfrac{10+8+19+6}{607} = \dfrac{43}{607} \approx 0.071$ or $\dfrac{33+18-8}{607} = \dfrac{43}{607} \approx 0.071$

E. $\dfrac{8+19+6+5}{607} = \dfrac{38}{607} \approx 0.063$ or $\dfrac{33+11-6}{607} = \dfrac{38}{607} \approx 0.063$

F. $\dfrac{10+8+6+5}{607} = \dfrac{29}{607} \approx 0.048$ or $\dfrac{18+11-0}{607} = \dfrac{29}{607} \approx 0.048$

G. $\dfrac{559}{607} \approx 0.921$

5. **Reactions to the Medication**

Medication	Drowsiness	Headache	Dizziness	Nausea	Totals
New Medication	116	110	32	41	299
Old Medication	67	11	3	4	85
Placebo	71	103	20	32	226
Totals	254	224	55	77	610

A. $\dfrac{254+226-71}{610} = \dfrac{409}{610} \approx 0.670$ B. $\dfrac{299+224-110}{610} = \dfrac{413}{610} \approx 0.677$

C. $\dfrac{77+85-4}{610} = \dfrac{158}{610} \approx 0.259$ D. $\dfrac{224+55-0}{610} = \dfrac{279}{610} \approx 0.457$

7. The Bureau of Labor Statistics of the US Department of Labor reported that the distribution of wages paid to hourly workers was

Gender and Age	Wage < $10/hour	Wage at least $10/hour	Totals
Male, 16 years and older	17,512,000	14,187,000	31,699,000
Female, 16 years and older	23,126,000	8,491,000	31,617,000
Totals	40,638,000	22,678,000	63,316,000

A. $\dfrac{31{,}699{,}000 + 40{,}638{,}000 - 17{,}512{,}000}{63{,}316{,}000} = \dfrac{54{,}825{,}000}{63{,}316{,}000} \approx 0.866$

B. $\dfrac{31{,}617{,}000 + 22{,}678{,}000 - 8{,}491{,}000}{63{,}316{,}000} = \dfrac{45{,}804{,}000}{63{,}316{,}000} \approx 0.723$

C. $\dfrac{31{,}699{,}000 + 22{,}678{,}000 - 14{,}187{,}000}{63{,}316{,}000} = \dfrac{40{,}190{,}000}{63{,}316{,}000} \approx 0.635$

D. $\dfrac{31{,}617{,}000 + 40{,}638{,}000 - 23{,}126{,}000}{63{,}316{,}000} = \dfrac{49{,}129{,}000}{63{,}316{,}000} \approx 0.776$

Exercise 6.7

1A. $\dfrac{6}{36} \times \dfrac{6}{36} = \dfrac{36}{1{,}296} \approx 0.028$ \qquad B. $\dfrac{6}{36} \times \dfrac{2}{36} = \dfrac{12}{1{,}296} \approx 0.009$

C. $\dfrac{6}{36} \times \dfrac{30}{36} = \dfrac{180}{1{,}296} \approx 0.139$ \qquad D. $\dfrac{3}{36} \times \dfrac{3}{36} = \dfrac{9}{1{,}296} \approx 0.007$

3A. $\dfrac{6}{31} \times \dfrac{5}{30} = \dfrac{30}{930} \approx 0.032$ \qquad B. $\dfrac{12}{31} \times \dfrac{6}{30} = \dfrac{72}{930} \approx 0.077$

C. $\dfrac{4}{31} \times \dfrac{25}{30} = \dfrac{100}{930} \approx 0.108$ \qquad D. $\dfrac{27}{31} \times \dfrac{26}{30} = \dfrac{702}{930} \approx 0.755$

5A. $\dfrac{109}{1630} \times \dfrac{108}{1629} = \dfrac{11{,}772}{2{,}655{,}270} \approx 0.004$ \qquad B. $\dfrac{1{,}521}{1630} \times \dfrac{1{,}520}{1629} = \dfrac{2{,}311{,}920}{2{,}655{,}270} \approx 0.871$

7A. $\dfrac{85}{1630} \times \dfrac{84}{1629} = \dfrac{7{,}140}{2{,}655{,}270} \approx 0.003$ \qquad B. $\dfrac{1{,}545}{1630} \times \dfrac{1{,}544}{1629} = \dfrac{2{,}385{,}480}{2{,}655{,}270} \approx 0.898$

9A. $(0.10)(0.10)(0.10) = (0.10)^3 = 0.001$ \qquad B. $(0.90)(0.90)(0.90) = (0.90)^3 = 0.729$

Tax Returns, Before and After Opening an IRA Account

C1

Tax Returns: Before IRA vs. After IRA

Single, full time student, part time worker's tax return. The student is claimed as a dependent on his parent's tax return and he has no IRA account.

Form 1040A — U.S. Individual Income Tax Return (2005)

Filing status: Single

Exemptions: Boxes checked on 6a and 6b: 0; Total number of exemptions claimed: 0

Income:

Line	Description	Amount
7	Wages, salaries, tips, etc. Attach Form(s) W-2.	10,209
8a	Taxable interest. Attach Schedule 1 if required.	569
8b	Tax-exempt interest. Do not include on line 8a.	
9a	Ordinary dividends. Attach Schedule 1 if required.	
9b	Qualified dividends (see page 25).	
10	Capital gain distributions (see page 25).	
11a	IRA distributions.	
11b	Taxable amount (see page 25).	
12a	Pensions and annuities.	
12b	Taxable amount (see page 26).	
13	Unemployment compensation and Alaska Permanent Fund dividends.	
14a	Social security benefits.	
14b	Taxable amount (see page 28).	
15	Add lines 7 through 14b (far right column). This is your **total income**.	10,778

Adjusted gross income:

Line	Description	Amount
16	Educator expenses (see page 28).	
17	IRA deduction (see page 28).	
18	Student loan interest deduction (see page 31).	
19	Tuition and fees deduction (see page 32).	
20	Add lines 16 through 19. These are your **total adjustments**.	
21	Subtract line 20 from line 15. This is your **adjusted gross income**.	10778

For Disclosure, Privacy Act, and Paperwork Reduction Act Notice, see page 58. Cat. No. 11327A Form 1040A (2005)

Form 1040A (2005) — Page 2

Tax, credits, and payments

Standard Deduction for—
- People who checked any box on line 23a or 23b or who can be claimed as a dependent, see page 32.
- All others:
  Single or Married filing separately, $5,000
  Married filing jointly or Qualifying widow(er), $10,000
  Head of household, $7,300

Line	Description	Amount
22	Enter the amount from line 21 (adjusted gross income).	10,778
23a	Check if: ☐ You were born before January 2, 1941, ☐ Blind; ☐ Spouse was born before January 2, 1941, ☐ Blind. Total boxes checked ▶ 23a	
23b	If you are married filing separately and your spouse itemizes deductions, see page 32 and check here ▶ 23b ☐	
24	Enter your standard deduction (see left margin).	5,000
25	Subtract line 24 from line 22. If line 24 is more than line 22, enter -0-.	5,778
26	If line 22 is over $109,475, or you provided housing to a person displaced by Hurricane Katrina, see page 33. Otherwise, multiply $3,200 by the total number of exemptions claimed on line 6d.	0
27	Subtract line 26 from line 25. If line 26 is more than line 25, enter -0-. This is your **taxable income.** ▶	5,778
28	**Tax,** including any alternative minimum tax (see page 34).	578
29	Credit for child and dependent care expenses. Attach Schedule 2.	
30	Credit for the elderly or the disabled. Attach Schedule 3.	
31	Education credits. Attach Form 8863.	
32	Retirement savings contributions credit. Attach Form 8880.	
33	Child tax credit (see page 38). Attach Form 8901 if required.	
34	Adoption credit. Attach Form 8839.	
35	Add lines 29 through 34. These are your **total credits.**	
36	Subtract line 35 from line 28. If line 35 is more than line 28, enter -0-.	
37	Advance earned income credit payments from Form(s) W-2.	
38	Add lines 36 and 37. This is your **total tax.** ▶	578
39	Federal income tax withheld from Forms W-2 and 1099.	624
40	2005 estimated tax payments and amount applied from 2004 return.	
41a	Earned income credit (EIC).	
41b	Nontaxable combat pay election.	
42	Additional child tax credit. Attach Form 8812.	
43	Add lines 39, 40, 41a, and 42. These are your **total payments.** ▶	624

If you have a qualifying child, attach Schedule EIC.

Refund

Direct deposit? See page 53 and fill in 45b, 45c, and 45d.

44	If line 43 is more than line 38, subtract line 38 from line 43. This is the amount you overpaid.	46
45a	Amount of line 44 you want **refunded** to you. ▶	46
45b	Routing number	
45c	Type: ☐ Checking ☐ Savings	
45d	Account number	
46	Amount of line 44 you want **applied to your 2006 estimated tax.**	

Amount you owe

| 47 | **Amount you owe.** Subtract line 43 from line 38. For details on how to pay, see page 54. ▶ | |
| 48 | Estimated tax penalty (see page 54). | |

Third party designee

Do you want to allow another person to discuss this return with the IRS (see page 55)? ☐ Yes. Complete the following. ☐ No

Designee's name ▶   Phone no. ▶ ( )   Personal identification number (PIN) ▶

Sign here

Joint return? See page 18. Keep a copy for your records.

Under penalties of perjury, I declare that I have examined this return and accompanying schedules and statements, and to the best of my knowledge and belief, they are true, correct, and accurately list all amounts and sources of income I received during the tax year. Declaration of preparer (other than the taxpayer) is based on all information of which the preparer has any knowledge.

Your signature | Date | Your occupation | Daytime phone number

Spouse's signature. If a joint return, both must sign. | Date | Spouse's occupation

Paid preparer's use only

Preparer's signature | Date | Check if self-employed ☐ | Preparer's SSN or PTIN

Firm's name (or yours if self-employed), address, and ZIP code | EIN | Phone no. ( )

Form 1040A (2005)

Single, full time student, part time worker's tax return. The student is claimed as a dependent on his parent's tax return and has an IRA account.

Form 1040A — U.S. Individual Income Tax Return (2005)

Label (See page 16.) — Use the IRS label. Otherwise, please print or type.

Presidential Election Campaign: Check here if you, or your spouse if filing jointly, want $3 to go to this fund (see page 18). ☐ You ☐ Spouse

Filing status (Check only one box.)
1. ☑ Single
2. ☐ Married filing jointly (even if only one had income)
3. ☐ Married filing separately. Enter spouse's SSN above and full name here.
4. ☐ Head of household (with qualifying person). (See page 19.) If the qualifying person is a child but not your dependent, enter this child's name here.
5. ☐ Qualifying widow(er) with dependent child (see page 19)

Exemptions
- 6a ☐ Yourself. If someone can claim you as a dependent, do not check box 6a.
- 6b ☐ Spouse
- Boxes checked on 6a and 6b: **0**
- Add numbers on lines above: **0**

Income

Line	Description	Amount
7	Wages, salaries, tips, etc. Attach Form(s) W-2.	10,209
8a	Taxable interest. Attach Schedule 1 if required.	569
8b	Tax-exempt interest.	
9a	Ordinary dividends.	
9b	Qualified dividends.	
10	Capital gain distributions.	
11a	IRA distributions.	
11b	Taxable amount.	
12a	Pensions and annuities.	
12b	Taxable amount.	
13	Unemployment compensation and Alaska Permanent Fund dividends.	
14a	Social security benefits.	
14b	Taxable amount.	
15	Add lines 7 through 14b. This is your **total income**.	10,778

Adjusted gross income

Line	Description	Amount
16	Educator expenses.	
17	IRA deduction.	2,000
18	Student loan interest deduction.	
19	Tuition and fees deduction.	
20	Add lines 16 through 19. These are your **total adjustments**.	2,000
21	Subtract line 20 from line 15. This is your **adjusted gross income**.	8,778

Cat. No. 11327A Form 1040A (2005)

Form 1040A (2005) Page 2

Tax, credits, and payments	22	Enter the amount from line 21 (adjusted gross income).	22	8,778
	23a	Check if: ☐ You were born before January 2, 1941, ☐ Blind ☐ Spouse was born before January 2, 1941, ☐ Blind } Total boxes checked ▶ 23a		
Standard Deduction for—	b	If you are married filing separately and your spouse itemizes deductions, see page 32 and check here ▶ 23b ☐		
• People who checked any box on line 23a or 23b or who can be claimed as a dependent, see page 32.	24	Enter your **standard deduction** (see left margin).	24	5,000
	25	Subtract line 24 from line 22. If line 24 is more than line 22, enter -0-.	25	3,778
	26	If line 22 is over $109,475, or you provided housing to a person displaced by Hurricane Katrina, see page 33. Otherwise, multiply $3,200 by the total number of exemptions claimed on line 6d.	26	0
• All others:	27	Subtract line 26 from line 25. If line 26 is more than line 25, enter -0-. This is your **taxable income.** ▶	27	3,778
Single or Married filing separately, $5,000	28	**Tax,** including any alternative minimum tax (see page 34).	28	378
	29	Credit for child and dependent care expenses. Attach Schedule 2. 29		
Married filing jointly or Qualifying widow(er), $10,000	30	Credit for the elderly or the disabled. Attach Schedule 3. 30		
	31	Education credits. Attach Form 8863. 31		
	32	Retirement savings contributions credit. Attach Form 8880. 32		
Head of household, $7,300	33	Child tax credit (see page 38). Attach Form 8901 if required. 33		
	34	Adoption credit. Attach Form 8839. 34		
	35	Add lines 29 through 34. These are your **total credits.**	35	
	36	Subtract line 35 from line 28. If line 35 is more than line 28, enter -0-.	36	
	37	Advance earned income credit payments from Form(s) W-2.	37	
	38	Add lines 36 and 37. This is your **total tax.** ▶	38	378
	39	Federal income tax withheld from Forms W-2 and 1099. 39 624		
	40	2005 estimated tax payments and amount applied from 2004 return. 40		
If you have a qualifying child, attach Schedule EIC.	41a	**Earned income credit (EIC).** 41a		
	b	Nontaxable combat pay election. 41b		
	42	Additional child tax credit. Attach Form 8812. 42		
	43	Add lines 39, 40, 41a, and 42. These are your **total payments.** ▶	43	624
Refund	44	If line 43 is more than line 38, subtract line 38 from line 43. This is the amount you **overpaid.**	44	246
Direct deposit? See page 53 and fill in 45b, 45c, and 45d.	45a	Amount of line 44 you want **refunded to you.** ▶	45a	246
	▶ b	Routing number ▶ c Type: ☐ Checking ☐ Savings		
	▶ d	Account number		
	46	Amount of line 44 you want **applied to your 2006 estimated tax.** 46		
Amount you owe	47	**Amount you owe.** Subtract line 43 from line 38. For details on how to pay, see page 54. ▶	47	
	48	Estimated tax penalty (see page 54). 48		

Third party designee
Do you want to allow another person to discuss this return with the IRS (see page 55)? ☐ Yes. Complete the following. ☐ No
Designee's name ▶   Phone no. ▶ ( )   Personal identification number (PIN) ▶

Sign here
Joint return? See page 18.
Keep a copy for your records.

Under penalties of perjury, I declare that I have examined this return and accompanying schedules and statements, and to the best of my knowledge and belief, they are true, correct, and accurately list all amounts and sources of income I received during the tax year. Declaration of preparer (other than the taxpayer) is based on all information of which the preparer has any knowledge.

Your signature | Date | Your occupation | Daytime phone number ( )
Spouse's signature. If a joint return, **both** must sign. | Date | Spouse's occupation |

Paid preparer's use only
Preparer's signature | Date | Check if self-employed ☐ | Preparer's SSN or PTIN
Firm's name (or yours if self-employed), address, and ZIP code | | EIN
| | Phone no. ( )

Form **1040A** (2005)

Comparison of the tax returns.

	No IRA account	With an IRA	Remarks
Wages, line 7	$10,209	$10,209	**Same Wages**
Interest Income, Line 8	$569	$569	Same interest income
IRA deduction, Line 17		$2,000	$2,000 contributed into an IRA account
Adjustments, Line 20		$2,000	Amount subtracted from wages
Adjusted Gross Income, Line 21	$10,778	$8,778	Different adjustable gross incomes
Adjusted Gross Income, Line 22	$10,778	$8,778	Copy Line 21 to Line 22
Standard Deduction, Line 24	$5,000	$5,000	Same deduction for being single
Line 25	$5,778	$3,778	Subtract Line 24 from Line 22.
Exemptions, Line 26	$0	$0	Student is claimed as a dependent on his parent's return and cannot claim himself.
Taxable Income, Line 27	$5,778	$3,778	Different taxable incomes
Tax from the Tax table, Line 28	$578	$378	Different tax obligations
Tax obligation Line 38	$578	$378	Different tax obligations
Withholdings, Line 39	$624	$624	Same amount of money withheld from paychecks
Total payments, Line 43	$624	$624	Same amount of money withheld from paychecks
Amount overpaid, subtract Line 38 from line 43.	$46	$246	Different refunds
Refund, Line 44	$46	$246	Different refunds

Index

INDEX

Action	481
Addition Rule for Prob.	510
Annuities	218
Amortization	114 - 124, 162 - 176, 237 - 239, 242, 248 - 249
Future Value	222 - 228, 242 - 243, 247 - 248
Present Value	232 - 237, 242 - 244
Sinking Fund	218 - 222, 242, 244 - 247
Appraisal	203
APR	106-111
APY	106-111
Circles	274
Area	292 - 296
Circumference	274 - 276, 292
Class Width	366, 368
Closing Costs	202 - 212
Closing Fee	204
Complementary Events	446 - 448
Credit Search	203
Decision Tree	481 - 488
Dependent Event	521 - 522
Diameter	275
Dimensional Analysis	265 - 266, 272 - 273, 284 - 289, 302 - 303, 317 - 319
Document Fee	203
Down Payment	203 - 212
Empirical Rule	401
Escrow	162 - 176
Event	438
Expected Value	472 – 477
Experiment	438
Graphs	349, 369
Bar	352 - 353, 356 - 358
Histogram	370 - 371, 373 - 375
Pie	354 - 355, 359 - 362
Polygon	371 - 373, 375 - 376

Hypotenuse	269
Independent Event	520
Insurance	204
Interest	204
Compound	94 - 106
Continuous	101 - 102, 105 - 106
Future Value	86 – 88, 97 - 102
Present Value	89 - 91, 103 - 106
Simple	82 - 91
IQR	389
IQR Boxplot	389 - 394
Leg	269
Legal Fee	204
Linear Equations	59 - 72
Loan Application Fee	203
Loan Payments	82
Amortization	114 - 124
Future Value Simple Interest	125 - 127
Zero Percent Financing	127 - 128
Mean	380
Median	380
Mixtures	53 - 55
Mortgages	161
Adjustable (ARM)	161
Balloon	161
Bi-weekly	165 - 167, 172 - 174
Fixed	161, 163 - 164, 170 - 172
Interest-only	161, 167 - 169, 174 - 176
Multiplication for Probability	520
Dependent	522
Independent	521
Mutually Exclusive	493, 504
No Prepayment Penalty	146 - 150
Normal Distribution	403
Nonstandard	415 - 421, 423 - 428
Standard	399 - 408, 410 - 414
Odds	455 - 462
Order of Operations	20 - 32
Outliers	388, 391
Payoff Table	481 - 482

Percent	37 - 43
Points	203, 205 - 206
Private Mortgage Insurance (PMI)	162 - 176, 204
Probability (Definition)	436
Probability Distribution	465 - 472
Proportion Applications	46 - 55
Blueprint Conversions	49 - 51
Burning Calories	51 - 53
Mixture Dilution	53 - 55
Monetary Conversions	46 - 49
Pythagorean Theorem	269 - 272
Qualitative Data	349, 366
Quantitative Data	349, 366
Quartiles	389
First	389 - 390
Third	389 - 391
Radius	274 - 275
Random Variable	465
Ratio and Proportion	35 - 44
Recording Fee	203
Rectangle	263
Area	283 - 285
Perimeter	263 - 265,
Refinancing	184 - 197
Remaining Balance	132
Amortization	136 - 140
Future Value	132 - 136
Rule of 78	140 - 144
Zero Percent Financing	144 - 145
Sample Space	438
Sector	277
Area	295 - 296
Perimeter	277 - 279
Signed Numbers	1
Addition	3 - 6
Multiplication/Division	11 - 17
Subtraction	6 - 11
Standard Deviation	394 - 397
Standard Normal Table	Inside Back Cover
Standard Score (z-score)	400, 416
State of Nature	481
Survey	204

D3

Tax Charge	204
Title Insurance	203
Title Search	203
Trial	438
Triangles	267
Area	290 - 292
Perimeter	267 - 272
Venn Diagram	493 - 504
Volume	312 - 319

The Standard Normal Curve

The numbers in the table represent the area under the curve starting at the mean, 0, and extending to the z score. This table is called the Standard Normal (z) Distribution Table.

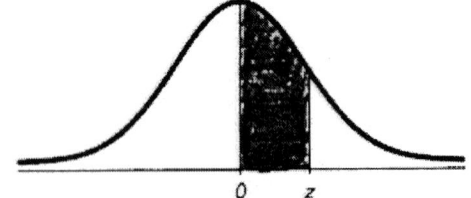

TABLE A-2 Standard Normal (z) Distribution

z	.00	.01	.02	.03	.04	.05	.06	.07	.08	.09
0.0	.0000	.0040	.0080	.0120	.0160	.0199	.0239	.0279	.0319	.0359
0.1	.0398	.0438	.0478	.0517	.0557	.0596	.0636	.0675	.0714	.0753
0.2	.0793	.0832	.0871	.0910	.0948	.0987	.1026	.1064	.1103	.1141
0.3	.1179	.1217	.1255	.1293	.1331	.1368	.1406	.1443	.1480	.1517
0.4	.1554	.1591	.1628	.1664	.1700	.1736	.1772	.1808	.1844	.1879
0.5	.1915	.1950	.1985	.2019	.2054	.2088	.2123	.2157	.2190	.2224
0.6	.2257	.2291	.2324	.2357	.2389	.2422	.2454	.2486	.2517	.2549
0.7	.2580	.2611	.2642	.2673	.2704	.2734	.2764	.2794	.2823	.2852
0.8	.2881	.2910	.2939	.2967	.2995	.3023	.3051	.3078	.3106	.3133
0.9	.3159	.3186	.3212	.3238	.3264	.3289	.3315	.3340	.3365	.3389
1.0	.3413	.3438	.3461	.3485	.3508	.3531	.3554	.3577	.3599	.3621
1.1	.3643	.3665	.3686	.3708	.3729	.3749	.3770	.3790	.3810	.3830
1.2	.3849	.3869	.3888	.3907	.3925	.3944	.3962	.3980	.3997	.4015
1.3	.4032	.4049	.4066	.4082	.4099	.4115	.4131	.4147	.4162	.4177
1.4	.4192	.4207	.4222	.4236	.4251	.4265	.4279	.4292	.4306	.4319
1.5	.4332	.4345	.4357	.4370	.4382	.4394	.4406	.4418	.4429	.4441
1.6	.4452	.4463	.4474	.4484	.4495 *	.4505	.4515	.4525	.4535	.4545
1.7	.4554	.4564	.4573	.4582	.4591	.4599	.4608	.4616	.4625	.4633
1.8	.4641	.4649	.4656	.4664	.4671	.4678	.4686	.4693	.4699	.4706
1.9	.4713	.4719	.4726	.4732	.4738	.4744	.4750	.4756	.4761	.4767
2.0	.4772	.4778	.4783	.4788	.4793	.4798	.4803	.4808	.4812	.4817
2.1	.4821	.4826	.4830	.4834	.4838	.4842	.4846	.4850	.4854	.4857
2.2	.4861	.4864	.4868	.4871	.4875	.4878	.4881	.4884	.4887	.4890
2.3	.4893	.4896	.4898	.4901	.4904	.4906	.4909	.4911	.4913	.4916
2.4	.4918	.4920	.4922	.4925	.4927	.4929	.4931	.4932	.4934	.4936
2.5	.4938	.4940	.4941	.4943	.4945	.4946	.4948	.4949 *	.4951	.4952
2.6	.4953	.4955	.4956	.4957	.4959	.4960	.4961	.4962	.4963	.4964
2.7	.4965	.4966	.4967	.4968	.4969	.4970	.4971	.4972	.4973	.4974
2.8	.4974	.4975	.4976	.4977	.4977	.4978	.4979	.4979	.4980	.4981
2.9	.4981	.4982	.4982	.4983	.4984	.4984	.4985	.4985	.4986	.4986
3.0	.4987	.4987	.4987	.4988	.4988	.4989	.4989	.4989	.4990	.4990

NOTE: For values of z above 3.09, use 0.4999 for the area.
*Use these common values that result from interpolation:

z score	Area
1.645	0.4500
2.575	0.4950